Handbook of

Enhanced Spectroscopy

Handbook of
Enhanced Spectroscopy

edited by
Pietro Giuseppe Gucciardi
Marc Lamy de la Chapelle
Nathalie Lidgi-Guigui

CRC Press
Taylor & Francis Group
Boca Raton London New York

CRC Press is an imprint of the
Taylor & Francis Group, an **informa** business

Handbook of Enhanced Spectroscopy

First published 2016 by Pan Stanford Publishing Pte. Ltd.

Published 2022 by CRC Press
Taylor & Francis Group
6000 Broken Sound Parkway NW, Suite 300
Boca Raton, FL 33487-2742

ISBN 13: 978-981-4613-32-3 (hbk)
ISBN 13: 978-981-4613-33-0 (ebk)

Visit the Taylor & Francis Web site at
http://www.taylorandfrancis.com

and the CRC Press Web site at
http://www.crcpress.com

British Library Cataloguing-in-Publication Data
A catalogue record for this book is available from the British Library.

Contents

Preface

Enhanced spectroscopies refer to any spectroscopical technique based on linear or nonlinear optical phenomena capable to amplify the cross-section of the process: Raman scattering, infrared absorption or fluorescence as well as second harmonic generation, coherent anti-Stokes Raman scattering . . . Such spectroscopies exhibit a substantial increase of the optical signal up to several orders of magnitude, thanks to the exploitation of the optical properties of metallic or semiconductor nanostructures or disordered nanostructured materials. More especially, plasmon-enhanced spectroscopies can be seen as a direct application of the resonant collective electron excitation (plasmons) inside nanostructures making them act as antennas for light, or nanoantennas.

Since the first observation of the surface-enhanced Raman scattering (SERS) in 1974 by M. Fleischmann et al., the enhanced spectroscopies field became a broad research field. Nearly 25,000 articles have been published about SERS (and more than 1500 only in 2014), nearly 8000 about metal-enhanced fluorescence (MEF), 600 about tip-enhanced Raman scattering (TERS), and more than 300 about surface-enhanced IR absorption (SEIRA).

Enhanced spectroscopies indeed are a very active research field which has experienced a revival in the nineties with the development of the nanotechnologies and the set up of new techniques and methodologies that allow a precise control of the shapes and geometries of the nanostructures' assemblies. In this way, a better control and a fine tuning of the optical properties of metallic nanosystems have paved the way to more reproducible, reliable, and controllable platforms for basic experiments and applications of the enhanced spectroscopies.

Theoretical, computational, and numerical simulation tools have been developed in recent years that allow for a better understanding of the enhancement mechanism. As the technological progress allows researchers to produce new systems and as the analytical systems are reaching better resolution, questions never ceased to emerge. The link between theory, nanofabrication, and analytical development is the engine of this innovative field that enhanced spectroscopies represents.

Another remarkable characteristic of the enhanced spectroscopies is the requirement to gather inter- and even transdisciplinarity around a research area. As such, chemists, physicists, and biologists are working together to produce a common knowledge.

This research field goes now from the fundamental research with the comprehension of the physical processes involved in the enhancement of the spectroscopic signal to the direct application of the enhanced spectroscopies in sensing, monitoring, or identifying chemical or biological agents and processes.

This *Handbook of Enhanced Spectroscopy* is intended for a broad audience: the fundamental knowledge is exposed, as well as the state of the art of the research performed by some of the most active teams in the world. We also were desirous of talking about the future of this field.

The book is organized in 15 chapters that follow this scope:

- The first four chapters are an introduction to enhanced spectroscopies: the mechanisms and theoretical background are exposed in Chapters 1 and 3. Chapter 2 is dedicated to the concept of plasmon resonance, while Chapter 4 describes the state of the art in the nanofabrication of plasmonic nanostructures dedicated to enhanced spectroscopies.
- The following five chapters are dedicated to SERS. The fundamentals are reviewed in chapter 5 to 7. Chapters 8 to 10 present applications and use of SERS.
- The principle and application of TERS are described in Chapter 11. The next two chapters present the applications to carbon nanomaterials (Chapter 12) and the study of stress profile in SiGe (Chapter 13) with this technique.

- The principles of mid infrared-enhanced spectroscopy are outlined in Chapter 14 as well as the sample preparation techniques
- Eventually, Chapter 15 is devoted to a new trend in enhanced spectroscopy which is based on the use of semi-conductor materials rather than plasmonics structures.

It was an honor to collaborate with such high quality of authors. We have no doubt that the ideas and principles exposed in this book will be a source of inspiration for the reader. Students or experts of the field will be able to use this book as a basis for further development in this continuously expanding field of enhanced spectroscopies.

Pietro Giuseppe Gucciardi
Marc Lamy de la Chapelle
Nathalie Lidgi-Guigui
Autumn 2015

Chapter 1

Enhanced Spectroscopy

Dominique Barchiesi
Automatic Mesh Generation and Advanced Methods (GAMMA3), University of Technology of Troyes (UTT), Institut National de Recherche en Informatique et Automatisme (INRIA), 12 rue Marie Curie, CS 42060, 10004 Troyes Cedex, France
dominique.barchiesi@utt.fr

1.1 Introduction

Spectroscopy is the characterization of matter through the measurement of its interaction with light. The discovery by Isaac Newton of the sprawl of colors (dispersion) of sunlight by a prism has been the foundation of further developments of spectroscopy. Joseph von Fraunhofer used a diffraction grating for the same goal and observed dark bands in the spectrum of the sunlight. Walter N. Hartley contributed to make the link between the position of atoms in the periodic table and the observed bands in spectra. The invention of the spectroscopy has induced a fantastic advanced in knowledge of universe due to its ability to determine the composition of the sun and stars. It has been then possible to analyze not only remote materials, but also the light sources, gas, and liquids in laboratories since the end of the nineteenth century. The twentieth century saw the development of new spectroscopic techniques and the

Handbook of Enhanced Spectroscopy
Edited by Pietro Giuseppe Gucciardi, Marc Lamy de la Chapelle, and Nathalie Lidgi-Guigui

ISBN 978-981-4613-32-3 (Hardcover), 978-981-4613-33-0 (eBook)
www.panstanford.com

improvement of the sensitivity of the devices in order to detect smaller and smaller concentrations of chemical components. The basis of spectroscopy is given considering the distinction between elastic and inelastic scattering, before investigating the way to improve the sensitivity of devices by field enhancement.

1.2 Spectroscopy: Elastic and Inelastic Scattering

Spectroscopy is based on the light analysis of the probing of matter by radiations of different wavelengths. It is convenient to rely on the wavelength of a radiation to the energy of the photon.

1.2.1 *Elastic Scattering*

The energy of photon E is related to the frequency of radiation ν by the Planck–Einstein law:

$$E = h\nu = \hbar\omega, \tag{1.1}$$

where $h = 6.62606957 \times 10^{-34}$ J.s is the Planck constant and $\hbar = h/(2\pi)$.

The angular frequency ω of a monochromatic radiation in vacuum is related to its frequency ν or its wavelength in vacuum λ_0 by:

$$\omega = 2\pi\nu = 2\pi\frac{c}{\lambda_0}, \tag{1.2}$$

c being the speed of light in the medium of propagation [20]. $c = 299\,792\,458$ m.s^{-1} in vacuum.

The energy of photon can therefore been expressed in joules (J), electron-volts (eV), or cm^{-1}. If $\tilde{E}$ is the energy of a photon expressed in joules, the associated wavenumber in vacuum is:

$$\frac{1}{\lambda_0} = \frac{\tilde{E}}{hc} \tag{1.3}$$

The rapid conversion between unities is obtained from:

$$\begin{aligned} 1\,\text{eV} &= 1.60217657 \times 10^{-19}\,J = 8.065544 \times 10^{5}\,\text{m}^{-1} \\ &= 8065.544\,\text{cm}^{-1} \end{aligned} \tag{1.4}$$

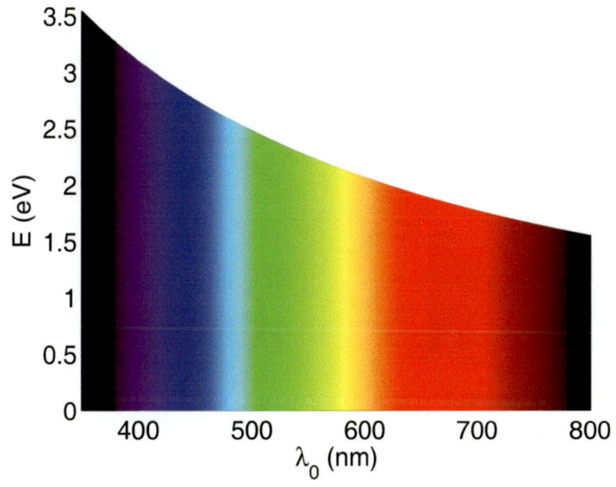

Figure 1.1 Energy of photons (eV) in the visible domain as a function of the wavelength (nm).

Example: a red radiation of wavelength 632.8 nm (wave number $1.5803 \times 10^6\ \text{m}^{-1} = 15803\ \text{cm}^{-1}$) corresponds to a photon of energy $3.1391 \times 10^{-19}\ \text{J} = 1.9593$ eV. Therefore, the commonly used unity depends on the domain of wavelength. In the visible domain, the electronvolt is convenient but the cm^{-1} is sometimes preferred for larger wavelengths. Figure 1.1 shows the energy of photons in the visible domain as a function of the wavelength.

If energy levels in matter are probed with photons with the same energy, the interaction of light with matter is elastic and results in Rayleigh scattering of light. Figure 1.2 shows the incoming white spectrum, corresponding to waves with periods between 400 nm and 800 nm. Two energies are characteristic of matter. The two corresponding wavelengths are in the scattered spectrum and are missing in the transmitted light (or extinction spectrum). Dark lines can be observed in the spectrum of the transmitted light (absorption spectrum), whereas bright lines appear on the dark ground (emission/scattered spectrum).

1.2.2 *Example of Hydrogen*

The energy levels are fingerprints of the matter, and therefore the composition of matter can be analyzed. They must be computed in

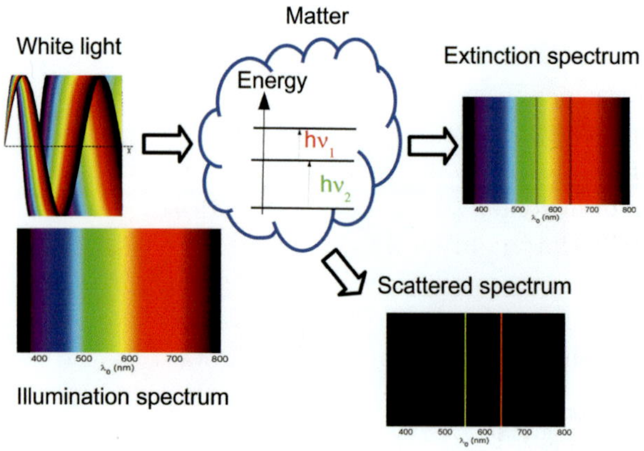

Figure 1.2 Principle of the elastic scattering of light by matter. The extinction spectrum is deprived of the two bands of energies characteristic of matter. On the contrary, the scattered spectrum exhibits the two bands.

the framework of quantum mechanics. Nevertheless for hydrogen, a simple mechanical model (introduced by Niels Bohr) of the proton and electron (mass m_e, charge q_e) in an hydrogen atom predicts an approximation of the energy levels E_n of electron, assuming the mass of the atomic nucleus m_p as infinite compared to that of electron (m_e). Actually $m_p \approx 1836\ m_e$. The underlying hypothesis are as follows:

- the constant speed of electron moving around the nuclear,
- the allowed orbits are quantified as an integer multiple of $\hbar = h/(2\pi)$,
- the electron radiates or absorbs energy only by change of orbits.

The energy of transition $E_{n\rightarrow m}$ between two discrete energy levels (n to m) is therefore approximately given by the Rydberg–Ritz formula:

$$\frac{E_{n\rightarrow m}}{hc} = \frac{1}{\lambda_0} = \underbrace{\frac{m_e q_e^4}{8h^3 c\epsilon_0^2}}_{R_\infty} \left|\frac{1}{n^2} - \frac{1}{m^2}\right|, \tag{1.5}$$

where ϵ_0 is the permittivity of vacuum, and n and m are positive integers. The (Johannes) Rydberg constant is $R_\infty = 1.0973731 \times$

10^7 m^{-1}. This formula was first deduced from experiments. This model can be refined by quantum mechanics considerations, but the present form is sufficient to illustrate that some specific wavelengths λ_0 can produce a jump of the electron from a low energy level to a higher one. This jump will produce a dark line in the extinction of spectrum of hydrogen (the cloud in Fig. 1.2). The excited electron comes back to its ground level by producing a photon associated to a radiation that is observable in the scattering spectrum (bright line). Both extinction (or absorption) and scattered (or emission) spectra calculated from the Rydberg formula are shown in Fig. 1.3.

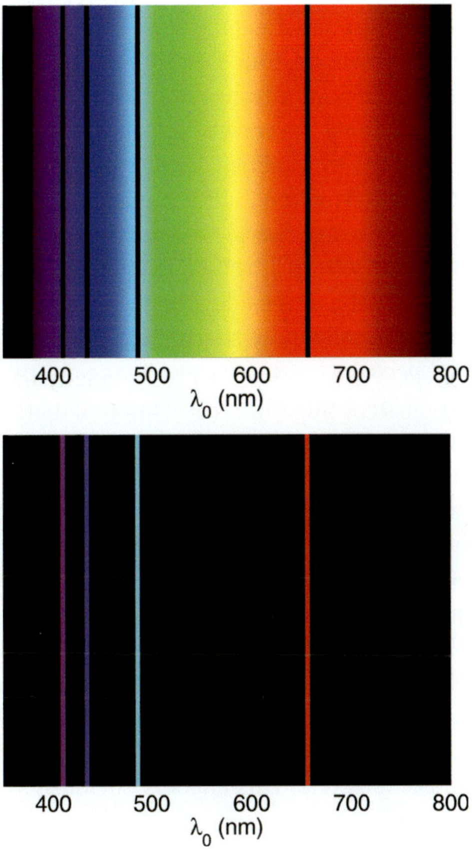

Figure 1.3 Extinction (left) and emission (right) visible spectra of hydrogen.

The elastic (Rayleigh) scattering of light has been introduced, but the inelastic scattering of light is also widely used.

1.3 Inelastic Scattering

In inelastic scattering, the scattered photon is at different energy from that of the incident photon. The energy difference corresponds to a transition between two states in the molecule [19].

In particular, the Raman spectroscopy involves transitions between rotational/vibrational energy levels. Full description and experimental results can be found in [10]: "Whenever light is diffused by the molecules of a transparent medium, the scattered radiations contain not only the wavelengths present in the incident light but also radiations of modified frequency. The effect is most striking when the scattering medium is a dust-free liquid, and the incident radiations consist of sharply defined spectral lines, e.g. the light of a quartz mercury lamp. It is then found that in the spectrum of the scattered light we have besides the incident lines also other new lines which are usually quite sharp." Explanation of the origin of this *secondary spectrum* (with reference to the elastic scattering) is given as: "the incident quantum of radiation is partly absorbed and partly scattered by the molecule. The shift of frequency would then correspond to some characteristic frequency of the molecule and would therefore be independent of the frequency of the exciting radiation." Chandrasekhara Venkata Raman was honored with Nobel Prize in Physics for his discovery in 1930.

If the scattered photon has less energy than the incident photon (the wavelength of the scattered wave is greater than that of the illumination) the associated effect is the Stokes process. If the scattered wave has higher energy (smaller wavelength), the anti-Stokes process corresponds to "stealing" energy of matter that was previously in an excited state. This effect is obviously weaker than the Stokes process (Fig. 1.4).

The advantages of the inelastic scattering are the possibility of probing small energy transitions (rovibrational) under visible illumination experimental setup, and the possibility of working in

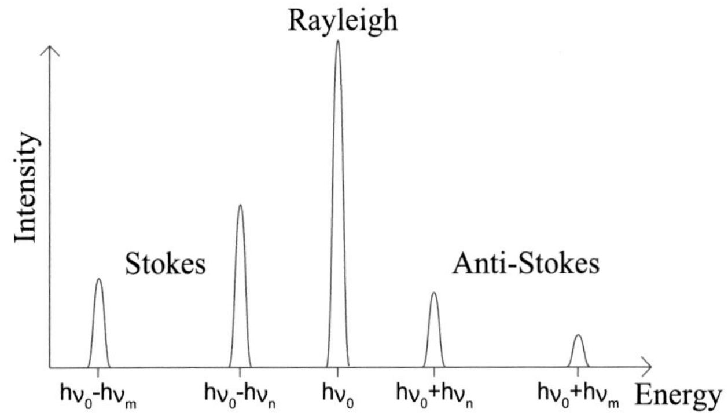

Figure 1.4 Schematic of a Raman spectrum as a function of energy of photons $h\nu$. The Rayleigh (elastic scattering), Stokes, and anti-Stokes (inelastic scattering) bands are shown.

water, of removing fluorescence and improving signal to noise ratio (compared to elastic spectroscopy) [12].

At this stage, it is clear that the concentration of the investigated material is a key point for the contrast of the lines in a spectrum, either in elastic or in inelastic spectroscopies. Therefore, to detect small concentrations of atoms or molecules, the increase of the sensitivity and of the resolution power of the measurement devices is basically searched. To this aim the local enhancement of the illumination and the decrease of its size are considered. The surface plasmon polariton resonance was proposed to achieve this goal [6]. To introduce the phenomenon of resonance, the historical point of view of Drude and Lorentz is often used. The practical cases of a metallic layer and of a metallic sphere are preferred in this chapter.

1.4 Field Enhancement by a Metallic Layer: The Surface Plasmon Polariton

In this section, we introduce a model of the interaction of a Gaussian beam with a metallic layer deposited on a glass substrate. The purpose of this model is to show how a field enhancement can be obtained by the excitation of a surface plasmon polariton (SPP), which is a resonance of the interaction between light and a thin

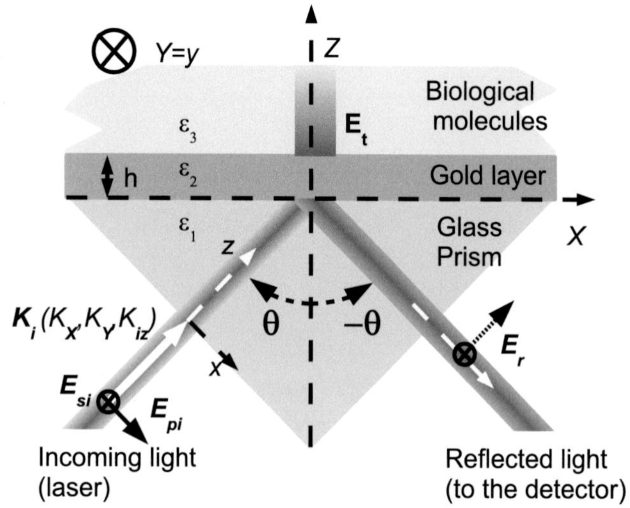

Figure 1.5 Principle of the surface plasmon resonance biosensor. The metal plate of thickness h and relative permittivity ϵ_2 is illuminated by a laser beam. θ is the angle of incidence. The relative permittivities of glass and of the surrounding medium are ϵ_1 and ϵ_3, respectively.

metallic plate. Such a field enhancement can be used to excite molecules in the vicinity of the metallic plate and therefore to enhance the spectroscopic signal (and to improve the sensitivity of the setup shown in Fig. 1.5).

The illumination must be described before introducing the model of the interaction between light and the multilayer, and finally the conditions of the field enhancement are given.

1.4.1 *The Illumination: A Gaussian Beam*

Any finite and polychromatic beam of light can be developed in a series of harmonic functions through the Fourier transform. The Fourier transform of the beam is a plane monochromatic wave expansion. The electric field of such a linearly polarized plane monochromatic wave is:

$$\mathbf{E}(\mathbf{r}, t) = \mathbf{E}^0 \exp(\imath(-\omega t + \mathbf{k}.\mathbf{r})), \tag{1.6}$$

where $\mathbf{E}^0$ is the amplitude, $\imath = (-1)^{\frac{1}{2}}$, ω is the angular frequency, $\mathbf{k} = (k_x, k_y, k_z)$ is the wave vector, and $\mathbf{r}$ is the position in the (x, y, z)

system of axis. The plane waves are basis functions of the Helmholtz equation, and therefore the wave vector verifies:

$$k_x^2 + k_y^2 + k_{iz}^2 = \epsilon_i \frac{\omega^2}{c^2}, \tag{1.7}$$

where ϵ_i is the relative permittivity of the medium i where the wave propagates. The relative permittivity is the square of the optical index of medium. It can be a complex number in an absorbing medium. In this case, the real part of the refractive index is related to the phase speed, while the imaginary part (the extinction coefficient) gives the amount of absorption loss when the electromagnetic wave propagates through the material.

Moreover, the electric field verifies the Maxwell equations in a medium free of additional electric charge:

$$\mathbf{k}.\mathbf{E} = 0. \tag{1.8}$$

Therefore, the three components of the electric field are not independent.

Light amplification by stimulated emission of radiation (LASER) sources are often used in experiments. In the following, the Gaussian beam is expanded in plane wave spectrum. The electric field of a Gaussian beam propagating along the z axis can be expanded in series of plane waves as follows [1]:

$$\mathbf{E}_i(x, y, z) = \iint E_0 \begin{pmatrix} \alpha_x \\ \alpha_y \\ -\frac{k_x\alpha_x + k_y\alpha_y}{k_z} \end{pmatrix} \times \pi a^2 \underbrace{e^{-a^2 \frac{k_x^2+k_y^2}{4}}}_{\text{Gaussian shape}} e^{ik_z z} e^{i(k_x x + k_y y)} dk_x dk_y \tag{1.9}$$

This equation corresponds to the 2D Fourier transform of a polarized field with Gaussian shape amplitude. The local polarization of each wave in the Fourier spectrum is given by $(\alpha_x, \alpha_y) = (1, 0)$ for (p) (or TM) polarization, or by $(\alpha_x, \alpha_y) = (0, 1)$ for (s) (or TE) polarization. If the Gaussian beam axis is rotated by an angle θ around the y axis, the projection of the Gaussian beam in the

$(X, Y = y, Z)$ can be written as:

$$\mathbf{E}_i(X, Y, Z) = \iint E_0 \begin{pmatrix} \frac{\alpha_x K_Z - K_Y \alpha_y \sin\theta}{K_Z \cos\theta + K_X \sin\theta} \\ \alpha_y \\ -\frac{K_X \alpha_x + K_Y \alpha_y \cos\theta}{K_Z \cos\theta + K_X \sin\theta} \end{pmatrix} \times \pi a^2 \underbrace{e^{-a^2 \frac{K_X^2 + K_Y^2}{4}}}_{\text{Gaussian shape}} e^{\imath K_Z Z} e^{\imath(K_X X + K_Y Y)} \cos\theta dK_X dK_Y, \tag{1.10}$$

where the wave vector is expressed in the (X, Y, Z) basis: $K_X = k_x \cos\theta + k_z \sin\theta$, $K_Y = k_y$, and $K_Z = k_z \cos\theta - k_x \sin\theta$. Each component of the electric field $\mathbf{E}_i(X, Y, Z)$ is modulated by the Gaussian shape function. Some typos errors in Ref. [1] prevent the reader to use directly the printed formula.

The beam is a sum of plane waves with a known modulation of its amplitude. Therefore, it is possible to consider the field that is transmitted by the metal plates under the above illumination, by considering a single plane monochromatic wave.

1.4.2 *The Local Transmitted and Reflected Fields*

The monochromatic plane wave illumination is considered in a general case. The incoming wave vector is $\mathbf{K}_1 = (K_X, K_Y, K_{1Z})$. The index 1 indicates the propagation in medium with relative permittivity ϵ_1. The transmitted and reflected field must be expressed in the (X, Y, Z) system of axis.

1.4.2.1 General Formulation of the Transmission Coefficients in Cartesian Coordinates

The reflection and transmission coefficients can be deduced from the classical Fresnel coefficients in (s) and (p) polarizations [9]:

- In (s) polarization, the electric field of the plane wave is in the (X, Y) plane. It is perpendicular to the plane of incidence $(\mathbf{K}, \mathbf{Z})$.
- In (p) polarization, the electric field is perpendicular to the (s) direction and to the wave vector. It is in the plane of

incidence. The magnetic field **B** of the plane electromagnetic wave is perpendicular to the plane of incidence (**K**, **Z**).

The transmission Fresnel coefficients are:

$$t_{(s)} = \frac{E_t}{E_i} = \frac{4K_{1Z}K_{2Z}e^{\imath(K_{2Z}-K_{3Z})h}}{(K_{1Z}+K_{2Z})(K_{2Z}+K_{3Z})+(K_{1Z}-K_{2Z})(K_{2Z}-K_{3Z})e^{2\imath K_{2Z}h}} \tag{1.11}$$

$$t_{(p)} = \frac{B_t}{B_i} = \frac{4\epsilon_2 K_{1Z}\epsilon_3 K_{2Z}e^{\imath(K_{2Z}-K_{3Z})h}}{(\epsilon_2 K_{1Z}+\epsilon_1 K_{2Z})(\epsilon_3 K_{2Z}+\epsilon_2 K_{3Z})+(\epsilon_2 K_{1Z}-\epsilon_1 K_{2Z})(\epsilon_3 K_{2Z}-\epsilon_2 K_{3Z})e^{2\imath K_{2Z}h}} \tag{1.12}$$

In the (X, Y, Z) system of axis, the transmitted electric field is a function of the impinging field $\mathbf{E}_t = T\,\mathbf{E}_i$:

$$\begin{pmatrix} E_{tX} \\ E_{tY} \\ E_{tZ} \end{pmatrix} = \begin{pmatrix} t_{(s)} & 0 & \frac{K_X(K_{Z1}\epsilon_3 t_{(s)}-K_{Z2}\epsilon_1 t_{(p)})}{(K_X^2+K_Y^2)\epsilon_3} \\ 0 & t_{(s)} & \frac{K_Y(K_{Z1}\epsilon_3 t_{(s)}-K_{Z2}\epsilon_1 t_{(p)})}{(K_X^2+K_Y^2)\epsilon_3} \\ 0 & 0 & t_{(p)}\frac{\epsilon_1}{\epsilon_3} \end{pmatrix} \begin{pmatrix} E_{iX} \\ E_{iY} \\ E_{iZ} \end{pmatrix} \tag{1.13}$$

The above formulation differs from that in Ref. [1] but can help to check that the divergences of the transmitted field vanish numerically. The Gaussian beam and the transmission and the reflection by a metal plate on a substrate can be used to compute the transmitted beam above the metal plate. Therefore, the conditions of the enhancement of the transmitted field can be discussed. The enhancement of this electric field is used to enhance the excitation of molecules above the metal plate and therefore can be used to increase the sensitivity of spectroscopic setups based on surface plasmon resonance (SPR). The field enhancement is obtained under specific conditions that are shortly investigated.

1.4.2.2 The Plasmon Polariton Resonance

Matrices of transmission (Eq. 1.13) have specific properties, which should be investigated to illustrate how resonance produces field enhancement. The denominator D of $t_{(p)}$ vanishes for a metallic plates, at a given angle of incidence and thickness h (Eq. 1.12). The transmission coefficients have therefore a pole in the complex plane [3, 4, 6]. The mathematical solution of the equation $D(K_X, e) = 0$ occurs for [5, 11]:

$$z_0 = e^{2\imath K_{2Z} h} = -\frac{(\epsilon_2 K_{1Z} + \epsilon_1 K_{2Z})(\epsilon_3 K_{2Z} + \epsilon_2 K_{3Z})}{(\epsilon_2 K_{1Z} - \epsilon_1 K_{2Z})(\epsilon_3 K_{2Z} - \epsilon_2 K_{3Z})}. \quad (1.14)$$

This transcendent equations can be solved numerically for h when the wavelength and the relative permittivities are fixed. Of course, h is a thickness and therefore must be a real number but the numerical solution gives a mathematical solution, i.e., a complex number in the general case. A matching between both solution can be found if the imaginary part of the numerical solution is negligible compared to its real part. A better approach consists of calculating the phase and the modulus of z_0, not forgetting well that h and K_{2Z} are complex numbers in the general case. Therefore, the complex number h can be evaluated for each θ, the problem of cutoff of logarithm (or of complex power elevation) function being avoided.

This best metal thickness is a function of the incidence angle θ of the impinging plane wave and corresponds to the SPP resonance conditions.

To go further, the complex residue of the transmission (t_p, Eq. 1.12) must be calculated [6]:

$$\lim_{z \to z_0} (z - z_0) t_{(p)} = -\frac{4\epsilon_2 K_{1Z} \epsilon_3 K_{2Z} ((\epsilon_2 K_{1Z} + \epsilon_1 K_{2Z})(\epsilon_3 K_{2Z} + \epsilon_2 K_{3Z}))^{\frac{K_{2Z} - K_{3Z}}{2K_{2Z}}}}{((\epsilon_2 K_{1Z} - \epsilon_1 K_{2Z})(\epsilon_3 K_{2Z} - \epsilon_2 K_{3Z}))^{-\frac{K_{2Z} + K_{3Z}}{2K_{2Z}}}}. \quad (1.15)$$

Let us illustrate numerically these results in a practical case of a gold layer.

1.4.2.3 Numerical Application to a Gold Layer on Glass at 632.8 nm

Let us consider a gold layer of thickness h and relative permittivity $-9.5099 + 1.2175\imath$ @632.8 nm [21, 25, 26] deposited on a glass substrate of relative permittivity $\epsilon_1 = 2.25$. The computation of h with Eq. 1.14 is plotted in Fig. 1.6. The physical value corresponds to the vanishing of the imaginary part of h (thin line) and to the maximum of the real part (thick line). The modulus is also indicated

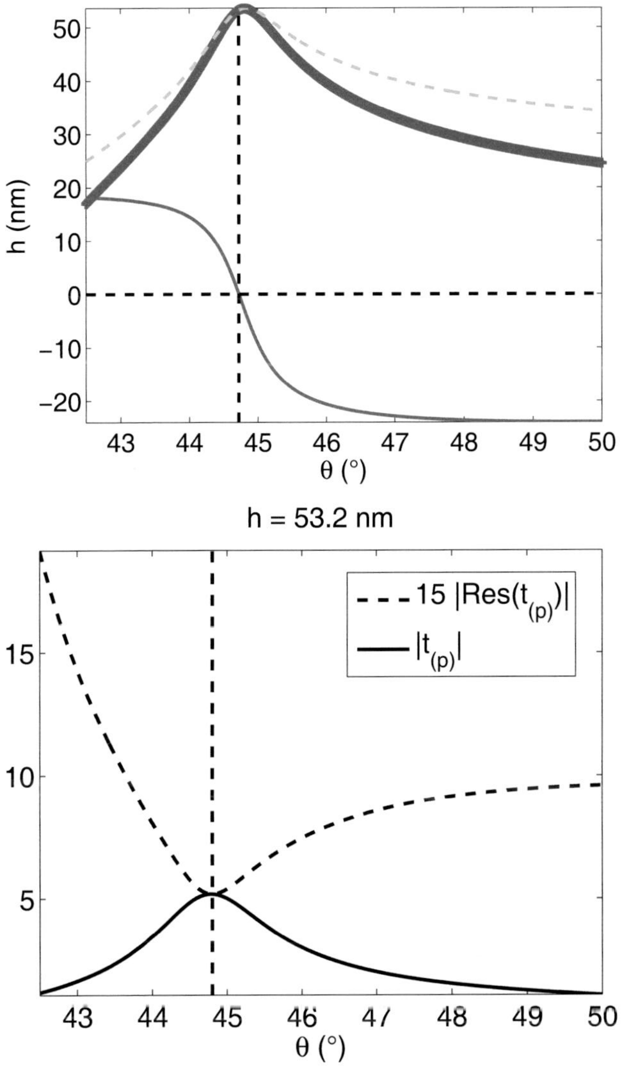

Figure 1.6 (Top) Thickness of the gold layer h as a function of the incidence angle θ (Eq. 1.14). The maximum value corresponds to the optimum thickness $h = 53.2$ nm to launch SPP. (Bottom) Residue of the transmission coefficient $t_{(p)}$ and field enhancement $|t_{(p)}|$ as a function of the incidence angle θ (Eq. 1.14) for $h = 53.2$ nm. The minimum value corresponds to the launch of the SPP. The field enhancement $|t_{(p)}|$ exhibits a maximum for the same incidence angle $\theta_p = 44.8^{\circ}$.

(- -) as well as the location of the zero and the optimum angle (horizontal and vertical dashed lines).

Figure 1.6 shows the high sensitivity of the SPP to the thickness and to the incidence angle. The field enhancement in this practical case is five times the amplitude of the incoming field. This result is obtained for a single plane wave impinging on the metal layer at a single wavelength. However, spectroscopy requires an analysis in a domain of wavelengths.

1.4.2.4 Numerical Application to a Gold Layer on Glass for Spectroscopy in the Visible Range

In the visible range of wavelengths (400–800 nm), the result of the optimization of the thickness of gold h for a better field enhancement $|E_t|/|E_0|$ with the above method is plotted in Fig. 1.7. The optimum thickness falls between 16 nm and 57 nm. The optimum angle of illumination (not shown here) varies from 42.88° to 45.81°. The corresponding field enhancement is an increasing function of the wavelength. This fact justifies the use of SPR for infrared experiments.

The results of this optimization suggest to consider the mean value of the optimized thickness h ($< h >= 47$ nm) to design an SPR for visible spectroscopy. Figure 1.8 shows the optimum angle of illumination and the field enhancement. Fortunately, the comparison between Figs. 1.7 and 1.8 shows that the thickness of gold is not a critical parameter.

The above investigation of the spectral behavior of the SPR setup has proved that a gold layer with adequate thickness, illuminated under a specific angle, can produce a field enhancement.

The models of the Gaussian beam and of the transmission by a metal layer being established, the field enhancement above the metal layer, can be fully described in a realistic case.

1.4.3 *The Enhancement of the Transmitted Electric Field by Surface Plasmon Excitation*

The transmission of the Gaussian beam can be computed from the transmitted field deduced from Eqs. 1.10 and 1.13 with Eqs. 1.12

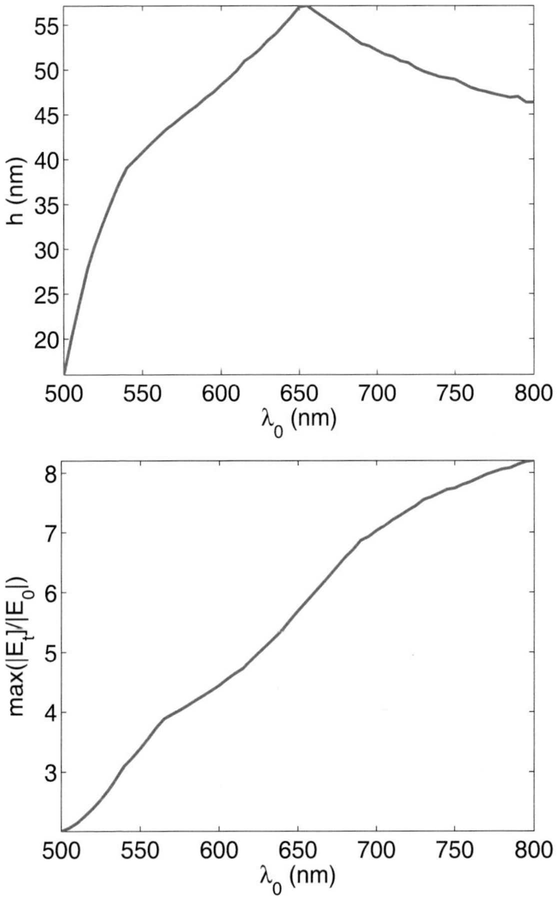

Figure 1.7 (Top) Thickness h of the gold layer for the best enhancement of the transmitted field gold layer h as a function of the wavelength λ_0. (Bottom) Transmitted field enhancement as a function of the wavelength λ_0.

and 1.11:

$$\mathbf{E}_t(X, Y, Z) = \iint E_0 T \begin{pmatrix} \frac{\alpha_x K_Z - K_Y \alpha_y \sin\theta}{K_Z \cos\theta + K_X \sin\theta} \\ \alpha_y \\ -\frac{K_X \alpha_x + K_Y \alpha_y \cos\theta}{K_Z \cos\theta + K_X \sin\theta} \end{pmatrix}$$

$$\times \pi a^2 \underbrace{e^{-a^2 \frac{K_X^2 + K_Y^2}{4}}}_{\text{Gaussian shape}} e^{\imath K_Z Z} e^{\imath(K_X X + K_Y Y)} \cos\theta dK_X dK_Y, \quad (1.16)$$

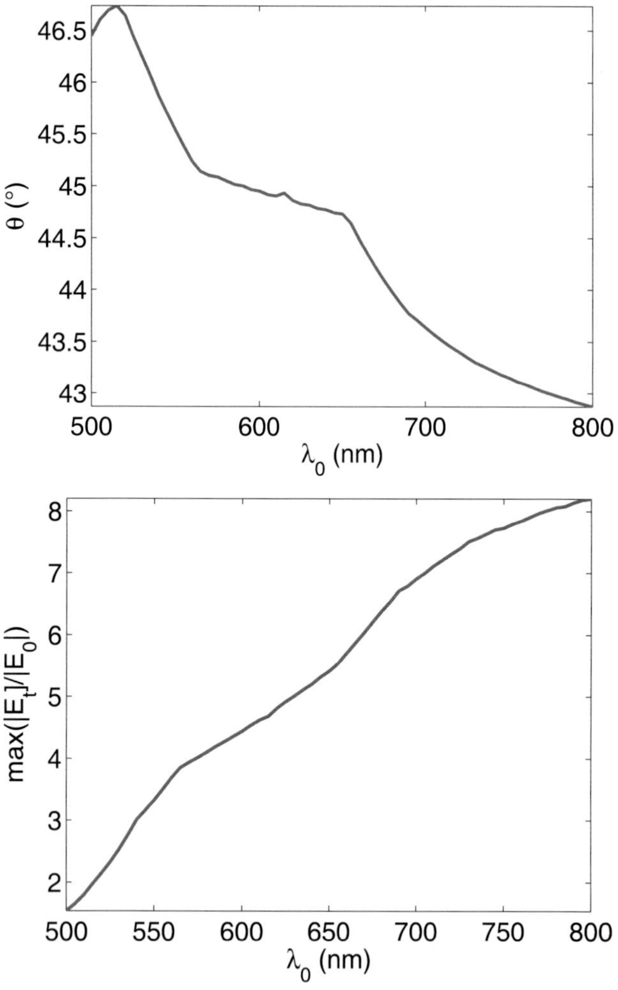

Figure 1.8 (Top) Optimum angle of illumination θ for the best enhancement of the transmitted field gold layer $h = 47$ nm as a function of the wavelength λ_0. (Bottom) Transmitted field enhancement as a function of the wavelength λ_0.

Only the propagating waves in glass are kept to compute the transmitted field. Therefore, $K_X^2 + K_Y^2 < \epsilon_1 \frac{\omega^2}{c^2}$ (see Eq. 1.7). The integral 1.16 is computed using the rectangle method with 128 terms. An example of application of the proposed model is the comparison with the experimental intensity enhancement observed on a silver layer by photon tunneling scanning microscopy [13, 14]. Figure 1.9 shows the comparison between experimental results and computations from Eq. 1.16.

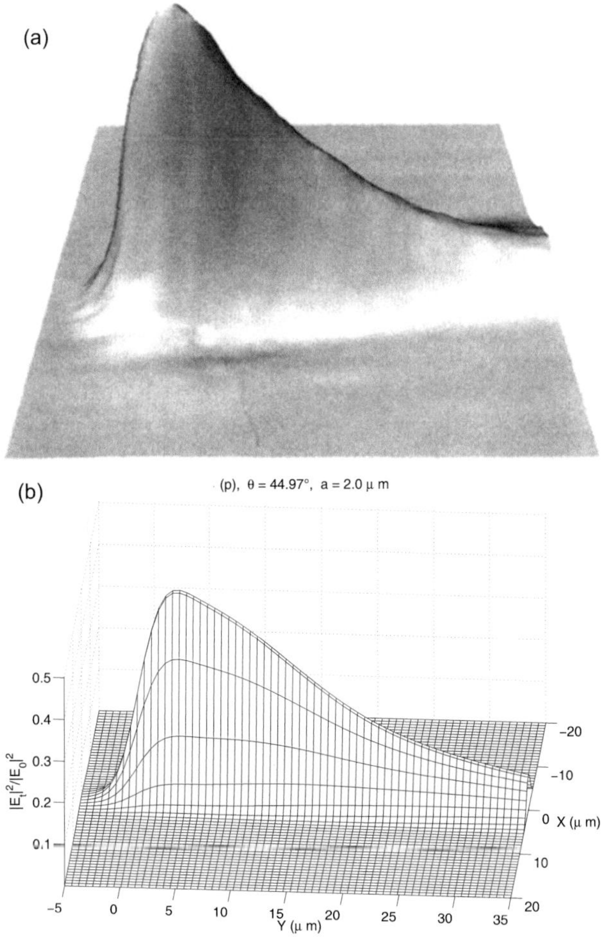

Figure 1.9 (a) Intensity of the electric field above the silver layer of thickness $h = 53$ nm and relative permittivity $\epsilon_2 = -17.9+0.7\imath$@632.8 nm. $\epsilon_3 = 1$ and the relative permittivity of the glass substrate is $\epsilon_1 = 2.123$. $\theta = 44.97^\circ$ and the width of the Gaussian beam is $a = 2$ μm. (b) Computation with Eq. 1.16. Figure reprinted with permission from Ref. [13], Copyright 1994 by the American Physical Society.

The SPP appears as an enhancement of the intensity above the metal film that can be used to excite molecules in its vicinity. However, the exponential decrease of the field amplitude deserves to be underlined. Actually, the Z dependance of the field lays in the $e^{\imath K_Z Z}$ term of Eq. 1.16. The exponential decay of the field is therefore $1/K_{3Z} \approx 100$ nm in this case. Only the molecules deposited at nanometric distance can make the most of the field enhancement.

Gold layer is easily deposited on glass substrate by sputtering [27]. Gold is malleable metal and is chemically unaffected by air and most corrosive reagents. A gold layer of thickness 48 nm is considered and Fig. 1.11 shows the field enhancement for a focused Gaussian beam: the field amplitude is multiplied by 3. The amplitude of a single plane wave impinging under the plasmon incidence angle is multiplied by about 5.

The field enhancement is highly sensitive to the angle of incidence. Varying the main angle of incidence of the beam of only 2° reduces the field enhancement by a factor of about 3. The field enhancement is also dependent on the optical properties of the metal. In the cases considered above, silver is more efficient than gold at $\lambda_0 = 632.8$ nm as it is less absorbing material: the imaginary part of the relative permittivity of gold is 1.7 times that of silver.

The enhancement of the transmitted field occurs at the expense of the reflected beam. The vanishing of the reflected plane wave when the SPP is launched was described in [2, 3, 6] using the rigorous approach of residues as above. Figure 1.10 shows the logarithm of the field enhancement $|E_t|(Z = h)$ and of the amplitude of the reflected field $|E_r|(Z = 0)$. The location of the minimum of the reflected field does not corresponds exactly to that of the maximum of the field enhancement, and therefore the detection of the minimum in the reflected light varying θ does not guarantee the maximum of the field enhancement [11].

Moreover comparison of the best parameters obtained by the complex analysis of the resonance ($h = 53.2$ nm, $\theta_p = 44.80^\circ$) differs slightly from the computed parameters for the maximum of $|E_t|(Z = h)$ ($h = 53.7$ nm and $\theta = 44.79^\circ$). This small shift is due to the influence of the numerator of $|E_t|(Z = h)$, which also depends on θ and h. The shape of the peak $\log|E_t|$ is sharp along the vertical

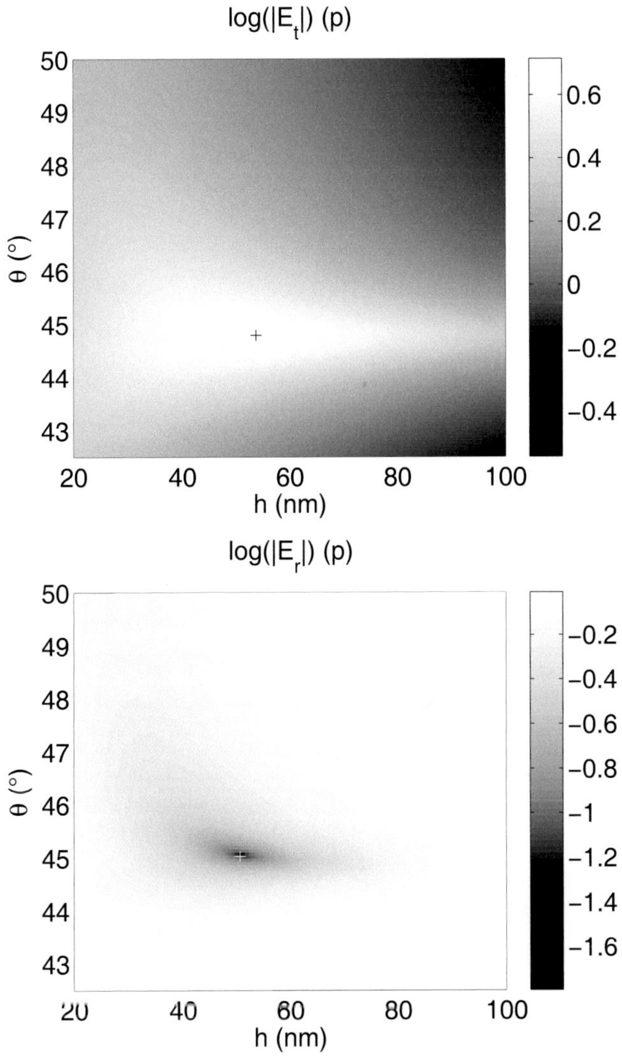

Figure 1.10 Transmitted field enhancement $|E_t|(Z = h)$ and reflected field $|E_r|(Z = 0)$ in logscale. The maximum of $|E_t|$ is obtained for $h = 53.7$ nm and $\theta = 44.79°$, whereas the minimum of $|E_r|$ is reached for $h = 50.6$ nm and, $\theta = 45.02°$.

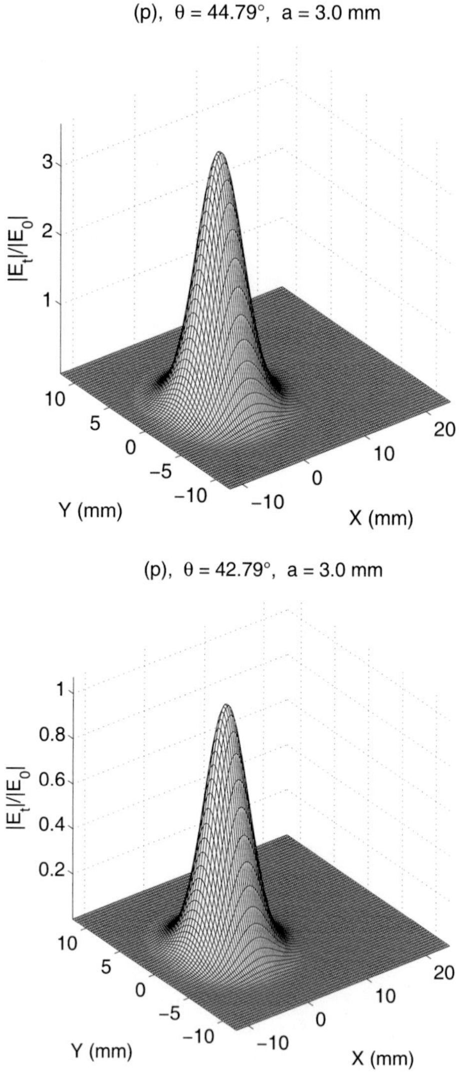

Figure 1.11 Amplitude of the electric field above the gold layer of thickness $h = 48$ nm and relative permittivity $\epsilon_2 = -9.510 + 1.217\imath$ at 632.8 nm. $\epsilon_3 = 1$ and the relative permittivity of the glass substrate is $\epsilon_1 = 2.25$. The width of the Gaussian beam is $a = 3$ mm. (a) $\theta = 44.80^\circ$, (b) $\theta = 42.80^\circ$.

axis showing the sensitivity of the field enhancement to the angle of incidence. The thickness appears to be a less critical parameter.

1.4.4 *Conclusion on the Surface Plasmon Polariton on a Gold Layer*

The field enhancement induced by SPP on a gold layer is about 5 in the investigated case; the intensity is therefore multiplied by 25. The use of a Gaussian beam slightly decreases the field enhancement, which is very sensitive to the angle of incidence of the beam. However, the thin layer of gold can be used to enhance locally the field just above the layer. The improvement of the sensitivity of a spectroscope based upon the SPR is therefore possible. However, the localized surface plasmon resonance (LSPR) using a metallic particle is more efficient.

1.5 Field Enhancement by a Spherical Particle

The model of the interaction between light and a spherical particle is known since 1908 [23] and tons of papers have been devoted to the applications of this model [7, 29]. The color changing of colloidal gold solutions was interpreted in terms of LSPRs [22, 29]. The localized plasmon resonance can be of great interest in surface-enhanced Raman spectroscopy, catalysis, biosensing, or in the treatment of skin cancer [15–17, 24, 28]. The problem of near-field calculation was addressed in [18] where it was shown that the calculation of the field near a metallic particle is possible with control of accuracy and that the electric field is enhanced locally. Therefore, this field enhancement around nanoparticles can be used to improve sensitivity of spectroscopic studies.

In this section, we use the classical results of the Mie theory without giving details that can be found elsewhere [8].

1.5.1 *The Mie Theory*

A spherical particle embedded in an infinite external medium is illuminated by a monochromatic plane wave of angular frequency ω.

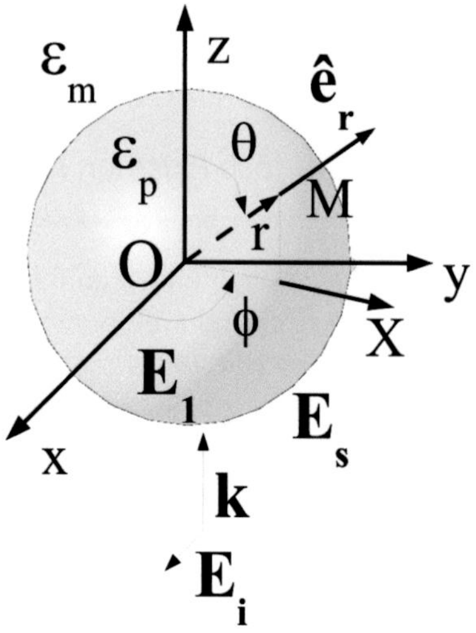

Figure 1.12 Spherical particle, system of coordinates, and material properties.

The central symmetry of the spherical particle of radius R suggests the appropriate system of coordinates: the spherical polar ones (r, θ, ϕ) (see Fig. 1.12). The corresponding orthonormal basis is denoted by $(\hat{e}_r, \hat{e}_\theta, \hat{e}_\phi)$.

The material property is the relative permittivity that differs in the surrounding medium and in the particle. For homogenous and isotropic spherical particles of radius R, the relative permittivity is a constant in and out the spherical particle. In the following, the ratio of the relative permittivities will be used:

$$m^2 = \frac{\epsilon_p}{\epsilon_m} \tag{1.17}$$

The field is expanded in series of spherical harmonics function that are solution of the wave equation in spherical coordinates. The size parameter is $x = 2\pi \frac{R}{\lambda_0}$ with R its radius and λ_0 the wavelength in vacuum. The following discussion is focused on the behavior of the coefficients a_n of the development in series of the electric field out of the sphere that are deduced from a system involving the boundary conditions at the surface of the sphere [8] [Eq. 4.52] for nonmagnetic materials:

$$a_n = \frac{1}{2}\frac{N}{D} \tag{1.18}$$

with

$$N = J_{n+\frac{1}{2}}(mx)\left((2m^2(n+1)-1)J_{n+\frac{1}{2}}(x) - 2m^2xJ_{n+\frac{3}{2}}(x)\right)$$
$$+ mxJ_{n+\frac{1}{2}}(x)\left(J_{n+\frac{3}{2}}(mx) - J_{n-\frac{1}{2}}(mx)\right) \quad (1.19)$$
$$D = J_{n+\frac{1}{2}}(mx)\left((m^2-1)(1+n)H^{(1)}_{n+\frac{1}{2}}(x) - m^2xH^{(1)}_{n+\frac{3}{2}}(x)\right)$$
$$+ mxH^{(1)}_{n+\frac{1}{2}}(x)J_{n+\frac{3}{2}}(mx) \quad (1.20)$$

This set of coefficients of the scattered field depends on both the material properties (through the relative permittivity of material $m^2 = \epsilon_p/\epsilon_m$) and the geometrical parameters (through $x = 2\pi R/\lambda_0$). The functions J_m and $H^{(1)}_m$ are respectively the Bessel and Hankel functions of the first kind.

1.5.2 *The Resonance of the Scattered Electric Field*

The above-mentioned coefficients of series of the electric field have denominator D that can vanish. If any, a pole of a_n and a resonance occur as in the case of the metal layer. The study is simpler considering small particles $x << 1$ ($R << \lambda_0/(2\pi)$). Series of D can be written:

$$D = 2\imath m^{n+\frac{1}{2}}\frac{m^2n+n+1}{2n+1} + o(x^2), \quad (1.21)$$

where $\imath = (-1)^{\frac{1}{2}}$. Therefore, the condition of vanishing of D is clearly $m^2 = -\frac{n+1}{n}$ [8]. The resonance can only occur for metallic particle of relative permittivity lower than -1. This mathematical solution corresponds to nonabsorbing metal. Each term (with index n) of the series expansion of the electric field can exhibit a resonance. The series of the first terms of a_n takes the following form for small values of x:

$$a_1 = \underbrace{-\imath\frac{2}{3}\frac{m^2-1}{m^2+2}x^3}_{\text{dipolar term}} - \imath\frac{2}{5}\frac{(m^2-1)(m^2-2)}{(m^2+2)^2}x^5 + o(x^6) \quad (1.22)$$

$$a_2 = -\imath\frac{2}{15}\frac{m^2-1}{2m^2+3}x^5 + o(x^6) \quad (1.23)$$

$$a_3 = -\imath\frac{4}{1575}\frac{m^2-1}{3m^2+4}x^7 + o(x^8) \quad (1.24)$$

The condition of resonance appears clearly in the denominators of each term of the series. The dominant term for small x is a_1, except for resonance conditions of higher order terms. The ratio $R_{a_2(5)/a_1(5)}$ of the term of order five of a_2 to that of a_1 is

$$R_{a_2(5)/a_1(5)} = \frac{(2+m^2)^2}{6(m^2-2)(3+2m^2)}. \tag{1.25}$$

Consequently, if $m^2 \approx -2$, a_1 is dominating. Considering a_1 the term of order five can only be neglected if:

$$\left|R_{a_1(5)/a_1(3)}\right| = \frac{3}{5}\left|\frac{m^2-2}{m^2+2}\right| x^2 << 1. \tag{1.26}$$

This condition cannot be fulfilled if $m^2 \approx -2$ except for very small nanometric particles ($x^2 << 5|m^2+2|/(3|m^2-2|)$) for which the relative permittivity of metal is obviously not the bulk one. Therefore, the study of the domain of validity of the dipolar approximation deserves to be investigated in view of its popularity in the basic course and its extensive use in models of field enhancement.

1.5.3 *Validity of the Dipolar Model*

Even if physically, any material is absorbing and therefore the relative permittivity of metal cannot be exactly -2, the dipolar term is meaningless, close to the resonance condition for a_1. The first terms of the series Eq. 1.22 suggest that all terms have singularities. Therefore, we are faced with an essential singularity of a_1 and the dipolar term alone cannot be considered for further discussions.

The optical property $m^2 \approx -2$ is further considered to focus on the conditions of resonance of a_1. *The correct handling of the singularity of a_1 requires to calculate the series of a_1 around $m = \imath\sqrt{2}$, before the series around $x = 0$.* The following series limited to order 1 is obtained:

$$a_1 = \frac{5}{6}\imath x + \frac{m-\imath\sqrt{2}}{\sqrt{2}}\left(-\frac{125}{54} + \frac{25}{18}\frac{1}{x} - \frac{2935}{1512}x + o(x^2)\right) + o(x^2)$$
$$+ o((m-\imath\sqrt{2})^2) \tag{1.27}$$

Consequently to describe accurately the resonance of a_1 for small x, the dipolar approximation is not relevant if $m^2 \approx -2$. Figures 1.13

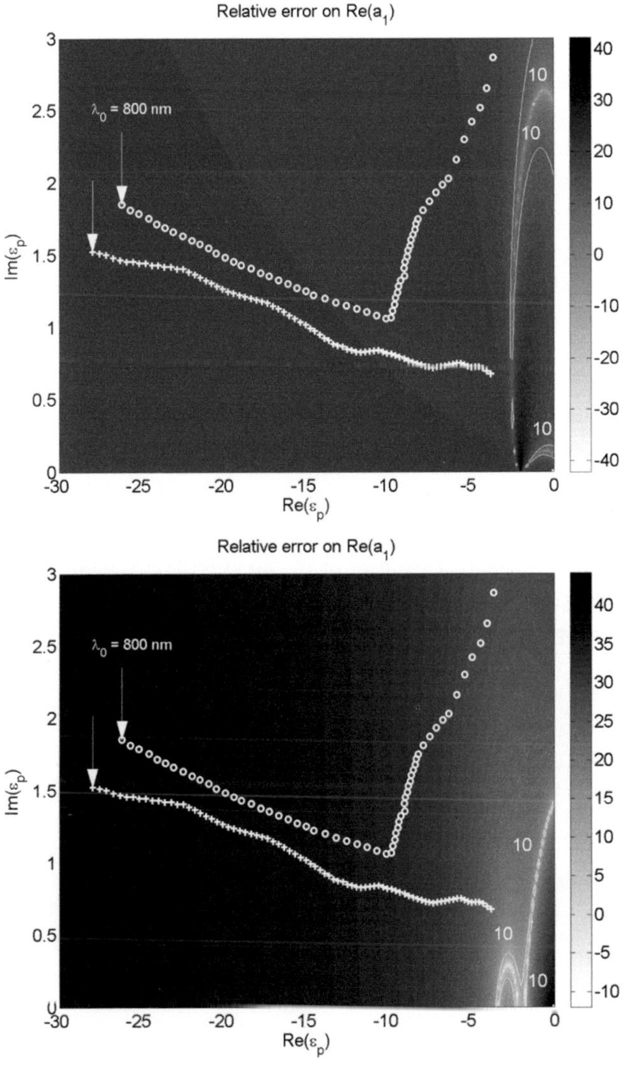

Figure 1.13 Relative error of the dipolar approximation on the real part of a_1 (top). Relative error of the series approximation on the real part of a_1 (bottom). The plots of the bulk relative permittivities of silver (+) and gold (o) are shown. An example of tolerance of 10% on the relative error is shown. All plots are in dB for readability.

and 1.14 show the relative error (in dB) of the dipolar approximation and the series approximation (Eq. 1.27) compared to a_1, in a wide domain of relative permittivity of metal. The gold (o) and silver (+) relative permittivities are also shown as references. The dipolar approximation is not valid near $m^2 = -2$ and the series (Eq. 1.27) are not valid far from $m^2 = -2$. The region of validity of the series is a limited zone near $m^2 = -2$ (see curves of the limit of zones of relative errors lower than 10%). These results illustrate the distance of action of the singularity of the dipolar model, i.e., the zone where the dipolar model cannot be used.

1.5.4 *Conclusion*

For small particles, the dominant term in the scattered electric field is a_1. This coefficient admits an essential singularity at $m^2 = -2$. The dipolar term is not relevant to evaluate the field enhancement. On the contrary, a rigorous treatment of the singularity gives a simple formula that enables a discussion on the size of the particle and the optical properties of the particle:

- For $m^2 = -2$ the result is:

$$a_1(m^2 - 2) = \frac{5}{6}\imath x + o(x^2) \tag{1.28}$$

- For $m^2 \approx -2$, the second-order correction for small x is:

$$\frac{m - \imath\sqrt{2}}{\sqrt{2}} \left(-\frac{125}{54} + \frac{25}{18}\frac{1}{x} - \frac{2935}{1512}x + o(x^2) \right)$$
$$+o(x^2) + o((m - \imath\sqrt{2})^2)$$

- For m^2 far from -2:

$$a_1 \approx -\imath\frac{2}{3}\frac{m^2 - 1}{m^2 + 2}x^3 + o(x^5) \tag{1.29}$$

Fortunately the optical properties of metals are often far from $m^2 = -2$, and therefore the dipolar model can be used for small particles as a good approximation of the amplitude of the electric field.

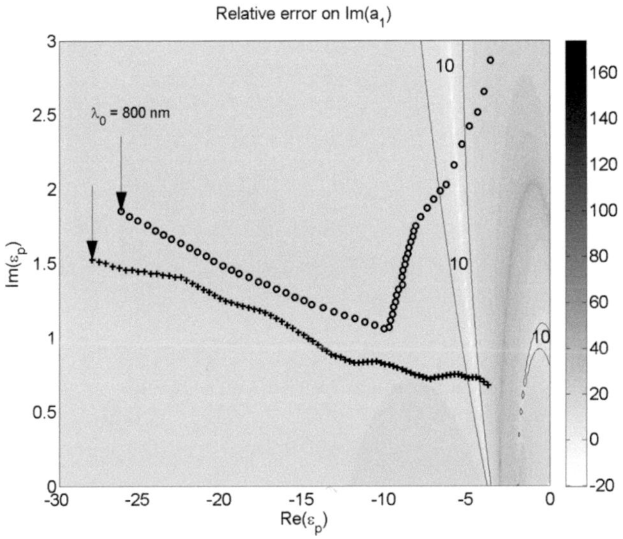

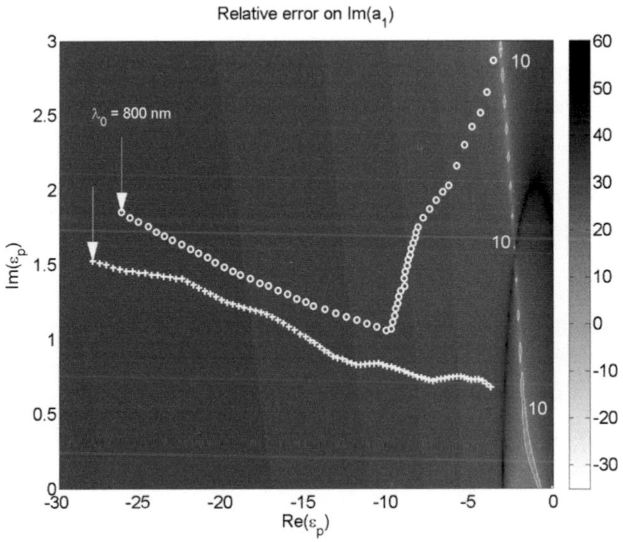

Figure 1.14 Relative error of the dipolar approximation on the imaginary part of a_1 (top). Relative error of the series approximation on the imaginary part of a_1 (bottom). The plots of the bulk relative permittivities of silver (+) and gold (o) are shown. An example of tolerance of 10% on the relative error is shown. All plots are in dB for readability.

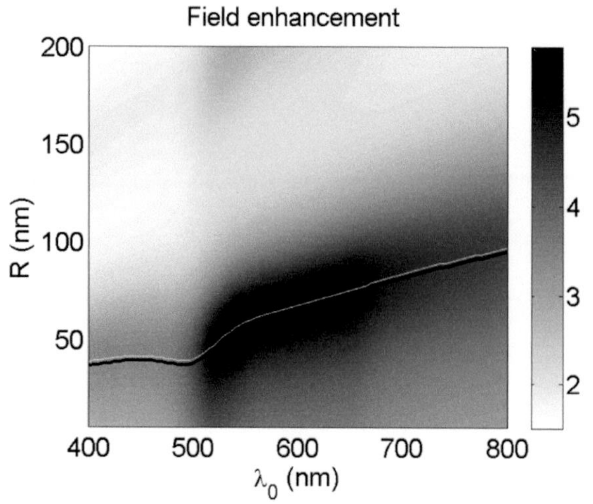

Figure 1.15 Field enhancement as a function of the radius R and the wavelength λ_0. The curve indicates the radius for each wavelength of the spectrum, leading to the best field enhancement. The field enhancement is the square root of the intensity in the vicinity of the sphere, in the direction of polarization of the impinging field (see Fig. 1.12). The full Mie theory is used to compute the field enhancement.

1.5.5 *The Field Enhancement in the Vicinity of a Small Spherical Gold Particle*

The case of a spherical gold nanoparticle of radius R, in the visible domain of wavelengths, is investigated in the following. An enhancement of the field around small metal particles is awaited for a given radius R and for each wavelength of the spectrum, in the direction of polarization of the incoming field (x). Let us note that the field enhancement does not occur exactly in this direction if the radius R increases. Therefore, the field enhancement is computed from the full Mie theory, without approximation, from only the first term of series a_1 and from the dipolar model (Eq. 1.29). Figure 1.15 shows the field enhancement as a function of the radius R and the wavelength λ_0. The best parameters follow the superimposed curve. The field enhancement is of about one order of magnitude. The optimum radius of nanoparticle is increasing with the wavelength, alike the field enhancement.

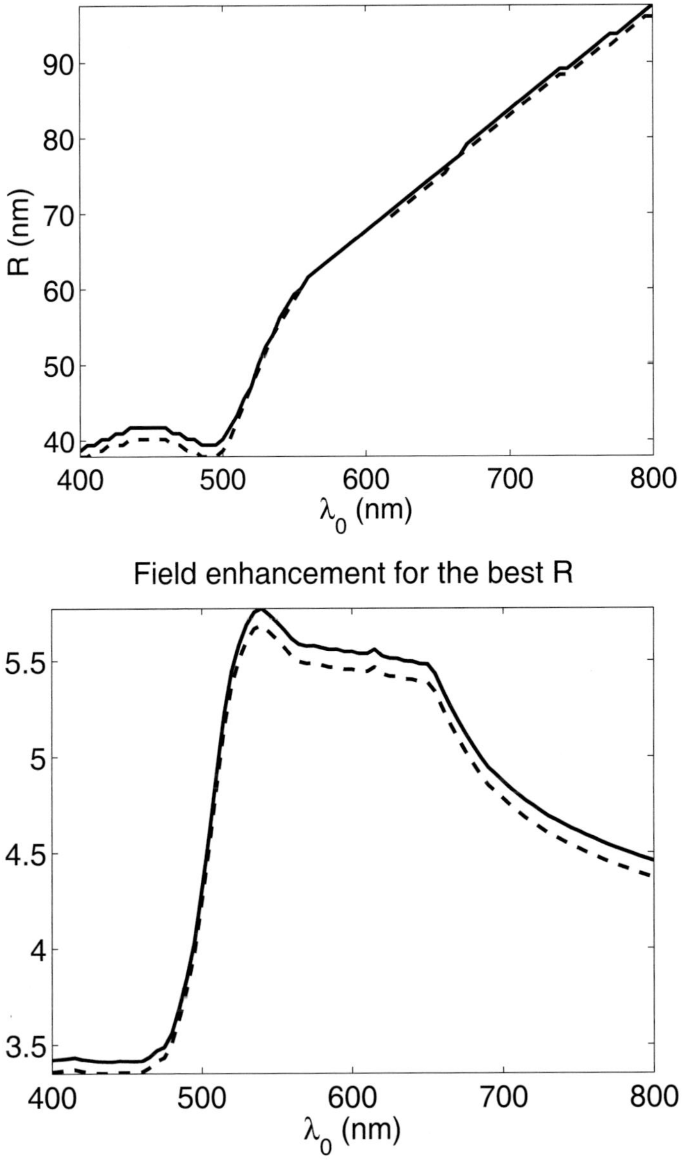

Figure 1.16 Best radius R as a function of the wavelength λ_0 from the full Mie theory and the first-term approximation (a_1) (top) and corresponding field enhancement (bottom).

The results obtained from considering only the first term of the Mie series (a_1) and the dipolar approximation are also plotted on Fig. 1.16. Only the first term of the series of the Mie coefficient of a_1 is sufficient to ensure a relative error of the recovered best radius lower than 4% and of the filed enhancement lower than 2% (Fig. 1.16).

A calculation from the dipolar approximation does not give the best radius and the best field enhancement at all. The radius is to small to preserve the result from the influence of the singularity of the dipole.

1.6 Conclusion

The dipolar model should be used with care in the domain of spectroscopy but remains the simplest one. The dipolar model can help to understand phenomenons but is not relevant for field enhancement calculations. Even if the concept of resonance is actually related to vanishing denominators, poles and essential singularities must be characterized to evaluate the field enhancement, which is obviously not infinite.

The field enhancement induced by specific illumination of a metallic layer is confined near the surface of the metal. This enhancement is not confined laterally and is less than one order of magnitude. The field enhancement induced by a sphere is about one order of magnitude and follows the same curve shape as the field enhancement by a metal layer. The advantage of the nanoparticle is the 3D confinement of the field enhancement in the direction of the incoming polarization and therefore these particles are able to increase the spatial resolution in spectroscopy. The field enhancement appears to vary as a function of the wavelength and a best field enhancement is awaited in infrared, and the metallic nanostructures are more efficient in this domain of wavelength than in the visible domain.

The surface plasmon resonance is able to increase locally the field and therefore is used to enhance the excitation of molecules and then the signal in spectroscopy, leading to enhanced spectroscopies.

References

1. Baida, F. I., Van Labeke, D. and Vigoureux, J. M. (2000). Near–Field surface plasmon microscopy. a numerical study of plasmon excitation, propagation and edge interaction using a 3d gaussian beam, *Phys. Rev. B.*
2. Barchiesi, D. (2011). *New perspectives in biosensors technology and applications*, chap. 5 (INTECH Open Access, Rijeka, Croatia), pp. 105–126, URL http://cdn.intechweb.org/pdfs/14870.pdf.
3. Barchiesi, D. (2012). A classroom theory of the surface plasmon polariton, *Eur. J. Phys.* **33**, 5, pp. 1345–1357.
4. Barchiesi, D., Kremer, E., Mai, V. P. and Grosges, T. (2008). A Poincaré's approach for plasmonics: The plasmon localization, *J. Microscopy* **229**, 2, pp. 525–532.
5. Barchiesi, D., Lidgi-Guigui, N. and Lamy de la Chapelle, M. (2012). Functionalization layer influence on the sensitivity of surface plasmon resonance (SPR) biosensor, *Optics Communications* **285**, 6, pp. 1619–1623.
6. Barchiesi, D. and Otto, A. (2013). Excitations of surface plasmon polaritons by attenuated total reflection, revisited, *La Rivista del Nuovo Cimento* **36**, 5, pp. 173–209.
7. Barchiesi, D. and Van Labeke, D. (1993). Application of Mie scattering of evanescent waves to scanning optical tunneling microscopy theory, *J. Mod. Opt.* **40**, 7, pp. 1239–1254.
8. Bohren, C. F. and Huffman, D. R. (1998). *Absorption and Scattering of Light by Small Particles* (John Wiley & Sons, Inc., New York).
9. Born, M. and Wolf, E. (1993). *Principles of Optics* (Pergamon Press, Oxford).
10. C. V. Raman, F. and Krishnan, K. S. (1928). A new class of spectra due to secondary radiation, *Indian J. Phys.* **2**, pp. 399–419.
11. Barchiesi D. (2013). *Principles and Optimization of Surface Plasmon Biosensor (SPR)*, chap. 11 (Pan Stanford Publishing Pte Ltd, Singapore), ISBN 13: 978-9814303613, pp. 333–357.
12. David, C. (2013). *Raman Spectroscopy Applied to Biomolecule Characterization*, chap. 1 (Pan Stanford Publishing Pte Ltd, Singapore), ISBN 13: 978-9814303613, pp. 1–33.
13. Dawson, P., de Fornel, F. and Goudonnet, J.-P. (1994). Imaging of surface plasmon propagation and edge interaction using a photon scanning tunneling microscope, *Phys. Rev. Lett.* **72**, 18, pp. 2927–2931.

14. Dawson, P., Smith, K. W., de Fornel, F. and Goudonnet, J.-P. (1995). Imaging of surface plasmon launch and propagation using photon scanning tunneling microscope, *Ultramicroscopy* **57**, 2–3, pp. 287–292.
15. Foteinopoulou, S., Vigneron, J. P. and Vandenbem, C. (2007). Optical near-field excitations on plasmonic nanoparticle-based structures, *Opt. Express* **15**, 7, pp. 4253–4267, URL http://www.opticsinfobase.org/oe/abstract.cfm?URI=oe-15-7-4253.
16. Grosges, T., Barchiesi, D., Kessentini, S., Gréhan, G. and Lamy de la Chapelle, M. (2011). Nanoshells for photothermal therapy: A monte-carlo based numerical study of their design tolerance, *Biomed. Opt. Express* **2**, 6, pp. 1584–1596, URL http://www.opticsinfobase.org/boe/abstract.cfm?URI=boe-2-6-1584.
17. Grosges, T., Barchiesi, D., Toury, T. and Gréhan, G. (2008). Design of nanostructures for imaging and biomedical applications by plasmonic optimization, *Opt. Lett.* **33**, 23, pp. 2812–2814.
18. Grosges, T., Borouchaki, H. and Barchiesi, D. (2007). Improved scheme for accurate computation of high electric near-field gradients, *Opt. Express* **15**, 3, pp. 1307–1321.
19. Le Ru, E. C. and Etchegoin, P. G. (2009). *Principles of Surface-Enhanced Raman Spectroscopy and related plasmonic effects*, 1st edn. (Elsevier, Amsterdam).
20. Logiurato, F. (2012). Relativistic derivations of de broglie and planck-einstein equations, *ArXiv e-prints* .
21. Luxpop (2013). http://www.luxpop.com.
22. Maier, S. A. (2007). *Plasmonics. Fundamentals and Applications* (Springer, New York, USA).
23. Mie, G. (1908). Beiträge zur Optik trüber Medien speziell kolloidaler Metallösungen (contributions to the optics of turbid media, especially colloidal metal solutions), *Ann. Phys.* **25**, pp. 377–445.
24. Mojarad, N. M., Sandoghdar, V. and Agio, M. (2008). Plasmon spectra of nanospheres under tightly focused beam, *J. Opt. Soc. Am. B* **25**, 4, pp. 651–658.
25. Palik, E. D. (1985). *Handbook of Optical Constants* (Academic Press Inc., San Diego USA).
26. Refractiveindex (2014). http://refractiveindex.info/
27. Salvi, J. and Barchiesi, D. (2014). Measurement of thicknesses and optical properties of thin films from surface plasmon resonance (SPR), *Applied Physics A* **115**, 1, pp. 245–255, doi:10.1007/s00339-013-8038-z.

28. Stilgoe, A. B., Nieminen, T. A., Kröner, G., Heckenberg, N. R. and Rubintsztein-Dunlop, H. (2008). The effect of Mie resonances on trapping in optical tweeters, *Opt. Express* **16**, 19, pp. 15039–15051.
29. Wriedt, T. (2012). *The Mie theory, Optical Sciences*, Vol. 169, chap. 2 (Springer-Verlag, Berlin), pp. 53–71.

Chapter 2

Plasmonic Imaging Systems and Dedicated Functionalized Biochips for Biosensing

Julien Moreau,[a] Jean-Pierre Cloarec,[b,c] Paul G. Charette,[c,d] Grégory Barbillon,[a] and Michael Canva[a]

[a]*Laboratoire Charles Fabry (LCF)—Institut d'Optique Graduate School (IOGS), Université Paris-Sud (UPS), CNRS, Palaiseau, France*
[b]*Université de Lyon, Institut des Nanotechnologies de Lyon (INL)site École Centrale de Lyon, CNRS, Ecully, France*
[c]*Laboratoire Nanotechnologies & Nanosystèmes, UMI 3463 CNRS, UdeS, INSA de Lyon, ECL, UJF, CPE Lyon. Université de Sherbrooke, Canada*
[d]*Département de Génie électrique et Génie Informatique, Université de Sherbrooke, Sherbrooke, Canada*
julien.moreau@institutoptique.fr

Surface plasmon resonance (SPR) imaging can be used as a multidimensional instrument to characterize surface (coordinates x and y) evolution (time t). In parallel, these surfaces can be functionalized and organized as biochips, provided they fulfill requirements specific to label free biomolecular sensing. Getting precise information about the thermodynamics of the probes and targets interactions also necessitates precise microfluidic conditions as well as enhanced image and data processing. Several

Handbook of Enhanced Spectroscopy
Edited by Pietro Giuseppe Gucciardi, Marc Lamy de la Chapelle, and Nathalie Lidgi-Guigui

ISBN 978-981-4613-32-3 (Hardcover), 978-981-4613-33-0 (eBook)
www.panstanford.com

experimental approaches, such as biomolecular interactions, whole cells analysis, or on-chip chemical synthesis, are presented herein, illustrating the diversity of bio-analytical strategies compatible with SPR imaging. Depending on the information needed, different experimental setups can be used by taking advantage of other dimensions such as incidence angle (θ), wavelength (λ), and polarization (P). We summarize here our different configurations, and their recent performances are reviewed. To further enhance the plasmonic biochip reader capabilities, it is anticipated that one will have to move from homogeneous gold film to other types of support such as gold nanodisks or micro/nanostructures.

2.1 Introduction

Plasmonics, as it is now called, has evolved as a field of its own, with many sub-fields in basics research as well as applications. SPR sensors are now widely used and/or studied for detection of chemical and biological species, and this point was addressed in some reviews, in particular by J. Homola and Dostálek [1] as well as the annual reviews by Rich and Myszka, since 1999 [2] up to 2011 [3] notably including study in Ref. [4]. A specific sub-field concerns the area of plasmonic imaging of surface structured into bioarrays or biochips, allowing the parallel monitoring of tens or hundreds of biomolecular interactions with all the advantages of label free characterization.

In this chapter, we will focus on some of our recent advances in the field of plasmonic imaging as used in biochip systems. The first section will be devoted to the plasmonic imaging instrumentation. We will briefly review the basics of the SPR and particularities of Surface Plasmon Resonance Imaging (SPRI) systems. This is part of a steadily growing field, as illustrated in Fig. 2.1. The emphasis here will be on showing how multidimensional they can be considered. We will mention the importance of image and data processing and will pay special attention to the specific microfluidic conditions that must be met in order to provide valuable dynamical information about the surface biomolecular interactions. In the next section, we will give examples of the biochip realization in terms

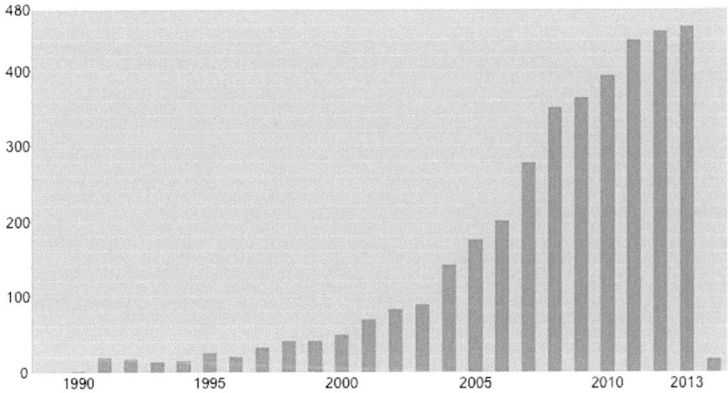

Figure 2.1 Illustration of the evolution of the steadily increasing number of publications per year in the "plasmon AND imaging" field, as obtained searching the Web of Science database (2014/01/20).

of chemical and biomolecular functionalization, which lies at the heart of such plasmonic label free sensor systems. Subsequently, in another section, we will emphasis different recent types of instrumentations specifically designed to gather data in respectively, the spectral and polarimetric dimensions. We will also highlight the necessity of calibration protocol in biochip experiments and how the temporal dimension can be used for easy and robust self-calibration procedures. Finally, before drawing conclusions, we will extrapolate the current designs and performances that seem to lead to other choices other than homogeneous gold film such as plasmonic substrate material.

2.2 Surface Plasmon Resonance Imaging as Multidimensional Instrument

2.2.1 *Basics of Plasmonic Imaging Sensing*

Surface plasmon resonance sensors are mainly based on plasmon resonance displacement monitoring to detect a modification of the index of refraction at the vicinity of the metallic layer. This can be achieved by a spectral, $R(\lambda)$, or an angular, $R(\theta)$, measurement.

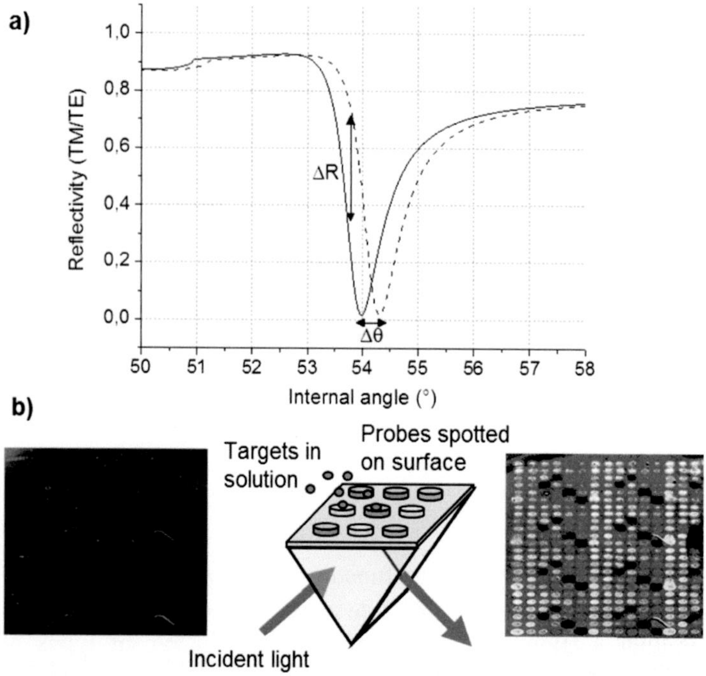

Figure 2.2 (a) Calculated reflectivity ($\lambda = 790$ nm) as a function of the internal incident angle on a 51 nm thick gold layer above a 3 nm chromium adhesion layer on a SF10 Prism. Full line: dielectric with $n = 1.329$; dashed line: step of index of $\Delta n = 5 \times 10^{-3}$. (b) Experimental reflectivity images (in false colors) obtained with SPRI before and after the injection of an oligonucleotide target on single strand DNA probes.

However, these two methods are not practical when one wants to add an imaging $R(x, y)$, capability to SPR sensors, while keeping a real-time capability $R(t)$. In SPRI systems, detection is realized by a local reflectivity measurement using a multiplexed detector array, such as a CCD camera. Indeed, it can be easily seen that a displacement of the plasmon resonance, expressed in terms of spectral or angular variation, can be detected as a reflectivity variation if one works at a fixed angle of incidence and for a fixed excitation wavelength (Fig. 2.2(a)).

With this imaging capability, the sensing surface can be functionalized with hundreds of different probe spots. Detection events such as a biomolecular targets binding to a probe spot will be directly seen in the acquired images as a local reflectivity variation. SPRI is a relatively simple multiplex sensor in the sense that it is a purely optical system where excitation and reading is done at the same wavelength, with the same light source. First SPRI system date from 1988 by Rothenhausler and Knoll [5] and numerous references can be found in the literature [6–8]. Commercial SPRI-based systems are sold by companies such as *Biacore* and *Horiba Scientific*. Figure 2.2(b) shows some typical images obtained with an SPRI setup before and after the injection of a target on a functionalized gold surface with various probes. Affinity between target and probes are directly interpreted from the local reflectivity variations observed. Most of our application examples will be taken using DNA–DNA interactions such as in Ref. [9].

2.2.2 *Optical Setup*

Figure 2.3 shows a general scheme of our SPRI system [10]. The Kretschmann configuration is used, where incident light passes through a prism before reaching the metallic layer. The light source can be the output of an optical fiber or directly the source itself such as an LED (λ = 790 nm, spectral width of 20 nm). After being collimated and TM polarized, the light is sent on a rotating motorized mirror, which is optically conjugated with the sensing surface. This ensures that the illumination of the surface will be

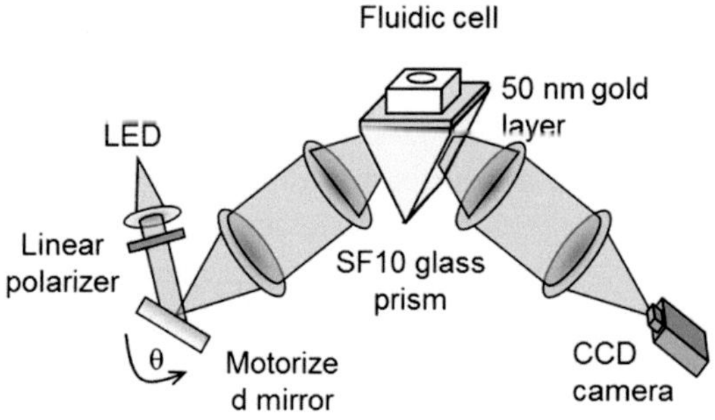

Figure 2.3 Scheme of the SPRI setup based on the Kretschmann configuration.

kept constant when the incident angle is changed. The surface is then conjugated with the CCD camera (12-bit, 1360 × 1024 pixels from Pixelfly). The simplest way to achieve this is to use an afocal system between mirror and sensing surface as well as one between the sensing surface and the detector. Changing the focal ratio for the two lenses that composed the latter afocal will change the magnification on the detector. A large field of view is usually chosen to be able to use the maximum number of probe spots on the surface. The imaging area is approximately 9 mm × 9 mm in this case. It is important to notice that in this configuration, the image of the light source is sent to infinity. For a perfect point source, this gives a perfectly homogenous illumination of the surface in terms of intensity and angle of incidence. For real light sources, illumination will never be strictly homogenous and some normalization procedures are mandatory to take into accounts both intensity and incident angle variation across the surface. Besides, unavoidable differences in probe surfacic densities, from one spot to another, will cause discrepancies in SPR responses. The normalization procedures are described in Section 2.4.2. The sensing surface itself is a thin gold layer of 48 nm thick with a 2 nm chromium adhesion layer, not deposited directly on the prism but on a glass coverslip. Although silver would be better for sensing, gold is usually preferred, as it does not suffer from oxidation processes. An index-matching oil is used to minimize secondary reflections between the prism and the glass coverslip. Chemical functionalization of the gold surface will be described in Section 2.3. Solutions are injected in a fluidic cell (about 10 μL) using a peristaltic pump, by placing against the gold surface.

Compared with SPR mono-sensors, some differences in the source used can be noted. First, a quasi-monochromatic light source such as a laser cannot be used in SPRI systems as speckle phenomena will strongly degrade the image quality and illumination stability over time. A source with a larger temporal coherence, such as LED, is therefore preferred. A drawback of LED sources is a much larger angular divergence compared with a laser beam. Reducing the effective size of the source with a pinhole in order to be as close as possible to a perfect point source and using some collimating optics can partially solve the problem. However, this rapidly reduces the amount of light available in the system. Typical angular divergence of 0.25° (full width at half maximum) can be achieved while keeping a sufficient amount of light falling on the detector. These trade-offs have a significant effect on the reflectivity minimum at the resonance (which is not equal to zero as expected for the ideal case) but the effect on the overall sensitivity, which proportional to the slope of the resonance, is surprisingly small: the loss is experimentally found to be less than 10%.

2.2.3 *Performances*

A sensor is for instance characterized by its sensitivity and its resolution. Optical index sensitivity, which is most commonly used, is defined as the reflectivity variation ΔR measured for a given step index Δn change in the buffer solution. It is directly related to the angular or spectral displacement of the SPR curve due to the step index as well as the width and depth of the resonance. Numerical simulation shows that the resonance becomes sharper as you increase the wavelength used to excite the surface plasmons. Therefore, it is expected that optical index sensitivity of SPR systems will increase linearly with the wavelength used. However, experimentally, this reduction in the width of the resonance is limited by the nonzero angular divergence of the illuminating beam. Sensitivity of the order of $\Delta R/\Delta n \sim 7500\%.\ \mathrm{RIU}^{-1}$ is obtained with our SPRI setup at the operating wavelength of 790 nm. Finally, the sensitivity is weakly dependent on the thickness of the gold layer around the optimal thickness of 48 nm (a difference of 5 nm in the thickness only reduce the sensitivity by 15%).

Resolution is the minimum step index of refraction detectable. It is directly related to the sensitivity previously defined and the noise level of the system. If we consider only the optical part of the setup, putting aside thermal effect and other chemical or biological-related drifts, shot noise is the dominant noise in an SPRI system. In this case, the noise of a reflectivity measurement on a given area of the sensing surface can be simply written as:

$$\text{Noise}\ (\%) = \frac{100}{\sqrt{R N_e N_{acc} n_{px}}}$$

where R is the mean reflectivity of the measured area, N_e is the full-well capacity of a camera's pixel, N_{acc} is the number of images accumulated in time, and n_{px} is the number of pixel averaged in the chosen measurement area. For the CCD camera used, N_e = 18,000 and N_{acc} is limited to 16 accumulations in order to keep the "real-time" capability. For a reflectivity around 30% at the working angle, this gives a theoretical noise level of 0.34% for one pixel, very close to what is observed experimentally. With the sensitivity given above, the resolution is then $r \sim 5 \times 10^{-5}$ RIU for a single pixel measurement. In practice, reflectivity measurements are made by averaging the signal on a whole probe spot image, a hundred pixel area, thus giving a resolution of $r \sim 5 \times 10^{-6}$ RIU.

Another way to characterize the sensitivity, certainly more useful in chemical and biological applications, is to calculate the minimum surface coverage for a given molecule, or **limit of detection**, that can be detected. In the case where the target molecular layer thickness bound on the surface is much smaller than the penetration depth of the plasmon evanescent wave ($e_b << L_z \approx 200$ nm), then there is a simple relationship between the surface coverage Γ and the variation of reflectivity ΔR detected:

$$\Gamma = \frac{\Delta R}{\frac{\partial n}{\partial C}\frac{\partial R}{\partial n}} L_z$$

where $\partial n/\partial C \sim 0.19$ cm^3/g for oligonucleotides and proteins and $L_z = 200$ nm. The minimum surface coverage is then related to the step index resolution r by:

$$\Gamma_{\min} = \frac{r}{\frac{\partial n}{\partial C}} L_z$$

with $r \sim 5 \times 10^{-6}$ RIU, the minimum surface coverage detectable is around 0.1 fmol/mm^2 for a 50 kDa target protein such as streptavidin. For a typical 300 μm diameter probe spot, this gives a theoretical limit of detection of about 10 attomol/spot. In practice however, in terms of the minimum detectable target concentration in the solution, going below 1 nmol/L concentration in an SPR experiment is usually extremely difficult without some amplification techniques. This is mainly due to the small volume of the microfluidic cell and the diffusion-limited analyte transport regime [11]. Also, because a large sensing area is usually preferred in

SPRI for chemical and biological applications, this limit of detection will be degraded by other sources of noise such as thermal drift, difference in fluidic across the sensing area, or inhomogeneous nonspecific adsorption.

The time resolution of the setup, defined as the time between two successive acquired images, is limited by the acquisition speed of the camera and the number of images accumulated in order to improve the signal to noise ratio (SNR) of the measurement. A trade-off between SNR and time resolution has to be chosen. A significant amount of time is also necessary for the software to capture and save the acquired images. Overall, the time resolution in our SPRI setup is around 3 seconds. In terms of spatial resolution, with the CCD camera used and a 9 mm × 9 mm sensing area, the spatial resolution was found to be about 14 microns. Spatial resolution can be in theory freely chosen by adjusting the focal lens used in the afocal, but two experimental aspects will limit the effective spatial resolution. First, because a large imaging area is preferred (a few millimeters to a centimeter size) in order to increase the number of different probes deposited on the surface, even with a camera having a few megapixel, the spatial resolution will be limited to a few microns at most. Using a camera with a larger number of pixels is possible, although it will significantly degrade the time resolution (lower acquisition speed) and sensitivity (smaller full-well capacity) of the setup. Second, as the imaging system is not working at a normal incidence from the gold layer, a perfect optical conjugation between the surface and the camera is not realized. Some defocus is always observed on the top and bottom part of the acquired images limiting further the benefit of increasing the spatial resolution.

2.2.4 *Image and Data Processing*

The sequence of events in a typical SPRI experiment is the following: first, the reflectivity of the whole sensing surface as a function of the incident angle is recorded by rotating the motorized mirror. This allows finding of the optimum working point where the reflectivity variation measured will be maximal. During this preliminary measurement, the fluidic cell is filled with the buffer solution. The incident angle is then fixed and reflectivity images of

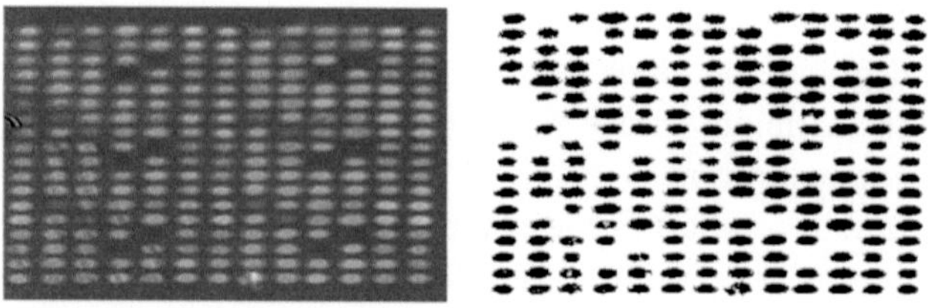

Figure 2.4 (Left) Reflectivity image obtained with the SPRI setup after a DNA–DNA interaction experiment and (Right) binary mask calculated using a threshold algorithm.

the sensing surface are recorded as a function of time while target solutions are injected. In order to extract relevant information from the stack of acquired images, some image processing is necessary. Mainly, a binary mask containing the probes spots matrix has to be applied on the stack of images (see Fig. 2.4). The simplest way is to use as a mask, a regular matrix of circular area using some prior knowledge on the position of the various spots. More complex solutions involved working directly on the experimental images obtained after a target–probes interaction has occurred and applied some threshold algorithms. The aim is to be able to measure a local variation of reflectivity only in the relevant areas on the biochip surface where probes have been bound. Measuring outside spots will only increase the noise and sometime can drastically reduce the precision of the measurement if some nonspecific interactions occurred. Numerous threshold algorithms exist [12, 13]; the main difficulty relies on the low contrast ($<1\%$) between the spots and the background in the raw reflectivity images recorded by the SPRI setup. Moreover, this background is usually heterogeneous across the image, which makes it impossible to find a common threshold for all spots. An efficient way to improve the homogeneity and increase the contrast is to define a background image and subtract it from the images stack. This background image can simply be one of the first images recorded, before the injection of any target. For long-lasting experiment, where the background changes with time due to some nonspecific absorption on the biochip surface, this background image must be regularly updated. In this case, more complex method to define the background image, such as morphological closing and opening algorithms, can be used.

Also, in order to better understand experimental results and better design the associated experimental conditions, some methods are needed to simulate the plasmonic resonance and its following shift due to the binding of a biomolecular layer. For any homogenous multilayer system, the Rouard extended method [14] can be used. It is based on multiple reflections on multilayered films that conserve the Snell invariant "$n \sin(\theta)$" in the complex dimension; that is when n is complex due to the absorption of a layer, $\sin(\theta)$ is also treated as a complex number. This basically allows simple and effective calculations of the reflectivity of an arbitrary large number of layers of any material. Such method was used to optimize plasmonic sensors substrate, in particular metal choice and associated thickness [15].

2.2.5 *The Other Possible Parameters R(x, y, t, λ, θ, P)*

The instrumental capability of SPRI sensors can be extended by adding some new dimensions in the system as schematically illustrated in Fig. 2.5.

The most obvious one is to add a spectral measurement capability as many molecules of biological interest have spectral signature in the visible and near IR. This can be done by using a spectrally tunable source instead of a simple LED as in the SPRI

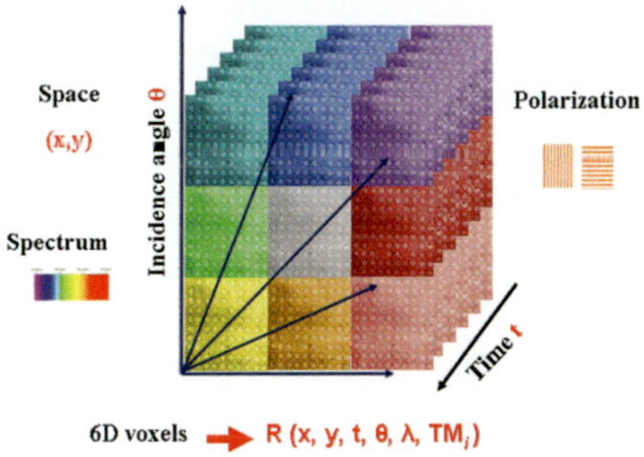

Figure 2.5 Schematic of the multi-parametric measurement capabilities of SPRI setup: $R(x, y, t, \theta, \lambda, \mathrm{TM})$.

optical setup describe above or using a white light source and a grating in front of the camera. Notice that adding a spectral measurement capability obliges to sacrifice another dimension, such as the real-time capability or one of the spatial dimensions. It is also possible to look at the polarimetric dimension with the aim of being sensitive to the anisotropic behavior on the sensing surface. These points will be discussed in the further section dealing especially with the recent advances in multidimensional SPRI systems.

2.2.6 *Choice of Metals*

We also started investigating another route to routinely be able to use silver material instead of gold, as explained thereafter, and overcoming the drawbacks the former materials exhibit with respect to the latter in particular in biofunctionalization properties as detailed in the functionalization section. Indeed, one of our recent work focused on the advantages and applicability of using silver instead of gold as metal to generate the plasmon wave. The most salient difference between the SPR responses of silver and gold as a function of angle is the shape of the resonance dip, which is narrower for silver. In fact, of all the noble metals that can be used for SPR, silver is theoretically capable of the highest sensitivity over the visible and near-infrared (NIR) wavelength ranges. This is illustrated in Fig. 2.6, which shows the sensitivity, in terms of change in reflectivity as a function of bulk refractive index at the point of maximum slope, as a function of wavelength for gold and silver.

However, even with lower sensitivity, gold remains the metal of choice in commercial SPR systems. Unlike silver, gold exhibits good chemical stability under ambient and aqueous conditions and can be chemically functionalized using several approaches. Indeed, none of the existing methods listed above for passivating silver are in widespread commercial use, presumably because of the perceived insufficient added value in performance at the cost of increased complexity of fabrication, compared with gold. However, the use of gold results in small losses of sensitivity at longer visible wavelengths and NIR, sensitivity at shorter wavelengths is severely affected. Excitation at short wavelengths is particularly interesting in applications of surface-plasmon field-

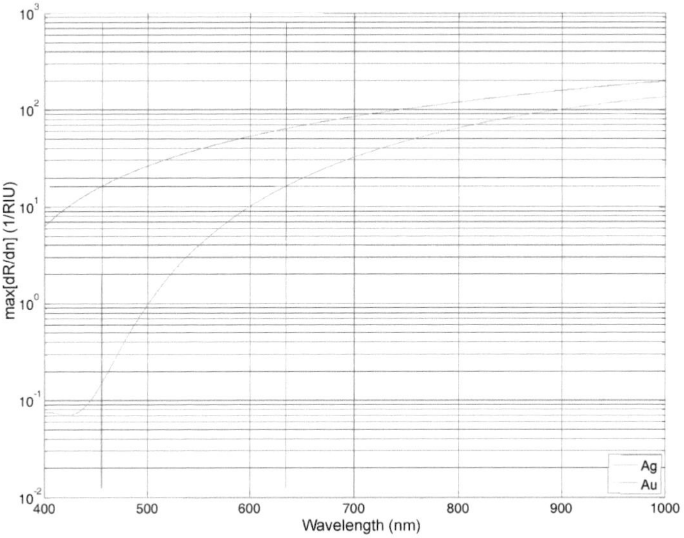

Figure 2.6 Maximum sensitivity to bulk index change, as defined by [1], for Au and Ag. The horizontal and vertical bars indicate the sensitivity correspondence between Au at 635 nm and Ag at 456 nm.

enhanced fluorescence spectroscopy [16], also known as surface plasmon-coupled fluorescence emission [17], owing to the greater range of fluorophores excited at short wavelengths.

2.2.7 *Microfluidic Issues*

Generally flow is laminar in microfluidic systems, there is little or no transport of ligands by self-mixing in the fluid. Hence, diffusion is the main mechanism by which target molecules migrate perpendicularly to the direction of fluid flow toward the sensing surface. As target molecules are consumed by the receptors on the sensing surface, the target supply in the immediate neighborhood of the surface may not be efficiently replenished, which will result in mass-transport-limited kinetics and the formation of a depletion zone. The depletion zone thickness is defined as the height above the surface where target replenishment by convection exactly balances consumption by diffusion to the sensor. The depletion zone thickness is typically less than 1 μm in microfluidic systems.

In a thin rectangular cross-section channel, an increase in volumetric flow rate will reduce the thickness of the depletion zone via an accentuation of the parabolic profile, thereby decreasing the distance over which ligands must diffuse to the sensing surface. However, the diffusive flux to the surface varies only weakly with the flow rate. Indeed, the volumetric flow rate must be increased by 10^3 to enhance the diffusive flux to the surface by a factor 10 [11]. Simply increasing flow rate, therefore, is a very inefficient method of increasing target diffusive flux to the sensor. Though recirculation eliminates target waste, capture efficacy and time-to-equilibrium are largely unchanged.

In addition to speed, the sensitivity of an SPR system may also be affected by the lack of adequate mixing. Indeed, high-performance SPR systems extract rate constants from parametric models of the experimental data. In the case of the well-known two-compartment model for mass-transport-limited kinetics [18], the model will fit the data poorly when there is too high a difference (over a factor 10 according to Ref. [19]) between the rate at which ligands diffuse to the surface and the rate at which they are consumed by the assay. If diffusion is too slow, a "fast" reaction will appear to be instantaneous and poor estimates of rate constants will result. Consequently, the range of measurable rate constants under mass-transport-limiting conditions can be extended in direct proportion to increasing the rate of mass transport to the surface by an alternative mechanism, such as mixing. We investigated several strategies for fluid mixing that can be combined with SPR [20, 21] and present thereafter the coupling of SPR biosensing with surface acoustic wave (SAW) mixing.

2.2.7.1 Coupled SPR biosensing and SAW-induced mixing

As a result, much work has been devoted to microfluidic mixing in order to both increase the speed and improve the sensitivity of biosensors in microfluidic devices. One particularly interesting review is the paper by Nguyen and Wu [22] on mixing techniques that presents an interesting graph of publications as a function of Péclet number versus Reynolds number. In our work, we have for instance investigated the use of SAW, which are mechanical oscillations that propagate on the surface of a material. They can be generated with interdigitated transducers (IDT) fabricated using thin-film metal deposition on a piezoelectric substrate. Shear-vertical (Rayleigh) SAW deform the substrate in the direction normal to the surface and transmit mechanical energy onto the fluid both by deformation of the surface and via the electric field, inducing turbulent flow within the fluid with acoustic streaming [23]. SAW have proven very effective for microfluidic mixing in the context of biosensing [24].

We have developed a device combining biosensing via SPR and micromixing via SAW on a single substrate [25]. Since both the SPR sensor and SAW transducer can be fabricated using low-cost microfabrication methods simultaneously on a single substrate, the design is well suited to chip-based biosensing applications. As an added bonus, high-intensity SAW has been shown to reduce and/or remove molecules nonspecifically bound to the surface, thereby increasing the measurement signal-to-noise ratio [26, 27]. Figure 2.7 shows a schematic diagram of the device with the SPR sensing surface and IDT electrodes on the common $LiNbO_3$ piezoelectric substrate. The IDT electrodes and SPR sensing surface

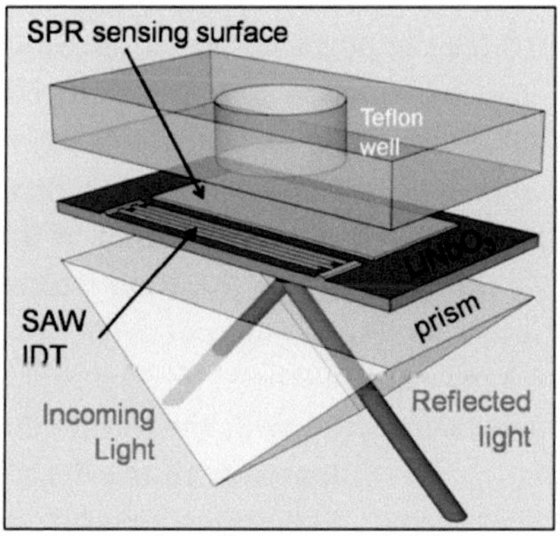

Figure 2.7 Schematic diagram showing the SPR sensing surface and IDT SAW electrodes on a common $LiNbO_3$ piezoelectric substrate. Also shown are the SPR excitation and reflected light paths that are coupled to the sensing surface via the prism, as well as the microfluidic cell. Reproduced from Ref. [25] by permission of The Royal Society of Chemistry.

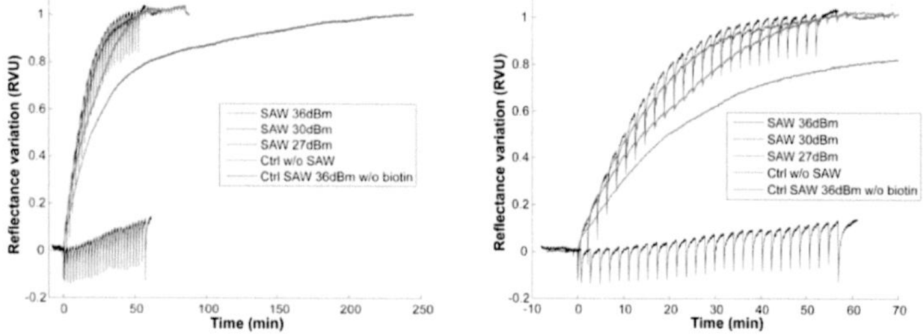

Figure 2.8 SPR reflectance variations at a fixed angle of interrogation as a function of time for a biotin–avidin assay in PBS (zoomed view on the right). Periodic SAW excitation was applied in 5 s duration pulses followed by 120 s relaxation intervals, for a range of SAW power levels (27, 30, 36, and dBm). A negative control without the application of SAW in the avidin–biotin assay is shown ("Ctrl w/o SAW"). A second negative control where unbiotinylated BSA was adsorbed onto the surface followed by injection of avidin under 36 dBm SAW is also shown ("Ctrl SAW 36dBm w/o biotin"). The downward spikes visible in the 36 dBm data are due to parasitic heat injection into the fluid by the SAW pulses. Results are given in reflectance variation units (RVU), where 1 RVU equals 100% reflectance change. Reproduced from Ref. [25] by permission of The Royal Society of Chemistry.

were fabricated from 48 nm-thick Au thin films with a 3 nm-thick Cr adhesion layer.

Figure 2.8 illustrates the performance of the SPR/SAW chip for measuring assay binding kinetics. SPR reflectance variations are shown as a function of time at a fixed angle of interrogation for a standard assay: biotin–avidin in phosphate buffer solution (PBS). Starting at the time of avidin introduction (time = 0 in the graphs), SAW excitation was applied in 5 s duration pulses followed by 120 s relaxation intervals, for a range of SAW power levels (27, 30, and 36 dBm). These experiments show that SAW action results in accelerated kinetics compared with the negative control due to microfluidic mixing ($\sim \times 5$ compared with the negative control for these particular experiments). Experiments performed with higher SAW power levels resulted in curves with an envelope that was identical to that of the 36 dBm curve, indicating that the chip had achieved a maximum mixing efficiency for that particular SAW/SPR design. Power levels of 30 and 27 dBm also resulted in accelerated kinetics but with intermediate levels of mixing efficiency.

The downward spikes in the SPR signal visible in the 36 dBm data are due to parasitic heat injection into the fluid by the SAW pulses, since the refractive index of the fluid changes with temperature (in the case of water, there is a 10^{-5} RIU refractive index change for every 0.1°C of temperature change, an easy measurement signal for SPR). The results also demonstrate that the chip nevertheless recovers well from the heat injection pulses with the binding curve maintaining an exponential envelope, as expected for this type of assay.

2.2.8 *SPRI Partial Conclusions*

Section 2.2 summarized the basis knowledge as well as the instrumental requirements for SPRI. Based on the metal–dielectric plasmonic properties, which can be simulated, the instrument essentially consists of an optical setup allowing to couple plane waves into and out of a propagating surface plasmon mode localized, with imaging capability. In order to provide useful information about the surface under study, image and data processing have to be performed. Special attention is paid to the microfluidic that conveys the targets biomolecules to the probes, especially if the seeked information is of dynamical nature. These main components give the possibility to monitor in parallel and real-time events on a surface organized into a biochip format. The following section addresses the special issue of functionalizing and structuring the biochip.

2.3 Biochip Functionalization

Surface plasmon resonance biosensing implies several advantages and constraints. The first interest of SPR is the ability to perform real-time detection of nonlabeled molecular species. This avoids any labeling step and enables to monitor molecular interactions without being hindered by any label. However, using nonlabeled species suffers an important drawback. The transducer signal changes can be trusted only if the functionalization is correctly, in fact close to perfectly, adapted to the biomolecular system to analyze. For instance, nucleic acid interactions do not imply the same requirements than proteins or carbohydrates, because these last molecules are more fragile or more difficult to synthesize. If the surface of SPR sensor is not correctly functionalized, nonspecific interactions of targets with various sites of the support may occur, and this could lead to false-positive or false-negative signals. Another requirement for SPR analysis is the stability of the surface layers. Progressive damaging of chemical and/or biological functionalization can lead to a gradual degradation of the quality of the analysis and add temporal drift components to the dynamical signals.

The goal of functionalization is therefore to implement a bioreceptor layer composed of biomolecular probes, aimed at specifically recognizing a (bio)molecular target. Functionalization can be schematically seen as a two-step operation: a first chemical modification of the initial metallic support (chemical functionalization); followed by the grafting of a bioreceptor, usually a layer of biomolecules (biofunctionalization). As described below, functionalization strategies are much more diverse than this simple two-step scheme. A few main strategies are reported in the literature for functionalizing SPR supports. The alkanethiol strategy is certainly the most studied and performed method, although other alternative approaches (e.g., polypyrrole electropolymerization) can be reported. Alkanethiol based functionalizations involve the affinity between thiol functional group and noble metals such as gold.

Whatever the chosen functionalization pathway, a wide diversity of biofunctionalization is reported. First, this diversity is due to the different (bio)molecular probe–target systems used for SPR biosensing, depending on the chosen biological model or biological application. Examples of SPR detection for nucleic acids, proteins, oligosaccharides, lipids, and living cells will be described below. For each class of biomolecular probes or targets, one can count several approaches for performing immobilization and biorecognition. Classical direct binding, use of additional linkers or biomolecules for tailoring surface properties, and DNA-directed immobilization (Zip-code sequences) are some of the strategies that will be presented.

Addressing molecular probes on sensor surfaces can be performed using technologies such as mechanical spotting, inkjet printing [28], and microcontact printing [29]. This problematic will only be slightly approached herein.

SPR analysis provides a powerful analysis technique for analyzing probe–target interactions, but it can also be used for monitoring other phenomena. SPR analysis can or could monitor functionalization steps, bio-oligomers on-chip direct synthesis, on-chip fishing and sorting, and molecular or cellular structural changes.

Chemical functionalization of gold surfaces with alkanethiol molecules and biofunctionalization are presented below in Section 2.3.1. Examples of biomolecular interactions to monitor with SPR are presented in Section 2.3.2. Alternative immobilization techniques, such as polypyrroleelectropolymerization [30–37], and direct chemical synthesis of biopolymers onto support [28, 38–43] cannot be presented in detail herein: the reader is invited to report to the cited references for more information.

2.3.1 *Chemical and Biomolecular Functionalizations of Gold Surfaces*

Gold substrates used for SPR measurement have to be modified in order to selectively bind target species. A first modification—named "chemical functionalization"—consists of grafting chemical functions onto gold. These chemical functions are then used to immobilize biomolecular probes, in a step we call herein "biomolecular functionalization."

2.3.1.1 Chemical functionalization of gold with alkanethiols

The most reported approach for functionalizing gold supports is to use the affinity of alkanethiols for gold, in order to form organic self-assembled monolayers (SAM) [29, 44, 45]. Alkanethiols are organic molecule bearing at least one terminal thiol (–SH), which is a chemical moiety spontaneously binding to gold. Self-assembly of alkanethiols on gold can yield, in appropriate conditions, a robust organic monolayer fully covering the metal surface. Though alkanethiol layers on gold are widely reported in the literature, their implementation and further chemical or biological functionalization can be either straightforward or uneasy. Love et al. [29] report how the implementation of the metal film itself can greatly influence the topography and organization of gold crystallites [46]. After deposition, gold is usually treated, prior to reaction with alkanethiols, in order to remove organic contaminants and rearrange the gold crystallites. For instance, flame annealing and chemical treatments can be used for both removal of contaminations and enlargement of the size of <111> gold crystallites, for typically yield micronic size terraces [29, 47–49].

The topography of substrate itself may strongly influence the dynamic organization (compacity, adjustment to the surface, defects, pin holes) of the alkanethiol layer. Considering the dynamic structuration of SAM on gold nanoparticles and clusters [29, 50], one can expect a strong influence of gold structuration on the next functionalization steps.

This may have consequences on the alkanethiol upper surface reactivity, thus influencing density of immobilized probes and biological recognition [51].

However a difficulty may arise from the fact that gold films of 200 nm thickness, which were widely used for studying at a nanometric scale the assembly of alkanethiols [52–55], may not be equivalent to the 50 nm. A 200 nm thick film is robust enough to stand a wide variety of gold cleaning and recrystallization prior to self-assembly. Films (48–50 nm thick) are less robust, and the conditions for gold coating (e.g., e-beam evaporation or sputtering coating) can strongly influence the final topography and the size of gold crystallites. Electrochemical reorganization, thermal, or flame annealing may be, in some cases, difficult to perform without damaging the sample, for instance, by diffusion to the surface of first anchoring layer (e.g., chromium).

Though macroscopic optical properties may seem to be identical, these discrepancies at the nanometric scale may cause important differences of behavior from a batch to another for alkanethiol monolayer assembly, thiols availability toward chemical grafting of biomolecules [50], and biomolecular probes ability to bind targets [51]. Understanding of alkanethiol films behavior toward ligand grafting is thus still a major field of study [56–60].

2.3.1.2 Biomolecular functionalization: immobilizing biomolecular probes onto SAM

After alkanethiol SAM implementation, the terminal moieties of the alkanethiols provide a surface enabling to bind biomolecular probes. These probes are aimed at further capturing biomolecular targets [28, 61]. Examples of SAM surface functional groups used for binding biomolecular probes are given in Table 2.1. Binding can be performed by chemical reaction with the surface, by physical interaction (hydrogen bonding, electrostatic interactions, hydrophobic interactions, complexation), or by a conjugation of both when several consecutive steps are involved.

Methyl (CH_3) terminated alkanethiols (Table 2.1(a)) yield a hydrophobic surface useful for adsorbing species through hydrophobic interactions. This type of surface is mainly adapted for binding lipidic membranes [62, 63], while it is not adapted to nucleic acids and proteins. **Primary amine** (NH2) terminated alkanethiols (Table 2.1(b)) can be used either for adsorption through electrostatic interactions at acidic pH, which causes amine protonation (thus providing $-NH_3^+$ charged moieties) [64] or for covalent binding [56, 59]. In the second case, the amine layer can bind proteins and peptides, whose carboxylic acid lateral chains were previously activated, for instance, through NHS/EDC procedure (see below).

Alkanethiols bearing a terminal **carboxylic acid** (COOH) moiety, such as 11-mercaptoundecanoic acid (MUA) (Table 2.1(c)), are classically used as a first functionalization monolayer, enabling further functionalization. In order to react with primary amines, carboxylic acid terminal groups need to be previously activated, for instance, by the N-hydroxysuccinimide/1-(3-dimethylaminopropyl)-N'-ethylcarbodiimide hydrochloride (NHS/EDC) procedure [65]. This yields an activated ester surface able to react with primary amines NH_2 encountered in proteins or modified oligodeoxyribonucleotides (ODN). ODN modified with a terminal primary amine connected through a linker to their 3′ or 5′ extremity are mainly bound to the surface through this terminal group. The primary amines on nitrogenous bases of nucleotides do not significantly bind to the activated ester groups. This simple approach enables to precisely choose the attachment point of an ODN on a surface.

Table 2.1 Examples of surface functional groups used for binding ligands. (a) Methyl, (b) primary amine, (c) carboxylic acid, and (d) terminal biotin

	Terminal functional available on surface after primary functionalization	Ligand and immobilization method (Chemical coupling, physical interactions, or complexation)	References
a	methyl —CH_3	Lipids, proteins	Duval et al. 2009 [62], Li et al. 2003 [70], Suraniti et al. 2007 [63], Tielens et al. [57]
b	Primary amine —NH_2	Nucleic acids; Poly(ethyleneimine) + extravidin	Balladur et al. 1997 [64]; Bassil et al. 2003 [66]
		—COOH moieties of peptide or protein (activated with NHS/EDC)	Boujday et al. 2008 [56]
		Linker + protein	Spadavecchia et al. 2009 [71], Boujday et al. 2009 [59]
c	Carboxylic acid —COOH (to activate with NHS/EDC for making it reactive)	—NH2 moieties of protein or peptides	Li et al. 2002 [72], Li et al. 2003 [70], Mazurczyk et al. 2008 [73]
		Polymer: carboxydextran; MAMVE; chitosan	• Mannelli et al. 2007 [74], Tombelli et al. 2002 [61], Lofas et al. 1990 [75]; El Khoury et al. 2010 [43]
d	• —Biotin	• Streptavidin, Neutravidin, Extravidin	• Maalouf et al. 2007 [76]

Alternatively, deprotonated carboxylic acid can be used for adsorbing positively charged polymer such as poly(ethyleneimine) (PEI) through electrostatic interactions. We also used to our advantage the fact that adsorbed PEI multilayer can bind negatively charged biomolecules such as extravidin [66, 67]. Similarly, dextran is another polymer commonly used for further functionalization of gold modified with MUA [67]. Maleic anhydride-alt-methyl vinyl ether (MAMVE) copolymer and chitosan, grafted on surface carboxylic acids, were also tested for protein/protein immobilization with protein microarrays [43, 68, 69].

Alkanethiols bearing a terminal biotin group (Table 2.1(d)) can be used for binding streptavidin-like proteins (streptavidin, extravidin, or neutravidin) [76, 77]. This first biomolecular layer is then used to bind biotinylated biomolecules (ODN, antibodies, or other proteins)—this approach is developed in Section 2.3.2.

The use of short chains of monoaminated alkanethiols (either cysteamine or cystamine) enabled efficient functionalization of gold in a shorter preparation time [51, 71] than classical thiols such as MUA. For instance, cysteamine-based functionalization enables immobilization of probes in one and a half days, while MUA-based functionalization implies 24–48 h for thiol self-assembly, plus further steps for grafting probes.

Besides, using mixed alkanethiols such as MUA/MOH [51] or biotin–thiol/MOH [76] can improve the efficiency of biofunctionalization, first by providing more degrees of freedom and chemical availability to the reactive sites of the surface; and second because a monolayer including hydrophilic surface groups (alcohol –OH, ethylene glycol) tends to decrease nonspecific adsorption [28, 78].

2.3.1.3 Biomolecular functionalization: immobilizing biomolecular probes onto gold

Biomolecular probes bearing accessible thiol moieties can be directly immobilized onto gold. Synthetic oligonucleotides can indeed be modified with sulfhydryl terminal groups [79], and proteins naturally contain thiol moieties in their cysteine lateral chain—as disulfide bridges, or as free thiol groups. The molecules can therefore be chemically anchored onto gold, through SH/gold interactions [80].

2.3.2 *Examples of Biomolecular Interactions Studied with SPR*

Different challenges may arise for performing biomolecular interactions on a support, more precisely at a liquid–solid interface. The first challenge is to ensure that the surfacic density of immobilized probes is high enough for capturing a detectable amount of targets. As indicated, a few 10^{11} probes/cm^2 is a common surfacic density for a monolayer of probes onto a flat support. Classically, the yield of chemical grafting of biomolecules onto a surface is far inferior to the yield obtained when the same reaction is performed in volume. This is due to surfacic steric hindrance, lower accessibility of reaction sites, inferior entropy variation, and diffusional mass transfer through boundary layer. Immobilizing a correct surfacic density of probe (i.e., a density yielding an optimal signal for biodetection) therefore usually requires a large concentration of probes. This is fairly easy for oligonucleotide probes, because such biomolecular species can be chemically synthesized on a large scale. However, it is not so easy to produce certain proteins (e.g., monoclonal antibodies) or oligosaccharides.

SPR microarrays imply another challenge: probe surfacic densities may vary from one spot to another—even for two probes with only slight differences, for instance, two oligonucleotides differing only with one nucleotide. This may be due to differences in drying behavior due to spotting buffer composition [81] or by differences of interactions of each probe at liquid–solid–air interface, in the spotted droplet before and during evaporation. Besides, when spotted probes are proteins, it is very plausible that biological activities of the different spotted proteins may vary from a spot to another, because of the diversity of interactions of proteins with the support. In all these cases, these discrepancies cause the spots ensemble to exhibit various surfacic densities of "useful" probes (i.e., biologically active and efficient for binding a molecular target). The comparison of the molecular interactions of spots with each other may therefore be a challenge for SPR analysis. Using calibration procedures may provide a solution for nucleic acid probes, as described in Section 2.4.2.

How to ensure the best possible biological activity of biomolecular probes? While ODN are, once again, comparatively easier to hybridize once protocols are correctly designed, each protein used on a support may need a specific chemical environment to avoid its denaturation. Avoiding nonspecific adsorption of undesired biomolecules while maximizing specific biomolecular affinity between probes and targets should be a main goal when designing a surface dedicated to analyze a probe–target interaction. Various approaches were explored for optimizing the bioreceptor layer. Section 2.3.2.1 introduces the use of **linkers or 3D polymers** for further tailoring of the surface properties, before immobilizing probes. Section 2.3.2.2 deals with **surface post-processing after probe immobilization**. This is a critical step that usually strongly influences the quality of biosensor measurements. In our experience, post-processing step should be taken into consideration irrespective of the method chosen for chemical and biomolecular functionalization.

Sections 2.3.2.3 to 2.3.2.5 introduce a few approaches for immobilizing molecular probes: direct grafting onto SAM, use of streptavidin and similar molecules, and DNA directed immobilization (or ZIP code immobilization). Section 2.3.2.6 introduces SPR measurements for analyzing living cells behavior. Examples of biomolecular interactions are summarized in Table 2.2.

2.3.2.1 Additional surface tailoring after SAM implementation: linkers and 3D polymers

Using a generic surface (e.g., carboxylic acid group of a MUA SAM) is not always adapted for optimizing the detection of probe–target interactions. The affinity between probe and target can be strongly influenced by the surrounding physicochemical environment. For example, secondary and tertiary structures of proteins can be modified by the surface electric charges, hydrogen bond donor or acceptor groups, or moieties exhibiting a hydrophobic behavior. Antibodies are usually considered as robust and nondenaturable proteins, keeping their affinity with their associated antigen even when submitted to (relatively) harsh conditions. But globular proteins (e.g., enzymes), either probes or targets, can on the contrary be denatured by contact with the sensor surface. It is therefore important to validate the influence of the surface chemical functionalization on proteins of interest [68, 69]. Moreover, it may be necessary to adapt the surface chemical functionalization when different protein probes are involved, which is classically the case in a biochip format. This can be done by modifying the first alkanethiol layer with additional molecules, either bifunctional (sometimes named “linkers” or “spacers”: PDC [71], Jeffamine (diamine PEG)

Table 2.2 Examples of biomolecular interactions onto supports

Chemical functionalization	Biomolecular probe	Biofunctionalization process	Molecular target to detect	Reference
MUA + PEI	biotinylated ODN	Electrostatic interactions (PEI) + extravidin/biotin complex	ODN	• Bassil et al. 2003 [66]
MUA + PEI; MUOH + dextran	ODN	Electrostatic interactions (PEI) or chemical grafting	DNA	• Mannelli et al. 2006 [67]
Alkylthiol + biotin	biotinylated antibody	Neutravidin/biotin complex	Bacteria	• Maalouf et al. 2007 [76]
Alkylthiol + biotin	biotinylated antibody	Neutravidin/biotin complex	Nanosome	• Vidic et al. 2007 [77]
Dodecanethiol	Lipid hybrid bilayer membrane	Lipid vesicles adsorption and fusion	Cytochrome C	• Suraniti et al. 2007 [63]
Dodecanethiol	Ni-DOGS phospholipidic-hemimembrane + PDNA	Hemimembrane fusion + membrane protein	PDNA	• Duval et al. 2009 [62]
Cysteamine + PDC + neutravidin	Biotinylated ODN	Neutravidin/biotin complex	ODN	• Spadavecchia et al. 2009 [71]

[28, 65, 73, 82] or polymeric (PEI [66], dextran [67, 74], MAMVE [43], and chitosan [68]. Such molecules have simultaneous purposes: (i) provide better availability of immobilized probes, by giving more degree of freedom; (ii) minimize nonspecific interactions between targets and surface, by adequately tuning surface energy (e.g., polyethylene glycol); and (iii) provide the less denaturing environment to the immobilized probes, in order to enable the best biological activity.

2.3.2.2 Surface post-processing after probe chemical grafting: capping and blocking steps

After chemical grafting of biomolecular probes onto a chemically reactive layer (either pure SAM, SAM + linkers, or SAM + 3D polymer), a fraction of reactive chemical moieties (e.g., primary amine, activated ester) is commonly still available onto the surface. During the analysis of a biological sample, these moieties would be susceptible to chemically react with nondesired species (e.g., proteins or nucleic acids that are different from the aimed target). They therefore have to be inactivated before any analysis. This **inactivation is a critical step** for avoiding aberrant behavior of the finalized biosensor. Inactivation can be performed in two ways, just after probe grafting: we name herein these post-processing steps "capping" and "blocking." These terms may vary from an article to another and are sometimes interchanged. The definitions we give below are therefore not universal. (1) **Capping** performs a reaction between the remaining reactive sites of the surface and small organic molecules (e.g., ethanolamine, succinic anhydride). Such capping molecule is used at high concentration. This yields an inactivation close to 100% of the surface remaining sites. (2) **Blocking** inactivates the surface remaining reactive sites, by using large "blocking molecules" (e.g., proteins or biomolecule mixtures, such as bovine serum albumin, milk proteins, casein, salmon sperm DNA, and other biomolecular polyelectrolytes). The blocking step provides two effects: (i) blocking molecules may inactivate the remaining reactive sites of the surface, like a capping step; (ii) additionally blocking molecules physisorb onto the surface, on sites that may have been otherwise available for physisorption, during the biological sample analysis. It therefore lowers nonspecific adsorption during the use the finalized biosensor. Blocking usually does not prevent complementary target to bind to the immobilized probes—of course the affinity of the probe with the target has to be significantly more important than nonspecific interaction of the probe with blocking molecules.

2.3.2.3 Direct immobilization of molecular probes on a functionalized support

The most straightforward approach for implementing a biofunctionalized surface is to directly couple the biomolecular probes onto the uppermost layer. ODN [83], peptides, and proteins [51, 56, 58, 60, 65, 73, 84] can all theoretically be chemically coupled to SAM modified gold, for instance, via carboxylic acid/amine reaction. As stated before, using a high concentration of probe for a covalent grafting (typically 25 μM) is rather easy for ODN, but can be problematic for proteins or others molecules. It was therefore important to elaborate others strategies for binding efficiently molecular probes. Some of these strategies are presented below.

2.3.2.4 Streptavidin and similar binding proteins

Streptavidin, extravidin, and neutravidin [85–88] are derived from the glycoprotein avidin. As protein, they can be immobilized onto a chemically functionalized surface. They can then be used for immobilizing other biomolecules, bearing biotin moieties. Biotin is a small organic molecule that can be incorporated into oligonucleotides, proteins, and alkanethiols. Streptavidin exhibits a strong affinity for biotin: the dissociation constant of a biotin–streptavidin complex is about 10^{-15} mol/L, which is comparable to a covalent binding. Each streptavidin molecule exhibits four biotin binding sites. It is then easy to bind other biomolecules bearing one or several biotins, such as ODN [66, 71], antibodies, peptides, or other proteins [65, 76, 77]. ODN or proteins (particularly antibodies) can be conjugated with biotin during their synthesis or in liquid medium. It is thus possible to bind these biomolecules to the surface with an excellent binding yield.

Streptavidin and similar molecules are to use in aqueous buffer solution. They are therefore easy to handle even with no experience in chemistry.

Avidin, as a positively charged glycoprotein, exhibits somehow nonspecific binding and has tendencies to form aggregates. Per comparison streptavidin exhibits an isoelectric point (IEP) of 5, and neutravidin an IEP of 6.3. These modified molecules are commonly expected to exhibit lower nonspecific adsorption than avidin—although deeper analysis show this is probably more complex [87]. User should adapt pH of the working buffer, depending on the chosen protein/streptavidin, extravidin, and neutravidin exhibit different optimal pH for decreasing nonspecific binding based on electrostatic interactions. As for proteins in general, it is recommended to limit denaturation of stored proteins (e.g., caused by freeze/thaw cycles) by aliquoting their stock solution, and do not freeze again a defrost streptavidin aliquot.

2.3.2.5 ZIP-code ODN: applications to ODN, protein and oligosaccharide immobilization

ZIP-code oligonucleotides [89] or "ZIP ODN" are more and more used for implementing microarrays. The ZIP strategy, also called DNA-directed immobilization (DDI), consists of immobilizing an ODN (A) of defined sequence (Fig. 2.9(A)) on a support.

By hybridization, this ODN is designed to bind another ODN (Fig. 2.9(B)), which is composed of two contiguous sequences: the first sequence (blue) is designed to hybridize specifically to the ZIP-code ODN (A). The second part (green) of the molecule corresponds to the probe itself, that is, the biomolecule aimed at binding the molecular target (Fig. 2.9(C) top): probe is an ODN; bottom: probe is glycoconjugate [90]. A library of ZIP-code ODN enables to reversibly address various probes, in the same binding conditions. It simplifies the elaboration of arrays, since a same batch of ZIP-code arrays can be used for various probes. The Zip-code strategy also enabled to analyze the interactions between ODN

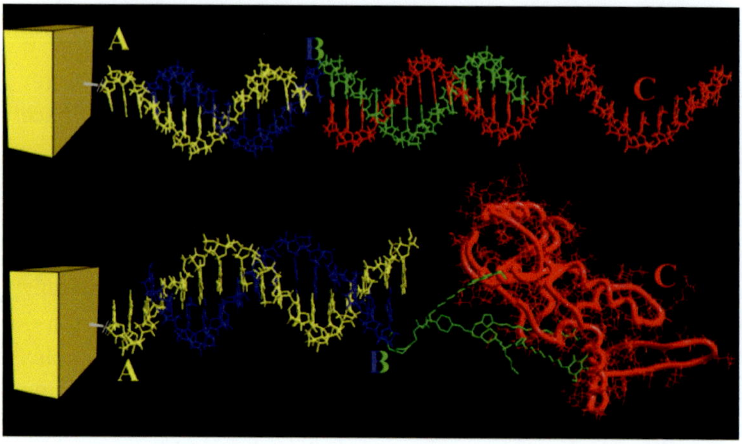

Figure 2.9 Two examples of ZIP code strategy. (Top) target (C, red) is a nucleic acid. The probe (B, green part) is an ODN. (Bottom) target (C, red) is a lectin. The probe (B, green part) is a glycomimetic molecule specifically binding to lectins.

and proteins. Maillart et al. studied the interactions between the p53 protein and oligonucleotide probes, while Corne et al. explored with SPR how damaged nucleic acids interact with Fpg, a protein recognizing and repairing damaged nucleic acids [31, 32]. ZIP-code strategy was also used for protein immobilization [91–93] and for glycomimetics [90, 94]. These classes of molecules cannot be easily chemically synthesized or enzymatically amplified. It can therefore be difficult to perform a chemical grafting of such molecules on a surface, because the concentration of molecules to bind is too low to attain a correct yield.

However, when conjugated with a nucleic acid probe, a protein or a glycomimetic can be immobilized with an excellent yield on a surface bearing complementary ZIP-code ODN. The yield of hybridization between ZIP-code ODN and its complementary sequence is indeed far higher than the yield of chemical coupling on the surface, for the same concentration of oligosaccharide. The ZIP-code approach therefore enables to synthesize a minute amount of probe for obtaining an adequate surface density of probes.

2.3.2.6 SPR and living cells

Surface plasmon resonance is commonly used for characterizing molecular interactions. It can also be used for monitoring phenomena at the cellular level. SPR was for instance tested for the detection of bacteria by Maalouf et al. [76]. In another perspective, Chabot et al. [95] used SPR for monitoring morphological changes of human cells forming a layer, under various stresses due to chemical or biochemical substances. Once again one can see that SPR can be useful not only for detecting mass transfer to a surface, but also for organizing changes of complex systems. SPR has been applied to the real-time monitoring of living cells to probe molecular processes [96, 97]. Recently, Chabot et al. [95] proposed a method for first-line broad-spectrum biosensing based on the combination of SPR with living cells. The method premise is that cellular activity induced by external agents is often associated with a change in cellular morphology, which in turn should lead to a variation of the effective refractive index at the interface between the cell membrane and the metal layer.

Three types of stimulation by injection of various agents known to influence cellular activity and cause morphological changes were evaluated: lipopolysaccharides (LPS), a bacterial endotoxin known to cause important cellular response often leading to cell death [95, 98]; sodium azide, a chemical toxin known to prevent cellular respiration by inhibiting the cytochrome c oxidase-respiratory chain complex IV [99]; pharmacological activation by thrombin, a physiological agonist involved in the modulation of cell layer integrity [100]. The proposed biosensing method could be applied to a large variety of cellular responses involving shape remodeling induced by external agents.

A comparison between the aggregated SPR responses over the entire cell population with phase contrast microscopy reveals that SPR signal variations are associated with the induction of cell death for LPS treatment and a contraction of the cell body for sodium azide. Thrombin-induced cellular response shows a rapid decrease of the measured laser reflectance over 5 minutes followed by a return to the original value. For this treatment, phase contrast micrographs relate the first phase of the SPR variation to cell contraction and increase of the inter-cellular gaps, whereas the recovery phase can be associated with a spreading of the cell on the sensing surface. Hence, the SPR signal is very consistent with the cellular response normally observed for these treatments.

As the cells have a higher refractive index than the surrounding liquid medium [101], changes in cell surface adhesion and/or morphology result in a corresponding change in the effective index of refraction on the sensing surface overall by changing the proportion in which the cells and the Hank's balanced salt solution (HBSS) solution affect the plasmon resonance. This in turn results in an observable SPR signal. Since the injection of a stimulating agent onto bare gold surfaces induces an increase in SPR reflectance and adding the same agent at the same concentration in the presence of cells leads to a decrease in reflectance, it is clear that the observed SPR reflectance changes are mainly due to cellular changes, while the surface adsorption of the agent does not have a significant effect.

2.3.3 *Biochip Partial Conclusions*

As described in the previous paragraphs, SPR can be used for studying a wide diversity of molecular interactions, with various functionalization approaches. The characteristics of the metallic layer, surface functionalization, biological activity of molecular probes, and nonspecific interactions strongly influence the analytical performances of an SPR biosensor and must be carefully combined for being adapted to a specific application. A classical application of SPR is real-time analysis of probe–target molecular interactions; it can be used for systems involving nucleic acids, peptides, proteins, carbohydrates, bacteria, and cells. SPR is also used for monitoring chemical reactions at solid–liquid interface, such as chemical functionalization or synthesis of molecular probes. Chemists and biologists are furthermore interested in studying structural changes with SPR, for instance, to monitor cell shape changes, or variation of orientation of anisotropic structures. This diversity of approach involves different conditions of use and implies to adapt SPR imagers for more and more modalities of analysis. In the following section, we will review some of our recent advances in the "multi-dimensional" SPRI systems enabling to further characterize with more detail the biochemical surfaces under investigation.

2.4 Recent Advances in Multidimensional SPRI Systems

2.4.1 *Spectral Dimension*

2.4.1.1 The different possible setups

There are many ways of going from a fixed wavelength system to a spectrally resolved system. It is not the scope of this section to review multiple wavelength systems solely allowing gain in precision using two or more fixed wavelength. This section discusses the acquisition and use of a continuous reflectivity spectrum. This requires the usage of a white light source coupled to a spectrometer capability.

In its simplest setup, the white light source is used to illuminate the sensor area and after being reflected in SPR mode, the light is directly collimated back to a spectrometer or a combination of a grating and a CCD camera (Fig. 2.10(a)). This provides the surface average reflectivity spectrum in real-time $R(\lambda, t)$, at fixed coupling angle θ, providing the spectrometer uses a parallel array of detectors [102]. An angular capability can be added allowing to provide $R(\lambda, \theta, t)$ with a significant loss in temporal resolution.

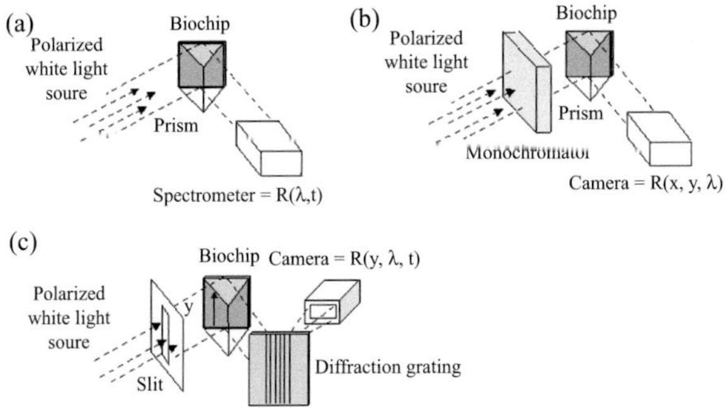

Figure 2.10 Schematic of the different setups for a spectrally resolved SPR sensor.

Another important configuration is obtained by placing a monochromator before the sensing area and scanning the wavelength (Fig. 2.10(b)), allowing using an imaging configuration to get $R(x, y, \lambda)$ or $R(x, y, \lambda, \theta)$, again with much lower temporal resolution actually depending on the number of wavelength and angle that are effectively monitored. In 2013, we demonstrated how to optimize a wavelength interrogation mode-based measurement that allows keeping both 2D imaging and real-time capabilities [104]. With only five wavelengths and an asymmetric pseudo-Lorentz fitting function, a resolution of 7 pm on the plasmon dip position determination is obtained. Moreover, the performances, both in terms of resolution and accuracy, are constant within a dynamic range of $\pm 5 \times 10^{-3}$ RIU of the surrounding media. This allows significant gain in data quality compared with the reflectivity interrogation mode, where the optimal conditions are strongly dependent on the buffer index variations and the surface chemistry inhomogeneity.

Finally, a third configuration (Fig. 2.10(c)) is based on a line-shaped illumination, and a grating, the line of light reflected by the sensor area is diffracted on the camera, giving $R(y, \lambda, t)$, allowing to keep the "real-time" capability of the system [105].

2.4.1.2 Spectro-imaging and gain in resolution

In this last configuration, both spectral information and temporal resolution can be obtained on a parallel set of biomolecular interactions. One dimension is used for spatial coded information and 1 dimension for spectral information, yet keeping the real-time capability of the camera acquisition dynamics. This was demonstrated in the case of a DNA/DNA interaction characterization experiment [105]. In this reference, an extended reflectivity spectrum was gathered allowing fitting of the spectral resonant shift $\Delta R(\lambda)$ with few picometer resolution, corresponding to a index sensitivity close to a monochannel sensor (3.5×10^{-7} RIU), and allowing discrimination of a single nucleotide polymorphism (SNP) on 20 bases long ssDNA sequence at concentration as low as 100 pM. The effective S/N is one of the closest to theoretical limit obtained so far [107]. This was made possible both because of the use of increase number of photons collected at the different wavelength and because of an enhanced data processing procedure.

2.4.1.3 Effect of chromophore dispersion

The advantages of the prior spectro-imaging system were demonstrated using transparent materials. As the SPR reflectivity is very sensitive to the minute changes in optical index, it can be anticipated that great effects will be induced by using absorbing materials. According to Kramers–Kronig relationship, in addition to their absorbing characteristics, these materials will also exhibit strong dispersion in the real part of their optical index. This is expected to affect both the shape and position of the plasmon resonance curves. We have conducted "spectral" experiments where the ssDNA materials used as target were labeled with organic dyes exhibiting strong absorption bands at different wavelengths, specifically the classical fluorescent biochip dyes Cy3 and Cy5, and as well Cy5.5, Rhodamine 6G, and Texas Red, spreading the absorption maximum position over nearly 200 nm in the mid region of the visible spectrum [102]. Starting with the absorption band, spectral index dispersion could be evaluated and SPR reflectivity simulations performed using the complex optical index of the materials. The calculated and experimental reflectivity spectra exhibited enhanced variations of reflectivity upon binding with the target material when their absorption spectrum matches the plasmon resonance, as in the case of the Cy5-labeled materials (Fig. 2.11, see [102]). A label will always improve the signal because of the extra mass it brings to the surface vicinity but when matching the plasmon resonance, an extra sensitivity enhancement is observed (160% in the case of Cy5 binded to 20 mer ssDNA). Together with spectro-angular mapping of the biosensing surface this would allow possible binded target discrimination if using several labels.

2.4.1.4 Characterizing the target spectral signature

Inversely, using the same experimental setup, one can start with the SPR reflectivity variations data induced by unknown dispersive

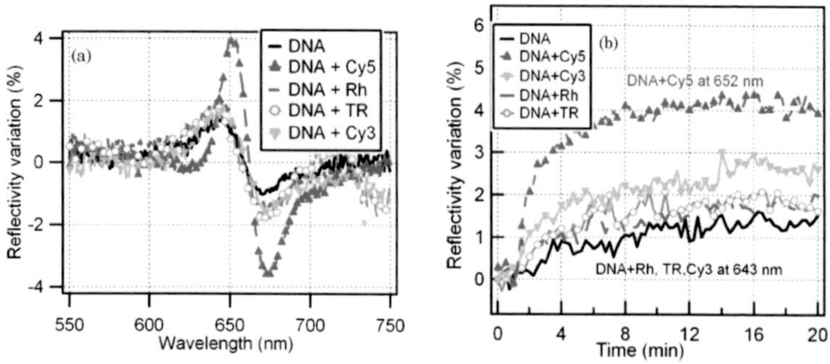

Figure 2.11 (Left) Reflectivity variation spectra after interaction with DNA target, without label, with Cy3, TR, Rh, and with Cy5. The reflectivity spectra are compared with the spectrum before interaction. (Right) Reaction kinetics at the maximum of sensitivity [102].

materials and extract the optical index dispersion, real and imaginary parts. This was effectively demonstrated using a dye solution and the setup described in Fig. 2.10(b). The principle is to acquire the complete angular and spectral reflectivity map of the sample. This set of data $R(\lambda,\theta)$ depends on the real and imaginary part of the sample optical index and on its spectral dispersion. The imaginary part is related to the absorption of the sample, while the real part, which is related to the speed of light in the medium, presents a large variation around the absorption band due to Kramers–Kronig relationships. Using the model of a multilayer system based on the Rouard method, the spectral refractive index of the sample can be extracted by fitting the experimental data. Figure 2.12 shows the experimental angulo-spectral surface plasmon reflectivity map, for a given point (x, y) on the sensor area, obtained with the dye solution above the gold surface.

As it can be seen in Figure 2.12, the influence of the dye solution is clearly visible in the plasmon spectrum shape. Figure 2.13 shows the reconstructed imaginary part of the dye ink, well in accordance with the absorption spectrum of the dye measured with a classical spectrometer. The reconstruction was only possible in the 600–700 nm band due to the fact that for other wavelengths, the plasmon resonance minimum of reflectivity falls outside the recorded angular data, making reconstruction of the sample optical index impossible.

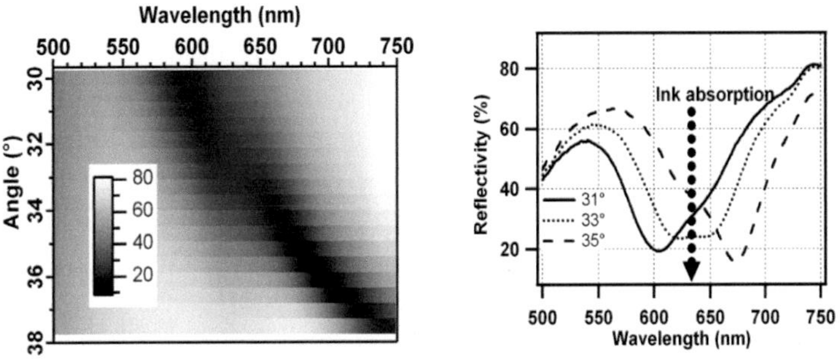

Figure 2.12 (Left) Angulo-spectral surface plasmon reflectivity map obtained with a dye solution, (Right) spectral reflectivity at 31°, 33°, and 35° extracted from the 2D surface reflectivity.

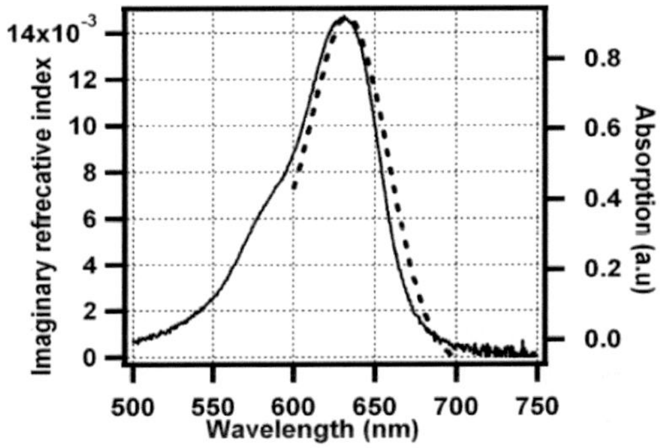

Figure 2.13 Absorption spectrum of the dye measured with a spectrometer (continues line) and reconstructed imaginary part of the refractive index using the SPR data (dashed line).

2.4.1.5 Angulo-spectral signal analysis

As already mentioned, the vast majority of published signal analysis algorithms for SPR are designed to either find the minimum in the reflectance curve [107] or fit a polynomial model to the entire reflectance curve so that the minimum or curve shift can then be estimated analytically. Indeed, most SPR systems are either angle-scanned or wavelength-scanned and hence produce 1D reflectance curves as a function of wavelength or angle. Spectro-angular SPR systems, however, are dual scanned, where both the angle and wavelength are scanned simultaneously. The output of such systems is a 2D image of the surface plasmon dip as in Fig. 2.11. Such systems require 2D signal analysis methods to take full advantage of the extra information. Some of our work [109] demonstrates a numerical technique that is capable of achieving a measurement resolution (5×10^{-9} RIU) over a large dynamic range. This is accomplished using linear algebra methods to analyze the 2D reflectance/dispersion images of the SPR response, as outlined schematically in Fig. 2.14 [103]. In this method, a set of orthogonal basis vectors, b_j, are calculated using principal component analysis from a "training set" of synthetic reflectance/dispersion images generated using rigorous coupled wave analysis (RCWA), which span the refractive index range of interest. The ensemble of basis is used to model the reflectance/dispersion image data over the refractive index domain of interest. In order to estimate the effective index of refraction, n_x, corresponding to a particular reflectance/dispersion image, the "inverse problem" is solved by projecting the image data against the basis using standard linear algebra techniques. The method

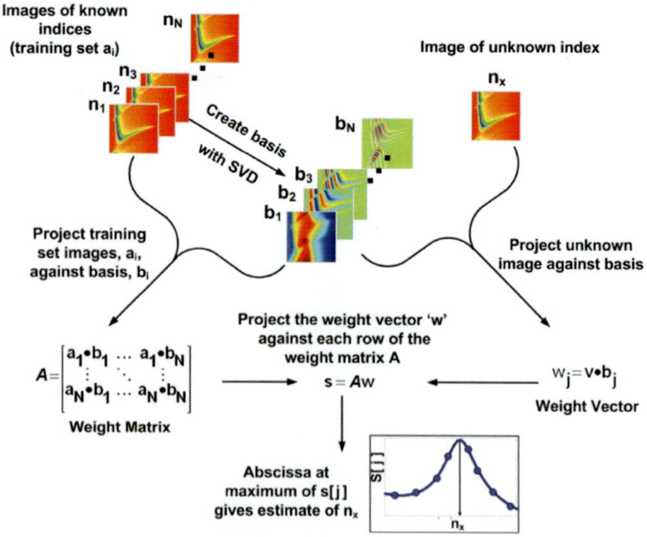

Figure 2.14 Schematic diagram of a linear algebra method to analyze the 2D reflectance/dispersion images of the surface plasmon resonance response [103].

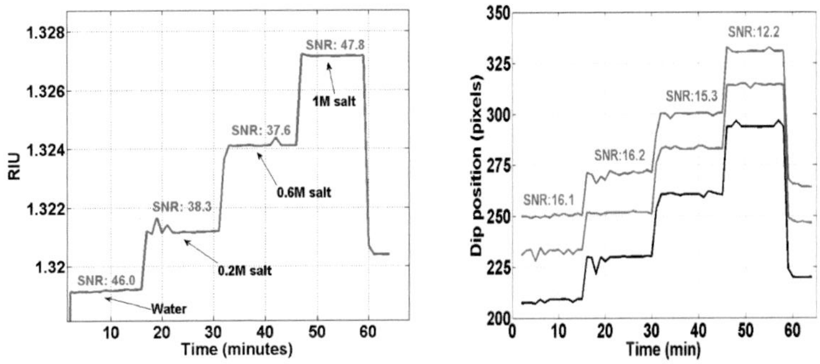

Figure 2.15 Index value jumps from salt solutions on a flat gold surface over time as measured by (Left) the 2D method [103] and (Right) the standard weighted centroid dip finding algorithm.

performs well in simulations with noisy data maintaining an error of less than 10^{-8} refractive index units with up to six bits of noise on 16 bit quantized image data.

Figure 2.15 shows experimental results obtained from index change steps brought about by the salt solutions (starting with water) for both the 2D method and a standard dip finding algorithm, as used for example in Refs. [105, 107]. Each solution was left in the chamber for approximately 15 minutes before a new solution was added. To compare the performance of each method, signal to noise ratios (SNR) were calculated at each step in refractive index.

Note that in the case of the 1D weighted-centroid dip finding algorithm; the SNR for a particular index step was averaged over three slices obtained from the corresponding dispersion/dispersion image.

2.4.2 *Temporal Dimension and Self-Calibration*

The obvious advantage of using an SPRI sensor compare to an SPR mono-detector is the possibility of simultaneously testing hundreds of different probes exposed to one or many target molecules. However, this only makes sense if the response of the system to a given perturbation, measured as a local reflectivity variation, is similar for all probes. Different factors can induce a nonhomogenous response of an SPRI sensor, a nonexhaustive list includes nonoptimal illumination in terms of intensity as well as angle of incidence, imperfect optical conjugation across the sensor area with the camera, inhomogeneity of the chemical functionalization leading to different probes effective surface concentration and difference in fluidic condition. For large sensing area of millimeter size, some calibration procedures are mandatory to correct these sources of error, especially in the case where minute differences between targets interaction signals have to be made to correctly and unambiguously discriminate.

2.4.2.1 TE normalization and step index calibration sequence

The first step is to normalize the different TM images with a TE image in order to compensate the nonuniform illumination of the sensing area. This TE image can be acquired once at the beginning of the experiment as illumination will usually not change during the experiment if incident angle and wavelength are kept constant—if source power stability is an issue, these referencing data have to be taken more often. To compensate for the variation in the SPR response across the sensor area, due to variation of angle of incidence or gold thickness, two buffer solutions with different index of refraction (such as PBS 1× /PBS 1.1×, $\Delta n = 2 \times 10^{-4}$ RIU) can be successively injected into the fluidic cell. This step index must be small enough to stay in the linear part of the SPR response. Figure 2.16 shows successive reflectivity variation images in the fluidic cell acquired with an SPRI setup before and after the injection of the PBS 1.1× solution. Some dispersion of the reflectivity variation during the plateau (images 4 and 5) can be observed.

From the previous stack of images, reflectivity curves as a function of time, on different points of the sensing area, can be extracted (Fig. 2.17, left). The dispersion in the variation of reflectivity measured appears clearly. The calibration procedure simply consists of normalizing to a mean value calculated on the plateau, all the previous raw reflectivity curves (Fig. 2.17, right). This correction factor will then be applied for all subsequent measurements.

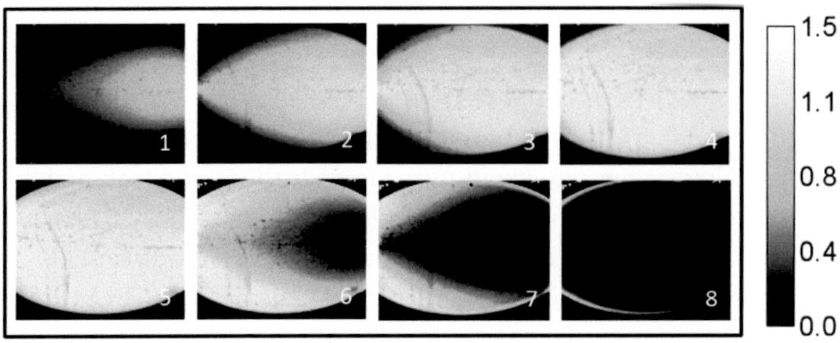

Figure 2.16 Time sequence of reflectivity variation images (TM/TE in %) for a step of index of 2 × 10^{-4} RIU. The PBS 1.1× buffer enters the fluidic cell on the right (images 1–3), a plateau is reached (images 4 and 5) and is then replaced by PBS 1× (images 6–8). The fluidic cell (diameter of 1 cm) does not appear circular due to the tilted imaging system.

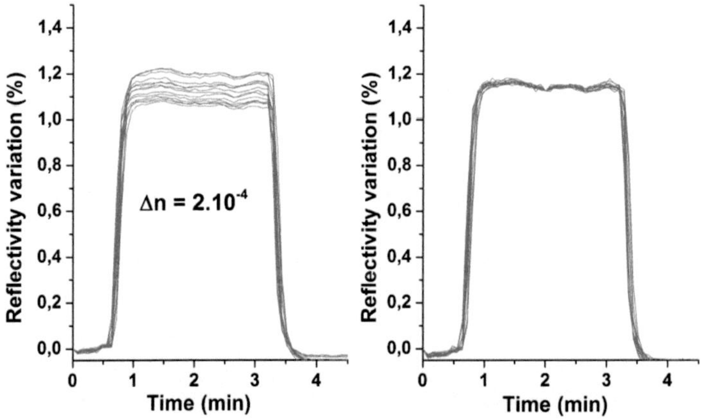

Figure 2.17 (Left) Raw reflectivity measured across the sensing area during the step index calibration sequence. (Right) Reflectivity curves obtained after normalization to a common mean value of the plateau.

2.4.2.2 Concept of self-calibration

It was found that using relatively stable surface chemistries, the reflectivity variation data for each spot were very reproducible as a function of "hybridization/rinsing/dehybridization/rinsing" cycle number. The data dispersion for many identical binding events on one single spot are found to be far narrower than the dispersion of one given event on many spots supposedly identical. This led us to attribute most of the data dispersion to inhomogeneity in the probe concentrations at the surface of the sensors—as discussed in Section 2.3.2. We have thereafter proposed and demonstrated a methodology that allows inline quantification and correction for such variations of biochemical sensitivity. It is based on co-functionalization of the different probes with a common so-called calibration probe. Such calibration probes was for our DNA/DNA experiment chosen as a 12 mer ssDNA sequence. Prior to quantifying the response of the biochip to the target solution under investigation, one quantifies the response to a calibration target containing solution. If all spot were homogeneously functionalized, such response would be the same for all, which is never experimentally the case. Such calibration dispersion is thereafter used to correct the subsequent measurements.

On the classical and sensitive case of SNP genotyping, standard deviations were calculated; without any correction, the discrimination can be possible in the best cases at 1σ with 68% of values, with the self-calibration correction method, discrimination increases from 1σ to 2σ for more than 95% of values when using the dynamical measurements and from less than 1σ to 3σ for 99% of values when using the quasi-equilibrium measurements. Using data segmentation in Fig. 2.18, one can see that genotyping cannot be unambiguously achieved prior to data normalization (top) and that 100% correct genotyping is clearly achievable once performed (bottom figure) [112]. Such patented methodology [113] could be generalized to different systems. As it is not specific to the reading system, it could be used identically in any dynamical reading system allowing sequential injection and reading of different solutions.

2.5 Toward Structured Substrates

2.5.1 *Conventional SPR Reaching Limits*

As recently suggested in an article from Piliarik and Homola [107], conventional SPRI system are close to the theoretical limits using today's optical components and propagating plasmons on

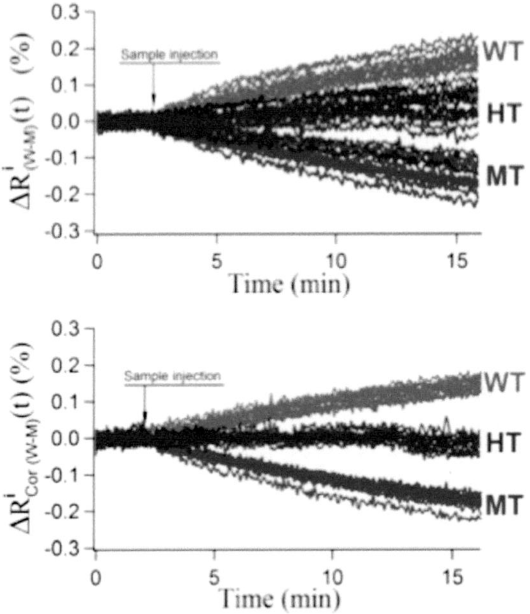

Figure 2.18 Superimposition of $\Delta R^{\mathrm{i}}(W - M)(t)$ for three sample's injections onto the biochip: WT, MT, and HT data. (Top) Data resulting from HT sample injection are, as is often the case, not separated from the two other sample's types. (Bottom) Same set of data using the calibration correction. The unambiguous genetic diagnostic, for the three possible genotypes, is clearly possible within minutes. Reprinted from Ref. [112] by permission of The Royal Society of Chemistry.

homogeneous gold films. That is one of the reasons why much research is nowadays also devoted to plasmon in structured materials at scale ranging from nano to micro. This includes studying the intermediate regime between the propagating plasmon, on which nearly all the current plasmonic biochips are based, and the localized plasmons, for which much effort has been devoted to nanoparticles, but mostly in solution format. Such intermediate regime includes explicitly what could be referred to as quasi-propagating plasmon and coupled localized plasmons. Some of our work in this specific sub-field is reported in the following paragraphs. Indeed clear preliminary data point out to the direction that nano- and micro- structured plasmonic sensor chips will gain substantial sensitivity compared with the homogeneous uniform ones.

Previously, while Malic et al. [114] have demonstrated an increase in sensitivity of biomolecular detection by employing nanopillar arrays, Kim et al. [115] have also shown theoretical calculations indicating that for nanostructured gold films a sensitivity increase is to be expected. Recently, Masson et al. [116] have shown increased sensitivity to localized refractive index changes by developing hexagonal arrays of nanoholes in gold films using self-assembled nanosphere masks.

Many theoretical calculations have been recently reported as in Refs. [117, 118], and a few publications [114–116] have reported remarkable results in this regime. It was also recently shown that propagative plasmon (PP) could be used to enhance the detection sensitivity when the target capture material is spatially organized [119]. In spite of these research activities, the optimal sensing structures, their dimensions, and properties are yet to be established and such materials have to be developed.

2.5.2 *LSPR Sensing and Imaging*

2.5.2.1 LSPR sensing

Since a few decades, the localized surface plasmon resonance (LSPR) nanosensors have also been widely used to monitor various biomolecular interactions including the adsorption of small molecules, ligand–receptor binding, protein adsorption on self-assembled monolayers, antibody–antigen binding, DNA and RNA hybridization, and protein–DNA interactions [120–122]. The principle of LSPR sensors transduction is similar to that of SPR sensors and is based on the measurements of local refractive index changes at or near the metallic nanoparticles surface [123–125]. The important parameters of these LSPR sensors are the refractive index sensitivity and the electromagnetic field decay length as for the SPR sensors. However, the values of these sensitivity and decay length for each sensors type are different. For SPR and LSPR sensors, sensitivity m related to a change of the surrounding dielectric index can reach 8800 nm/RIU [126] and is mainly around 200 nm/RIU

[127]. The SPR sensors are more sensitive than the LSPR sensors with a factor 44, but the electromagnetic (EM) field decay length l_d of LSPR sensors (5–30 nm) [128] is smaller than those of SPR sensors (200–350 nm) [126, 129]. This decay length provides to LSPR nanosensor an enhanced sensitivity. Consequently, SPR and LSPR exhibit similar effective sensitivities. The parameters m (sensitivity) and l_d (EM field decay length) depend on the size, shape, and the composition of metallic nanoparticles. The response of these two types of plasmonic sensors can be described using the simple model of Campbell group [126]:

$$\Delta\lambda = m\Delta n[1 - \exp(-2d/l_d)]$$

where $\Delta\lambda$ is wavelength shift, m is the refractive index sensitivity, $\Delta n = n_{adsorbate} - n_{surrounding\ medium}$ is the local refractive index change induced by an adsorbate, d is the effective thickness of adsorbate layer, and l_d is the evanescent EM field decay length.

To realize these nanosensors (particles) on substrates (glass) or in solution, different techniques, such as chemical synthesis, electron beam lithography (EBL), nanosphere lithography (NSL), and soft UV nanoimprint lithography (UV-NIL) are used. Several groups use the chemical synthesis to fabricate nanoparticles with different shapes (Fig. 2.19): spheres [130], rods [108, 131], prisms [106, 132], disks [111, 133], cubes [110], and core–shell nanoparticles [134, 135]. For all of the shapes, the fabrication principle is based on the reduction of a metallic salt in order to realize nanoparticles with varied sizes and shapes. Another technique uses the electrochemical reduction of metallic salts with a surfactant presence [136].

The two methods are very efficient to create nanoparticles with a large aspect ratio. For the core–shell nanoparticles, one of most used systems is a dielectric core with a metallic shell as nanorices [137, 138]. Usually, the dielectric core (silica) is carried out by the Stöber method, which allows obtaining a range in size from 80 to 500 nm [139]. The metallic shell is obtained by growth from metallic salts in solution and in the presence of NH_4OH to produce a preferential deposition onto the surface [140].

Previously, the fabrication methods were used to realize nanoparticles in solution. Other techniques exist to obtain nanoparticles bound at substrate: the lithographic techniques. We will

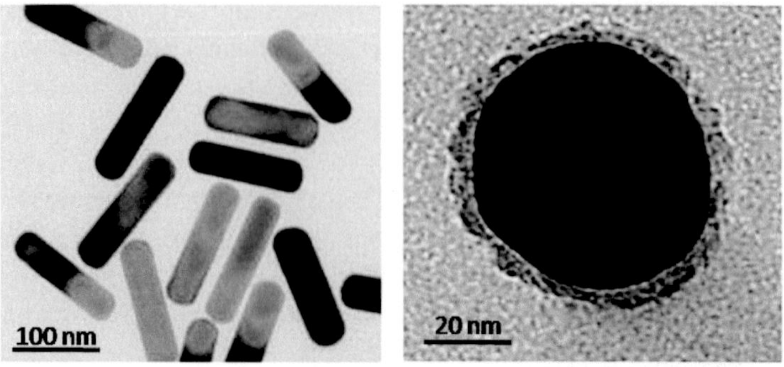

Figure 2.19 TEM images of gold nanoparticles, (Left) nanorods and (Right) core (Au)/shell (silica) nanoparticle.

focus on three types of lithography, for example, electron beam lithography (EBL) [120, 141], nanosphere lithography (NSL) [116, 119], and soft UV nanoimprint lithography (UV-NIL) [142], to produce metallic nanoparticles on substrates. The most known technique is EBL. The principle of this technique is to expose a thin layer of photoresist (e.g., PMMA) to high-energy electrons beam in order to obtain the desired pattern in series. This exposure is followed by a chemical development, then the evaporation of the metallic layer. To finish, a lift-off process is used to withdraw the remaining photoresist and we obtain the metallic nanoparticles with the shape, size, and the periodicity desired. EBL offers the possibility to realize different shapes, sizes, and periodicities of patterns with a high degree of accuracy on large zones. However, EBL is usually time consuming and expensive.

Two alternative techniques enable to implement metallic nanoparticles arrays: the nanosphere lithography (Fig. 2.10) and the soft UV nanoimprint lithography. First, the principle of NSL is based on the fabrication of a 2D mask obtained by the self-assembly of nanospheres monodispersed in size. Then, a step of metallic layer deposition is realized by evaporation or similar techniques. The last step is to remove the nanosphere mask in order to obtain the triangular nanoparticles [143, 144]. NSL allows controlling the size and the interparticle spacing. Concerning the soft UV-NIL, the principle is based on the realization of a mold transparent with UV

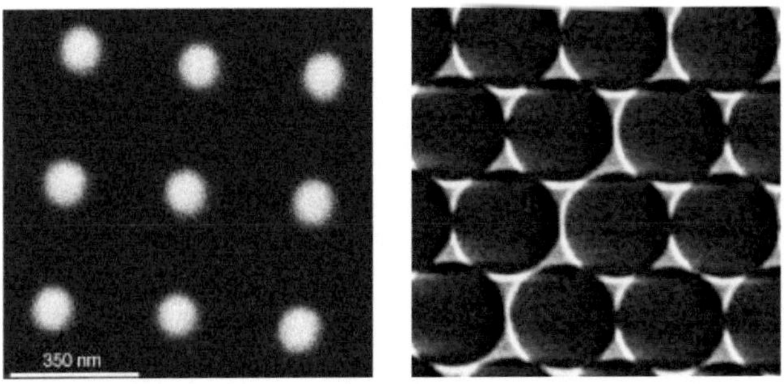

Figure 2.20 SEM images of gold nanoparticles: (Left) nanodisks obtained by EBL, (Right) nanotriangles obtained by NSL.

on which the desired pattern is realized. To create this flexible mold, polydimethylsiloxane (PDMS) is generally used [142, 145, 146]. Then, the flexible mold is used to imprint the pattern in a polymer sensitive to UV (AMONIL) [146]. The AMONIL is deposited by spin coating on the substrate. Then, the PDMS mold is deposited on the AMONIL with (or without) external pressure and the sample is exposed to UV lamp in order to imprint the pattern in AMONIL. Thus, AMONIL is cured and the PDMS mold is released from the sample. The next step is to remove the residual thickness of AMONIL with suitable reactive ion etching (RIE) process after imprint [145]. The RIE process is stopped at the level of substrate surface. After RIE, steps of metallic layer deposition and lift-off process are realized as for EBL technique. UV-NIL permits to realize large surfaces (some square millimeter) on several types of substrates with a weak time of fabrication compared with EBL. With the same PDMS mold, several samples can be realized.

Thus, these metallic nanoparticles fixed on the substrates are used as biological sensors. The important parameters of these sensors are the sensitivity *m* and EM field decay length l_d. Several groups already demonstrated that the sensitivity *m* depends on the geometry of nanoparticles. Sau et al. observed that the sensitivity varied in significant manner with the shape and the size of nanoparticles. For example, they found a sensitivity of 83 nm/RIU for Au nanocubes (length = 44 nm) and 703 nm/RIU for Au branched nanoparticles (length = 80 nm) [147]. For Au nanorods of length of 40 nm and 74 nm with a diameter of 17 nm, they found a sensitivity of 195 nm/RIU and 288 nm/RIU, respectively [147]. Tam et al. demonstrated for core–shell nanoparticles (silica/gold) immobilized on glass substrates that the SPR sensitivity of these nanoparticles (nanoshells) increased with the total particle size and to a lesser extent with increasing r_1/r_2, where r_1 is the radius of nanoshell core (silica) and r_2 is the total radius of nanoshell. Precisely, this SPR sensitivity of nanoshells may be enhanced when the nanoshells are larger with thinner gold shells [137].

Some examples of plasmonic biodetection are presented below. First, several groups have studied the biotin–streptavidin interaction—this complex is presented in Section 2.3.2.4. Barbillon et al. demonstrated that gold nanodisks were very sensitive to the detection streptavidin (SA) and anti-biotin (AB) (Fig. 2.21) with a detection limit of 7 pM and 3 nM, respectively [148]. The dimensions of used gold nanodisks were 90 nm of diameter, 50 nm of height, and 250 nm of interdisks spacing and these Au nanoparticles were carried out on glass substrates by EBL (Fig. 2.20). Au nanodisks were functionalized in a suitable way to detect SA and AB. The limit of detection was determined by

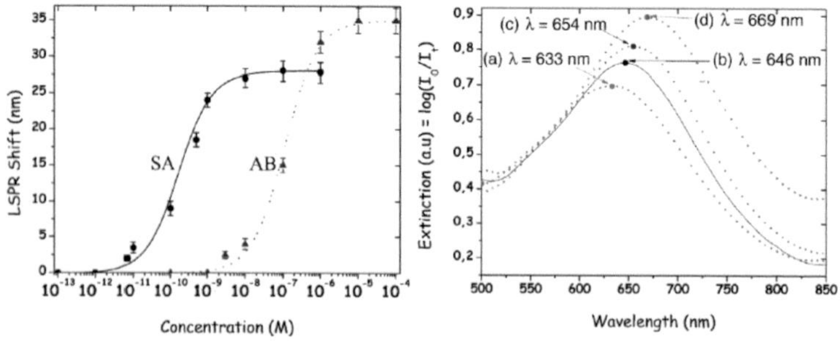

Figure 2.21 (Left): LSPR shifts of Au nanodisks following adsorption from 1mM MUA (thiol) and 1 mM Biotin as a function of SA or AB concentration. (Right): an example of smoothed LSPR spectra at each step of the surface modification of Au nanodisks. (a) Au nanodisks before chemical modification, $\lambda = 633$ nm. (b) Au nanodisks after modification with 1 mM MUA $\lambda = 646$ nm. (c) Au nanodisks after modification with 1 mM Biotin $\lambda = 654$ nm. (d) Au nanodisks after modification with 100 nM anti-Biotin $\lambda = 669$ nm.

extinction spectroscopy, while measuring the wavelength LSPR shift. Marinakos et al. performed measurements with a plasmonic sensing (LSPR shift) [149] yielding a limit of detection of 94 pM for SA. Moreover, they demonstrated that the dimensions and the mass of the molecule affect the magnitude of the LSPR shift. The saturation of the two studied systems SA (60 kDa) and AB (150 kDa) yielded shifts of 28 nm and 35 nm, respectively. Thus, it appears clearly that a bigger mass density at the nanoparticle surface results in a larger LSPR shift. Faure et al. demonstrated that the sensitivity (LSPR shift) could be increased by a fourfold in terms of detected molecules (SA) using the grafting of a luminescent nanoparticle (nanohybrid) to the streptavidin (the target molecule to be detected) [150]. This system was thus composed of SA and nanohybrid. Its dimensions were increased, as well as its refractive index. These new dimensions enabled to measure LSPR shifts, by extinction spectroscopy, for smaller incubation times, therefore decreasing the inferior limit of detection of SA molecules, by comparison with analysis free SA (1 SA by nanohybrid) [150]. The used gold nanoparticles realized by EBL are of the following dimensions: 140 nm of diameter, 70 nm of height, and 340 nm of periodicity. From Fig. 2.22, the limits of detection (incubation time) were determined at 10 min for SA + nanohybrid and 16 min for only SA corresponding to 3×10^5 and 1.2×10^6 detected molecules on a real zone of $30 \times 30\ \mu m^2$, respectively. The increasing of the size and refractive index of the SA/nanohybrid system explain the LSPR shift enhancement.

In addition, the Van Duyne's group demonstrated that the LSPR nanosensor could enable an ultrasensitive detection of amyloid β-derived diffusible ligands (ADDLs)/anti-ADDLs, molecules possibly involved in the development of Alzheimer's disease. From LSPR shifts measurements for this interaction of ADDLs and anti-ADDLs, a limit of the anti-ADDLs detection of 10 nM was measured by varying the concentration of anti-ADDL antibody [117]. By using this technology, they can detect very weak concentrations. With improvements in the selectivity and the sensitivity, the LSPR nanobiosensor could be an accurate and economical alternative tool.

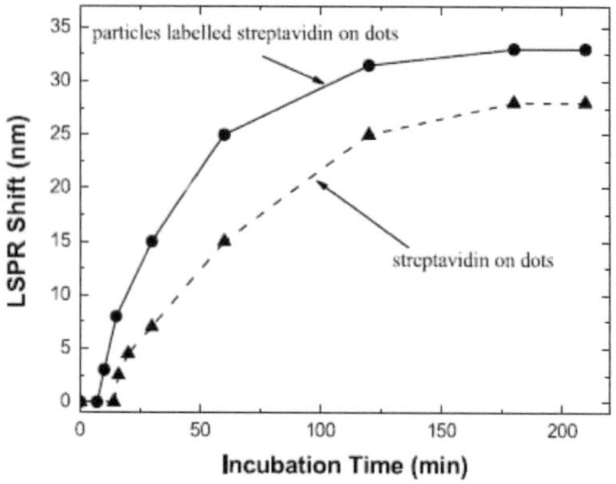

Figure 2.22 LSPR shifts of binding of free streptavidin (dotted line) and particle-labeled streptavidin (solid line) on biotinylated dot arrays.

2.5.2.2 Multiplexed LSPR imaging

LSPR imaging is based on the use of optical properties of plasmonic nanostructures. The advantages of these single plasmonic nanostructures are the potential increasing of biological detection sensitivity [151–153] and their nonphotobleaching. The single plasmonic nanoparticles (or nanostructures) and the detection instrumentation can allow for multiplexing [154–156]. This multiplexed imaging gives an important advantage for biological systems, because the disease detection involves several components. Moreover, multiple sizes of single plasmonic nanoparticles can also be used for therapeutic applications in which specific cells can be targeted by several drugs for an optimal efficiency.

Taton et al. demonstrated a first study for multiplexed biological detection [157]. Various plasmonic nanoparticles of distinct LSPR wavelength were used and functionalized with oligonucleotides in order to obtain a multiplexed detection of DNA microarrays. Plasmonic multiplexing can also be used for in vivo imaging of biological tissue and cells in order to obtain a medical diagnostic [158–160]. Recently in 2013, Ruemmele et al. have described an LSPR imaging instrument and presented proof-of-principle studies to demonstrate homogeneity of sensing on a large surface (1.1 cm^2), and the ability to perform multiple biosensing simultaneously [161].

Optical properties of plasmonic nanoparticles, such as the nonphotobleaching (or the very weak photobleaching), and large scattering cross-section are ideally enabled to perform LSPR imaging for in vivo sensing. They also facilitate combined imaging and targeted drug delivery for therapeutic applications. The use of these single plasmonic nanoparticles as multiplexed LSPR imaging system opens up the interesting way for the improvement of the diagnoses and therapies. Indeed, these single plasmonic nanoparticles may be enabled to transport and release drugs at a precisely targeted location.

2.5.3 *Numerical Simulations*

Because of the cost and time needed to fabricate nanostructured samples for SPR imaging system, an exhaustive experimental characterization of plasmonic nanostructures is not conceivable. Some preliminary numerical studies must be done in order to understand the different propagating plasmon and localized surface plasmon modes and their coupling. This allows choosing the suitable parameters sizes of structuration. RCWA calculations can be used to numerically calculate the reflectance from different 2D periodic gold structures as the wavelength, angle of incidence of the incident radiation, and the refractive index of the media around the SPRI sensor chips are varied [162]. However, for more complex nanostructured samples such as the metallic nanoparticles arrays described above, a model based on the hybridization of two classical methods (see Fig. 2.23), the Fourier modal method and the finite element method, was developed by Lalanne et al. [163, 164], which allows precise calculation with reasonable computation speeds and memory used. Such a numerical tool allows among other things to compute the distribution of the electrical field around the nanostructures. As the SPR sensitivity is directly related to the intensity of the plasmonic electrical field close (a few nanometers) to the metallic surface, the impact of the nanostructure size and shape on the final sensitivity can be predicted [165, 166]. This is also crucial for choosing the appropriate areas for the biomolecule probe localization.

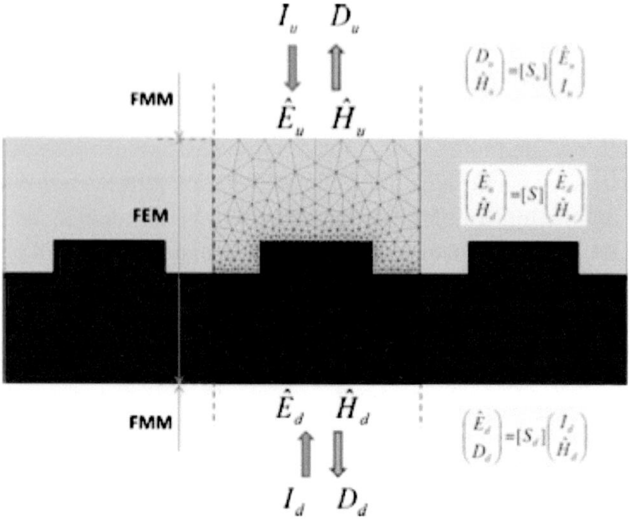

Figure 2.23 Sketch of the zone handled with FEM and FMM in the hybrid model and the associated S-matrix for different layers.

2.6 Conclusions

In summary, plasmonic imaging applied to the field of biochip systems is a fast growing research topic, due to the many advantages it offers; in particular evanescent sensing, label-free, access to interactions dynamics as well as ease of operation and low cost, to name a few. To perform at its best, it however requires many expertise to be assembled into a unique system, ranging from physics, in particular plasmonics, experimental, and modeling, to chemistry, biology, fluidics, data processing, etc. In this chapter, we have summarized the features and performances of enhanced SPRI systems. These biochips readers are capable of monitoring hundreds of molecular interactions, as small as binding of short oligonucleotides sequences to complex events such as cells adhesion. Different instrumental developments have extended the multidimensional capacity of SPRI systems, such as giving access to spectral information on the dielectric medium or on the bound molecule targets and/or molecular conformational states. Importance of microfluidics and data processing was also studied.

Such systems allow precise biomolecular surface characterization for both study and use of biochip functionalization and operation. The specific surface chemical and biochemical functionalizations, as necessitated by the presence of a metallic sensing layer, usually gold, and the label free constraint, have been reviewed and different possibilities are illustrated in detail. The vast diversity of approaches and applications of SPR imaging implies to carefully and globally adapt the analytical chain (implementation of support, surface modifications, biomolecular recognition, physical measurements, data processing) to optimize each specific type of analysis. Further enhancement of the capabilities of SPRI readers will require moving from simple inhomogeneous gold film to more dedicated substrates. This specific field of research will most certainly grow in the coming years.

References

1. Homola, J. and J. Dostálek, *Surface plasmon resonance based sensors*. Springer series on chemical sensors and biosensors. 2006, Berlin, New York: Springer. xii, 251p.
2. Myszka, D. G., Survey of the 1998 optical biosensor literature. *Journal of Molecular Recognition*, 1999. **12**(6): pp. 390–408.
3. Rich, R. L. and D. G. Myszka, Survey of the 2009 commercial optical biosensor literature. *Journal of Molecular Recognition*, 2011. **24**(6): pp. 892–914.
4. Rich, R. L. and D. G. Myszka, Grading the commercial optical biosensor literature-Class of 2008: "The Mighty Binders." *Journal of Molecular Recognition*, 2010. **23**(1): pp. 1–64.
5. Rothenhausler, B. and W. Knoll, Surface-plasmon microscopy. *Nature*, 1988. **332**(6165): pp. 615–617.
6. Jordan, C.E., et al., Surface plasmon resonance imaging measurements of DNA hybridization adsorption and streptavidin/DNA multilayer formation at chemically modified gold surfaces. *Analytical Chemistry*, 1997. **69**(24): pp. 4939–4947.
7. O'Brien, M. J., et al., A surface plasmon resonance array biosensor based on spectroscopic imaging. *Biosensors & Bioelectronics*, 2001. **16**(1/2): pp. 97–108.

8. Zhang, T., et al., Measuring particle-substrate distance with surface plasmon resonance microscopy. *Journal of Optics a-Pure and Applied Optics*, 2001. **3**(5): pp. 333–337.
9. Hottin, J., et al., Plasmonic DNA: Towards genetic diagnosis chips. *Plasmonics*, 2007. **2**(4): pp. 201–215.
10. Lecaruyer, P., et al., Surface plasmon resonance imaging as a multidimensional surface characterization instrument—Application to biochip genotyping. *Analytica Chimica Acta*, 2006. **573**: pp. 333–340.
11. Squires, T. M., R. J. Messinger, and S. R. Manalis, Making it stick: convection, reaction and diffusion in surface-based biosensors. *Nature Biotechnology*, 2008. **26**(4): pp. 417–426.
12. Yang, Y. H., et al., Normalization for cDNA microarray data: A robust composite method addressing single and multiple slide systematic variation. *Nucleic Acids Research*, 2002. **30**(4): p. e15.
13. Quackenbush, J., Microarray data normalization and transformation. *Nature Genetics*, 2002. **32**: pp. 496–501.
14. Lecaruyer, P., et al., Generalization of the Rouard method to an absorbing thin-film stack and application to surface plasmon resonance. *Applied Optics*, 2006. **45**(33): pp. 8419–8423.
15. Lecaruyer, P., et al., Metallic film optimization in a surface plasmon resonance biosensor by the extended Rouard method. *Applied Optics*, 2007. **46**(12): pp. 2361–2369.
16. Dostalek, J. and W. Knoll, Biosensors based on surface plasmon-enhanced fluorescence spectroscopy. *Biointerphases*, 2008. **3**(3): pp. Fd12–Fd22.
17. Gryczynski, I., et al., Effects of sample thickness on the optical properties of surface plasmon-coupled emission. *Journal of Physical Chemistry B*, 2004. **108**(32): pp. 12073–12083.
18. Myszka, D. G., et al., Extending the range of rate constants available from BIACORE: Interpreting mass transport-influenced binding data. *Biophysical Journal*, 1998. **75**(2): pp. 583–594.
19. Roper, D. K., Enhancing lateral mass transport to improve the dynamic range of adsorption rates measured by surface plasmon resonance. *Chemical Engineering Science*, 2006. **61**(8): pp. 2557–2564.
20. Raynal, F., et al., Study of a chaotic mixing system for DNA chip hybridization chambers. *Physics of Fluids*, 2004. **16**(9): pp. L63–L66.
21. Raynal, F., et al., Towards better DNA chip hybridization using chaotic advection. *Physics of Fluids*, 2007. **19**(1): p. 017112.

22. Nguyen, N. T. and Z. G. Wu, Micromixers—a review. *Journal of Micromechanics and Microengineering*, 2005. **15**(2): pp. R1–R16.
23. Sritharan, K., et al., Acoustic mixing at low Reynold's numbers. *Applied Physics Letters*, 2006. **88**(5): p. 054102.
24. Kuznetsova, L. A. and W. T. Coakley, Applications of ultrasound streaming and radiation force in biosensors. *Biosensors & Bioelectronics*, 2007. **22**(8): pp. 1567–1577.
25. Renaudin, A., et al., Integrated active mixing and biosensing using surface acoustic waves (SAW) and surface plasmon resonance (SPR) on a common substrate. *Lab on a Chip*, 2010. **10**(1): pp. 111–115.
26. Cular, S., et al., Removal of nonspecifically bound proteins on microarrays using surface acoustic waves. *IEEE Sensors Journal*, 2008. **8**(3/4): pp. 314–320.
27. Meyer, G. D., et al., Nonspecific binding removal from protein microarrays using thickness shear mode resonators. *IEEE Sensors Journal*, 2006. **6**(2): pp. 254–261.
28. Cloarec, J. P., et al., *A multidisciplinary approach for molecular diagnostics based on biosensors and microarrays. IRBM*, 2008. **29**(2/3): pp. 105–127.
29. Love, J. C., et al., Self-assembled monolayers of thiolates on metals as a form of nanotechnology. *Chemical Reviews*, 2005. **105**(4): pp. 1103–1169.
30. Livache, T., et al., Biosensing effects in functionalized electroconducting conjugated polymer layers—addressable DNA matrix for the detection of gene-mutations. *Synthetic Metals*, 1995. **71**(1/3): pp. 2143–2146.
31. Corne, C., et al., SPR-imaging based assays on an oligonucleotide-array to analyze DNA lesions recognition and excision by repair proteins. Science & Mathematics, 2008. **52**(1): pp. 249–250.
32. Maillart, E., et al., Versatile analysis of multiple macromolecular interactions by SPR imaging: Application to p53 and DNA interaction. *Oncogene*, 2004. **23**(32): pp. 5543–5550.
33. Grosjean, L., et al., A polypyrrole protein microarray for antibody-antigen interaction studies using a label-free detection process. *Analytical Biochemistry*, 2005. **347**(2): pp. 193–200.
34. Haddour, N., et al., Electrogeneration of a poly(pyrrole)-NTA chelator film for a reversible oriented immobilization of histidine-tagged proteins. *Journal of the American Chemical Society*, 2005. **127**(16): pp. 5752–5753.

35. Cherif, B., et al., Clinically related protein-peptide interactions monitored in real time on novel peptide chips by surface plasmon resonance imaging. *Clinical Chemistry*, 2006. **52**(2): pp. 255–262.

36. Mercey, E., et al., Polypyrrole oligosaccharide array and surface plasmon resonance imaging for the measurement of glycosaminoglycan binding interaction. *Analytical Chemistry*, 2008. **80**(9): pp. 3476–3482.

37. Gondran, C., et al., Detection of carbohydrate-binding proteins by oligosaccharide-modified polypyrrole interfaces using electrochemical surface plasmon resonance. *Analyst*, 2008. **133**(2): pp. 206–212.

38. Maskos, U. and E. M. Southern, Oligonucleotide hybridizations on glass supports—A novel linker for oligonucleotide synthesis and hybridization properties of oligonucleotides synthesized in situ. *Nucleic Acids Research*, 1992. **20**(7): pp. 1679–1684.

39. Blanchard, A. P., et al., High-density oligonucleotide arrays. *Biosensors & Bioelectronics*, 1996. **11**(6/7): pp. 687–690.

40. Lipshutz, R. J., et al., High density synthetic oligonucleotide arrays. *Nature Genetics*, 1999. **21**: pp. 20–24.

41. Lausted, C., et al., POSaM: A fast, flexible, open-source, inkjet oligonucleotide synthesizer and microarrayer. *Genome Biology*, 2004. **5**(8): p. R58.

42. Wieczerzak, E., et al., Monitoring of native chemical ligation on solid substrate by surface plasmon resonance. *Peptide Science*, 2008. **90**(3): pp. 415–420.

43. El Khoury, G., et al., Development of miniaturized immunoassay: Influence of surface chemistry and comparison with enzyme-linked immunosorbent assay and Western blot. *Analytical Biochemistry*, 2010. **400**(1): pp. 10–18.

44. Laibinis, P. E., et al., Comparison of the structures and wetting properties of self-assembled monolayers of normal-alkanethiols on the coinage metal-surfaces, Cu, Ag, Au. *Journal of the American Chemical Society*, 1991. **113**(19): pp. 7152–7167.

45. Ulman, A., Formation and structure of self-assembled monolayers. *Chemical Review*, 1996. **96**: pp. 1533–1554.

46. Skaife, J. J., et al., Influence of nanometer-scale topography of surfaces on the orientational response of liquid crystals to proteins specifically bound to surface-immobilized receptors. *Langmuir*, 2001. **17**(18): pp. 5448–5457.

47. Ahn, W., et al., Electroless gold island thin films: Photoluminescence and thermal transformation to nanoparticle ensembles. *Langmuir*, 2008. **24**(8): pp. 4174–4184.

48. Boussert, S., et al., An intramolecular O-N migration reaction on gold surfaces: Toward the preparation of well-defined amyloid surfaces. *ACS Nano*, 2009. **3**(10): pp. 3091–3097.

49. Lauer, M. E., et al., Formation and healing of micrometer-sized channel networks on highly mobile Au(111) surfaces. *Langmuir*, 2007. **23**(10): pp. 5459–5465.

50. Hostetler, M. J., et al., Dynamics of place-exchange reactions on monolayer-protected gold cluster molecules. *Langmuir*, 1999. **15**(11): pp. 3782–3789.

51. Briand, E., et al., Functionalisation of gold surfaces with thiolate SAMs: Topography/bioactivity relationship - A combined FT-RAIRS, AFM and QCM investigation. *Surface Science*, 2007. **601**(18): pp. 3850–3855.

52. Poirier, G. E., et al., Molecular-scale characterization of the reaction of ozone with decanethiol monolayers on Au(111). *Journal of the American Chemical Society*, 1999. **121**(41): pp. 9703–9711.

53. Poirier, G. E. and M. J. Tarlov, Molecular ordering and gold migration observed in butanethiol self-assembled monolayers using scanning-tunneling-microscopy. *Journal of Physical Chemistry*, 1995. **99**(27): pp. 10966–10970.

54. Poirier, G. E. and M. J. Tarlov, The C(4 × 2) superlattice of N-alkanethiol monolayers self-assembled an Au(111). *Langmuir*, 1994. **10**(9): pp. 2853–2856.

55. Poirier, G. E., et al., 2-Dimensional liquid-phase and the Px-root-3-phase of alkanethiol self-assembled monolayers on Au(111). *Langmuir*, 1994. **10**(10): pp. 3383–3386.

56. Boujday, S., et al., In-depth investigation of protein adsorption on gold surfaces: Correlating the structure and density to the efficiency of the sensing layer. *Journal of Physical Chemistry B*, 2008. **112**(21): pp. 6708–6715.

57. Tielens, F., et al., Characterization of omega-functionalized undecanethiol mixed self-assembled monolayers on Au(111): A combined polarization modulation infrared reflection-absorption spectroscopy/X-ray photoelectron spectroscopy/periodic density functional theory study. *Journal of Physical Chemistry C*, 2008. **112**(1): pp. 182–190.

58. Tielens, F., et al., Stability of binary SAMs formed by omega-acid and alcohol functionalized thiol mixtures. *Langmuir*, 2009. **25**(17): pp. 9980–9985.

59. Boujday, S., et al., Innovative surface characterization techniques applied to immunosensor elaboration and test: Comparing the efficiency of Fourier transform-surface plasmon resonance, quartz

crystal microbalance with dissipation measurements, and polarization modulation-reflection absorption infrared spectroscopy. *Analytical Biochemistry*, 2009. **387**(2): pp. 194–201.

60. Boujday, S., et al., Detection of pathogenic Staphylococcus aureus bacteria by gold based immunosensors. *Microchimica Acta*, 2008. **163**(3/4): pp. 203–209.
61. Tombelli, S., et al., Improved procedures for immobilisation of oligonucleotides on gold-coated piezoelectric quartz crystals. *Biosensors & Bioelectronics*, 2002. **17**(11/12): pp. 929–936.
62. Duval, A., et al., Polarimetric surface plasmon resonance imaging biosensor. *Optics Letters*, 2009. **34**(23): pp. 3634–3636.
63. Suraniti, E., et al., Construction of hybrid bilayer membrane (HBM) biochips and characterization of the cooperative binding between cytochrome-c and HBM. *Langmuir*, 2007. **23**(12): p. 6835.
64. Balladur, V., et al., Determination of the main forces driving DNA oligonucleotide adsorption onto aminated silica wafers. *Journal of Colloid and Interface Science*, 1997. **194**(2): p. 408.
65. Soultani-Vigneron, S., et al., Immobilisation of oligo-peptidic probes for microarray implementation: Characterisation by FTIR, atomic force microscopy and 2D fluorescence. *Journal of Chromatography B-Analytical Technologies in the Biomedical and Life Sciences*, 2005. **822**(1/2): pp. 304–310.
66. Bassil, N., et al., One hundred spots parallel monitoring of DNA interactions by SPR imaging of polymer-functionalized surfaces applied to the detection of cystic fibrosis mutations. *Sensors and Actuators B-Chemical*, 2003. **94**(3): pp. 313–323.
67. Mannelli, I., et al., Surface plasmon resonance imaging (SPRI) system and real-time monitoring of DNA biochip for human genetic mutation diagnosis of DNA amplified samples. *Sensors and Actuators B-Chemical*, 2006. **119**(2): pp. 583–591.
68. Yang, Z., et al., Characterization of three amino-functionalized surfaces and evaluation of antibody immobilization for the multiplex detection of tumor markers involved in colorectal cancer. *Langmuir*, 2013. **29**(5): pp. 1498–1509.
69. Yang, Z., et al., Improvement of protein immobilization for the elaboration of tumor-associated antigen microarrays: Application to the sensitive and specific detection of tumor markers from breast cancer sera. *Biosensors and Bioelectronics*, 2013. **40**(1): pp. 385–392.

70. Li, L. Y., et al., Protein adsorption on alkanethiolate self-assembled monolayers: Nanoscale surface structural and chemical effects. *Langmuir*, 2003. **19**(7): pp. 2974–2982.
71. Spadavecchia, J., et al., New cysteamine based functionalization for biochip applications. *Sensors and Actuators B: Chemical*, 2009. **143**(1): p. 139.
72. Li, L. Y., et al., In situ single-molecule detection of antibody-antigen binding by tapping-mode atomic force microscopy. *Analytical Chemistry*, 2002. **74**(23): pp. 6017–6022.
73. Mazurczyk, R., et al., Low-cost, fast prototyping method of fabrication of the microreactor devices in soda-lime glass. *Sensors and Actuators B: Chemical*, 2008. **128**(2): p. 552.
74. Mannelli, I., et al., DNA immobilisation procedures for surface plasmon resonance imaging (SPRI) based microarray systems. *Biosensors & Bioelectronics*, 2007. **22**(6): pp. 803–809.
75. Lofas, S. and B. Johnsson, A novel hydrogel matrix on gold surfaces in surface-plasmon resonance sensors for fast and efficient covalent immobilization of ligands. *Journal of the Chemical Society-Chemical Communications*, 1990(21): pp. 1526–1528.
76. Maalouf, R., et al., Label-free detection of bacteria by electrochemical impedance spectroscopy: Comparison to surface plasmon resonance. *Analytical Chemistry*, 2007. **79**(13): pp. 4879–4886.
77. Vidic, J., et al., Gold surface functionalization and patterning for specific immobilization of olfactory receptors carried by nanosomes. *Analytical Chemistry*, 2007. **79**(9): p. 3280.
78. Li, L. Y., et al., Protein adsorption on oligo(ethylene glycol)-terminated alkanethiolate self-assembled monolayers: The molecular basis for nonfouling behavior. *Journal of Physical Chemistry B*, 2005. **109**(7): pp. 2934–2941.
79. Marie, R., et al., Adsorption kinetics and mechanical properties of thiol-modified DNA-oligos on gold investigated by microcantilever sensors. *Ultramicroscopy*, 2002. **91**(1): pp. 29–36.
80. Lalander, C. H., et al., DNA-directed self-assembly of gold nanoparticles onto nanopatterned surfaces: controlled placement of individual nanoparticles into regular arrays. *ACS Nano*, 2010. **4**(10): pp. 6153–6161.
81. Dugas, V., et al., Droplet evaporation study applied to DNA chip manufacturing. *Langmuir*, 2005. **21**(20): pp. 9130–9136.

82. El Khoury, G., et al., Acid deprotection of covalently immobilized peptide probes on glass slides for peptide microarrays, in *2007 Annual International Conference of the IEEE Engineering in Medicine And Biology Society, Vols. 1–16*. 2007. pp. 2242–2246.
83. Guedon, P., et al., Characterization and optimization of a real-time, parallel, label-free, polypyrrole-based DNA sensor by surface plasmon resonance imaging. *Analytical Chemistry*, 2000. **72**(24): pp. 6003–6009.
84. Briand, E., et al., Immobilization of Protein A on SAMS for the elaboration of immunosensors. *Colloids And Surfaces B—Biointerfaces*, 2006. **53**(2): pp. 215–224.
85. Helm, C. A., et al., Measurement of ligand-receptor interactions. *Proceedings of the National Academy of Sciences*, 1991. **88**(18): pp. 8169–8173.
86. Häussling, L., et al., Direct observation of streptavidin specifically adsorbed on biotin-functionalized self-assembled monolayers with the scanning tunneling microscope. *Angewandte Chemie International Edition in English*, 1991. **30**(5): pp. 569–572.
87. Nguyen, T. T., et al., Comparison of the energetics of avidin, streptavidin, neutrAvidin, and anti-biotin antibody binding to biotinylated lipid bilayer examined by second-harmonic generation. *Analytical Chemistry*, 2011. **84**(1): pp. 201–208.
88. Green, N. M., Avidin and streptavidin. *Methods in Enzymology*, 1990. **184**: pp. 51–67.
89. Gerry, N. P., et al., Universal DNA microarray method for multiplex detection of low abundance point mutations. *Journal of Molecular Biology*, 1999. **292**(2): pp. 251–262.
90. Chevolot, Y., et al., DNA-based carbohydrate biochips: A platform for surface glyco-engineering. *Angewandte Chemie—International Edition*, 2007. **46**(14): pp. 2398–2402.
91. Boozer, C., et al., DNA-directed protein immobilization for simultaneous detection of multiple analytes by surface plasmon resonance biosensor. *Analytical Chemistry*, 2006. **78**(5): pp. 1515–1519.
92. Boozer, C., et al., DNA directed protein immobilization on mixed ssDNA/oligo (ethylene glycol) self-assembled monolayers for sensitive biosensors. *Analytical Chemistry*, 2004. **76**(23): pp. 6967–6972.
93. Ladd, J., et al., DNA-directed protein immobilization on mixed self-assembled monolayers via a streptavidin bridge. *Langmuir*, 2004. **20**(19): pp. 8090–8095.

94. Chevolot, Y., et al., Multiplexed binding determination of seven glycoconjugates for Pseudomonas aeruginosa Lectin I (PA-IL) using a DNA-based carbohydrate microarray. *Chemical Communications*, 2011. **47**(31): pp. 8826–8828.
95. Chabot, V., et al., Biosensing based on surface plasmon resonance and living cells. *Biosensors and Bioelectronics*, 2009. **24**(6): p. 1667.
96. Hide, M., et al., Real-time analysis of ligand-induced cell surface and intracellular reactions of living mast cells using a surface plasmon resonance-based biosensor. *Analytical Biochemistry*, 2002. **302**(1): pp. 28–37.
97. Yanase, Y., et al., The SPR signal in living cells reflects changes other than the area of adhesion and the formation of cell constructions. *Biosensors and Bioelectronics*, 2007. **22**(6): pp. 1081–1086.
98. Heumann, D. and T. Roger, Initial responses to endotoxins and Gram-negative bacteria. *Clinica Chimica Acta*, 2002. **323**(1): pp. 59–72.
99. Leary, S. C., et al., Chronic treatment with azide in situ leads to an irreversible loss of cytochrome c oxidase activity via holoenzyme dissociation. *Journal of Biological Chemistry*, 2002. **277**(13): pp. 11321–11328.
100. Coughlin, S. R., Protease-activated receptors in hemostasis, thrombosis and vascular biology. *Journal of Thrombosis and Haemostasis*, 2005. **3**(8): pp. 1800–1814.
101. Beuthan, J., et al., The spatial variation of the refractive index in biological cells. *Physics in Medicine and Biology*, 1996. **41**(3): pp. 369–382.
102. Nakkach, M., et al., Absorption and related optical dispersion effects on the spectral response of a surface plasmon resonance sensor. *Applied Optics*, 2008. **47**(33): pp. 6177–6182.
103. Alleyne, C. J., et al., Numerical method for high accuracy index of refraction estimation for spectro-angular surface plasmon resonance systems. *Optics Express*, 2008. **16**(24): pp. 19493–19503.
104. Sereda, A., et al., High performance multi-spectral interrogation for surface plasmon resonance imaging sensors. *Biosensors and Bioelectronics*, 2014. **54**(0): pp. 175–180.
105. Bardin, F., et al., Surface plasmon resonance spectro-imaging sensor for biomolecular surface interaction characterization. *Biosensors & Bioelectronics*, 2009. **24**(7): pp. 2100–2105.
106. Jin, R. C., et al., Controlling anisotropic nanoparticle growth through plasmon excitation. *Nature*, 2003. **425**(6957): pp. 487–490.

107. Piliarik, M. and J. Homola, Surface plasmon resonance (SPR) sensors: Approaching their limits? *Optics Express*, 2009. **17**(19): pp. 16505–16517.
108. Murphy, C. J. and N. R. Jana, Controlling the aspect ratio of inorganic nanorods and nanowires. *Advanced Materials*, 2002. **14**(1): pp. 80–82.
109. Johansen, K., et al., Surface plasmon resonance: instrumental resolution using photo diode arrays. *Measurement Science & Technology*, 2000. **11**(11): pp. 1630–1638.
110. Sun, Y. G. and Y. N. Xia, Shape-controlled synthesis of gold and silver nanoparticles. *Science*, 2002. **298**(5601): pp. 2176–2179.
111. Maillard, M., et al., Tuning the size of silver nanodisks with similar aspect ratios: Synthesis and optical properties. *Journal of Physical Chemistry B*, 2003. **107**(11): pp. 2466–2470.
112. Hottin, J., et al., Biochip data normalization using multifunctional probes. *Analyst*, 2012. **137**(13): pp. 3119–3125.
113. Canva, M., Procédé d'auto-calibration de biopuces, 2005: France extended.
114. Malic, L., et al., Enhanced surface plasmon resonance imaging detection of DNA hybridization an periodic gold nanoposts. *Optics Letters*, 2007. **32**(21): pp. 3092–3094.
115. Kim, H. M., et al., Detection of biomolecular binding through enhancement of localized surface plasmon resonance (LSPR) by gold nanoparticles. *Sensors*, 2009. **9**(4): pp. 2334–2344.
116. Live, L. S., et al., Localized and propagating surface plasmons in gold particles of near-micron size. *Journal of Physical Chemistry C*, 2009. **113**(1): pp. 40–44.
117. Byun, K. M., et al., Sensitivity enhancement of surface plasmon resonance imaging using periodic metallic nanowires. *Journal of Lightwave Technology*, 2008. **26**(9/12): pp. 1472–1478.
118. Yoon, S. J. and D. Kim, Target dependence of the sensitivity in periodic nanowire-based localized surface plasmon resonance biosensors. *Journal of the Optical Society of America a—Optics Image Science and Vision*, 2008. **25**(3): pp. 725–735.
119. Lisboa, P., et al., Sensitivity enhancement of surface-plasmon resonance imaging by nanoarrayed organothiols. *Advanced Materials*, 2008. **20**(12): p. 2352–2358.
120. Riboh, J.C., et al., A nanoscale optical biosensor: Real-time immunoassay in physiological buffer enabled by improved nanoparticle

adhesion. *Journal of Physical Chemistry B*, 2003. **107**(8): pp. 1772–1780.

121. Haes, A. J., et al., A localized surface plasmon resonance biosensor: First steps toward an assay for Alzheimer's disease. *Nano Letters*, 2004. **4**(6): pp. 1029–1034.
122. Sannomiya, T., et al., In situ sensing of single binding events by localized surface plasmon resonance. *Nano Letters*, 2008. **8**(10): pp. 3450–3455.
123. Zhao, J., et al., Localized surface plasmon resonance biosensors. *Nanomedicine*, 2006. **1**(2): pp. 219–228.
124. Barbillon, G., et al., Electron beam lithography designed chemical nanosensors based on localized surface plasmon resonance. *Surface Science*, 2007. **601**(21): pp. 5057–5061.
125. Barbillon, G., et al., How nanoparticles encapsulating fluorophores allow a double detection of biomolecules by localized surface plasmon resonance and luminescence. *Nanotechnology*, 2008. **19**(3): p. 035705.
126. Jung, L. S., et al., Quantitative interpretation of the response of surface plasmon resonance sensors to adsorbed films. *Langmuir*, 1998. **14**(19): pp. 5636–5648.
127. Barbillon, G., et al., Detection in near-field domain of biomolecules adsorbed on a single metallic nanoparticle. *Journal of Microscopy*, 2008. **229**(2): pp. 270–274.
128. Barbillon, G., et al., Detection of chemical molecules with integrated plasmonic glass nanotips. *Surface Science*, 2008. **602**(16): pp. L119–L122.
129. Sagle, L. B., et al., Advances in localized surface plasmon resonance spectroscopy biosensing. *Nanomedicine*, 2011. **6**(8): pp. 1447–1462.
130. Frens, G., Controlled nucleation for regulation of particle-size in monodisperse gold suspensions. *Nature—Physical Science*, 1973. **241**(105): pp. 20–22.
131. Jana, N. R., et al., Wet chemical synthesis of high aspect ratio cylindrical gold nanorods. *Journal of Physical Chemistry B*, 2001. **105**(19): pp. 4065–4067.
132. Jin, R. C., et al., Photoinduced conversion of silver nanospheres to nanoprisms. *Science*, 2001. **294**(5548): pp. 1901–1903.
133. Maillard, M., et al., Silver nanodisks. *Advanced Materials*, 2002. **14**(15): pp. 1084–1086.
134. Ung, T., et al., Controlled method for silica coating of silver colloids. Influence of coating on the rate of chemical reactions. *Langmuir*, 1998. **14**(14): pp. 3740–3748.

135. Ung, T., et al., Optical properties of thin films of Au@SiO2 particles. *Journal of Physical Chemistry B*, 2001. **105**(17): pp. 3441–3452.

136. Yun, M. H., et al., Electrochemically grown wires for individually addressable sensor arrays. *Nano Letters*, 2004. **4**(3): pp. 419–422.

137. Tam, F., et al., Geometrical parameters controlling sensitivity of nanoshell plasmon resonances to changes in dielectric environment. *Journal of Physical Chemistry B*, 2004. **108**(45): pp. 17290–17294.

138. Wang, H., et al., Nanorice: A hybrid plasmonic nanostructure. *Nano Letters*, 2006. **6**(4): pp. 827–832.

139. Westcott, S. L., et al., Construction of simple gold nanoparticle aggregates with controlled plasmon-plasmon interactions. *Chemical Physics Letters*, 1999. **300**(5/6): pp. 651–655.

140. Jackson, J. B. and N. J. Halas, Silver nanoshells: Variations in morphologies and optical properties. *Journal of Physical Chemistry B*, 2001. **105**(14): pp. 2743–2746.

141. Barbillon, G., et al., Two examples of nanostructured gold surfaces as biosensors. Surface-enhanced chemiluminescence and double detection by surface plasmon resonance and luminescence. *Gold Bulletin*, 2008. **41**(2): pp. 174–186.

142. Barbillon, G., et al., Gold nanoparticles by soft UV nanoimprint lithography coupled to a lift-off process for plasmonic sensing of antibodies. *Microelectronic Engineering*, 2010. **87**(5/8): pp. 1001–1004.

143. Hulteen, J.C. and R. P. Vanduyne, Nanosphere lithography—A materials general fabrication process for periodic particle array surfaces. *Journal of Vacuum Science & Technology a—Vacuum Surfaces and Films*, 1995. **13**(3): pp. 1553–1558.

144. Dimitrov, A. S., et al., A comparison between the optical properties of amorphous and crystalline monolayers of silica particles. *Langmuir*, 1999. **15**(16): pp. 5257–5264.

145. Hamouda, F., et al., Nanoholes by soft UV nanoimprint lithography applied to study of membrane proteins. *Microelectronic Engineering*, 2009. **86**(4/6): pp. 583–585.

146. Hamouda, F., et al., Sub-200 nm gap electrodes by soft UV nanoimprint lithography using polydimethylsiloxane mold without external pressure. *Journal of Vacuum Science & Technology B*, 2010. **28**(1): pp. 82–85.

147. Sau, T. K., et al., Properties and applications of colloidal nonspherical noble metal nanoparticles. *Advanced Materials*, 2010. **22**(16): pp. 1805–1825.

148. Barbillon, G., et al., Sensitive detection of biological species through localized surface-plasmon resonance on gold nanodisks. *Thin Solid Films*, 2009. **517**(9): pp. 2997–3000.

149. Marinakos, S. M., et al., Plasmonic detection of a model analyte in serum by a gold nanorod sensor. *Analytical Chemistry*, 2007. **79**(14): pp. 5278–5283.

150. Faure, A. C., et al., Core/shell nanoparticles for multiple biological detection with enhanced sensitivity and kinetics. *Nanotechnology*, 2008. **19**(48): p. 485103.

151. Klar, T., et al., Surface-plasmon resonances in single metallic nanoparticles. *Physical Review Letters*, 1998. **80**(19): pp. 4249–4252.

152. Mock, J. J., et al., Local refractive index dependence of plasmon resonance spectra from individual nanoparticles. *Nano Letters*, 2003. **3**(4): pp. 485–491.

153. Lal, S., et al., Nano-optics from sensing to waveguiding. *Nature Photonics*, 2007. **1**(11): pp. 641–648.

154. Jain, P. K., et al., Calculated absorption and scattering properties of gold nanoparticles of different size, shape, and composition: Applications in biological imaging and biomedicine. *Journal of Physical Chemistry B*, 2006. **110**(14): pp. 7238–7248.

155. Rosi, N. L. and C. A. Mirkin, Nanostructures in biodiagnostics. *Chemical Reviews*, 2005. **105**(4): pp. 1547–1562.

156. Schultz, S., et al., Single-target molecule detection with nonbleaching multicolor optical immunolabels. *Proceedings of the National Academy of Sciences of the United States of America*, 2000. **97**(3): pp. 996–1001.

157. Taton, T. A., et al., Two-color labeling of oligonucleotide arrays via size-selective scattering of nanoparticle probes. *Journal of the American Chemical Society*, 2001. **123**(21): pp. 5164–5165.

158. El-Sayed, I. H., et al., Surface plasmon resonance scattering and absorption of anti-EGFR antibody conjugated gold nanoparticles in cancer diagnostics: Applications in oral cancer. *Nano Letters*, 2005. **5**(5): pp. 829–834.

159. Alivisatos, P., The use of nanocrystals in biological detection. *Nature Biotechnology*, 2004. **22**(1): pp. 47–52.

160. Hu, R., et al., Metallic nanostructures as localized plasmon resonance enhanced scattering probes for multiplex dark-field targeted imaging of cancer cells. *Journal of Physical Chemistry C*, 2009. **113**(7): pp. 2676–2684.

161. Ruemmele, J. A., et al., A localized surface plasmon resonance imaging instrument for multiplexed biosensing. *Analytical Chemistry*, 2013. **85**(9): pp. 4560–4566.

162. Li, L. F., New formulation of the Fourier modal method for crossed surface-relief gratings. *Journal of the Optical Society of America a—Optics Image Science and Vision*, 1997. **14**(10): pp. 2758–2767.

163. Besbes, M., et al., Numerical analysis of a slit-groove diffraction problem. *Journal of the European Optical Society—Rapid Publications*, 2007. **2**: p. 07022.

164. Hugonin, J. P., et al., Hybridization of electromagnetic numerical methods through the G-matrix algorithm. *Optics Letters*, 2008. **33**(14): pp. 1590–1592.

165. Sarkar, M., et al., Introducing 2D confined propagating plasmons for surface plasmon resonance sensing using arrays of metallic ribbons. *Sensors and Actuators B—Chemical*, 2014. **191**: pp. 115–121.

166. Chamtouri, M., et al., Enhanced SPR sensitivity with nano-micro-ribbon gratings—An exhaustive simulation mapping. *Plasmonics*, 2014. **9**(1): pp. 79–92.

Chapter 3

Modeling of Enhanced Electromagnetic Fields in Plasmonic Nanostructures

Maria Antonia Iatì,[a] Elefterios Lidorikis,[b] and Rosalba Saija[c]

[a] *CNR-Istituto per i Processi Chimico-Fisici, Viale F. Stagno D'Alcontres 37, 98158 Messina, Italy*
[b] *Department of Materials Science and Engineering, University of Ioannina, Ioannina 45110, Greece*
[c] *Dip. di Fisica e di Scienze della Terra, Viale F. Stagno D'Alcontres 31, 98166 Messina, Italy*
iati@me.cnr.it

This chapter provides a description of some of the main approaches suitable for modeling enhanced electromagnetic fields in plasmonic nanostructures, ranging from the quasi-static approximation to the exact description in terms of higher order multipole field expansions and the consideration of nonlocal and quantum effects. The strengths and limitations of each method will be discussed in terms of efficiency, accuracy, and applicability to nanoplasmonic systems of interest.

3.1 Introduction

The interaction processes between electromagnetic radiation and conduction electrons in metal nanoparticles (MNPs) produce

Handbook of Enhanced Spectroscopy
Edited by Pietro Giuseppe Gucciardi, Marc Lamy de la Chapelle, and Nathalie Lidgi-Guigui

ISBN 978-981-4613-32-3 (Hardcover), 978-981-4613-33-0 (eBook)
www.panstanford.com

coherent localized plasmon oscillations, termed localized surface plasmon resonances (LSPRs), with a resonant frequency depending on the composition, size, geometry, dielectric environment, and particle–particle distance in the presence of aggregation [80]. As a result, electric fields near the particle are greatly enhanced and the particle optical extinction (scattering + absorption) shows a maximum at the plasmon resonance frequency [60]. This important behavior of MNPs finds a broad range of applications in electronics, optical sensing, biomedicine, and nanotechnology [57]. Nanostructures supporting plasmon oscillations also play a significant role in surface-enhanced techniques, such as surface-enhanced Raman spectroscopy (SERS) and surface/metal-enhanced fluorescence (SEF or MEF) [57]. The enhanced electromagnetic fields extend beyond the nanoparticle surface by some nanometers, creating a sensing volume in which molecule detection is possible [80].

LSPRs appear to be strongly dependent on MNP size and shape. For small particles, the extinction is dominated by absorption, while with increasing particle size, scattering takes over [60]. As particle size increases, higher order multipole modes appear and a description in terms of dipole contributions becomes inaccurate. Furthermore, in presence of MNP aggregates, significantly larger field enhancements in comparison with individual nanoparticles can be observed at the inter-junction between adjacent particles, due to nearest-neighbor interaction between the MNPs. These inter-junction regions characterized by highly intense and localized electromagnetic fields are called "hot spots" [30] and are critical to enhanced sensing applications. It is therefore fundamental to get a full comprehension of the particle–particle interaction at the nanoscale.

The recent advances in synthesis and characterization of plasmonic nanostructures made it possible to realize nanoparticles with many different shapes, such as nanoshells, nanowires, nanocubes, nanostars, nanotriangles, just to mention a few of them [18, 70, 77, 98]. This implies the need for simultaneous advances in accurate simulation techniques capable of computing the optical properties of nonspherical, eventually aggregated, nanoparticles. Several analytical and numerical methods have been proposed in the literature for simulating the optical behavior of nonspherical MNPs and actually there is no single universal method providing the best results in all cases. The choice of a particular technique is suggested by the specific particle parameters. It is not the purpose of this work to give an exhaustive summary of the different methods. Detailed information on specific computational techniques can be found in the literature and comprehensive reviews appear in Myroshnychenko et al. [69] and Zhao et al. [119].

Currently, the most commonly used numerical approaches for describing the optical response of MNPs are the discrete dipole approximation (DDA), the finite-difference time domain (FDTD) method, and the boundary (finite) element method (BEM, FEM). The DDA, originally introduced by Purcell and Pennypacker [83] for describing light scattering by interstellar dust grains, was subsequently developed by Draine [21] and Goodman, Draine, and Flatau [29]. The DDA simulates the metal particle as an array of N point dipoles of known polarizability tensor organized on a lattice to represent the nanoparticle shape. This method has found many applications in modeling extinction spectra of MNPs with different shapes and in simulating spectra of plasmonic bands [40, 81]. The accuracy of DDA calculations critically depends on the discretization step and so the main drawback of this method is that it can become time consuming. A comprehensive review of the DDA can be found in Yurkin and Hoekstra [118].

Another approach that has been extensively applied to model nanostructure optical properties is the FDTD method [31, 102]. This approach is based on the numerical integration of the Maxwell equations in the time domain and therefore can be very advantageous when time-resolved fields are required [69, 99, 117]. The space and time derivatives of the electric and magnetic fields are approximated using a finite-difference scheme with space and time discretizations selected to minimize computational errors and ensure numerical stability [66]. The fields are specified at spatial grid points. Values at the grid points for the previous and current time steps are used to calculate the values at the next time step. The FDTD method, although time-domain based, can also be used to calculate the nanoparticle frequency domain spectra using Fourier transforms [119]. The popularity of the FDTD method stems from its simple implementation, flexibility, and versatility. Disadvantages in its use may be due to some limitations in terms of efficiency and accuracy, and in terms of the maximum particle size parameter [66].

A powerful approach when dealing with MNP aggregates and dispersions of randomly oriented particles is the transition matrix (T-matrix) technique [8, 66, 112]. In many applications, the T-matrix allows a rigorous description of the scattering process, surpassing other commonly used techniques in terms of efficiency and size parameter range. Moreover, the transformation properties under rotation of the T-matrix make it possible to perform an analytical evaluation of the scattering characteristics averaged over particle orientation [9, 12, 27]. This unique feature of the T-matrix method has relevant effects in saving computational time. Given a particle reference frame, the T-matrix needs to be computed only once, then the average scattering characteristics of a particle ensemble can be computed through simple analytical formulas [8, 56, 66]. Therefore, the computational effort required by the T-matrix method turns out to be much lighter than the one required by other numerical approaches, such as the DDA and the FDTD, which need *ex novo* calculation of the scattered field whenever the orientation of the scattering particle changes. A comprehensive review of the applications of the T-matrix method for modeling MNPs optical behavior and their applications in sensing, surface-enhanced Raman scattering, and manipulation with laser tweezers is given in Khlebtsov [42].

This chapter is organized into three main parts. In the beginning, we will describe the particle field enhancement within the quasi-static approximation, discussing particle size and shape effects, and particle–particle and particle–substrate interaction. Then, we will introduce an exact approach to the description of the optical behavior of single and aggregated MNPs through the multipole field expansion and the T-matrix approach. Finally, we will present an extension of the theory to nonlocal and quantum effects, which become important as the particle size gets smaller than the electron mean free path in the bulk material and when MNP pair separations enter the sub-nanometer scale.

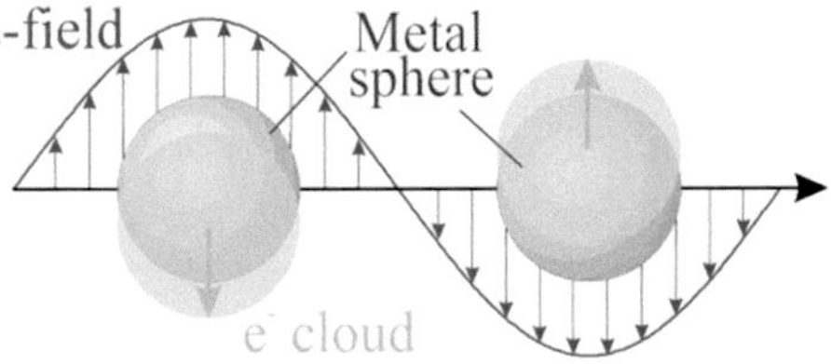

Figure 3.1 Schematic of plasmon oscillation for a sphere, showing the displacement of the conduction electron charge cloud relative to the nuclei. Reprinted with permission from Ref. [40], Copyright 2003, American Chemical Society.

3.2 Field Enhancement in the Quasi-Static Limit

A great deal of understanding and intuition is gained in the quasi-static approximation, where closed form expressions are possible. When the radius R of a spherical MNP is much smaller than the wavelength, there is no significant variation of the incident field across the particle volume and electrostatics can be used. When illuminated, the conduction electrons in the MNP will oscillate coherently under the action of the exciting light field, causing surface charges to develop at opposite particle boundaries, as shown schematically in Fig. 3.1. These, however, result into a restoring force, defining an oscillation eigenfrequency that is dependent on the density and effective mass of the electrons, the size and shape of the particle, as well as on the dielectric environment. When the MNP is illuminated at this eigenfrequency, the electron cloud coherent oscillation becomes resonant, giving rise to the so-called LSPR, with attributes such as very high polarization fields and scattering/absorption cross sections. To first order the surface charge distribution defines a dipole moment, and thus in the quasi-static limit the response of the MNP is that of a dipole polarization $\mathbf{p}(\omega)$:

$$\mathbf{p}(\omega) = \varepsilon_0 \varepsilon_h \alpha(\omega) \mathbf{E}_0(\omega) \tag{3.1}$$

where ε_0 is the vacuum permittivity, ε_h the host dielectric constant (assumed real), and $\mathbf{E}_0$ the incident field. The dipole polarizability $\alpha(\omega)$ is [7, 50, 57]:

$$\alpha(\omega) = 4\pi R^3 \frac{\varepsilon(\omega) - \varepsilon_h}{\varepsilon(\omega) + 2\varepsilon_h} \tag{3.2}$$

where $\varepsilon(\omega)$ is the dielectric function of the metal. The LSPR appears at $\mathrm{Re}\{\varepsilon(\omega)\} = -2\varepsilon_h$. Analytical expressions also exist for ellipsoidal MNP in the quasi-static limit, while for more complex shapes a numerical approach is required [40]. The polarizability of a MNP relates to its absorption and scattering cross sections [7, 50, 57]

$$C_{\mathrm{abs}} = k\mathrm{Im}\{\alpha\},\ C_{\mathrm{sca}} = k^4|\alpha|^2/6\pi. \tag{3.3}$$

An oscillating dipole re-radiates, so then the total field at point $\mathbf{r}$ from the MNP center is the sum of the incident and re-radiated fields [37]

$$\mathbf{E} \cong \mathbf{E}_0 + \frac{e^{ikr}}{4\pi\varepsilon_0\varepsilon_h r^3}\{k^2r^2(\hat{\mathbf{n}}\times\mathbf{p})\times\hat{\mathbf{n}} + (1-ikr)[3\hat{\mathbf{n}}\cdot(\hat{\mathbf{n}}\cdot\mathbf{p}) - \mathbf{p}]\} \tag{3.4}$$

where $\hat{\mathbf{n}} = \mathbf{r}/r$ is the unit vector in the direction of $\mathbf{r}$. For plasmonic applications where high field enhancement is needed, such as in SERS [1, 23, 44–46, 64, 68, 71, 111], only the near-field term $\propto r^{-3}$ of Eq. 3.4 is important. These near-fields are related to enhanced light–matter interactions such as absorption and emission. In particular, absorption in a material is given by $\sigma|\mathbf{E}|^2$ [37], so the absorption enhancement is given by:

$$F_{\mathrm{abs}} = |\mathbf{E}(\mathbf{r},\omega)|^2/|\mathbf{E}_0(\mathbf{r},\omega)|^2 \tag{3.5}$$

while in the electromagnetic contribution to SERS, both excitation at frequency ω and emission at the Stokes-shifted frequency ω_s are enhanced, yielding a SERS enhancement:

$$F_{\mathrm{SERS}} = |\mathbf{E}(\mathbf{r},\omega)|^2|\mathbf{E}(\mathbf{r},\omega_s)|^2/(|\mathbf{E}_0(\mathbf{r},\omega)|^2|\mathbf{E}_0(\mathbf{r},\omega_s)|^2). \tag{3.6}$$

We can then use Eq. 3.4 in its near-field limit to evaluate Eq. 3.5 and 3.6. Furthermore, and depending on the dimensionality of the understudy system, Eqs. 3.5 and 3.6 can be integrated around the MNP to yield expressions for absorption and for SERS (including tip-enhanced) in one-dimensional [15], two-dimensional [53, 59, 93], and three-dimensional [52, 53] systems. In particular it is found that the SERS enhancement scales as a function of distance Δ of the MNP surface as $\propto R^{12}|\eta(\omega)|^2|\eta(\omega_s)|^2/(R+\Delta)^n$, where $n = 11$ for a 1D system (e.g., carbon nanotube, see Fig. 3.2a,b) and $n = 10$ for a 2D system (e.g., graphene, see Fig. 3.2c) in the incoherent case, while the exponent changes to 10 and 8 respectively

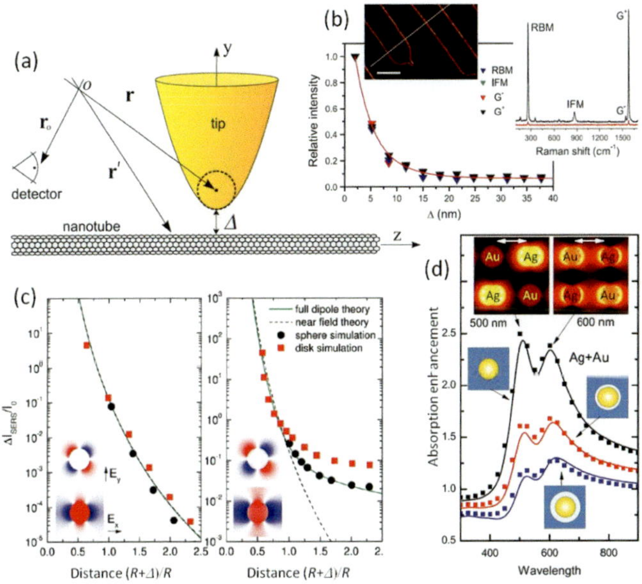

Figure 3.2 (a) Schematic for spatially resolved Raman scattering of carbon nanotubes. The dashed circumference represents a small sphere of radius R_{tip} centered at the tip apex. (b) Approach curve $I/I_{\text{max}} = M^{-1} + C(R_{\text{tip}} + \Delta)^{-10}$ for the measured radial breathing mode (RBM), intermediate frequency mode (IFM), G^-, and G^+ Raman bands of a carbon nanotube. M and C were treated as adjustable parameters. Measurements in right inset show the far-field ($\Delta = 38$ nm, red curve) and near-field ($\Delta = 2$ nm, black curve) Raman spectra. In the top inset, the tip-enhanced G-band Raman image of a carbon nanotube is shown (adapted from Ref. [15]). (c) FDTD simulations for SERS enhancement in 2D graphene by suspended nanospheres and nanodisks, compared with the near-field model and a full-field model that takes both near- and far-fields into account. The NP diameter is 10 nm (left) and 140 nm (right), with 23 nm and 320 nm periodicities, respectively. The insets show the fields on a plane running through the NP center, at the corresponding LSPR of a 10 nm sphere (left) and a 140 nm disk (right), showing the characteristic dipole field distributions (adapted from Ref. [93]). (d) Total 3D absorption enhancement caused by a Au and Ag NP composite ($R = 4$ nm, $f = 5.5\%$). Three coatings are considered with thicknesses 0, 1, and 2 nm ($n_c = n_h$). Lines for analytical (Eq. 3.7) and symbols for numerical (FDTD) results. The insets plot the field intensity at $\lambda = 500$ nm (Ag resonance) and $\lambda = 600$ nm (Au resonance) for the uncoated case (adapted from Ref. [53]).

in the coherent case [59]. In the above expression, $\eta = \alpha/4\pi R^3$ is the normalized polarizability. Finally, the effect of MNP near-fields on the absorption enhancement in the surrounding material (e.g., for photovoltaic applications) is [52, 53]:

$$F_{\text{abs}} \cong \beta \left(1 - f + 2f|\eta|^2 + 2f\text{Re}\{\eta\}\right) \tag{3.7}$$

where $\beta = 1 - C_{\text{abs}}/L^2$ is a phenomenological parameter incorporating the effect of resistive losses in the MNPs, L is the average distance between NPs, and $f = 4\pi R^3/3L^3$ is the NP average volume filling ratio. Equation (3.7) assumes that the absorption of the bare surrounding material is low without the MNP, while it is also valid for mixtures of different metal NP (see Fig. 3.2d). More detailed expressions about absorption and Raman enhancements in systems with different dimensionalities can be found in the above-referenced papers.

3.2.1 *Particle Shape and Size Effects*

From the previous section, it gets clear that the particle polarizability plays a central role in the study of plasmonic-enhanced near-fields. Analytical expressions for the polarizability of MNP with shapes different from spherical are mainly restricted to ellipsoids. For an ellipsoid with semi-axes $a_1 > a_2 > a_3$ and light illumination along one of the principal axes, the corresponding polarizability is [7]

$$\alpha_i = 4\pi a_1 a_2 a_3 \frac{\varepsilon - \varepsilon_h}{3\varepsilon_h + 3L_i(\varepsilon - \varepsilon_h)} = V\left(\frac{\varepsilon_h}{\varepsilon - \varepsilon_h} + L_i\right)^{-1} \tag{3.8}$$

where $V = 4\pi a_1 a_2 a_3/3$ is the particle volume and L_i is the geometrical factor when the field is polarized along the principal axis i:

$$L_i = \frac{3V}{8\pi} \int_0^\infty \frac{dq}{(a_i^2 + q)f(q)} \tag{3.9}$$

with $f(q) = \sqrt{(q + a_1^2)(q + a_2^2)(q + a_3^2)}$. They satisfy $\sum L_i = 1$ and $L_1 \leq L_2 \leq L_3$, while $L_i = 1/3$ for a sphere. From Eq. 3.8, we now find the LSPR condition:

$$\text{Re}\{\varepsilon\} = \varepsilon_h(1 - L_i^{-1}) \tag{3.10}$$

A special class of ellipsoids is the spheroids, which have two of the three semi-axes equal. If the smaller two axes ($a_2 = a_3$) are equal we have a prolate spheroid (cigar-like), in which case:

$$L_1 = \frac{1-e^2}{e^2}\left(-1 + \frac{1}{2e}\ln\frac{1+e}{1-e}\right) \tag{3.11}$$

while when the larger two axes are equal ($a_1 = a_2$), we have an oblate spheroid (disk-like):

$$L_1 = \frac{g(e)}{2e^2}\left[\frac{\pi}{2} - \tan^{-1} g(e)\right] - \frac{g^2(e)}{2},$$
$$g(e) = \left(\frac{1-e^2}{e^2}\right)^{1/2} \tag{3.12}$$

where $e = 1 - a_3^2/a_1^2$ is the particle eccentricity. The other geometrical factors are found using $\sum L_i = 1$. The prolate spheroid ranges from a needle ($e = 1$) to a sphere ($e = 0$), while the oblate spheroid ranges from a disk ($e = 1$) to a sphere ($e = 0$). Finally, the polarizabilities of a finite cylinder have been shown to have a close relationship with those of an ellipsoid [110].

Expressions are also derived for ellipsoids coated by shell of constant thickness and dielectric function ε_c. Assuming the semi-axes of the total particle structure to be a_1', a_2', and a_3', the polarizability is [7]:

$$\alpha_i = V\frac{(\varepsilon_c - \varepsilon_h)\left[\varepsilon_c + (\varepsilon - \varepsilon_c)(L_i^{(1)} - fL_i^{(2)})\right] + f\varepsilon_c(\varepsilon - \varepsilon_c)}{\left[\varepsilon_c + (\varepsilon - \varepsilon_c)(L_i^{(1)} - fL_i^{(2)})\right]\left[\varepsilon_h + (\varepsilon_c - \varepsilon_h)L_i^{(2)}\right] + \varepsilon_c(\varepsilon_c - \varepsilon_h)} \tag{3.13}$$

where now $V = {}^{4\pi a_1' a_2' a_3'}/_3$ is the total volume of the particle and here $f = {}^{a_1 a_2 a_3}/_{a_1' a_2' a_3'}$ is the volume fraction of the core. $L_i^{(1)}$and $L_i^{(2)}$ are the geometrical factors of the inner and outer ellipsoids respectively. For a sphere, $L_i^{(1)} = L_i^{(2)} = 1/3$.

For larger nanoparticles, the quasi-static approximation fails due to retardation effects, and higher order corrections must be considered. An expansion of the first TM mode of Mie theory for a sphere yields [51, 61]

$$\alpha = V\frac{1 - \left(\frac{1}{10}\right)(\varepsilon + \varepsilon_h)x^2 + O(x^4)}{\left(\frac{\varepsilon_h}{\varepsilon - \varepsilon_h} - \frac{1}{3}\right) - \frac{1}{30}(\varepsilon + 10\varepsilon_h)x^2 - i\frac{4\pi^2\varepsilon_h^{3/2}}{3}\frac{V}{\lambda_0^3} + O(x^4)} \tag{3.14}$$

where $x = 2\pi R/\lambda_0$ is the so-called size parameter. The x^2 term in the numerator incorporates retardation effects, while the x^2 term in the denominator incorporates retardation effects in the depolarization field [61]. The overall effect is the shifting of the LSPR to lower frequencies. Finally, the third term in the denominator is purely imaginary and incorporates the effect of radiation damping, caused by the direct radiative decay of the LSPR into photons [47, 115], and results into a significant broadening and weakening of the LSPR as particle size increases.

Equation (3.14) was been extended empirically to apply for ellipsoid particles as well [51]

$$\alpha_i = V\left(\frac{\varepsilon_h}{\varepsilon - \varepsilon_h} + L_i + A\varepsilon_h x_i^2 + B\varepsilon_h^2 x_i^4 - i\frac{4\pi^2\varepsilon_h^{3/2}}{3}\frac{V}{\lambda_0^3}\right)^{-1} \tag{3.15}$$

where the size parameter is now defined according to the relevant semi-axis along the field polarization direction and the A, B are treated as adjustable parameters. This expression was fitted to simulated data and they found [51]

$$\begin{aligned} A(L_i) &= -0.4865L_i - 1.046L_i^2 + 0.8481L_i^3 \\ B(L_i) &= 0.01909L_i + 0.1999L_i^2 + 0.6077L_i^3 \end{aligned} \tag{3.16}$$

3.2.2 *Particle–Particle and Particle–Substrate Interactions*

If the MNPs are not isolated from each other, then the average field acting on each particle is the sum of the incident one and of the one re-radiated by all the other MNPs. Accounting for this in case of sub-wavelength separations is most simply done within the dipole approximation by considering the macroscopic polarization $P = N \cdot p$, where N is the number density of the MNPs, $p = 4\pi\varepsilon_0\varepsilon_h R^3\eta(E_0 + E_i) = 4\pi\varepsilon_0\varepsilon_h R^3\tilde{\eta}E_0$ is the induced particle dipole moment, E_0 is the incident field, E_i is the average additional field due to scattering from the other MNPs, and $\eta = \alpha/4\pi R^3$ is the normalized polarizability. It is shown [37] that $E_i = P/3\varepsilon_0\varepsilon_h$, which, when combined with the above two equations, yields that the effect of the MNP re-radiated fields is accounted for by the simple adjustment of the normalized polarizability [52, 53]

$$\tilde{\eta}^{-1} = \eta^{-1} - f \tag{3.17}$$

where $f = 4\pi NR^3/3$ is the MNP volume filling ratio. For ellipsoids with semi-axes a, b, and c, we replace $R^3 \rightarrow abc$. Equation (3.17) applies for coated particles as well, and is also valid for 2D MNP arrays (for normal incidence at least [53]).

A more convenient form, applicable to uncoated particles only, is obtained in terms of the geometric factors, after combining Eq. 3.17 with Eq. 3.8 to find

$$\tilde{L}_i = L_i - f/3 \tag{3.18}$$

for a regular array of ellipsoid particles aligned to each other with the incident field parallel to their principal axis i. The LSPR condition is now $\varepsilon = \varepsilon_h(1 - \tilde{L}_i^{-1})$, that is, shifted to more negative values of ε and thus to longer wavelengths. The main effect of the scattered fields is to reduce the restoring force on the MNP free electrons leading into a decreased plasmon frequency and thus the red-shift observed in MNP arrays. Note that the above approximation breaks down for MNPs in close proximity, as it ignores the detailed surface charge distributions and the resulting high fields created within the narrow gap in between the MNPs. More detailed expressions for MNP pairs and aggregates have been described by Kreibig and Vollmer [50], while pairs of MNPs with different shapes were considered by Gluodenis et al. [28] and comparisons to experiments were made by Urich et al. [106].

For large-period MNP arrays, when the inter-particle separation is comparable to the wavelength, diffractive coupling between the plasmon resonances becomes active giving rise to the so-called lattice plasmons (LPs), which have characteristically narrow line shapes. These are best understood within the coupled-dipole approach that again incorporates the fields scattered by all nanoparticles as an effective change of the polarizability [20, 58, 96, 120, 121]:

$$\tilde{\eta}^{-1} = \eta^{-1} - G \tag{3.19}$$

where for normal incidence and a square array of dipoles (particles)

$$G = R^3 \sum_{j \neq i} r_{ij}^{-3} e^{ikr_{ij}} \left[(1 - ikr_{ij})(3\cos^2\theta_{ij} - 1) + k^2 r_{ij}^2 \sin^2\theta_{ij} \right] \tag{3.20}$$

with θ_{ij} the angle between $\mathbf{r}_{ij}$ and the polarization direction. At normal incidence, the onset of diffraction occurs at $\lambda = \varepsilon_h^{1/2} D$, the so-called Rayleigh anomaly, where G becomes maximum (diverges for an infinite array). If now there is a frequency at which $\mathrm{Re}\{\eta^{-1} - G\} = 0$, a lattice resonance appears in addition to the LSPR resonance, albeit sharper and stronger than the LSPR. Excitation of lattice resonances has been demonstrated experimentally [3, 16, 48, 88, 107] and has been shown to hold great promise for both bulk [73] and surface [96] refractive index sensing, and for application in label-free affinity biosensing [33].

The above expression assumes that the MNPs are embedded inside an infinite uniform material ε_h. If the MNPs are resting on top of a substrate with dielectric constant ε_s, the LSPR is modulated due to the close interaction with the substrate. This has been treated within a dipole-image approach by Yamaguchi et al. [116] and Bobbert and Vlieger [5], where in case of small spherical MNPs and normal incidence we find in the spherical dipole approximation

$$\tilde{\eta}^{-1} = \eta^{-1} - F \tag{3.21}$$

where $F = \frac{R^3}{(2d)^3}\frac{\varepsilon_s - \varepsilon_h}{\varepsilon_s + \varepsilon_h}$, d is the distance from the MNP center to the substrate (for light polarized perpendicular to the substrate surface this becomes $\tilde{\eta}^{-1} = \eta^{-1} - 2F$). For particle arrays on a substrate, the cumulative effect of Eqs. 3.17 and 3.21 can be used [116].

For spherical particles ($R = d$) and typical dielectric substrates ($\varepsilon_s \neq -\varepsilon_h$), the image charge correction is generally small ($F \sim 0.1$). For spheroids, however, the R^3 term in the expression for F must be replaced by the product of the three semi-axes and d is the semi-axis perpendicular to the surface. It is easily seen that for flat oblate spheroids, as the particle center comes closer to the surface and thus to its image charge, the correction becomes increasingly large. In such cases, however, consideration of higher order multipoles becomes important [7, 89]. Bobbert and Vlieger [6] considered this case by solving the problem in the appropriate spheroidal coordinates and derived the appropriate adjustment for the geometric depolarization factors L_i, which compare well with experiments by Valamanesh et al. [108]. Finally, polarizability of truncated spheres on a substrate were presented by Wind et al. [113, 114].

3.3 Multipole Field Expansion and T-Matrix in Plasmonics

In this section we follow the approach by Borghese et al. [8], which proved to be very suitable for an analytical description of the field scattered by MNPs in the near- and far-field regions, in order to go beyond the dipole approximation.

We assume that the particle is embedded into a homogeneous, isotropic, indefinite medium of real refractive index n. The incident field is the polarized plane wave

$$E_I = E_0 \hat{e}_I \exp(ik_I \cdot r), \tag{3.22}$$

with unit polarization vector $\hat{\mathbf{e}}_I$ and propagation vector $\mathbf{k}_I = k\hat{\mathbf{k}}_I$, where $k = nk_v$ and $k_v = \omega/c$.The scattered field, at a large distance from the particle (within which the origin of the coordinates is assumed to lie), has the asymptotic form

$$E_I = E_0(e^{ink_v r}/r) f(k_S, k_I) \tag{3.23}$$

in order to satisfy the radiation condition at infinity [36]. The vector function $\boldsymbol{f}$ is the normalized scattering amplitude that, for particles of general morphology, depends on the direction of the incident wavevector and on the direction of observation $\mathbf{k}_S = \hat{\mathbf{r}}$. Since $\nabla \cdot \mathbf{E}_S = 0$ in the external region, the scattering amplitude satisfies the transversality condition:

$$\hat{\mathbf{r}} \cdot \mathbf{f} = 0. \tag{3.24}$$

Equation (3.24) gives the well-known result that in the far zone the scattered field is transverse to the direction of observation $\hat{r}$.

Usually, the field polarization is referred to the plane of scattering. Therefore, we define two pairs of mutually orthogonal unit vectors $\hat{u}_{I\eta}$ and $\hat{u}_{S\eta}$, where $\eta = 1, 2$. The orientations are chosen in order to satisfy:

$$\hat{u}_{I1} \times \hat{u}_{I2} = \hat{k}_I, \hat{u}_{S1} \times \hat{u}_{S2} = \hat{k}_S$$

with $\hat{\mathbf{u}}_{I2} = \hat{\mathbf{u}}_{S2}$ orthogonal to the plane of scattering. We denote an incident field polarized along $\hat{\mathbf{u}}_{I\eta}$ as $\mathbf{E}_{I\eta}$ and the corresponding scattered field as $\mathbf{E}_{S\eta}$; $\mathbf{f}_\eta$ will then indicate the scattering amplitude.

3.3.1 *The Transition Matrix Method*

The scattered field can be obtained by applying to the fields the boundary conditions across the surface of the particle. To solve this problem, Waterman introduced the transition matrix (T-matrix) method, based on the definition of a linear operator that relates the incident field to the scattered field [112]. Hereafter, we summarize the essentials of the T-matrix method. A comprehensive review of the method can be found in Borghese et al. (2007) [8] and Mishchenko et al. (2002) [66].

The starting point of the method is the field expansion in terms of a complete set of vector solutions of the Helmholtz equation. A suitable basis for the expansion is represented by the spherical multipole fields, that is, the vector solutions of the Maxwell equations in a homogeneous medium that are simultaneous eigenfunctions of the square and of the z-component of the angular momentum, as well as of the parity.

The multipole expansion of the electric field of a plane wave and of the corresponding scattered wave is:

$$\begin{aligned}\mathbf{E}_I &= E_0 \sum_{plm} \mathbf{J}_{lm}^{(p)}(\mathbf{r}, \mathbf{k})\, W_{lm}^{(p)}(\hat{\mathbf{e}}_I, \hat{\mathbf{k}}_I) \\ \mathbf{E}_S &= E_0 \sum_{plm} \mathbf{H}_{lm}^{(p)}(\mathbf{r}, \mathbf{k})\, A_{lm}^{(p)}(\hat{\mathbf{e}}_I, \hat{\mathbf{k}}_I)\end{aligned} \tag{3.25}$$

where $\mathbf{J}_{lm}^{(p)}(\mathbf{r}, \mathbf{k})$ denotes vector multipole fields that are defined by:

$$\begin{aligned}J_{lm}^{(1)}(r, k) &= j_l(kr) X_{lm}(\hat{r}) \\ J_{lm}^{(2)}(r, k) &= \frac{1}{k} \nabla \times j_l(kr) X_{lm}(\hat{r}).\end{aligned} \tag{3.26}$$

Superscript $p = 1, 2$ refers to the values of the parity index, distinguishing magnetic multipoles ($p = 1$) from electric ones ($p = 2$). In the above equations, $j_l(kr)$ are spherical Bessel functions, and $\mathbf{X}_{lm}$ are vector spherical harmonics [37]. The $\mathbf{H}_{lm}^{(p)}$ denotes multipole fields that satisfy the radiation condition at infinity. They are identical to the $\mathbf{J}_{lm}^{(p)}$ multipole fields except for the substitution of the Hankel function of the first kind $h_l(kr)$ in place of the spherical Bessel functions $j_l(kr)$. In Eq. 3.25, the amplitudes $W_{lm}^{(p)}$ are given as [26]:

$$\begin{aligned}W_{lm}^{(1)}(\hat{e}_I, \hat{k}_I) &= 4\pi i^l \hat{e}_I \cdot X_{lm}(\hat{k}_I), \\ W_{lm}^{(2)}(\hat{e}_I, \hat{k}_I) &= 4\pi i^l \hat{e}_I \cdot [X_{lm}(\hat{k}_I) \times \hat{k}_I].\end{aligned} \tag{3.27}$$

It is interesting to observe that in Eq. 3.25 the multipole fields $\mathbf{H}_{lm}^{(2)}$ are not orthogonal to $\mathbf{r}$. They can be decomposed into their radial and transverse component according to the following equation:

$$H_{lm}^{(2)} = \frac{1}{kr}\left[l\left(l+1\right)\right]^{1/2} h_l Y_{lm}\hat{r} - \frac{1}{kr}\left[krh_l\right]' X_{lm}(\hat{r}) \times \hat{r}.$$

This means that the field scattered by a particle always has a radial component that, due to the properties of the Hankel functions, may become rather big at the particle surface or near the particle. For large values of kr, the radial component decays faster than the transverse component, so that in the far zone the field appears as purely transverse to $\hat{\mathbf{r}}$ according to the properties of the scattering amplitude. In the near zone, the scattered field is not transverse but includes a nonvanishing radial component, that is, a component in the direction of observation $\hat{\mathbf{r}}$.

In common experimental set-ups, scattering and absorption measurements are performed in the far-field region. However, the advances in surface-enhanced spectroscopies (such as SERS) and scanning near-field optical microscopy (SNOM) meant a growing interest toward the analysis of local electromagnetic fields on the particle surface or in proximity of the particle [84]. A theoretical comprehension of the different field behavior in the far and near zone appears extremely important for optimizing device design and surface-enhanced techniques.

From a formal point of view, the scattering process can be described by a linear operator S defined by the following equation:

$$\mathbf{E}_S = S\mathbf{E}_I.$$

The operator S, called transition operator, can be introduced by the linearity of the Maxwell equations and of the equations expressing the boundary conditions across the surface of the particle. The representation of the operator S on the basis of the spherical multipole fields gives the T-matrix defined by the following equation [112]:

$$A_{lm}^{(p)} = \sum_{p'l'm'} S_{lml'm'}^{(pp')}(r,k)\, W_{l'm'}^{(p')}(\hat{u}_{I\eta}, \hat{k}_I) \qquad (3.28)$$

relating the multipole amplitudes of the incident field to those of the scattered field. The elements of the T-matrix contain all the information about the scattering process, but are independent on the state of polarization of the incident field.

The scattering amplitude can be expressed in terms of the T-matrix through the equation:

$$f_{\eta\eta'} = -\frac{1}{4\pi k} \sum_{pp'll'mm'} W_{lm}^{(p)*}(\hat{u}_{S\eta}, \hat{k}_S) S_{lml'm'}^{(pp')} W_{l'm'}^{(p')}(\hat{u}_{I\eta}, \hat{k}_I), \quad (3.29)$$

which can be considered the most important equation in the T-matrix approach, since all the observable quantities, such as the extinction cross section, are given in terms of the scattering amplitude.

Borghese et al. showed that the transformation properties of the multipole fields under rotation of the coordinate frame imply corresponding transformation properties of the T-matrix under rotation of the scattering particle [12]. A similar idea had already been suggested by Varadan [109], but the procedure devised by Borghese proved to yield the first correct formula [41], which is applicable to the calculation of the orientational average of all the quantities of interest for any choice of the orientational distribution function of the particles. The rotation transformation properties for the T-matrix can be used to develop efficient analytical procedures for averaging scattering characteristics over particle orientations [66]. The random orientation averaging is used when the aggregate is rotated randomly during the observation time, or when we deal with an ensemble of randomly oriented particles [67, 74]. When the analytical solution for the orientation averaging is used, the averaging procedure can be carried out much more efficiently. As a result, once the T-matrix has been calculated for an arbitrary chosen orientation of one of the particles in an ensemble with respect to the incident field, it is straightforward to calculate all the scattering characteristics for any other particle orientation [9].

3.3.2 *Aggregates of Spheres*

In this section, we describe how the T-matrix approach solves the scattering problem by an aggregate (cluster) of spheres without resorting to any approximation [8]. The cluster model turns out to be very useful since it allows to simulate many situations of practical interest where we deal with nonspherical scatterers. The first progress in devising a technique to calculate the optical properties of linear chains of spherical particles is by Bruning and Lo [14], successively extended by Peterson and Strom for general aggregation geometry [79]. A relevant contribution to the solution of the problem is due to Borghese et al. [10].

Consider a cluster of spheres numbered by the index α, of radius ρ_α, and refractive index n_α, whose centers lie at $\mathbf{R}_\alpha$. The spheres may have different refractive index and different radii. The field scattered by the whole aggregate is written as the superposition of the fields scattered by the single spheres,

$$\mathbf{E}_{S\eta} = E_0 \sum_\alpha \sum_{plm} \mathbf{H}_{lm}^{(p)} (r_\alpha, k)\, A_{\eta\alpha lm}^{(p)} \tag{3.30}$$

where $\mathbf{r}_\alpha = \mathbf{r} - \mathbf{R}_\alpha$. The field within each sphere is taken in the following form:

$$\mathbf{E}_{T\eta\alpha} = E_0 \sum_{plm} \mathbf{J}_{lm}^{(p)} (r_\alpha, k)\, C_{\eta\alpha lm}^{(p)} \tag{3.31}$$

with $k_\alpha = n_\alpha k_v$, so that it is regular everywhere inside the sphere. The amplitudes $A_{\eta\alpha lm}^{(p)}$ and $C_{\eta\alpha lm}^{(p)}$ are determined by applying on the field the boundary conditions across the surface of each of the spheres. Now, the scattered field is given by a linear combination of multipole fields with different origins, whereas the incident field is given by a combination of multipole fields centered at the origin of the coordinates. Since the boundary conditions must be imposed at the surface of each of the spheres, for example, of the α-th sphere, we use the *addition theorem* [8] to write the whole field in terms of multipole fields centered at $\mathbf{R}_\alpha$. With the help of this theorem, the scattered field at the surface of the α-th sphere turns out to be

$$\mathbf{E}_{S\eta} = E_0 \left[\sum_{plm} \mathbf{H}_{lm}^{(p)} (\mathbf{r}_\alpha, k)\, A_{\eta\alpha lm}^{(p)} + \sum_{\alpha'} \sum_{p'l'm'} \mathbf{J}_{lm}^{(p)} (r_\alpha, k)\, H_{\alpha lm\alpha' l'm'}^{(pp')} A_{\eta\alpha' l'm'}^{(p)} \right].$$

Analogously, the incident field at the surface of the α-th sphere is as given below:

$$\mathbf{E}_{I\eta} = E_0 \sum_{plm} \sum_{p'l'm'} \mathbf{J}_{lm}^{(p)} (\mathbf{r}_\alpha, k)\, J_{\alpha lm0l'm'}^{(pp')} W_{I\eta l'm'}^{(p')} \tag{3.32}$$

where $\mathbf{R}_0 = 0$ is the vector position of the origin. Now, applying the boundary conditions, we get for each α four equations among which the elimination of the amplitudes of the internal field is trivial. Once this elimination is done, we get a system of linear nonhomogeneous equations:

$$\sum_{\alpha'} \sum_{p'l'm'} M^{(pp')}_{\alpha lm\alpha' l'm'} A^{(p)}_{\eta\alpha lm} = -\mathcal{W}^{(p)}_{I\eta\alpha lm}, \tag{3.33}$$

where we define

$$\mathcal{W}^{(p)}_{I\eta\alpha lm} = \sum_{p'l'm'} J^{(pp')}_{\alpha lm0l'm'} W^{(p')}_{I\eta\alpha lm} \tag{3.34}$$

and

$$M^{(pp')}_{\alpha lm\alpha' l'm'} = (R^{(p)}_{\alpha l})^{-1} \delta_{\alpha\alpha'} \delta_{pp'} \delta_{ll'} \delta_{mm'} + H^{(pp')}_{\alpha lm\alpha' l'm'}. \tag{3.35}$$

The quantities $R^{(p)}_{\alpha l}$ are the Mie coefficients for the scattering from the α-th sphere. In turn, the quantities $H^{(pp')}_{\alpha lm\alpha' l'm'}$ which come from the addition theorem, describe the multiple scattering processes occurring among the spheres in the aggregate. The occurrence of these processes is a manifestation of the coupling between spheres. The elements $H^{(pp')}_{\alpha lm\alpha' l'm'}$ of the transfer matrix couple multipole fields of any parity, with origin on different spheres. The formal solution to system (3.33)

$$A^{(p)}_{\eta\alpha lm} = -\sum_{p'l'm'} \left[M^{-1}\right]^{(pp')}_{\alpha lm\alpha' l'm'} \mathcal{W}^{(p')}_{I\eta\alpha' l'm'} \tag{3.36}$$

relates the multipole amplitudes of the incident field to those of the field scattered by the whole object. In order to define the T-matrix for the whole aggregate, it is necessary to express the scattered field in terms of multipole fields with the same origin. Because of the addition theorem, the scattered field can be written as

$$\begin{aligned} \mathbf{E}_{S\eta} &= \left[\sum_{plm} \sum_{\alpha'} \sum_{p'l'm'} \mathbf{H}^{(p)}_{lm} (r_\alpha, k) J^{(pp')}_{0lm\alpha' l'm'} A^{(p')}_{\eta\alpha' l'm'} \right] \\ &= \sum_{plm} \mathbf{H}^{(p)}_{lm} (\mathbf{r}, k) \mathcal{A}^{(p')}_{\eta\alpha' l'm'}, \end{aligned}$$

which is valid in the region external to the smallest sphere including the whole aggregate, and thus gives the scattered field in the far region. The preceding equation shows that the field scattered by the whole cluster can be expanded as a series of vector multipole fields with a single origin (monocentered expansion) provided that the amplitudes are given as follows:

$$\mathcal{A}^{(p)}_{\eta\alpha lm} = \sum_{\alpha' p' l' m'} J^{(pp')}_{0lm\alpha' l' m'} A^{(p')}_{\eta\alpha' l' m'}.$$

Substituting the amplitudes $\mathcal{A}$ in the expression (3.36), we can define the T-matrix of the aggregate as:

$$S^{(pp')}_{lml'm'} = - \sum_{\alpha\alpha' qq' ll' mm'} J^{(pq)}_{0lm\alpha LM} \left[M^{-1}\right]^{(qq')}_{\alpha LM\alpha' L' M'} J^{(q'p')}_{\alpha L' M' 0l'm'}. \tag{3.37}$$

In order to calculate the T-matrix of a cluster, we have to solve the system (3.33), which has, in principle, infinite order. Of course, the system must be truncated to some finite order by including the multipole expansions terms up to order l_{M}, chosen so to ensure the convergency of the calculations. For a cluster of N spheres, this implies solving a system of order $d_{\mathrm{M}} = 2Nl_{\mathrm{M}}(l_{\mathrm{M}} + 2)$, which may grow rather big. Actually, the inversion of the matrix M is responsible for most of the CPU time required for the calculation. The CPU time scales as d_{M}^3. As a consequence, the computational demand increases with the cube of the number of the spheres.

3.3.3 *An Application of T-Matrix to Plasmonic Nanoaggregates*

The study of the optical behavior of clusters of MNPs has recently raised a growing interest due to their unique properties stemming from the strong plasmon coupling between the spheres. MNP aggregates find interesting applications in many fields such as nanobiotechnology and nanomedicine [43, 100], optical manipulation at the nanoscale [62], and sensing enhancement applications [60]. A nice overview on coupled plasmonics can be found in Stockman (2011) [97], Pelton et al. (2008) [78], and Jain and El-Sayed (2010) [38].

As MNPs are brought in close proximity to each other with a separation much smaller than the wavelength of light, their individual plasmon oscillations couple due to the near-field interaction, resulting in coupled plasmon resonance modes [38, 72, 80, 86]. Because of this interaction, near fields in the gaps between MNPs can be enhanced by several additional orders with respect to single nanoparticles, giving origin to "hot spots" [78]. Plasmon coupling has been explained with an intuitive picture in terms of a plasmon hybridization model, in analogy with atomic orbital hybridization and formation of bonding and antibonding orbitals in molecules [72, 82]. The strong plasmon coupling determines many interesting optical effects that cannot be observed in densely packed dielectric aggregates [43].

As a result of the aggregation, the LSPR tends to red shift, and the shift can reach near-infrared regions. This red shift is an indication of the favorable coupling of the plasmon oscillations, which means that a lower energy is required to drive the coupled plasmon oscillation modes [38]. The amount of the red shift is determined by the interparticle distance. The closer the particles, the larger the red shift of the plasmon resonance [38].

The enhanced fields at the junction of nanoparticles can be used to strongly enhance the Raman scattering of molecules lying in these junctions [65, 101]. A decrease in gap distance between MNPs results in an increase in SERS enhancement [35, 65, 76]. Further reading on approaches to describe the optical behavior of metal aggregates can be found in Refs. [32, 42, 78, 84].

The T-matrix approach shows its validity when computational results are compared with experiments (see Fig. 3.3). As an example, we model the MNPs as clusters of gold spherical subunits and choose the dimensions of each subunit so to reproduce the range measured from the TEM image of the experimental sample (see insert in Fig. 3.3) and so that the wavelength of the lower energy plasmon resonance of the cluster is close to the measured value [62].

In the experimental sample, the aggregation process is obtained through a coating with pyridines molecules. To simulate the experimental aggregate, in our models the single spheres are not in mutual contact, but separated by a very small distance of about 0.5 nm. We adopt the optical constants for gold measured by Johnson and Christy [39]. The use of these optical constants raises several critical issues due to the lack of convergence in the multipole expansion [8]. This is the reason why we introduce the longitudinal dielectric function described by Lindhard [54], as reported and simplified by Pack et al. [75]. We investigate the optical properties

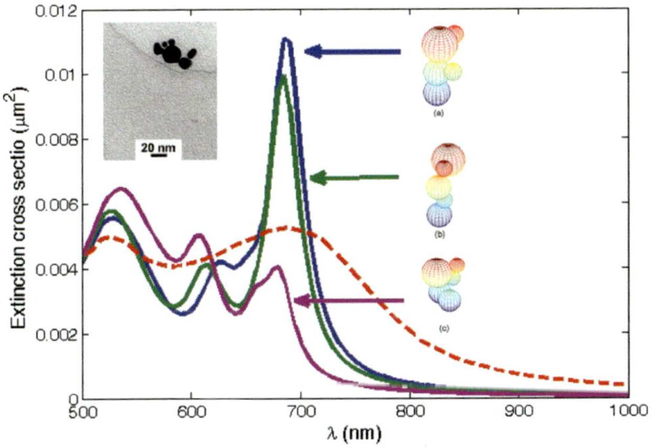

Figure 3.3 Models of three gold nanoaggregates built so to reproduce the sample TEM image (inset). Computed extinction cross sections (solid lines) are compared with the experimental extinction (dashed line). Adapted from Ref. [63].

of several gold aggregates, which were built to satisfy the above requirements. As an example, we show in Fig. 3.3 the extinction cross sections for three different models of clusters simulating the TEM image, and we compare them with the experimental extinction spectrum (dashed line).

It is important to observe that the measured extinction is an ensemble average resulting from the different optical properties of the gold nanoparticles in the sample, thus we would need to average over all possible clusters configurations to have a match with the measured UV–Vis spectrum [61].

For a better understanding of the optical properties of gold nanoaggregates we study the extinction properties of aggregates with increasing size [62]. We keep the overall aggregate geometrical configuration fixed (model cluster (a) in Fig. 3.3) and change the dimensions of the spherical subunits to build clusters with a progressively growing overall size. Each cluster is composed of five spheres. In Fig. 3.4a, we consider a cluster made of one sphere with a radius of 10 nm, two spheres with a radius of 7.5 nm, and two spheres with a 5 nm radius. For comparison, we also consider the smallest sphere containing the whole aggregate that has a radius

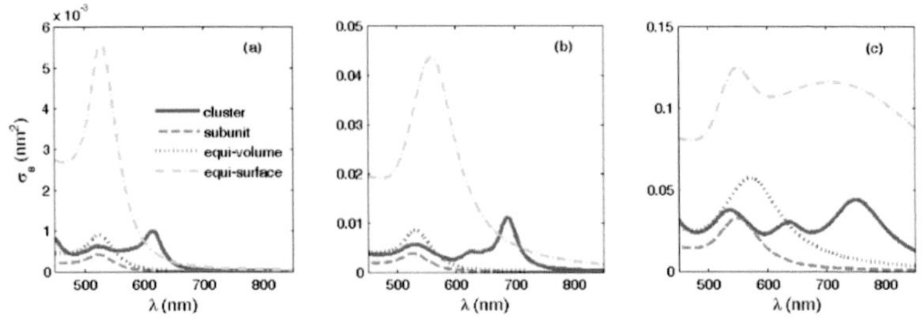

Figure 3.4 Modeling of the extinction properties of gold nanoaggregates with increasing size. The model cluster has the configuration (a) shown in Fig. 3.3. In (a–c) we plot the calculated extinction cross section for the model cluster (blue solid line) and compare it with the extinction cross section of the largest cluster subunit (dashed green), the gold sphere with equivolume (dotted red), and the smallest gold sphere containing all the cluster (dot-dash cyan). In (a), the five subunits are one sphere with a radius of 10 nm, two spheres with a radius of 7.5 nm, and two spheres with a 5 nm radius. In (b), the subunits have double size with respect to (a). In (c), the subunits have four times the size than in (a). Adapted from Ref. [62].

$r_S = 22.5$ nm, and the equivalent volume sphere that has a radius $r_V = 12.8$ nm. In Fig. 3.4b, we double the size of the cluster, that is, one sphere with a radius of 20 nm, two spheres with a radius of 15 nm, and two spheres with a 10 nm radius. The smallest sphere containing the cluster has a radius $r_S = 45$ nm. The equivalent volume sphere has a radius $r_V = 25.6$ nm. In Fig. 3.4c, the cluster is made of one sphere with a radius of 40 nm, two spheres with a radius of 30 nm, and two spheres with a 20 nm radius. The smallest sphere containing the cluster has a radius $r_S = 90$ nm. The equivalent volume sphere has a radius $r_V = 51.2$ nm [62].

The pictures show how the long wavelength plasmonic resonance, arising from the subunit aggregation, red shifts with the increase of the aggregate size. The results in the pictures also confirm the dependence of the extinction on the aggregation process. The extinction by a nanoaggregate is strongly different from the extinction by the volume equivalent sphere or by the smallest sphere containing the whole aggregate. This confirms the importance of using computational methods capable of a detailed

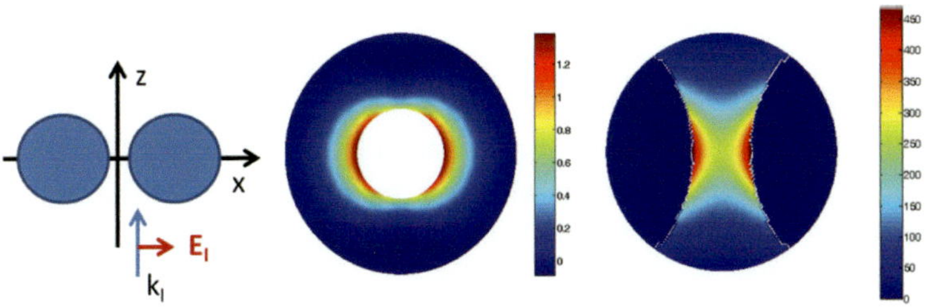

Figure 3.5 (Left) Sketch of the dimer geometry. The spheres have a radius of 10 nm and a top-to-top distance $d = 4$ nm. (Center) Color map of the electric field intensity around a spherical gold nanoparticle with radius 10 nm at $\lambda = 532$ nm. (Right) Color map of the electric field intensity at the inter-particle gap for the gold dimer at $\lambda = 553$ nm. All the quantities are normalized to the electric field intensity of the incident field.

description of the true particle morphology and aggregation stage.

In the color maps in Fig. 3.5, we plot the electric field intensity distribution around a spherical gold nanoparticle with radius 10 nm (central panel) and at the interjunction between two gold spheres in a dimer (right panel). The dimer configuration is the one depicted at the left in the figure. The spheres in the dimer are identical and have a radius of 10 nm. We choose to plot the electric field intensity at the wavelength for which we maximize the field enhancement, that is, at $\lambda = 532$ nm for the sphere, and at $\lambda = 553$ nm for the dimer. The electric field intensity distribution for the dimer gives a clear evidence of the hot spot at the interjunction, where, due to the interaction between the near fields of the coupled particles, we get a maximum field enhancement.

In conclusion, the T-matrix method has found many applications in plasmonics, ranging from the study of plasmon coupling in nanoaggregates to SERS enhancement calculations [2, 13], and optical trapping and manipulations of single and aggregated particles [11, 90, 91]. Despite the relevant recent advances in computational techniques, there are several fields where the T-matrix still appears as the best choice in terms of efficiency, especially when one deals with randomly oriented particles [42].

3.4 Extension for Nonlocal and Quantum Effects

Up to now, the metallic dielectric response has been assumed local, that is, $\varepsilon(\omega)$ being only a function of frequency. As the particle size approaches the electron mean free path in the bulk material (~13 nm for Au [39]), however, the microscopic electron interactions become important and this assumption starts breaking down. For particles smaller than 10 nm in particular, there are significant deviations from the classical result [50]. To lowest order, quantum confinement effects can be simply considered assuming an increased electron scattering rate at the particle surface [34, 49, 50]. This can be quantified as an increased damping rate $A\upsilon_F/R$, where υ_F is the Fermi velocity and $A \sim 1$ is a phenomenological parameter depending on the surface chemistry and can be used as an adjustable parameter. The size-dependent dielectric function is then:

$$\varepsilon(\omega, R) = \varepsilon_{\text{core}}(\omega) - \frac{\omega_p^2}{\omega(\omega + i\gamma + iA\upsilon_F/R)} \tag{3.38}$$

where $\varepsilon_{\text{core}}(\omega) = \varepsilon(\omega) + \omega_p^2/(\omega^2 + i\gamma\omega)$, that is, the bulk dielectric function minus the free electron contribution, is the contribution of interband transitions and core electron polarizability. However, while this method is successful in capturing the major effects of plasmon damping, it overestimates the plasmon broadening and underestimates the plasmon blue-shift [50, 95].

For better results, a nonlocal description for the metal permittivity is required. More than just for isolated small metal particles, nonlocality brings in important implications when MNPs are in close proximity, irrespective of their size, and/or when their shape includes sharp features. When the relevant minimum feature size (separation or apex curvature) is smaller than 10 nm, a large charge and near-field build up is predicted by local theory at the particle surface (hot spot). An accurate treatment must take into account the electron–electron interactions beyond the delta-function distribution predicted by local theory. In particular, Pauli exclusion and Coulomb repulsion result into a pressure term in the equations of motion that counters the compression of the electron gas at the nanoparticle surface. The scale of this effect is inferred by the electron Fermi wavelength, which is around 0.5 nm for both gold and silver. The electron pressure has been typically described within a linearized hydrodynamical approach for the electron collective motion in the metal [4, 19, 17, 24, 104]

$$\beta^2\nabla(\nabla\cdot\mathbf{J})+(\omega^2+i\gamma\omega)\mathbf{J}=i\omega\omega_p^2\varepsilon_0\mathbf{E} \tag{3.39}$$

where $\mathbf{J}$ is the electron polarization current and the pressure coefficient is $\beta=\sqrt{3/5}\upsilon_F$ [25], proportional to the Fermi velocity. Coupled with Maxwell's wave equation

$$\nabla\times\nabla\times\mathbf{E}=\varepsilon_{\text{core}}(\omega^2/c^2)\mathbf{E}+i\omega\mu_0\mathbf{J} \tag{3.40}$$

it describes the retarded nonlocal response of the nanoplasmonic system. The above equations in a uniform medium predict a spatially nonlocal longitudinal dielectric function $\varepsilon_L(\omega,q)=\varepsilon_{\text{core}}-\omega_p^2/(\omega^2+i\gamma\omega-\beta^2q^2)$, enabling longitudinal plasmon oscillations when $\varepsilon_L(\omega,q)=0$, which is satisfied at $q_0=\sqrt{\omega^2+i\gamma\omega-\omega_p^2/\varepsilon_{\text{core}}}/\beta$. The transverse dielectric function, on the other hand, remains local and equal to the bulk dielectric function $\varepsilon(\omega)=\varepsilon_{\text{core}}-\omega_p^2/(\omega^2+i\gamma\omega)$. The induced surface charges are then forced to spread over a thickness of $\sim 1/\text{Im}\{q_0\}$, which is the decay length of the excited longitudinal plasmons in the metal. Values of $\hbar\omega_p$, $\hbar\gamma$, and υ_F for the noble metals (Au, Ag, Cu) are (9, 9.1, 8.9) eV, (0.07, 0.02, 0.09) eV, and (1.4, 1.39, 1.57) $\times$ 10^6 m/s respectively [39]. At frequencies much smaller than the plasma frequency, the decay length converges to $l_0\cong\varepsilon_{\text{core}}^{1/2}\beta/\omega_p$, which for Au ($\varepsilon_{\text{core}}(\omega<<\omega_p)\approx 10$) is of the order of $l_0\cong 0.25$ nm, while at higher frequencies it reaches up to 0.5 nm [19]. The main effect of the nonlocal response, thus, is that the electron gas cannot be squeezed into an infinitesimally small thickness on the particle surface, as dictated by a classical local electromagnetic treatment, but rather it spreads out by a distance of the order of l_0, putting an upper limit to the field enhancement achievable by metallic nanostructures. For example in SERS, where an $\sim|E|^4$ dependence is expected [68], the estimated maximum enhancement is found to be limited to $\sim 10^{10}$ [105].

For small isolated particles, the effect of nonlocality is the observed blue-shift and quenching of the plasmon resonance [50, 95]. An analytical solution for the polarizability of small spheres that includes the nonlocal effects that can be derived [19, 85] by providing a nonlocal extension to Mie theory. In the quasi-static dipole limit, it becomes:

$$\alpha_{NL} = 4\pi R^3 \frac{\varepsilon - \varepsilon_h(1+\delta)}{\varepsilon + 2\varepsilon_h(1+\delta)}, \tag{3.41}$$

where $\delta = (\varepsilon - \varepsilon_{\text{core}}) j_1(x_0)/(\varepsilon_{\text{core}} x_0 j_1'(x_0))$ and $x_0 = q_0 R$ [85]. j_1 is the spherical Bessel function and the prime denotes differentiation with respect to the argument. It is easy to see that the magnitude of δ quickly diminishes as the particle size increases. In the most simple manner, thus, the nonlocal effects enter as a mere scaling of the dielectric function, either that of the host $\varepsilon_h \to \varepsilon_h(1+\delta)$ or that of the metal $\varepsilon \to \varepsilon(1+\delta)^{-1}$. The understanding and intuition of standard Mie theory is now carried into the nonlocal regime.

In more complex geometries involving particles in close separations or particles with sharp features, the explicit solution to Eqs. 3.39–3.40 becomes demanding. Proposals to alleviate this include a transformation optics approach [24] or more recently the Local Analogue Model (LAM) [55]. The latter is particularly simple and transferable to any standard numerical local Maxwell solver: in the LAM approach the nonlocal effects are mapped onto a thin (local) dielectric cover layer of thickness Δd and dielectric function ε_t on the particle surface, extending inwards the metal as depicted in Fig. 3.6a. The dielectric function and thickness of this equivalent layer are related as

$$\frac{\varepsilon_t}{\Delta d} = \frac{\varepsilon \varepsilon_h}{\varepsilon - \varepsilon_h} q_0. \tag{3.42}$$

The model can be employed in two ways: either by assuming a fixed layer thickness Δd and adjusted dielectric function ε_t or by assuming a fixed dielectric function $\varepsilon_t = \varepsilon_h$ and adjusted layer thickness. The latter basically comprises a mere geometrical redefinition of the particle surface. In the case of the Au particle dimer studied in Ref. [55], the fixed thickness approach provides an excellent agreement with the full calculation, while the fixed dielectric approach works adequately only well below the plasma frequency. This method enables the quantitative analysis of nonlocal effects in complex plasmonic structures using only local dielectric functions, and thus can be integrated within any standard Maxwell solver offering simplicity and physical insight.

A weakness of the hydrodynamical model is that it ignores the possible charge density spill out outside the metal particle boundaries. This has been shown to be an important factor in explaining the red-shifting of plasmon spectra for small free-electron metals such as sodium [87, 103], while it is absolutely essential for incorporating quantum tunneling effects between the particles when their separation distance reaches the sub-nanometer scale [92, 94]. The latter is the main mechanism limiting the monotonous build-up of surface charges in particles with decreasing separations and thus responsible for providing an upper limit to light concentration in plasmonic structures. In particular, it is found that tunneling imposes a confinement limit of $10^{-8}\lambda^3$ for visible light [103], setting fundamental limits to applications where extreme field concentration is required. Taking into account quantum tunneling is thus essential when designing metallic structures with sub-nanometric separations, targeting strong charge and field build-ups.

To go beyond the hydrodynamical model into the quantum tunneling regime requires a first-principles quantum treatment, such as the time-dependent density functional theory (TDDFT) [122] that addresses the plasmonic optical response in the quantum level and offers a full account of the nonlocal effects. However, due to its computational complexity this is practical only for the smallest of the particles containing only a few thousand electrons. To solve this problem, a classical local model that effectively incorporates the quantum tunneling effects, the so-called quantum-corrected model (QCM) has been recently implemented [22]. In this approach, the junction between two particles is represented by a fictitious material whose dynamical conductivity is adjusted so it effectively mimics the effects of electron tunneling. The quantum-corrected calculation can then be performed within any standard Maxwell solver, providing excellent agreement to full quantum mechanical calculations [22]. The fictitious material is shown schematically in Fig. 3.6b. Because tunneling probability depends very sensitively on separation distance, when treating curved surfaces (e.g., two metal spheres) different points (on a vertical plane) between the surfaces will be characterized by different distance l and thus different tunneling transmission $T(l)$. To remedy this, a numbers of shells with different effective conductivity can be employed, as shown in

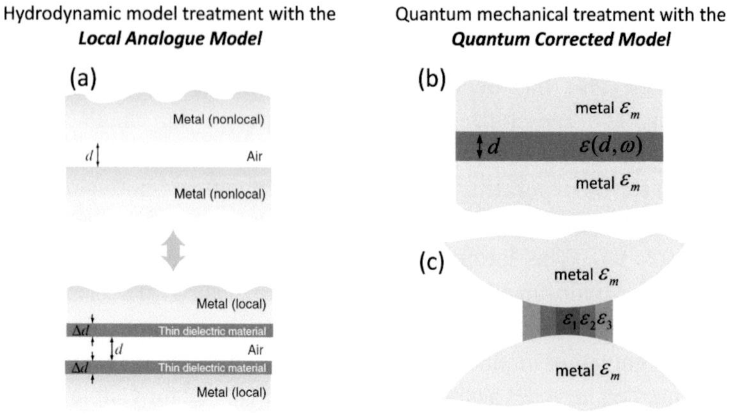

Figure 3.6 Hydrodynamic model treatment: (a) A gold-air-gold geometry within the nonlocal picture (top) and the Local Analogue Model (bottom). (b) A simple plasmonic system is viewed as composed by two semi-infinite planar interfaces separated by a distance l and characterized by tunneling transmission $T(l)$. The tunneling induced conductivity between the surfaces defines a separation-dependent dielectric material $\varepsilon Q(l, \omega)$. (c) In case of curved surfaces such as a sphere dimer, this can be employed in the Quantum Corrected Model as shells of fictitious material (adapted from Ref. [22]).

Fig. 3.6b. The tunneling-induced conductivity between the surfaces then defines a separation-dependent dielectric material

$$\varepsilon_Q(l, \omega) = 1 - \frac{\omega_g^2}{\omega(\omega + i\gamma_g(l))} \tag{3.43}$$

where $\gamma_g(l) = \varepsilon_0\omega_g^2/\sigma_0(l)$ and $\sigma_0(l)$ is the static conductivity calculated from TDDFT and follows an exponential dependence with separation distance (see [22] for the calculation details and characteristic values of $\gamma_g(l)$ for Au and Na). For simplicity, a value of $\omega_g = \omega_p$, the plasma frequency of the metal under study, can be used.

The quantum modeling reveals three distinct regimes regarding the dimer's optical response [22, 94]: (a) *non-contact regime*, for separations larger than 0.5 nm, where electron tunneling is practically inactive and quantum mechanical and classical results are generally in agreement regarding number of plasmon modes, resonance shifts, and field enhancements, (b) *tunneling regime*, where the electron transfer neutralizes the high charge densities on the particles' surface and leads to a drastic reduction of field enhancement compared to what predicted by a classical approach. For even further decreasing separation the extinction spectra progress smoothly to the (c) *contact regime*, where the dimer behaves as a single nonspherical particle. The modes in this region are the so-called charge transfer plasmons, where considerable difference with the classical treatment is found for all separations. It is important to note that due to excessive tunneling, the contact region is actually reached at finite positive separation distances.

3.5 Conclusions

The study of the unique optical properties of MNPs is a fast growing field due to the many applications that they find in several research, technological, and biomedical fields. A full theoretical comprehension of the electromagnetic processes involved is therefore fundamental to optimize surface-enhanced spectroscopic techniques. Approximated descriptions are often successfully used to get an insight into the physics of the problem, but in the presence of complex or aggregated nanostructures they may hide important effects. In such situations, accurate computational approaches are required to get a good agreement with experiments and a real understanding of the electromagnetic processes at the nanoscale.

Acknowledgments

M. A. I. and R. S. warmly thank Prof. Ferdinando Borghese and Prof. Paolo Denti for many years of enjoyable and fruitful collaboration and for sharing their deep knowledge of electromagnetic scattering.

M. A. I. acknowledges funding from the European Union Seventh Framework Programme (FP7/2007-2013) under Grant Agreement No. 241818 (FP7-HEALTH-F5-2009-241818-NANOANTENNA).

References

1. Alonso-González, P., Albella, P., Schnell, M., Chen, J., Huth, F., Garcıá-Etxarri, A., Casanova, F., Golmar, F., Arzubiaga, L., Hueso, L. E., Aizpurua, J., and Hillenbrand, R. (2012). Resolving the electromagnetic mechanism of surface-enhanced light scattering at single hot spots, *Nature Commun.*, **3**, 684.
2. Arya, K. (2006). Scattering T-matrix theory in wave-vector space for surface-enhanced Raman scattering in clusters of nanoscale spherical metal particles, *Phys. Rev. B*, **74**, 195438.
3. Auguié, B., and Barnes, W. L. (2008). Collective resonances in gold nanoparticle arrays, *Phys. Rev. Lett.*, **101**, 143902.
4. Boardman, A. D. (1982). *Electromagnetic Surface Modes* (Wiley, New York).
5. Bobbert, P. A., and Vlieger, J. (1986). Light scattering by a sphere on a substrate, *Physica A*, **137**, pp. 209–242.
6. Bobbert, P. A., and Vlieger, J. (1987). The polarizability of a spheroidal particle on a substrate, *Physica A*, **147** pp. 115–141.
7. Bohren, C. F., and Huffman, D. R. (1983). *Absorption and Scattering of Light by Small Particles* (Wiley, New York).
8. Borghese, F., Denti, P., and Saija, R. (2007). *Scattering from Model Nonspherical Particles*, 2nd ed. (Springer-Verlag, Berlin Heidelberg).
9. Borghese, F., Denti, P., Saija, R., Iatì, M. A., and Sindoni, O. I. (2001). Optical properties of a dispersion of anisotropic particles with non-randomly distributed orientations. The case of atmospheric ice crystals, *J. Quant. Spectrosc. Radiat. Transfer*, **70**, pp. 237–251.
10. Borghese, F., Denti, P., Saija, R., Toscano, G., and Sindoni, O. I. (1984). Multiple electromagnetic scattering from a cluster of spheres. I. Theory, *Aerosol Sci. Technol.*, **3**, pp. 227–235.
11. Borghese, F., Denti, P., Saija, R., Iatì, M. A., and Maragò, O. M. (2008). Radiation force and torque on optically trapped linear nanostructures, *Phys. Rev. Lett.*, **100**, 163903.
12. Borghese, F., Denti, P., Saija, R., Toscano, G., and Sindoni, O. I. (1984). Macroscopic optical constants of a cloud of randomly oriented nonspherical scatterers, *Nuovo Cim.*, **B 81**, pp. 29–50.
13. Boyack, R., and Le Ru, E. C. (2009). Investigation of particle shape and size effects in SERS using T-matrix calculations, *Phys. Chem. Chem. Phys.*, **11**, pp. 7398–7405.

14. Bruning, J. H., and Lo, Y. T. (1971). Multiple scattering of EM waves by spheres. I. Multipole expansion and ray-optical solutions, *IEEE Trans. Antennas Propag.*, **19**, pp. 378–390

15. Cançado, L. G., Jorio, A., Ismach, A., Joselevich, E., Hartschuh, A., and Novotny, L. (2009). Mechanism of near-field Raman enhancement in one-dimensional systems, *Phys. Rev. Lett.*, **103**, 186101.

16. Chu, Y., Schonbrun, E., Yang, T., and Croziera, K. B. (2008). Experimental observation of narrow surface plasmon resonances in gold nanoparticle arrays, *Appl. Phys. Lett.*, **93**, 181108.

17. Ciracì, C., Hill, R. T., Mock, J. J. , Urzhumov, Y., Fernández-Domínguez, A. I., Maier, S. A., Pendry, J. B., Chilkoti, A., and Smith, D. R. (2012). Probing the ultimate limits of plasmonic enhancement, *Science*, **337**, pp. 1072–1074.

18. Daniel, M. C., and Astruc, D. (2004). Gold nanoparticles: Assembly, supramolecular chemistry, quantum-size-related properties, and applications toward biology, catalysis, and nanotechnology, *Chem. Rev.*, **104**, pp. 293–346.

19. David, C., and García de Abajo, F. J. (2011). Spatial nonlocality in the optical response of metal nanoparticles, *J. Phys. Chem. C*, **115**, pp. 19470–19475.

20. de Abajo, F. J. G. (2007). Light scattering by particle and hole arrays, *Rev. Mod. Phys.*, **79**, pp. 1267–1290.

21. Draine, B. T. (1988). The discrete-dipole approximation and its application to interstellar graphite grains, *Astrophys. J.*, **333**, pp. 848–872.

22. Esteban, R., Borisov, A. G., Nordlander, P., and Aizpurua, J. (2012). Bridging quantum and classical plasmonics with a quantum-corrected model, *Nature Commun.*, **3**, 825.

23. Fazio, B., D'Andrea, C., Bonaccorso, F., Irrera, A., Calogero, G., Vasi, C., Gucciardi, P. G., Allegrini, M., Toma, A., Chiappe, D., Martella, C., and de Mongeot, F. B. (2011). Re-radiation enhancement in polarized surface-enhanced resonant Raman scattering of randomly oriented molecules on self-organized gold nanowires, *ACS Nano*, **5**, pp. 5945–5956.

24. Fernández-Domínguez, A. I., Wiener, A., García-Vidal, F. J., Maier, S. A., and Pendry, J. B. (2012). Transformation-optics description of nonlocal effects in plasmonic nanostructures, *Phys. Rev. Lett.*, **108**, 106802.

25. Fetter, A. L. (1973). Electrodynamics of a layered electron gas. I. Single layer, *Ann. Phys. (N.Y.)*, **81**, pp. 367–393.

26. Fucile, E., Borghese, F., Denti, P., Saija, R., and Sindoni, O.I. (1997). General reflection rule for electromagnetic multipole fields on a plane surface, *IEEE Trans. Antennas Prop.*, **45**, pp. 868–875.
27. Fucile, E., Borghese, F., Denti, P., and Saija, R. (1993). Theoretical description of dynamic light scattering from an assembly of large axially symmetric particles, *J. Opt. Soc. Am. A,* **10**, pp. 2611–2617.
28. Gluodenis M. M., and Foss, Jn. C. A. (2002). The effect of mutual orientation on the spectra of metal nanoparticle rod-rod and rod-sphere pairs, *J. Phys. Chem. B*, **106**, pp. 9484–9489.
29. Goodman, J. J., Draine, B. T., and Flatau, P. J. (1991). Application of fast-Fourier-transform techniques to the discrete-dipole approximation, *Opt. Lett.*, **16**, pp. 1198–1200.
30. Halas, N. J., Lal, S., Chang, W.-S., Link, S., and Nordlander, P. (2011). Plasmons in strongly coupled metallic nanostructures, *Chem. Rev.*, **111**, pp. 3913–3961.
31. Hao, F., Nehl, C. L., Hafner, J. H., and Nordlander, P. (2007). Plasmon resonances of a gold nanostar, *Nano Lett.*, **7**, pp. 729–732.
32. Hellmers, J., Riefler, N., Wriedt, T., Eremin, Y. A. (2008). Light scattering simulation for the characterization of sintered silver nanoparticles, *J. Quant Spectrosc Radiat Transfer,* **109**, pp. 1363–1373.
33. Homola, J. (2008). Surface plasmon resonance sensors for detection of chemical and biological species, *Chem. Rev.*, **108**, pp. 462–493.
34. Hövel, H., Fritz, S., Hilger, A., Kreibig, U., and Vollmer, M. (1993). Width of cluster plasmon resonances—Bulk dielectric functions and chemical interface damping, *Phys. Rev. B*, **48**, pp. 18178–18188.
35. Huang, X., El-Sayed, I. H., Qian, W., and El-Sayed, M. A. (2007). Cancer cells assemble and align gold nanorods conjugated to antibodies to produce highly enhanced, sharp, and polarized surface Raman spectra: A potential cancer diagnostic marker, *Nano Lett.*, **7**, pp. 1591–1597.
36. Ishimaru, A. (1978). *Wave Propagation and Scattering in Random Media* (Academic, New York).
37. Jackson, J. D. (1999) *Classical Electrodynamics*, 3rd ed. (Wiley, New York).
38. Jain, P. K., and El-Sayed, M. A. (2010). Plasmon coupling in noble metal nanostructures, *Chem. Phys. Lett.*, **487**, pp. 153–164.
39. Johnson, P. B., and Christy, R. W. (1972). Optical constants of the noble metals, *Phys. Rev. B*, **6**, 4370.

40. Kelly, K. L., Coronado, E., Zhao, L. L., and Schatz, G. C. (2003). The optical properties of metal nanoparticles: The influence of size, shape, and dielectric environment, *J. Phys. Chem. B*, **107**, pp. 668–677.
41. Khlebtsov, N. G. (1992). Orientational averaging of light-scattering observables in the J-matrix approach, *Appl. Opt.*, **31**, pp. 5359–5365.
42. Khlebtsov, N. G. (2013). T-matrix method in plasmonics: An overview, *J. Quant. Spectrosc. Radiat Transfer*, **123**, pp. 184–217.
43. Khlebtsov, N. G., and Dykman, L. A. (2010). Optical properties and biomedical applications of plasmonic nanoparticles, *J. Quant. Spectrosc. Radiat. Transfer*, **111**, pp. 1–35.
44. Kneipp, K., Kneipp, H., Itzkan, I., Dasari, R. R., and Feld, M. S. (2002). Surface enhanced Raman scattering and biophysics, *J. Phys. Cond. Mat.*, **14**, pp. R597–R624.
45. Kneipp, K., Moskovits, M., and Kneipp, H., editors. (2006) *Surface-Enhanced Raman Scattering: Physics and Applications* (Springer-Verlag, Berlin, Heidelberg).
46. Kneipp, K., Wang, Y., Kneipp, H., Perelman, L. T., Itzkan, I., Dasari, R. R., and Feld, M. S. (1997). Single molecule detection using surface-enhanced Raman scattering (SERS), *Phys. Rev. Lett.*, **78**, pp. 1667–1670.
47. Kokkinakis, Th., and Alexopoulos, K. (1972). Observation of radiative decay of surface plasmons in small silver particles, *Phys. Rev. Lett.*, **28**, pp. 1632–1634.
48. Kravets, V. G., Schedin, F., and Grigorenko, A. N. (2008). Extremely narrow plasmon resonances based on diffraction coupling of localized plasmons in arrays of metallic nanoparticles, *Phys. Rev. Lett.*, **101**, 087403.
49. Kreibig, U., and Fragstein, C. V. (1969). The limitation of electron mean free paths in small silver particles, *Z. Phys.*, **224**, pp. 307–323.
50. Kreibig, U., and Vollmer, M. (1995). *Optical Properties of Metal Clusters* (Springer-Verlag, Berlin).
51. Kuwata, H., Tamaru, H., Esumi, K., and Miyano, K. (2003). Resonant light scattering from metal nanoparticles: Practical analysis beyond Rayleigh approximation, *Appl. Phys. Lett.*, **83**, pp. 4625–2627.
52. Lagos, N., Sigalas, M. M., and Lidorikis, E. (2011). Theory of plasmonic near-field enhanced absorption in solar cells, *Appl. Phys. Lett.*, **99**, 063304.
53. Lidorikis, E. (2012). Modeling of enhanced absorption and Raman scattering caused by plasmonic nanoparticle near fields, *J. Quant. Spectr. Rad. Transf.*, **113**, pp. 2573–2584.

54. Lindhard, J. (1954). On the properties of a gas of charged particles, *R Danish Acad. Sci. Lett: Math-Fys. Medd.*, **28**, 8.
55. Luo, Y., Fernandez-Dominguez, A. I., Wiener, A., Maier, S. A., and Pendry, J. B. (2013). Surface plasmons and nonlocality: A simple model, *Phys. Rev. Lett.*, **111**, 093901.
56. Mackowski, D. W., and Mishchenko, M. I. (1996). Calculation of the T matrix and the scattering matrix for ensembles of spheres, *J. Opt. Soc. Am. A*, **13**, pp. 2266–2278.
57. Maier, S. A. (2007) *Plasmonics: Fundamentals and Applications* (Springer, New York).
58. Markel, V. A. (1993). Coupled-dipole approach to scattering of light from a one-dimensional periodic dipole structure, *J. Mod. Opt.*, **40**, pp. 2281–2291
59. Maximiano, R. V., Beams, R., Novotny, L., Jorio, A., and Cançado, L. G. (2012). Mechanism of near-field Raman enhancement in two-dimensional systems, *Phys. Rev. B*, **85**, 235434.
60. Mayer, K. M., and Hafner, J. H. (2011). Localized surface plasmon resonance sensors, *Chem. Rev.*, **111**, pp. 3828–3857.
61. Meier, M., and Wokaun, A. (1983). Enhanced fields on large metal particles: Dynamic depolarization, *Opt. Lett.*, **8**, pp. 581–583.
62. Messina, E., Cavallaro, E., Cacciola, A. Saija, R., Borghese, F., Denti, P., Fazio, B., D'Andrea, C., Gucciardi, P. G., Iatì, M. A., et al. (2011). Manipulation and Raman spectroscopy with optically trapped metal nanoparticles obtained by pulsed laser ablation in liquids, *J. Phys. Chem. C*, **115**, pp. 5115–5122.
63. Messina, E., Cavallaro, E., Cacciola, A., Iatì, M. A., Gucciardi, P. G., Borghese, F., Denti, P., Saija, R., Compagnini, G., Meneghetti, M., Amendola, V., and Maragò, O. M. (2011). Plasmon-enhanced optical trapping of gold nanoaggregates with selected optical properties, *ACS Nano*, **5**, pp. 905–913.
64. Messinger, B. J., von Raben, K. U., Chang, R. K., and Barber, P. W. (1981). Local fields at the surface of noble-metal microspheres, *Phys. Rev. B*, **24**, pp. 649–657.
65. Michaels, A. M., Jiang, J., and Brus, L. (2000). Ag nanocrystal junctions at the site for surface-enhanced Raman scattering of single Rhodamine 6G molecules, *J. Phys. Chem. B*, **104**, pp. 11965–11971.
66. Mishchenko, M. I., Travis, L. D., and Lacis, A. A. (2002). *Scattering, Absorption, and Emission of Light by Small Particles* (Cambridge University Press).

67. Mishchenko, M. I., Wiscombe, W. J., Hovenier, J. W., and Travis, J. D. (2000). *Light Scattering by Nonspherical Particles: Theory, Measurements, and Geophysical Applications*, eds. Mishchenko, M. I., Hovenier, J. W., and Travis, J. D. "Overview of scattering by nonspherical particles" (Academic, San Diego) pp. 29–60.

68. Moskovits, M. (1985). Surface-enhanced spectroscopy, *Rev. Mod. Phys.*, **57**, pp. 783–826.

69. Myroshnychenko, V., Rodriguez-Fernandez, J., Pastoriza-Santos, I., Funston, A. M., Novo, C., Mulvaney, P., Liz-Marzan, L. M., and Garcia de Abajo, F. J. (2008). Modelling the optical response of gold nanoparticles, *Chem. Soc. Rev.*, **37**, pp. 1792–1805.

70. Nehl, C. L., Liao, H., and Hafner, J. H. (2006). Optical properties of star shaped gold nanoparticles, *Nano Lett.*, **6**, pp. 683–688.

71. Nie, S., and Emory, S. R. (1997). Probing single molecules and single nanoparticles by surface-enhanced Raman scattering, *Science*, **275**, pp. 1102–1106.

72. Nordlander, P., Oubre, C., Prodan, E., Li, K., and Stockman, M. I. (2004). Plasmon hybridization in nanoparticle dimers, *Nano Lett.*, **4**, pp. 899–903.

73. Offermans, P., Schaafsma, M. C., Rodriguez, S. R. K., Zhang, Y. C., Crego-Calama, M., Brongersma, S. H., and Gómez Rivas, J. (2011). Universal scaling of the figure of merit of plasmonic sensors, *ACS Nano*, **5**, pp. 5151–5157.

74. Okada, Y. (2010). *Light Scattering Reviews*, 5th ed. Kokhanovsky, A. A., Chapter 1 "Numerical simulations of light scattering and absorption characteristics of aggregates" (Springer-Verlag, Berlin, Heidelberg) pp. 3–35.

75. Pack, A., Hietschold, M., and Wannemacher, R. (2001). Failure of local Mie theory: Optical spectra of colloidal aggregates, *Opt. Commun.*, **194**, pp. 277–287.

76. Pal., A., and Pal, T. (1999). Silver nanoparticle aggregate formation by a photochemical method and its application to SERS analysis, *J. Raman Spectrosc.*, **30**, pp. 199–204.

77. Pastoriza-Santos, I., and Liz-Marzan, L. M. (2008). Colloidal silver nanoplates. State of the art and future challenges, *J. Mater Chem.*, **18**, pp. 1724–1737.

78. Pelton, M., Aizpurua, J., and Bryant, G. (2008). Metal-nanoparticle plasmonics, *Laser Photon. Rev.*, **2**, pp. 136–159.

79. Peterson, B., and Strom, S. (1973). T-matrix for electromagnetic scattering from an arbitrary number of scatterers and representations of E(3)*, *Phys. Rev. D*, **8**, pp. 3661–3678.
80. Petryayeva, E., and Krull, U. J. (2011). Localized surface plasmon resonance: Nanostructures, bioassays and biosensing—A review, *Anal. Chimi. Acta*, **706**, pp. 8–24.
81. Prescott, S., and Mulvaney, P. (2006). Gold nanorod extinction spectra, *J. Appl. Phys.*, **99**, pp. 123504 1–7.
82. Prodan, E., Radloff, C., Halas, N. J., and Nordlander, P. (2003). A hybridization model for the plasmon response of complex nanostructures, *Science*, **302**, pp. 419–422.
83. Purcell, E. M., and Pennypacker, C. R. (1973). Scattering and absorption of light by nonspherical grains, *Astrophys. J.*, **186**, pp. 705–714.
84. Quinten, M. (2011). *Optical Properties of Nanoparticle Systems* (Wiley-VCH).
85. Raza, S., Yan, W., Stenger, N., Wubs M., and Mortensen, N. A. (2013). Blueshift of the surface plasmon resonance in silver nanoparticles: Substrate effects, *Opt. Exp.*, **21**, pp. 27344–27355.
86. Rechberger, W., Hohenau, A., Leitner, A., Krenn, J. R., Lamprecht, B., and Aussenegg, F. R. (2003). Optical properties of two interacting gold nanoparticles, *Opt. Commun.*, **220**, pp. 137–141.
87. Reiners, T., Ellert, C., Schmidt, M., and Haberland, H. (1995). Size dependence of the optical response of spherical sodium clusters, *Phys. Rev. Lett.*, **74**, pp. 1558–1561.
88. Rodriguez, S. R. K., Abass, A., Maes, B., Janssen, O. T. A., Vecchi, G., and Gómez Rivas, J. (2011). Coupling bright and dark plasmonic lattice resonances, *Phys. Rev. X*, **1**, 021019.
89. Roman-Velazquez, C. E., Noguez, C., and Barrera, R. G. (2000). Substrate effects on the optical properties of spheroidal nanoparticles, *Phys. Rev. B*, **61**, pp. 10427–10436.
90. Saija, R., Denti, P., Borghese, F., Maragò, O. M., and Iatì, M. A. (2009). Optical trapping calculations for metal nanoparticles. Comparison with experimental data for Au and Ag spheres, *Opt. Exp.*, **17**, pp. 10231–10241.
91. Saija, R., Denti, P., Borghese, F., Maragò, O. M., and Iatì, M. A. (2007). Optical trapping of nonspherical particles in the T-matrix formalism, *Opt. Exp.*, **15**, pp. 11984–11998. Errata (2007), **15**, 6946.

92. Savage, K. J., Hawkeye, M. M., Esteban, R., Borisov, A. G., Aizpurua, J., and Baumberg, J. J. (2012). Revealing the quantum regime in tunnelling plasmonics, *Nature (London)*, **491**, pp. 574–577.

93. Schedin, F., Lidorikis, E., Lombardo, A., Kravets, V. G., Geim, A. K., Grigorenko, A. N., Novoselov, K. S., and Ferrari, A. C. (2010). Surface enhanced Raman spectroscopy of graphene, *ACS Nano*, **4**, pp. 5617–5626.

94. Scholl, J. A., García-Etxarri, A., Koh, A. L., and Dionne, J. A. (2013). Observation of quantum tunneling between two plasmonic nanoparticles, *Nano Lett.*, **13**, pp. 564–569.

95. Scholl, J. A., Koh, A. L., and Dionne, J. A. (2012). Quantum plasmon resonances of individual metallic nanoparticles, *Nature*, **483**, pp. 421–428.

96. Špačková, B., and Homola, J. (2013). Sensing properties of lattice resonances of 2D metal nanoparticle arrays: An analytical model, *Opt. Exp.*, **21**, pp. 27490–27502.

97. Stockman, M. I. (2011). Nanoplasmonics: Past, present, and glimpse into future, *Opt. Exp.*, **19**, pp. 22029–22106.

98. Sun, Y., and Xia, Y. (2002). Shape-controlled synthesis of gold and silver nanoparticles, *Science*, **298**, pp. 2176–2179.

99. Taflove, A. (1995). *Computational Electrodynamics: The Finite-Difference-Time-Domain Method* (Artech House, Boston).

100. Tam, J. M., Tam, J. O. , Murthy, A., Ingram, D. R., Ma, L. L. , Travis, K., et al. (2010). Controlled assembly of biodegradable plasmonic nanoclusters for near-infrared imaging and therapeutic applications, *ACS Nano*, **4**, pp. 2178–2184.

101. Taylor, R. W., Lee, T. C., Scherman, O. A., Esteban, R., Aizpurua, J., Huang, F. M., Baumberg, J. J., and Mahajan, S. (2011). Precise subnanometer plasmonic junctions for SERS within gold nanoparticle assemblies using cucurbit[*n*]uril "glue", *ACS Nano*, **5**, pp. 3878–3887.

102. Teixeira, F. L. (2007). FDTD/FETD methods: A review on some recent advances and selected applications, *J. Microwaves Optoelectron.*, **6**, pp. 83–95.

103. Teperik, T. V., Nordlander, P., Aizpurua, J., and Borisov, A. G. (2013). Robust subnanometric plasmon ruler by rescaling of the nonlocal optical response, *Phys. Rev. Lett.*, **110**, 263901.

104. Toscano, G., Raza, S., Jauho, A.-P., Mortensen, N. A., and Wubs, M. (2012). Modified field enhancement and extinction by plasmonic nanowire dimmers due to nonlocal response, *Opt. Exp.*, **20**, pp. 4176–4188.

105. Toscano, G., Raza, S., Xiao, S., Wubs, M., Jauho, A.-P., Bozhevolnyi, S. I., and Mortensen, N. A. (2012). Surface-enhanced Raman spectroscopy: Nonlocal limitations, *Opt. Lett.*, **37**, pp. 2538–2540.

106. Urich, A., Pospischil, A., Furchi, M. M., Dietze, D., Unterrainer, K., and Mueller, T. (2012). Silver nanoisland enhanced Raman interaction in graphene, *Appl. Phys. Lett.*, **101**, 153113.

107. Väkeväinen, A. I., Moerland, R. J., Rekola, H. T., Eskelinen, A.-P., Martikainen, J.-P., Kim, D.-H., and Törmä, P. (2013). Plasmonic surface lattice resonances at the strong coupling regime, *Nano Lett.*, DOI: 10.1021/nl4035219.

108. Valamanesh, M., Borensztein, Y., Langlois, C., and Lacaze, E. (2011). Substrate effect on plasmon resonance of supported flat silver nanoparticles, *J. Phys. Chem. C*, **115**, pp. 2914–2922.

109. Varadan, V. K. (1980). *Acoustic Electromagnetic and Elastic Wave Scattering—Focus on the T-Matrix Approach,* eds. Varadan, V. K., and Varadan, V. V., "Multiple scattering of acoustic, electromagnetic and elastic waves" (Pergamon, New York) pp. 103–134.

110. Venermo, J., and Sihvola, A. (2005). Dielectric polarizability of circular cylinder, *J. Electrost.*, **63**, pp. 101–117.

111. Wang, D.-S., and Kerker, M. (1981). Enhanced Raman scattering by molecules adsorbed at the surface of colloidal spheroids, *Phys. Rev. B*, **24**, pp. 1777–1790.

112. Waterman, P. C. (1971). Symmetry, unitarity, and geometry in electromagnetic scattering, *Phys. Rev. D*, **3**, pp. 825–839.

113. Wind, M. M., Bobbert, P. A., Vlieger, J., and Bedeaux, D. (1987). The polarizability of a truncated sphere on a substrate II, *Physica A*, **143**, pp. 164–182.

114. Wind, M. M., Vlieger, J., and Bedeaux, D. (1987). The polarizability of a truncated sphere on a substrate I, *Physica A*, **141**, pp. 33–57.

115. Wokaun, A., Gordon, J. P., and Liao, P. F. (1982). Radiation damping in surface-enhanced Raman scattering, *Phys. Rev. Lett.*, **48**, pp. 957–960.

116. Yamaguchi, T., Yoshida, S., and Kinbara, A. (1974). Optical effect of the substrate on the anomalous absorption of aggregated silver films, *Thin Solid Films*, **21**, pp. 173–187.

117. Yee, K. S. (1966). Numerical solution of initial boundary value problems involving Maxwell's equations in isotropic media, *IEEE Trans. Antennas Propag.,* **14**, pp. 302–307.

118. Yurkin, M. A., and Hoekstra, A. G. (2007). The discrete dipole approximation: An overview and recent developments, *J. Quant. Spectrosc. Radiat. Transfer*, **106**, pp. 558–589.

119. Zhao, J., Pinchuk, A. O., Mcmahon, J. M., Li, S., Ausman, L. K., Atkinson, A. L., and Schatz, G. C. (2008). Methods for describing the electromagnetic properties of silver and gold nanoparticles, *Accounts of Chemical Research*, **41,** pp. 1710–1720.

120. Zou, S. N., and Schatz, G. C. (2004). Narrow plasmonic/photonic extinction and scattering line shapes for one and two dimensional silver nanoparticle arrays, *J. Chem. Phys.*, **121**, pp. 12606–12612.

121. Zou, S., Janel, N., and Schatz, G. C. (2004). Silver nanoparticle array structures that produce remarkably narrow plasmon lineshapes, *J. Chem. Phys.*, **120**, pp. 10871–10875.

122. Zuloaga, J., Prodan, E., and Nordlander, P. (2009). Quantum description of the plasmon resonances of a nanoparticle dimer, *Nano Lett.*, **9**, pp. 887–891.

Chapter 4

Design and Fabrication of Nanostructures for Plasmon-Enhanced Spectroscopies

Remo Proietti Zaccaria,[a] Simone Panaro,[a] Andrea Toma,[a] Manohar Chirumamilla,[a] Andrea Giugni,[b] Gobind Das,[a] Roman Krahne,[a] and Enzo Di Fabrizio[b,c]

[a] *Istituto Italiano di Tecnologia, via Morego 30, 16163 Genova, Italy*
[b] *Physical Science and Engineering and BESE Divisions, KAUST, King Abdullah University of Science and Technology, Thuwall, 23955-6900, Kingdom of Saudi Arabia*
[c] *BIONEM Lab, University of Magna Graecia, Campus S. Venuta, Germaneto, viale Europa, I-88100 Catanzaro, Italy*
remo.proietti@iit.it

4.1 Introduction

"I would like to describe a field, in which little has been done, but in which an enormous amount can be done in principle. This field is not quite the same as the others in that it will not tell us much of fundamental physics but it is more like solid-state physics in the sense that it might tell us much of great interest about the strange phenomena that occur in complex situations.

Handbook of Enhanced Spectroscopy
Edited by Pietro Giuseppe Gucciardi, Marc Lamy de la Chapelle, and Nathalie Lidgi-Guigui

ISBN 978-981-4613-32-3 (Hardcover), 978-981-4613-33-0 (eBook)
www.panstanford.com

Furthermore, a point that is most important is that it would have an enormous number of technical applications. What I want to talk about is the problem of manipulating and controlling things on a small scale." It is with these words that R. P. Feynman, on the 29 December 1959, at the annual meeting of the American Physical Society, introduced for the first time the concept of nanotechnology (the word "nano" dates back to the ancient Greece, with *nano* ($= \nu\alpha\nu o\varsigma$) indicating something of small dimensions). Since Feynman's speech, this intriguing discipline has achieved incredible goals going even beyond Feynman's expectations. For example, the 24 million volumes mentioned by Feynman in his "There's Plenty of Room at the Bottom" would fit in about 30 Tbyte hard drive, which has dimensions much lower than the 3 square yards initially estimated by Feynman.

When nanotechnology joins other disciplines such as photonics, plasmonics, electronics, or biology, the number of technical applications becomes almost countless. For example, CPU approaching the 10 GHz frequency clock, liquid crystal or plasma screens, solar cells, nanocatalysis, micro/nanofibers for surgical purposes, nanoparticles for optical tagging of biological molecules, and devices for detection of single/few molecules are just few of many examples where nanotechnology plays a fundamental role.

Here we shall introduce some examples of plasmonics-oriented nanotechnology. In particular, we shall start by showing how optical modes, known as dark modes, can be excited with far-field radiation by opportune design of metallic nanosystems. We will proceed by considering three-dimensional standing plasmonic nanostructures with the purpose of single/few molecule detection. We shall demonstrate the amazing effectiveness of these devices when compared with the two-dimensional counterpart. Afterwards, we will move to the Raman characterization of a specific kind of two-dimensional metallic devices known as APA, which, similarly to the previous case, have been finding strong utilization for single/few molecules detection applications. Finally, by using photocurrent spectroscopy, we shall introduce semiconductor and metallic/semiconductor structures used as optical sensors and for photovoltaic applications.

4.2 Plasmon Hybridization: Main Features and Implications in Near- and Far-Field Optical Response

Plasmon hybridization is an electromagnetic (EM) phenomenon, extensively studied in the last decade [1, 2], concerning the combination of interacting plasmonic modes. When two plasmonic charge distributions interact with each other, the total system admits two plasmonic modes corresponding to low energy (bonding) and high energy (antibonding) configurations, respectively. The smaller is the mutual distance between the charge distributions (stronger interaction), the larger is the energy gap between bonding and antibonding modes [3].

One of the simplest cases of plasmonic hybridization that can be theoretically studied consists of two identical plasmonic spheres (Fig. 4.1(a)). Each sphere supports different order modes, labeled by their angular momentum number "*l*" [1]. By applying the variational principle to the system, it is possible to show the energy splitting of unperturbed levels into bonding and antibonding ones and the influence of the interparticle distance "D" on the energy gap [1].

Despite the long-range effectiveness of EM fields, plasmonic hybridization becomes appreciable for interparticle gaps much smaller than the system scale. In fact, the strength of the near-field coupling between two plasmonic fields is determined by the overlapping of their evanescent tails [3] (Fig. 4.1(b)), whose characteristic length is a small fraction of the system size [3]. Therefore, the maximum interparticle gap required for plasmon hybridization varies according to the spectral region in which the structures resonate. For example, we can find 100 nm gaps in asymmetric slits [4] (Fig. 4.1(c)) working in the infrared range and 20 nm gaps in disk heptamer [5] (Fig. 4.1(d)) presenting hybridization in the visible (VIS) range.

In order to study the EM response of complex plasmonic nanoassemblies, it is convenient to consider systems characterized by simple elemental units. For this reason, in the following part, examples of rod-like nanoantenna dimers and trimers, showing the effects of gap-coupling in their near- and far-field responses, will be presented.

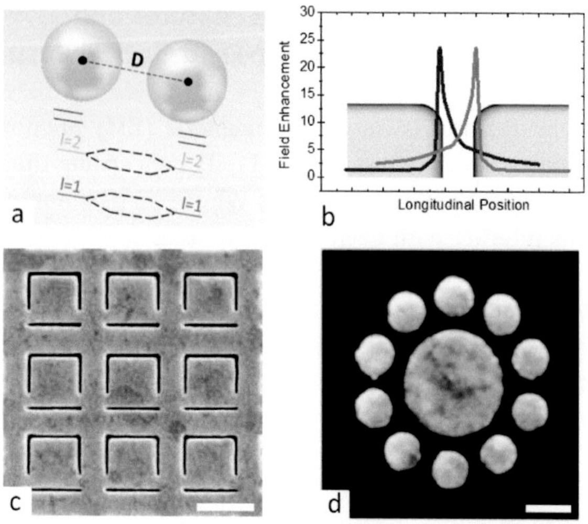

Figure 4.1 (a) sketch showing the system of two plasmonic spheres and the splitting of "*l*" energy levels; (b) sketch showing the overlapping of plasmon evanescent tails; (c) SEM image of an array of asymmetric slits (scale bar: 500 nm); (d) SEM image of gold disks eptamer (scale bar: 100 nm). Reprinted with permission from Ref. [5], Copyright 2012 American Chemical Society.

Rod-like antennas can support short and long axis *localized surface plasmons* (*LSPs*) and both of them can be hybridized in complex systems. For example, the fabricated L-shape nanoantenna dimer shown in Fig. 4.2(a) presents, in the near-infrared (NIR) region, the hybridization of the long axis LSPs into in-phase and out-of-phase collective oscillations [6]. The investigation of the charge current densities induced in the dimer by a plane wave impinging at normal incidence leads to a clear distinction between the two hybridized modes supported by the system. In the first mode (bonding mode, Fig. 4.2(c)), the charge barycenters of the two antennas keep their mutual distance fixed (in-phase oscillation). Whereas in the second mode (antibonding mode, Fig. 4.2(e)), the charges in both arms contemporarily move away from the gap region (out-of-phase oscillation). Therefore, in the bonding mode the currents try to neutralize the net charge in the gap, while in the antibonding mode the currents tend to unbalance the gap charge

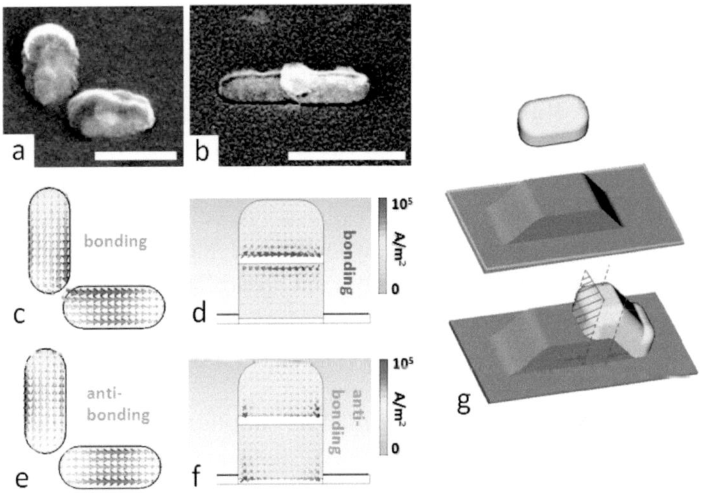

Figure 4.2 (a) SEM image of gold L-shape nanoantenna dimer (scale bar: 200 nm); (b) SEM image of gold stacked optical antennas (scale bar: 300 nm); (c, e) 2D charge current density plots evaluated on a plane parallel to substrate, passing through the center of the structures: respectively L-shape bonding and antibonding modes are shown; (d, f) SOAs 2D charge current density plots evaluated on a plane perpendicular to the long axis of lower antenna, passing through the center of the gap region: respectively bonding and anti-bonding modes are shown; (g) sketch depicting the three steps of SOAs fabrication: (i) lower antenna fabrication by *Electron Beam Lithography*; (ii) capping with dialectric spacer; (iii) *overlayer exposure* for upper antenna fabrication (red arrows depict the lower/upper antenna overlapping region).

[Note that the attributes of in-phase and out-of-phase collective oscillations are not univocally related to bonding and antibonding modes and the system geometry plays an essential role in such relation.

We realized *stacked optical antennas* (*SOAs*) (Fig. 4.2(b)) by a three-step process: first we fabricated a single antenna on a dielectric substrate (upper sketch in Fig. 4.2(g)), then we capped it with a thin dielectric spacer (central sketch in Fig. 4.2(g)) and finally we fabricated a second antenna on the top of the first one. During the last step, we shifted the upper antenna along its main axis, so that the two antennas were only partially overlapped (the overlapping region is between the two dashed lines in the lower sketch of Fig. 4.2(g)). By considering the charge density currents of SOAs on a plane perpendicular to the antennas long axis (red plane in lower sketch of Fig. 4.2(g)), we can observe the hybridization of the short axis LSPs. In fact, by simulating the charge current densities induced by a normally incident plane wave polarized along the short axis of SOAs, we can recognize both out-of-phase (Fig. 4.2(d)) and in-phase LSP modes (Fig. 4.2(f)). Different from planar L-shape system, SOAs support an out-of-phase mode more energetically convenient than the in-phase one. In fact, in the SOAs system, the LSPs are forced to flow in parallel along the dielectric gap (spacer) and not in series, as in the L-shape configuration.

The phenomenon of hybridization has implications not only in the near-field response of plasmonic systems but also in their far-field response. The charge current densities previously defined for upper and lower antennas in SOAs can be directly associated to dipolar momenta. By considering the system in terms of a global dipolar momentum, the antibonding mode is a strongly radiative (*bright*), while the bonding mode is expected to weakly couple to far-field radiation (the two dipoles—top and bottom side of the spacer—tend to cancel each other). We performed far-field simulations with the same EM source employed for near-field analysis, reporting the extinction efficiency spectrum of 10 nm gap SOAs (Fig. 4.3(d)). The spectrum shows an intense peak around 600 nm, which can be ascribed to the strongly radiating antibonding mode, and a small peak around 900 nm related to the bonding mode. By changing the gap between 20 nm and 5 nm, a high variability of bonding resonance wavelength has been appreciated (a 600 nm span). For this reason, we defined the bonding resonance as *fundamental* and plotted both bonding and antibonding resonance wavelengths as function of the fundamental one (Fig. 4.3(b)). The corresponding gap variation is reported in the plot of Fig. 4.3(a). Different from the bonding plot (red contour circles), the antibonding curve (blue contour squares) does not show an appreciable evolution. In order to understand the reason of such behavior, we remember that the energy gap depends on the intensity of the near-field interaction. So we performed near-field simulations on 10 nm gap SOAs, evaluating the *field enhancement* in bonding and antibonding condition. We

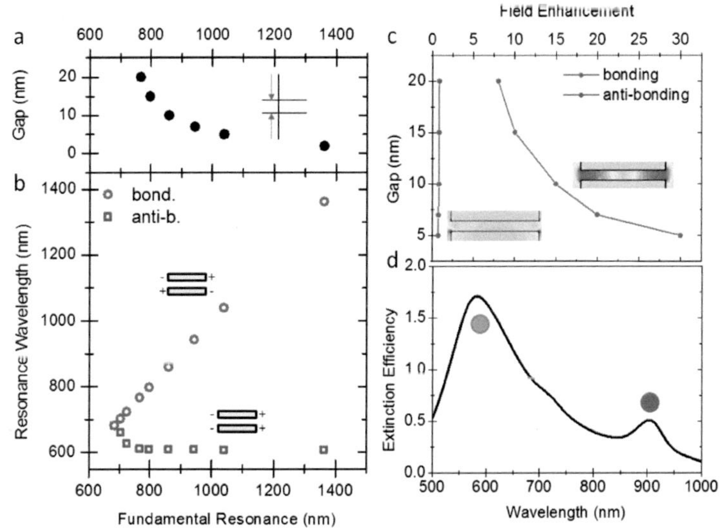

Figure 4.3 (a) Plot showing the relation between the fundamental resonance wavelength and the interparticle gap in SOAs; (b) plot showing the variation of SOAs bonding (red circles) and antibonding (blue squares) resonance wavelengths as function of the fundamental resonance wavelength; (c) plot showing the dependency of SOAs field enhancement on interparticle gap, in bonding (red trace) and antibonding (blue trace) condition (right/left inset: 2D plot of field enhancement inside SOAs gap in bonding/antibonding condition); (d) simulated extinction efficiency spectrum of 10 nm gap SOAs in normal incidence condition and polarization parallel to SOAs short axis.

observed that the bonding mode induces the generation of two intense *hot-spots* in the dielectric spacer (see right sketch in Fig. 4.3(c)), while the antibonding mode strongly suppresses the electric field inside the gap (see left sketch in Fig. 4.3(c)). Therefore, we plotted the bonding and antibonding field enhancement, in the hot-spot position, as function of interparticle gap (Fig. 4.3(c)). It is quite evident how, in bonding condition, the electric field increases for smaller gaps (red plot). Conversely, in the antibonding case, the electric field does not show a significant evolution, remaining strongly suppressed. We can relate the low variability of antibonding wavelength to the absence of field enhancement that the system experiences in antibonding condition.

By studying this system, we could clarify some aspects related to plasmonic hybridization. In particular, we could highlight the main implications that bonding and antibonding charge configurations have both into the near- and far-field responses of a coupled plasmonic system.

The small peak observed in Fig. 4.3(d) around 900 nm has been associated to a mode that weakly couples to the far-field radiation. Interestingly, this peak shows a sharper line-shape with respect to usual plasmonic resonances in the far-field, due to its reduced *radiation damping* [7]. If ideally we kept on reducing the radiation damping of a bright mode, we would reach the *dark* mode limit case. Despite the high spectral selectivity of dark modes, they cannot be directly exploited in far-field optical applications [8].

In the last few years, new plasmonic devices have been conceived presenting high spectral *figure of merit* in the far-field [5], by inducing the hybridization of bright and dark modes at the so-called *plasmonic Fano resonance* [9]. In strong gap-coupling regimes, the spectral overlapping of a bright and a dark mode can promote an energy transfer between the two plasmonic modes, leading to an interferential effect [10]. When light interacts with such systems, two different transition channels can be defined: (i) the incoming photon is converted into a bright plasmon and the same plasmon is reconverted into an outgoing photon; (ii) the incoming photon is converted into a bright plasmon which, via near-field channel, becomes a dark plasmon and the inverse transition creates an outgoing photon. During these processes (time scale in the order of fsec), the probability amplitudes related to each transition *path* destructively interfere and reduce the light–matter interaction [10]. For this reason, it is possible to observe, around the Fano resonance, a characteristic dip in the extinction spectrum, critically dependent on the near-field coupling.

In perspective of realizing a device with such a property, the first step is the hybridization of the LSP of a linear antenna with the out-of-phase dark mode of an *aligned antenna dimer*. The fabricated device is a *T-shape nanoantenna trimer* (Fig. 4.4(a)) consisting of three equivalent rod-like antennas. By exciting such a system with normally incident light polarized along the central antenna (Fig. 4.4(b)), it is possible to induce the LSP mode "$|B\rangle$"

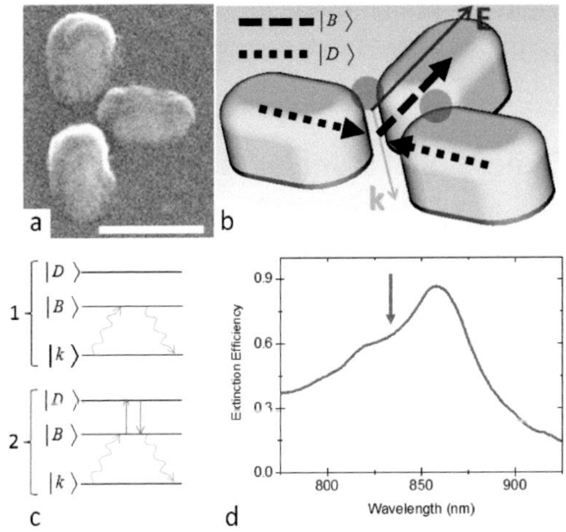

Figure 4.4 (a) SEM image of T-shape nanoantenna trimer (scale bar: 200 nm); (b) sketch of T-shape antennas, depicting the bright (long dash arrow) and dark (short dash arrows) modes induced inside the system for light impinging under normal incidence condition and polarized parallel to the central antenna long axis (the regions of gap-coupling have been depicted with two transparent red circle areas); (c) diagrams showing the two transition paths that light can select: (1) $|k\rangle$-$|B\rangle$-$|k\rangle$; (2) $|k\rangle$-$|B\rangle$-$|D\rangle$-$|B\rangle$-$|k\rangle$); (d) experimental extinction efficiency spectrum of T-shape antennas collected for light impinging at normal incidence and polarized parallel to the central antenna long axis (the Fano resonance is indicated by the red arrow).

(long dash arrow). For geometrical reasons, light cannot couple to the remaining antenna arms. At this point, only the gap-coupling between the central antenna and the other two arms can induce the excitation of the dark mode "$|D\rangle$" along the aligned arms (short dash arrows). For granting the first transition channel selected by free radiation "$|k\rangle$" (upper diagram in Fig. 4.4(c)), the single antenna can be sufficient. Instead, for the second channel (lower diagram in Fig. 4.4(c)), it is necessary to introduce the aligned antenna dimer. By performing far-field transmission spectroscopy on a T-shape nanoantenna array, the Fano resonance phenomenon was observed in the collected extinction efficiency spectrum (Fig. 4.4(d)). Although the estimated interparticle gap of T-shape antennas is around 10 nm (strong coupling regime), only a slight reduction in the extinction spectrum was noticed, probably due to *convolution effects* among the far-field contributions of all the elements in the array.

4.3 Three-Dimensional Standing Nanostar Structures

The fascinating properties exhibited by nanostructures when elevated from the bulk substrate through dielectric pillars have opened up a wide range of applications in the field of plasmon-enhanced spectroscopy [11–15]. In particular, standing plasmonic nanostars are a novel class of intriguing structures in the field of plasmon-enhanced spectroscopy, in fact they posses sharp tips where strongly localized electric fields are focalized. This feature, combined with the standing characteristic, allows the detection of single/few molecules by surface-enhanced Raman spectroscopy (SERS) [16, 17]. The strength of the local EM field on the nanostructure is mainly dependent on structure shape, size, interparticle separation, and dielectric constant. Top-down approaches, such as electron beam lithography (EBL) and reactive ion etching (RIE) techniques can provide the capability of patterning nanostructures with homogeneity and uniformity. In the present section, fabrication and characterization of 3D plasmonic nanostar (standing nanostars) by EBL and RIE techniques are investigated for field-enhanced spectroscopy by means of numerical calculations and SERS measurements.

Schematic illustration of 3D Au plasmonic nanostar structures with size L, inter-star separation I_{SS}, pillar height h, branch width B_W and length B_L are shown in Fig. 4.5(a). Nanostars are fabricated by the combination of EBL and RIE techniques [11]. Normal-incidence scanning electron microscopy (SEM) image of 3D nanostar structures with geometrical features of B_W and B_L of 65 and 40, 150 nm of I_{SS}, L and h, respectively, is shown in Fig. 4.5(b). The corresponding magnified image of a single standing nanostar with tilted view is represented in the inset, which underlines the morphological quality in terms of sharp edges. In order to understand the electric field enhancement nature of 2D and 3D plasmonic nanostar structures, numerical calculations (CST-Microwave) are performed. In order to properly represent the

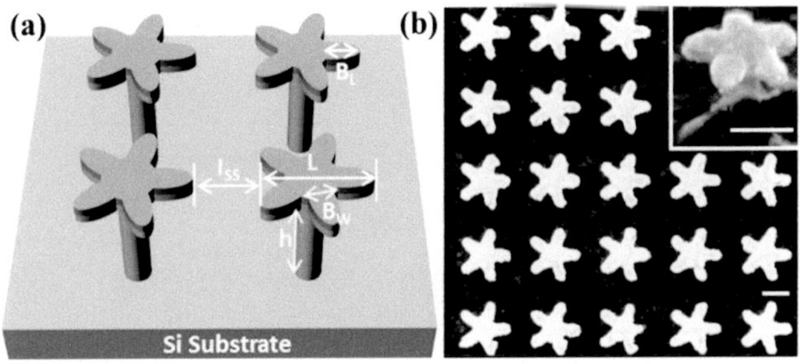

Figure 4.5 (a) Schematic representation of 3D Au plasmonic nanostar structures. (b) Normal-incidence SEM image of 3D nanostar structures. Magnified SEM image of a single nanostar with 52° tilted view is shown in the inset. Length of the scale bar is 100 nm.

fabricated structures, the tetragonal mesh resolution of 0.3 nm was chosen. The dielectric constant of Si and Au employed in the present simulation were $\varepsilon_{Si} = 15.21$ and $\varepsilon_{Au} = -9.79 + i\,1.97$ [18, 19].

Figures 4.6(a) and 4.6(b) show the contour plots of the near electric field distribution in 2D (no-standing) nanostars for *xy*- and *xz*-planes, respectively. The corresponding plots for 3D nanostructures are illustrated in Figs. 4.7(c) and 4.7(d). The 2D nanostar shows concentrated electric field around the star tips with an enhancement factor, with respect to the incident light, equal to 2. This value is due to the strong interaction between the nanostar and the dielectric substrate, since the local electric field tends to extend deeply into the dielectric substrate, therefore a reduced electric field enhancement is observed. According to previous reports [11, 20], an increment in the local electric field enhancement and sensitivity can be observed by elevating the nanostructure height, where the localized electric field associated to the nanostructure is completely decoupled from the substrate. In particular, the contour profile for the 3D nanostar (Figs. 4.6(c) and 4.6(d)) shows a maximum enhancement value of 16, which is around eight times higher than the 2D nanostar case. The localized electric field is in fact completely decoupled from the substrate (Fig. 4.6(d)), therefore fully available for molecule sensing in localized surface plasmon resonance (LSPR) or SERS-based measurements. By considering that the SERS enhancement [21] is proportional to $|E|^4$, these 3D

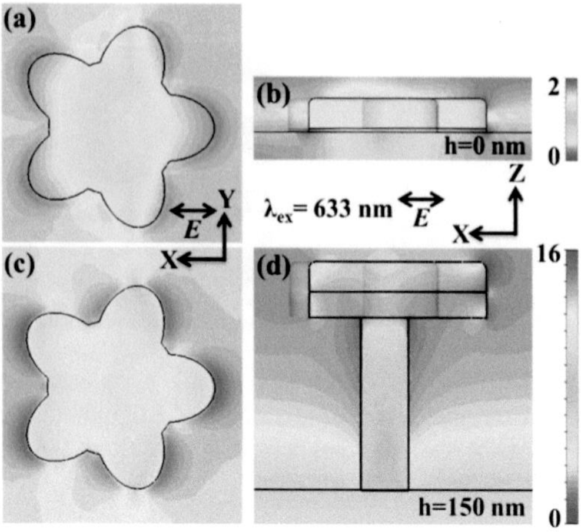

Figure 4.6 Electric field distribution for a single nanostar structure in (a, c) *xy*-plane, respectively 2D and 3D. The corresponding plots for the *xz*-plane are shown in (b) and (d), respectively.

plasmonic nanostars can provide an enhancement of 6.5×10^4 by considering only the electromagnetic enhancement. If the chemical enhancement is also accounted [22], these nanostructures can provide an average enhancement in the order of 10^6–10^7 that allows the detection of molecules at ultralow concentrations.

To properly understand the enhancing capability of the 3D nanostar structures, SERS measurements are performed by using Rhodamine-6G (R6G) as a probe molecule. Renishaw inVia micro-Raman spectroscopy in backscattering geometry mode is used to analyze the substrate. Thermoelectrically cooled charge-coupled device (CCD) is used as detector. The instrument is equipped with an excitation laser of 633 nm wavelength and 150× objective lens (NA = 0.95) of Leica microscope. Its spectral resolution is 1.1 cm^{-1}. R6G molecules were deposited on the nanostar structures by means of chemisorption technique, where the nanostar substrate was immerse in R6G (10 μM) solution for 20 min (which allows the adsorption of R6G molecules on the metal surface). Afterwards, the sample was extracted from the R6G solution and rinsed with de-ionized water in order to remove excess molecules. Finally, the substrate was dried with nitrogen gas and then SERS measurements

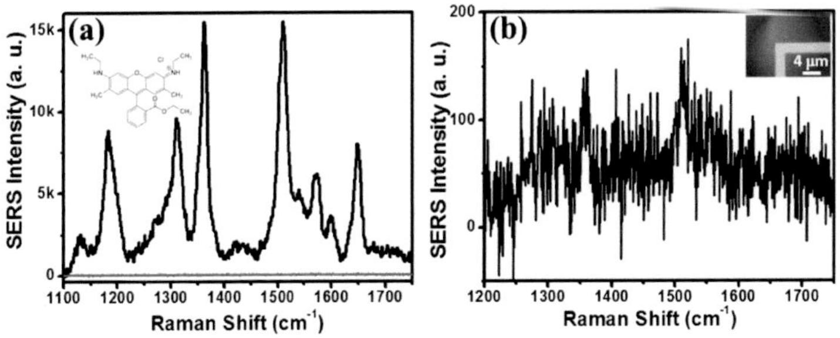

Figure 4.7 (a) Typical SERS spectrum of R6G molecule (black) and background Raman spectra over 3D plasmonic nanostar structures (red). (b) Reference Raman spectrum of R6G molecule over flat gold marker. The inset represents the optical image of the gold marker.

were performed. For Raman and SERS measurements, x-polarized 633 nm laser source with 50 μW power and 30 s accumulation time were used.

Typical Raman spectrum (red trace) of 3D plasmonic nanostar structures before chemisorption of R6G molecules in the range of 1100–1800 cm^{-1} (in order to avoid c-Si first- and second-order peaks centered at 521 and 965 cm^{-1}, respectively) is shown in Fig. 4.7(a). No significant Raman features within the spectral range are observed, thus confirming the absence of surface contamination, which highlights the fabrication quality of the nanostar substrates for sensing applications. Typical SERS spectrum of R6G molecule (black trace) over 3D plasmonic nanostar structures is also shown in Fig. 4.7. The characteristic peaks of R6G are clearly distinguishable and the major bands centered at 1360, 1510, and 1649 cm^{-1} can be assigned to C–H bending, a combination of C–N stretching, C–H and N–H bending and combination of ring stretching of the C–C vibration and C–H_x bending of the xanthenes ring, respectively. The positions of the peaks agree well with previous findings [23]. In order to calculate the SERS enhancement factor, the reference Raman measurement is taken on flat gold marker (Fig. 4.7(b)), which shows low intense peaks of R6G. The optical image of gold marker is shown in the inset. The SERS enhancement factor is calculated by assuming that the molecules (closely packed) are deposited only on the metal surface. The reference band centered at 1510 cm^{-1} is used to calculate the SERS enhancement factor due to its high signal-to-noise ratio. By considering a laser beam focal radius of 500 nm, the average SERS enhancement factor for the reference band centered at 1510 cm^{-1} is about 10^6 with respect to flat Au marker, which well agrees with the enhancement factor calculated through numerical simulations.

4.4 Gold-Coated Anodic Porous Alumina Substrates

Here we introduce anodic porous alumina (APA) substrates produced by means of electrochemical technique. In this regard, with the purpose of obtaining APA templates with different pore size and wall thickness, aluminum substrates were anodized in three different acidic solutions, namely phosphoric (APA1), oxalic (APA2), and sulfuric (APA3) acids. Thereafter, the gold deposition (thickness: 25 nm) was carried out over all the substrates to empower the plasmonic behavior. Three substrates (APA1→AuAPA1, APA2→AuAPA2, APA3→AuAPA3) were employed to investigate their SERS characteristics. SEM images of these substrates before and after gold deposition are shown in Fig. 4.8 and all the related parameters are reported in Table 4.1.

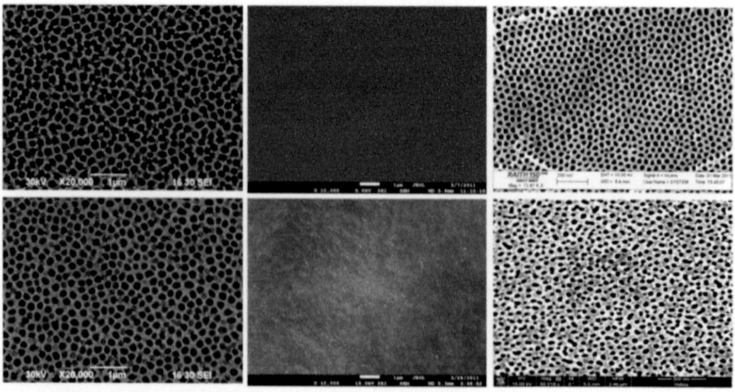

Figure 4.8 SEM images of APA substrate before (top) and after (bottom) gold deposition for APA1, APA2, and APA3 samples (from left to right).

Table 4.1 Various APA physical parameters before and after gold

Parameters	Substrates					
	APA1		APA2		APA3	
	As prepared	Au-coated	As prepared	Au-coated	As prepared	Au-coated
Pore diameter, *d* (nm)	181 ± 45	157 ± 35	82 ± 7	59 ± 9	29 ± 4	15 ± 3
Wall thickness, *w* (nm)	82 ± 35	104 ± 35	21 ± 7	40 ± 9	13 ± 4	36 ± 3
Porosity, *P* (%)	43	15	49	26	35	6

The SERS analysis of Cresyl violet (CV) molecules, chemisorbed on the AuAPA substrates, was performed by using a 633 nm laser line in the spectral range of 300–1300 cm^{-1} (Fig. 4.9). The chosen geometries show pore diameters from 15 nm to 160 nm and wall thicknesses from 36 to 100 nm. The Raman spectra were also collected from AuAPA substrates before depositing any molecules (background measurement to check any possible contamination) and from chemisorbed Au substrates (for reference) [24].

In all the collected SERS spectra, the characteristic vibrational bands of CV are observed. The intense Raman bands centered at around 348, 591, 675, and 1186 cm^{-1} can be attributed to the out-of-plane skeleton deformation, to the combination of in-plane $N–H_2$ and ring bending, to the ring deformation and combination of $N–H_2$ rocking and $C–H_x$ rocking, respectively [25]. From the vibrational bands observed in Fig. 4.9, it is most probable that the CV is oriented to have the $N–H_2$ group very close to the gold film, leading to a strong Raman scattering and, consequently, to higher SERS signal. It has been observed [26] that CV molecules, drop deposited over Si substrates, show a fluorescence background with a low intense peak at around 591 cm^{-1}. Therefore, it is also noteworthy to emphasize that significant fluorescence quenching is also established, as already observed for gold nanoparticles [27].

When the pore size/wall thickness of AuAPA substrate decreases from around 160 nm/100 nm to around 60 nm/40 nm, as reported in Table 4.1, a significant increase in the SERS intensity for the CV bands is observed, as illustrated in Fig. 4.9. Figure 4.10 shows the reference band intensity of CV, centered at around 591 cm^{-1},

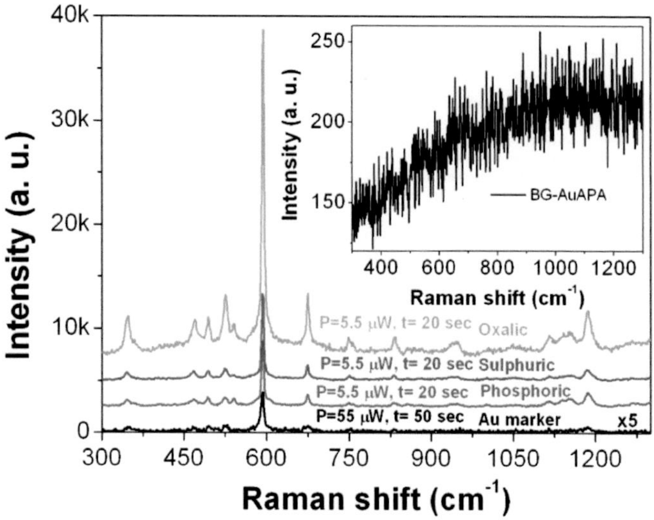

Figure 4.9 SERS spectra of CV deposited over all the substrates (Au, AuAPA1-AuAPA3). In the inset, the background signal (from Analyst).

for different SERS substrates. The figure shows an increase in intensity from around 2,000 counts to more than 30,000 counts, while keeping all the measurement parameters fixed but varying the pore size and wall thickness (see Table 4.1). In addition, a further decrease in d and w reduces the band intensity. Hence, it is very important to choose the value of d and w with extreme care. Finally, by noticing that the molecules are attached only on the top of the AuAPA walls, the number of molecules under illumination is the highest for AuAPA3 (lowest porosity of 6%) and the lowest for AuAPA2 (highest porosity of 26%), with an intermediate value for AuAPA1 (porosity of 15%).

The SERS enhancement factor G can be estimated from the CV SERS spectra measured from the three different gold-coated APA substrates,

$$G = \left(\frac{I_{\text{SERS}}/A_{\text{SERS}}}{I_{\text{ref}}/A_{\text{ref}}} \right)$$

In the equation, I and A are the intensity and the active area over which the molecules are deposited, respectively. The enhancement factor G is calculated for the AuAPA substrates with respect to the flat gold substrate. The results show G values in the range from 0.98 $\times 10^3$ to 1×10^4 for all the three cases, with the highest value of 1×10^4 appearing for AuAPA2. Interestingly, this substrate is not the one showing the smallest pores (and, therefore, also the highest pore density) and wall thickness, that is actually AuAPA3. For the G calculation, only the molecules attached to the AuAPA surface were considered. The molecules attached into the pores were instead neglected since the interior part of the pores was out of focus, namely negligible signal from this area was collected. Here, we propose that the highest enhancement factor G for AuAPA2 with respect to AuAPA1 and AuAPA3 is not solely due to the pore size or wall thickness, but it also depends on their mutual combination (d and w).

The present findings were found to be highly reproducible, since the SERS spectra of CV taken at different points of each substrate appeared very similar one another, as reported in Fig. 4.10(a). This result is an intrinsic confirmation of the good uniformity of the APA geometry as well as of the gold coating. In this regard, SERS mapping measurements were also performed in a large area and, thereafter, the analysis was performed for the reference peak centered at

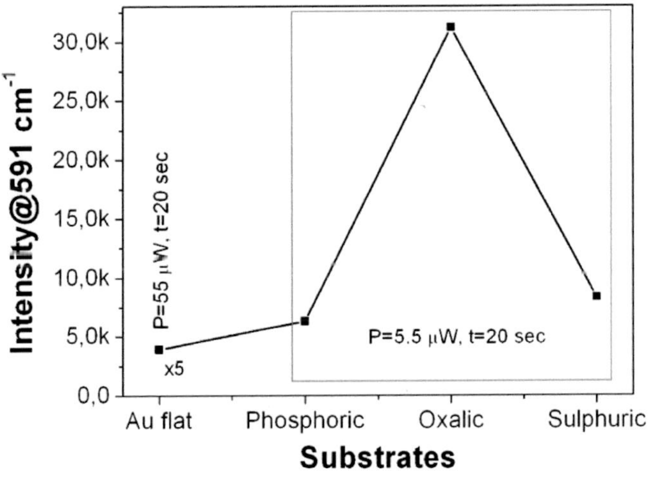

Figure 4.10 Intensity curve for Raman band centered at 591 cm^{-1} for different substrates.

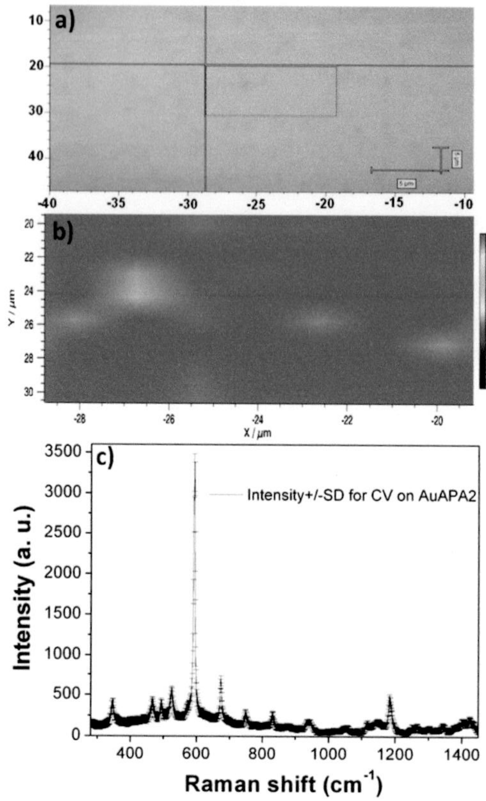

Figure 4.11 2D SERS mapping over AuAPA2 substrate to demonstrate the reproducibility features.

591 cm^{-1}. The statistical analysis, such as standard deviation over the whole SERS spectra, was also performed for all the substrates. The mapping area, its image analysis, and thereafter the SERS spectrum with standard deviation for the AuAPA2 substrate are shown in Figs. 4.11(a–c). The figures show very small variation in intensity throughout the mapping region. In addition, Fig. 4.9 demonstrates that with the AuAPA substrate it is possible to obtain a strongly enhanced Raman signal even for fluorescent molecules.

In order to provide support for the experimental results shown in Fig. 4.9 and Fig. 4.10, we have also performed a series of simulations with the intent of calculating the near-field electromagnetic distribution on the APA substrates. This kind of analysis can, in fact, provide information on the level of excitation of any sample deposited on an SERS substrate. In Fig. 4.12, simulations of the electric field due to the LSP generation on the SERS substrates at $z = 0$ nm (i.e., on the front surface of the substrate) using CST software are presented. In particular, a 25 nm thick golden slab showing a honeycomb pattern with a varying periodicity of 260, 100, and 55 nm, and sub-wavelength air holes with a diameter of 160, 60, and 15 nm, respectively, was simulated. The source light was taken at 633 nm with linear polarization along the x-axis. The overall simulated structure is shown in Fig. 4.12(a). For all the periodicity patterns, the results show an electric field strongly affected by the material discontinuity. In fact, the absolute value of the total electric field shows some nodes at the gold–air interface. This is well explained by the z-component of the electric field as seen in Fig. 4.12(b). In fact, the x-polarization creates antiphase hot-spots along the x-axis, which determines the abrupt change of the electric field as in Figs. 4.12(c–e).

The relative electric field distribution profile over the gold surface area for all the AuAPA substrates are shown in Figs. 4.12(f–h). It is easily understood that the intensity of the field is proportional to the coupling strength among the holes, in fact the coupling decreases when the holes are far from one another. This aspect is especially stressed when we look at the relative value of electric field for all the samples, where AuAPA2 shows a value up to five times higher than the AuAPA3 substrate. These simulation findings were found in good agreement with the experimental results shown in Fig. 4.10.

From the experimental measurements of CV molecules deposited on the different APA substrates (Fig. 4.10), the substrate providing the best enhancement resulted to be "AuAPA2", as confirmed by the respective simulations (Fig. 4.12). In fact, the total enhancement of Raman intensity in the SERS phenomena is a combination of strong electric field (due to the metal surface nanostructure size, shape, and interparticle gap) and strong molecular polarizability (i.e., chemical enhancement). It is found that the major part of the signal enhancement is due to the electric field enhancement,

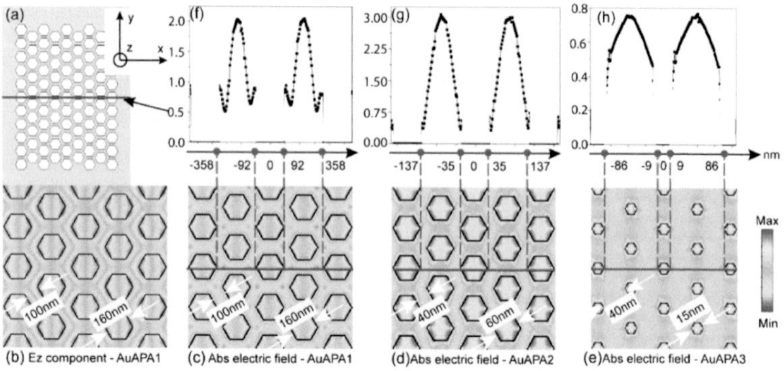

Figure 4.12 CST simulations of honeycomb structures by varying pore size *d* and wall thickness *w*. (a) The overall simulated structure; (b) amplitude of the *z*-component of the electric field in the *x*–*y* plane of AuAPA1; total electric field in the *x*–*y* plane for (c) AuAPA1 with $d = 160$ nm and $w = 80$ nm, (d) AuAPA2 with $d = 60$ nm and $w = 40$ nm, (e) AuAPA3 with $d = 15$ nm and $w = 40$ nm; (f), (g), and (h) relative electric field profile over the gold surface along the highlighted line for all the SERS structures. The excitation light is $\lambda = 633$ nm on the SERS surface with the dielectric function of the gold slab $\varepsilon_{\text{real}} = -9.83$ and $\varepsilon_{\text{imaginary}} = 1.97$. The gold thickness is always $h = 25$ nm.

whereas the maximum chemical enhancement can be around 100 [28].

4.5 Periodic Nanocuboids as Plasmonic Substrates

Nanocuboid SERS substrates with different matrix dimensions (1 × 1, 2 × 1, 2 × 2, 2 × 3, and 4 × 4) and interparticle gaps IPG (20–500 nm) were fabricated by using top-down EBL technique, followed by gold deposition through e-beam evaporation. Various molecules, such as rhodamine 6G (R6G), myoglobin protein, and wild/mutated type W1837R peptides were examined after chemisorption over the SERS devices [23, 29].

SEM and atomic force microscopy (AFM) images of nanocuboids with matrix size of 2 × 2, 3 × 3, and 4 × 4 are shown in Fig. 4.13. The SEM images of the devices show the periodic layout of the nanocuboid structures for different matrix arrays, whereas AFM

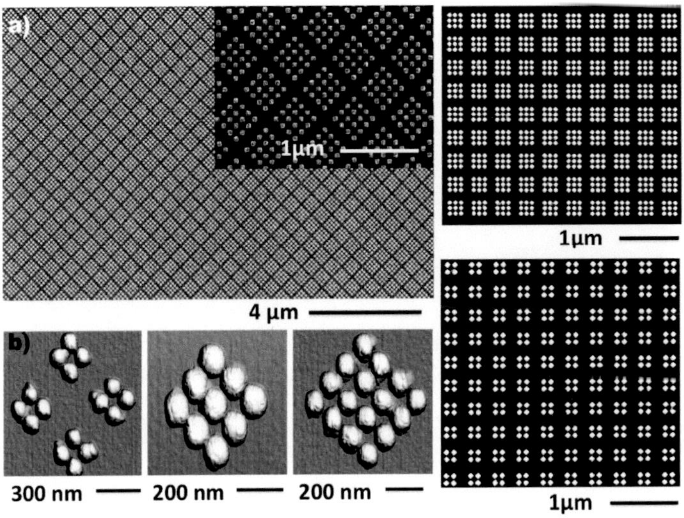

Figure 4.13 (a) SEM and (b) AFM images of nanocuboid structures with different matrix array sizes (2 × 2, 3 × 3, and 4 × 4).

results confirm the 25 nm structures height. The images show cuboids with strongly rounded edges because of the convolution between the relatively large AFM tip and the nanostructures geometrical features.

In order to investigate the electric field distribution on these substrates, numerical calculations were performed by finite integration methods using CST commercial software. In particular, the gold nanocuboids were modeled with a geometry as close as possible to the fabricated structures. In fact, as can be observed in Fig. 4.14, the corners of the cuboids were slightly smoothed to match the experimental geometrical features. The side length of the cuboid structure and IPG were fixed at 80 nm and 20 nm, respectively. The distribution of the electric field, at $\lambda = 633$ nm, was calculated for various array sizes from 1 × 1 to 5 × 5 cuboids. In order to avoid any electric field dependence on the mesh density, we have constructed the same mesh density per cuboid for all the array sizes. In Fig. 4.14, the electric field distributions for the samples with cuboid matrices 1 × 1 and 3 × 3 are shown. The electric field polarization is fixed along the x-axis. It is clearly observed

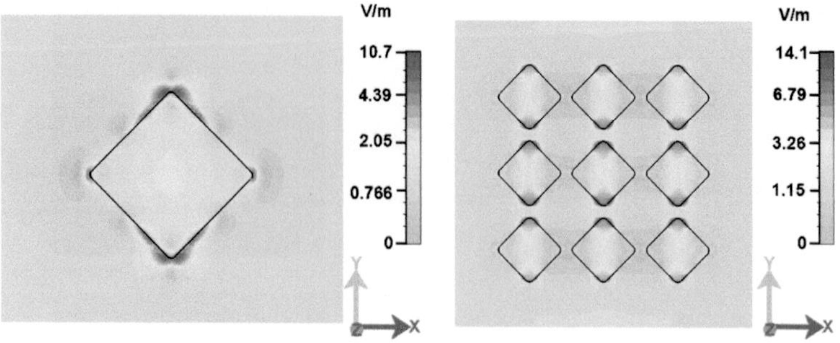

Figure 4.14 Electric field distribution image for nanocuboids matrix 1 × 1 (a) and 3 × 3 (b) SERS device. The corners of the structures are slightly rounded to make the design more realistic. The light is *x*-polarized. The interparticle gap between two nanocuboids is 20 nm.

that the electric field is mainly confined on the corners along the *y*-axis, however, lower electric field can also be revealed on the corners along the *x*-axis. This behavior is ought to the mismatching resonant conditions between the wavelength and the dimensions of the cuboids [30]. To be noticed that the field enhancement obtained by using a realistic nanocuboid (rounded corners) is expected to be lower than for an ideal structure (sharp corners).

The SERS measurements carried out for R6G dye in the range of 1080–2000 cm^{-1} are shown in Fig. 4.15. In particular, R6G was deposited over different gold-based cuboid SERS substrates using chemisorption technique. The featureless Raman spectrum shown in the inset clearly demonstrates the substrate being perfectly clean without any contamination. In the spectrum the characteristics intense peaks are visible at around 1182, 1310, 1361, 1510, and 1649 cm^{-1}, attributed to the out of plane C–H_x bending, the combination of C–H and N–H bending, the combination of ring stretching of C–C vibration, N–H bending and C–H_x wagging, C–H bending, the combination of C–N stretching, C–H and N–H bending, and the combination of ring stretching of C–C vibration and C–H_x bending, respectively [31–33]. A shoulder at around 1275 cm^{-1} is also observed, related to the C–O–C stretching vibration of xanthene group. The spectrum was obtained by using an SERS device

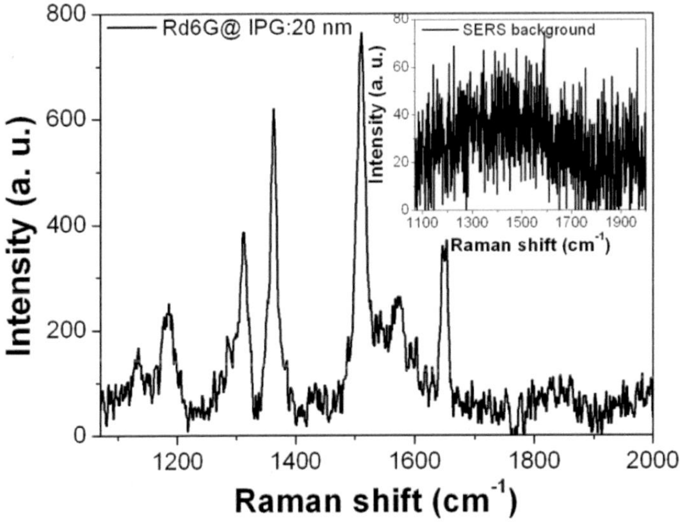

Figure 4.15 SERS spectrum in the range of 1080–2000 cm^{-1}, for R6G dye molecule, deposited on nanocuboids (4 × 4 cuboids matrix size with IPG 20 nm) by means of chemisorption technique. Raman spectrum of bare SERS device (without any molecule on it) is shown in the inset.

characterized by a 4 × 4 cuboids matrix with IPG 20 nm and cuboids side of 80 nm.

The variation in the reference peak height of R6G SERS spectrum, centered at 1510 cm^{-1}, for all the SERS substrates with IPG in the range of 20–500 nm is plotted in Fig. 14.16(a). The reference intensity is found to be decreasing with the increase in IPG of the nanocuboid SERS device. The trend in the variation of the intensity can be categorized in two steps: (a) the device with IPG less than 250 nm and (b) the device for IPG larger than 250 nm. It is observed that the SERS intensity diminishes quickly (from around 800 counts to 150 counts) when IPG is increased from 20 to 250. Afterwards, the SERS intensity remains roughly constant.

Furthermore, a series of SERS measurements were performed for devices with varying cuboids matrix sample and metrology, keeping the IPG fixed (i.e., around 20 nm). Cuboids matrix such as 1 × 1, 2 × 1, 2 × 2, 3 × 3, and 4 × 4 were investigated to understand their

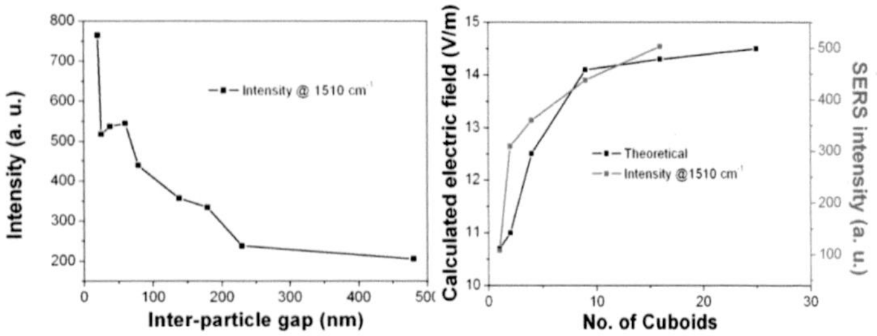

Figure 4.16 (a) SERS intensity of rhodamine 6G for (a) the band centered at 1510 cm^{-1} for nanocuboid substrates with varying interparticle gap and (b) SERS intensity for different matrix arrays. The comparison with the calculated electric field is also reported.

influence on the SERS intensity. In Fig. 4.16(b), the SERS intensity (blue line) is plotted against the number of cuboids (e.g., in case of 2 × 1, the number of cuboids is 2 while for 4 × 4, it is 16). It is revealed from Fig. 4.16(b) that the increase in the number of cuboids leads to a sharp increase in SERS intensity reaching to 500 counts for the device with 4 × 4 nanostructures. Any further increase in the array size has, however, a little influence on the SERS intensity. This trend can be easily understood considering that a 4 × 4 cuboids matrix covers an area of 0.3 μm^2, which corresponds to the surface area illuminated by the source light. Hence, an increase in dimension of the nanostructure array will not influence the SERS intensity. The calculated electric field with varying matrix size is also plotted for comparison finding a good consistency between them.

The enhancement factor is calculated for the above nanocuboid structures and was found to be 1.4×10^5 with respect to a nonpatterned gold substrate.

To demonstrate the cuboids matrix capability as biosensor, we have also investigated both myoglobin (Mb) and the labeled-free W1837R peptide of BRCT domain of the BRCA1 tumor protein, deposited over a 4 × 4 cuboids nanostructure. Various measurements were performed for Mb at different locations of the SERS device. Mb SERS spectrum with standard deviation is shown in Fig. 4.17. The standard deviation error bars throughout the spectral range indicate the reproducible behavior of the SERS device and thus the hot-spots generation at different locations.

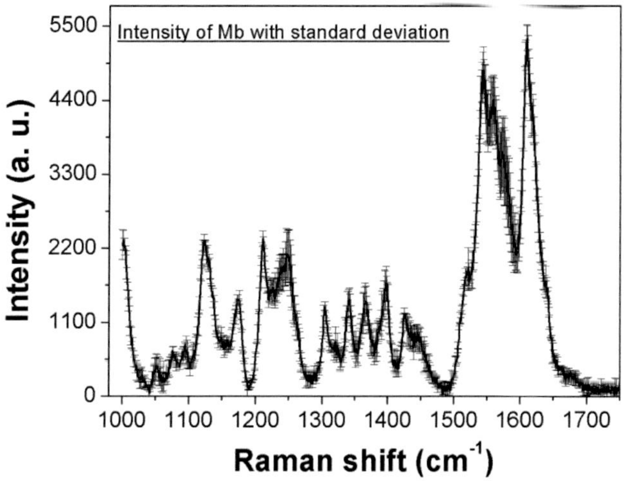

Figure 4.17 SERS spectrum of myoglobin protein chemisorbed over nanocuboid substrate. Standard deviation error bar is also shown in order to confirm the reproducibility at different positions on the substrate.

4.6 Scanning Photocurrent Microscopy

The photoconductivity of nanomaterials can be exploited for photovoltaics and photodetection applications. Toward the integration of novel material into practical devices, the photoconductive properties of the active material need to be studied and optimized, but equally important is to know how this material is behaving in a specific device configuration. Here, the information on where the photocurrent is actually generated can be very helpful for designing efficient device architectures. In this respect scanning photocurrent microscopy (SPCM) is a powerful tool for local probing of photocurrent, and has been applied to carbon nanotubes [34], semiconductor nanowires [35–38] and sheets [39], and recently also to thin films of colloidal nanocrystals [40, 41]. The experimental setup for SPCM can be realized in two ways: (i) the laser beam is fixed and the sample is scanned by a piezo-controlled stage. This approach gives large freedom on the scanning speed and can be adapted to the time response of the electrical circuit. In practice the photocurrent recorded from single nanoobjects can be quite

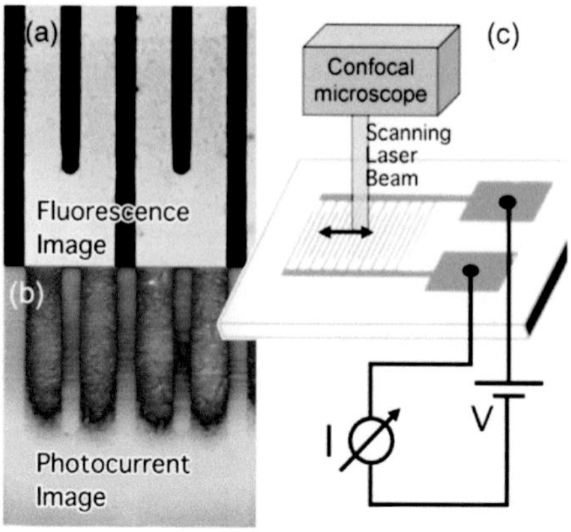

Figure 4.18 The use of a confocal fluorescence microscope synchronized with the recording of the photocurrent allows to record fluorescence and photocurrent images simultaneously. (a) Grey scale image of the fluorescence of a thin film consisting of fluorescent CdSe/CdS core–shell nanocrystals deposited on top of interdigitated electrodes. The electrodes appear black in the image. (b) Photocurrent image recorded at fixed voltage bias of the same region as in (a), blue (red) color corresponds to low (high) photocurrent signal. (c) Scheme illustrating the experimental setup formed by the scanning laser beam of the confocal microscope and the electrical circuit to record the photocurrent in situ. The spatial resolution is defined by the diffraction limited spot size of the laser beam. Figure reprinted with permission from Ref. [41], Copyright 2012 by the American Physical Society.

low and slow scanning enables larger integration times. (ii) In the second approach (illustrated in Fig. 4.18) a commercial confocal fluorescence microscope is used for the light excitation (here the laser beam is typically scanned via galvanometer mirrors), which includes the versatility of this instrument (recording of fluorescence, spectral response etc.), but limits the pixel dwell time for the photocurrent recording to the parameters imposed by the confocal microscope. In this case the scanning speed is typically quite high (with dwell times in the range from few to some hundred of μs), and the electrical circuit needs to be configured to have a low inherent noise level. A possible approach uses short electrical connections to the first current amplifier stage. In the following we will discuss the results obtained with the latter approach, recorded from either homogeneous nanoparticle films or from individual nanowire bundles.

4.6.1 SPCM on Colloidal Semiconductor Nanocrystal Films

Thin films of colloidal semiconductor nanocrystals are a very promising material as active layers in photovoltaic devices and in photodetectors [42–45]. Here, the densely packed nanocrystals in the active layer form a so-called "nanocrystal solid", where they are separated by organic surfactant molecules, which result in tunneling barriers for the transport of charges between the particles. Therefore, the individual nanocrystals can be regarded as weakly coupled islands, and transport of the charge carriers occurs via tunneling. In such systems the photoconduction has been described either on the local level, by considering the exciton dissociation in two adjacent nanocrystals [46], or by competing phenomenological models which describe the photo response of thin nanocrystal films by macroscopic quantities such as sheet resistance and voltage barrier [40, 47–50].

The work of Kudera et al. [41] integrated the local model that describes the charge dissociation with the phenomenological macroscopic treatment that is dominated by the pathways involved in the harvesting of the photo-generated charges. This was achieved by modeling the spatial photocurrent distribution (displayed in Figs. 4.19(a,b)) by assuming exciton ionization according to the model of Leatherdale et al. [46], followed by a random walk of the photo-generated electrons and holes to their respective electrodes similar to Mott's theory of transport in amorphous materials. In this random walk model, the mobility of the charges can be expressed by a parameter $x_{0,e/h}$ that can be evaluated by fits to the photocurrent profiles by the probability distribution

$$P_h(x) = \frac{1}{1+\exp\left(\frac{x-x_{0,h}}{\Delta x_h}\right)};\ P_e(x) = \frac{1}{1+\exp\left(\frac{L-x-x_{0,e}}{\Delta x_e}\right)}$$

for electrons traveling to the electrode located at $x = L$ and P_h for holes traveling to $x = 0$. The x_0 parameter is correlated to the diffusion length of the electrons and holes within the nanocrystal film by $x_0e/x_0h = (L_De/L_Dh)^2$, therefore the analysis of the spatial profiles of the photocurrent amplitude in the SPCM images can provide deeper insight in the film properties (Figs. 4.19(c, d)).

4.6.2 SPCM on Au-Decorated CdSe Nanowires

Hybrid metal–semiconductor systems promise to enhance the performance of photovoltaic or optical sensor devices with respect to their bare semiconductor counterparts. Here the semiconductor portion can absorb the light, while the metal parts facilitate the electron–hole dissociation and charge transport [51]. Furthermore, the scattering and plasmonic properties of the metal parts can lead to an enlarged absorption cross section of the hybrid nanomaterial. For example, metal NPs have been embedded in solar cell materials to improve their performance [52]. In blended materials, the mixing of metal and semiconductor nanoparticles can improve the performance either by broad band light scattering or, in a specific spectral band matching the absorber, by plasmon resonance. However, well-defined all-inorganic metal–semiconductor interfaces could have additional advantages due to the intrinsic Schottky fields at the metal semiconductor contact, which can be expected to improve charge dissociation of the photo-generated electron–hole pairs. In this respect, gold decoration of semiconductor nanowires is thus of particular interest. Chakraborty et al. [38] observed strongly enhanced photoconduction from Au-decorated CdSe nanowires with respect to the bare CdSe wires. The question then was: what is at the origin of the photocurrent enhancement? Is it the light scattering, the plasmonic properties, or the Schottky fields induced by the metal portions? The enhancement due to the plasmonic properties could be excluded by comparing the absorption with the photocurrent spectra, where the plasmonic resonance band appeared solely in the optical spectrum. Light scattering played a significant role, which could be evaluated by comparing the photocurrent of the Au-decorated nanowires with that of physical mixtures of CdSe nanowires and Au nanoparticles. The impact of the Schottky fields could be elucidated by the SPCM experiments shown in Fig. 4.20. Here bare CdSe nanowires in Figs. 4.20(a,b) show the expected pinning of the photocurrent maximum at the electrode contact due to the enhanced dissociation caused by the Schottky field. For

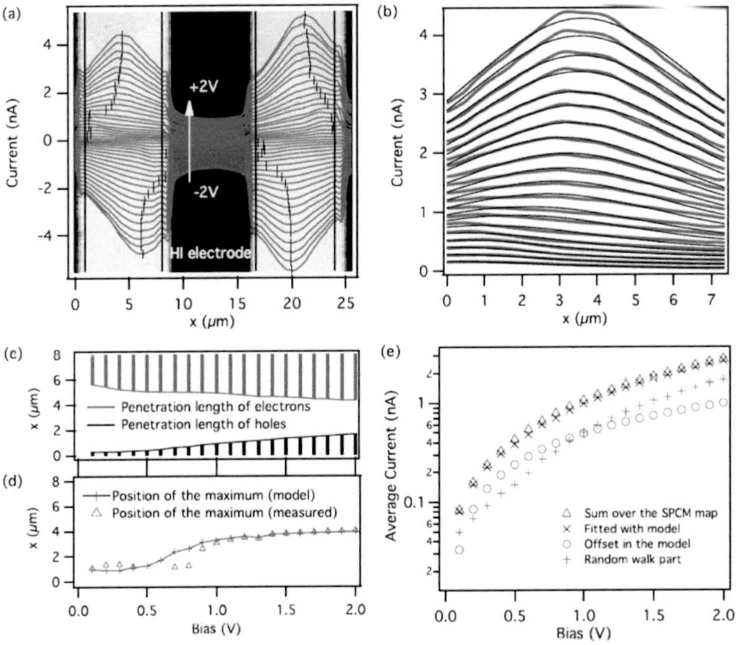

Figure 4.19 (a) Photocurrent amplitude along a profile perpendicular to the electrodes for bias voltages from −2 V to 2 V in steps of 100 mV. A fluorescence image of the corresponding area is displayed as background in order to visualize the position of the electrodes . The photocurrent signal at a given position was obtained by averaging over all current values on a line parallel to the electrodes on the corresponding current maps. The investigated area corresponds to two electrodes gaps in Figs. 4.18(a, b). The small black marks on the individual curves indicate the position of the maxima as obtained by fitting a Gaussian to the segments of the profiles limited by the vertical black lines. (b) Photocurrent profiles from the first gap (red lines) and the corresponding fits (black) at bias voltages from 0.1 V to 2 V on an enlarged scale. (c) Penetration length x_0 of the sigmoidal function from eq. 3 versus bias voltage. The displayed range of x represents the left gap in (a), therefore, the cathode is located at $x = 0$, the anode at $x = 8$ µm. The plot displays $L - x_0$ for the electrons, x_0 for the holes. (d) Position of the spatial current maximum as determined from the fits. (e) Current–voltage curves obtained by integrating over the current values of the different spatial current maps and normalized by the number of data points at different bias voltages (triangles) and as reproduced with the model (crosses). The different terms of the model are also displayed separately. Figure reprinted with permission from Ref. [41], Copyright 2012 by the American Physical Society.

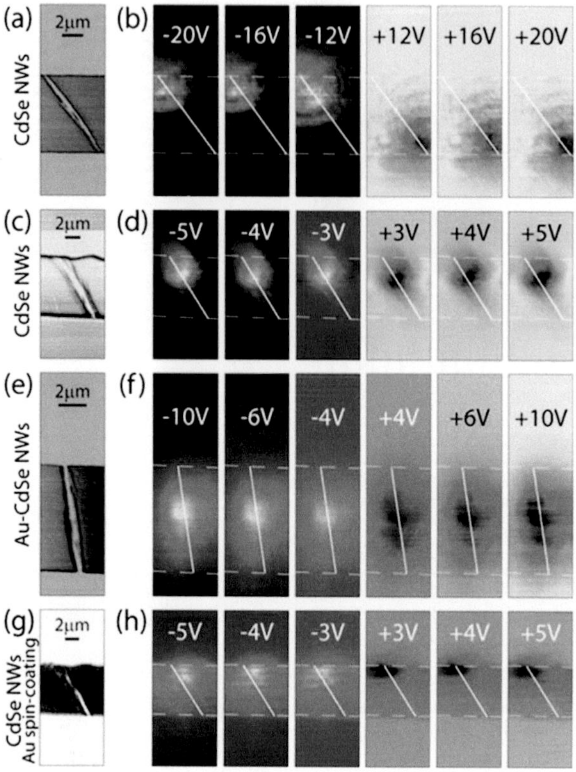

Figure 4.20 SPCM applied to CdSe and Au-decorated CdSe nanowires that bridge the gap between two parallel electrodes. The panels show optical reflection (a, c, e, g) and SPCM (b, d, f, h) measurements on bare CdSe NWs (a–d), Au-CdSe NWs (e–f), and of the sample from (c–d) spin coated with Au NPs (e–f). Bundles of NWs that bridge the electrode gap can be identified in the optical reflection images of the investigated areas. (b, d, f, h) Sequences of SPCM images recorded at different bias voltages. The bright (dark) color contrast corresponds to large photocurrent at negative (positive) bias voltage. All SPCM images were recorded at a wavelength of 515 nm with power of 1 μW. As a guide to the eye, the NW bundles are indicated by white solid lines in the SPCM plots, and the electrode borders by dashed lines. Reproduced from Ref. [38] with permission from The Royal Society of Chemistry.

Au-decorated CdSe nanowires, the spatial photocurrent distribution is very different, where the photocurrent maxima extended much more into the electrode gap, or were even localized within the center of the gap (Fig. 4.20(f)). The localization of the photocurrent maximum within the center of the electrode gap corroborates the impact of the many localized Schottky fields related to the Au decoration along the wire. In a microscopic picture, the Schottky field related to the Au NPs results in an upward bending of the band toward nanowire surface, resulting in an increased space-charge region and to a radial separation of electrons and holes (with respect to the NW axis). Therefore, the Au decoration of semiconductor nanowires by wet chemical synthesis methods can provide a promising alternative for the fabrication of efficient materials for photovoltaic devices, due to large spectral range of enhanced absorption and more efficient charge separation.

4.7 Conclusions

In summary, we have introduced different examples based on nanotechnology and plasmonic-based spectroscopy. We started with planar metallic device having the purpose of exciting the so-called dark modes by means of far-field radiation. This study was faced both theoretically, through numerical simulations, and experimentally, through fabrication and characterization. Afterwards, we moved to 3D plasmonic Au nanostar structures fabricated by EBL and RIE techniques. Numerical simulations were carried out to reveal the physical mechanism behind the electric field enhancement. R6G at 10 μM concentration was used to test the substrate efficiency. The use of this kind of structures is motivated by their peculiarity in forming very high electric field enhancements with respect to nanostars laying down on the bulk substrate. An average SERS enhancement factor of $\sim 10^6$ was observed for 3D plasmonic nanostar structures with respect to a flat Au marker. 3D nanostar structures appear as promising SERS substrates for detecting biological and chemical species even at ultralow concentrations.

The next device was an array of Au-nanocuboids fabricated by using a top-down nanofabrication technique. This technique offers a great control over the size and IPG of the metallic nanocuboids, hence a strict control over plasmon resonance generation. This procedure was used to fabricate nanocuboids matrix of different dimensions (1×1, 2×2, 3×3, and 4×4) and with various IPG (cuboid corner-to-corner from 20 to 500 nm). Morphological assessments were carried out by means of SEM and AFM, while SERS measurements were employed to verify the device reliability as a biosensor. Numerical simulations were also performed and finding consistency with the experimental SERS results. After depositing R6G, SERS measurements were performed for all the samples by varying the matrix dimensions (from 1×1 to 4×4) and by varying the IPG (in the range of 20–500 nm). SERS enhancement was estimated to be 1.4×10^5 with respect to a gold unpatterned substrate. Finally, we studied the photoconductivity of semiconductor and metal/semiconductor structures for sensor and solar cells applications. In particular, colloidal semiconductor nanocrystals and hybrid metal–semiconductor systems where carefully analyzed with scanning photocurrent technique.

References

1. Nordlander, P., Oubre, C., Prodan, E., Li, K., and Stockman, M. I. (2004). Plasmon Hybridization in Nanoparticle Dimers, *Nano Lett.*, **4**, pp. 899–903.
2. Funston, A. M., Novo, C., Davis, T. J., and Mulvaney, P. (2009). Plasmon Coupling of Gold Nanorods at Short Distances and in Different Geometries, *Nano Lett.*, **9**, pp. 1651–1658.
3. Biagioni, P., Huang, J. S., and Hecht, B. (2012). Nanoantennas for visible and infrared radiation, *Rep. Prog. Phys.*, **75**, p. 024402.
4. Nikolaenko, A. E., De Angelis, F., Boden, S. A., Papasimakis, N., Ashburn, P., Di Fabrizio, E., and Zheludev, N. I. (2010). Carbon Nanotubes in a Photonic Metamaterial, *Phys. Rev. Lett.*, **104**, p. 153902.
5. Lassiter, J. B., Sobhani, H., Knight, M. W., Mielczarek, W. S., Nordlander, P., and Halas, N. J. (2012). Designing and Deconstructing the Fano Lineshape in Plasmonic Nanoclusters, *Nano Lett.*, **12**, pp. 1058–1062.

6. Panaro, S., et al. (2013). Design and Top-Down Fabrication of Metallic L-Shape Gap Nanoantennas Supporting Plasmon-Polariton Modes, *Microelectron. Eng.*, **111**, pp. 91–95.
7. Hao, F., Sonnerfraud, Y., Van Dorpe, P., Maier, S. A., Halas, N. J., and Nordlander, P. (2008). Symmetry Breaking in Plasmonic Nanocavities: Subradiant LSPR Sensing and a Tunable Fano Resonance, *Nano Lett.*, **8**, pp. 3983–3988.
8. Gómez, D. E., Teo, Z. Q. , Altissimo, M., Davis, T. J., Earl, S., and Roberts, A. (2013). The Dark Side of Plasmonics, *Nano Lett.*, **13**, pp. 3722–3728.
9. Luk'yanchuk, B., Zheludev, N. I., Maier, S. A., Halas, N. J., Nordlander, P., Giessen, H., and Chong, C. T. (2010). The Fano Resonance in Plasmonic Nanostructures and Metamaterials, *Nat. Mater.*, **9**, pp. 707–715.
10. Zhang, S., Genov, D. A., Wang, Y., Liu, M., and Zhang, X. (2008). Plasmon-Induced Transparency in Metamaterials, *Phys. Rev. Lett.*, **101**, p. 047401.
11. Hatab, N. A., Hsueh, C.-H., Gaddis, A. L., Retterer, S. T., Li, J.-H., Eres, G., Zhang, Z., and Gu, B., (2010). Free-Standing Optical Gold Bowtie Nanoantenna with Variable Gap Size for Enhanced Raman Spectroscopy, *Nano Lett.*, **10**, pp. 4952–4955.
12. Oh, Y.-J., Jeong, K.-H. (2012). Glass Nanopillar Arrays with Nanogap-Rich Silver Nanoislands for Highly Intense Surface Enhanced Raman Scattering, *Adv. Mater.*, **24**, pp. 2234–2237.
13. Pecora, E. F., Lawrence, N., Gregg, P., Trevino, J., Artoni, P., Irrera, A., Priolo, F., and Negro, L. D. (2012). Nanopatterning of Silicon Nanowires for Enhancing Visible Photoluminescence, *Nanoscale*, **4**, pp. 2863–2866.
14. Seo, K., Wober, M., Steinvurzel, P., Schonbrun, E., Dan, Y., Ellenbogen, T., and Crozier, K. B. (2011). Multicolored Vertical Silicon Nanowires, *Nano Lett.*, **11**, pp. 1851–1856.
15. Jeong, S., Garnett, E. C., Wang, S., Yu, Z., Fan, S., Brongersma, M. L., McGehee, M. D., and Cui, Y. (2012). Hybrid Silicon Nanocone–Polymer Solar Cells, *Nano Lett.*, **12**, pp. 2971–2976.
16. Rodríguez-Lorenzo, L., Álvarez-Puebla, R. N. A., Pastoriza-Santos, I., Mazzucco, S., Stéphan, O., Kociak, M., Liz-Marzán, L. M., and García de Abajo, F. J. (2009). Zeptomol Detection Through Controlled Ultra-sensitive Surface-Enhanced Raman Scattering. *Journal of the American Chemical Society*, **131**, pp. 4616–4618.
17. Hrelescu, C., Sau, T. K., Rogach, A. L., Jäckel, F., Laurent, G., Douillard, L., and Charra, F. (2011). Selective Excitation of Individual Plasmonic Hotspots at the Tips of Single Gold Nanostars, *Nano Lett.*, **11**, pp. 402–407.

18. Rakic, A. D., Djuriic, A. B., Elazar, J. M., and Majewski, M. L. (1998). Optical Properties of Metallic Films for Vertical-Cavity Optoelectronic Devices, *Appl. Opt.*, **37**, pp. 5271–5283.

19. Alabastri, A., Tuccio, S., Giugni, A., Toma, A., Liberale, C., Das, G., De Angelis, F., Di Fabrizio, E., and Proietti Zaccaria, R. (2013). Molding of Plasmonic Resonances in Metallic Nanostructures: Dependence of the Non-Linear Electric Permittivity on System Size and Temperature, *Materials,* **6**, pp. 4879–4910.

20. Dmitriev, A., Hägglund, C., Chen, S., Fredriksson, H., Pakizeh, T., Käll, M., and Sutherland, D. S. (2008). Enhanced Nanoplasmonic Optical Sensors with Reduced Substrate Effect, *Nano Lett.*, **8**, pp. 3893–3898.

21. Kneipp, K. M. M., and Kneipp, H. (2007). *Surface-Enhanced Raman Scattering. Physics and Applications*. Springer: New York.

22. Reilly, T. H., Chang, S.-H., Corbman, J. D., Schatz, G. C., and Rowlen, K. L. (2006). Quantitative Evaluation of Plasmon Enhanced Raman Scattering from Nanoaperture Arrays, *J. Phys. Chem. C,* **111**, pp. 1689–1694.

23. Chirumamilla, M., Das, G., Toma, A., Gopalakrishnan, A., Zaccaria, R. P., Liberale, C., De Angelis, F., and Di Fabrizio, E. (2012). Optimization and Characterization of Au Cuboid Nanostructures as a SERS Device for Sensing Applications, *Microelectronic Engineering*, **97**, pp. 189–192.

24. Das, G., Patra, N., Gopalakrishanan, A., Proietti Zaccaria, R., Toma, A., Thorat, S., Di Fabrizio, E., Diaspro, A., and Salerno, M. (2012). Surface Enhanced Raman scattering Substrate Based on Gold-Coated Anodic Porous Alumina Template, *Microelectronic Engineering*, **97**, pp. 383–386.

25. Kudelski, A. (2005). Raman Studies of Rhodamine 6G and Crystal Violet Sub-monolayers on Electrochemically Roughened Silver Substrates: Do Dye Molecules Adsorb Preferentially on Highly SERS-Active Sites?, *Chem. Phys. Lett.*, **414**, pp. 271–275.

26. Das, G., Patra, N., Gopalakrishnan, A., Proietti Zaccaria, R., Toma, A., Thorat, S., Di Fabrizio, E., Diaspro, A., and Salerno, M. (2012). Fabrication of Large-Area Ordered and Reproducible Nanostructures for SERS Biosensor Application, *Analyst,* **137**, pp. 1785–1792.

27. Dulkeith, E., Morteani, A. C., Niedereichholz, T., Klar, T. A., Feldmann, J., Levi, S. A., van Veggel, F. C. J. M., Reinhoudt, D. N., Möller, M., and Gittins, D. I. (2002). Fluorescence Quenching of Dye Molecules Near Gold Nanoparticles: Radiative and Nonradiative Effects, *Phys. Rev. Lett.*, **89**, 203002.

28. Liu, Y. (2002). Evidence of Chemical Effect on Surface-Enhanced Raman Scattering of Polypyrrole Films Electrodeposited on Roughened Gold Substrates, *Langmuir*, **18**, pp. 174–181.
29. Das, G., Chirumamilla, M., Toma, A., Gopalakrishnan, A., Proietti Zaccaria, R., Alabastri, A., Leoncini, M., and Di Fabrizio, E. (2013). Plasmon Based Biosensor for Distinguishing Different Peptides Mutation States, *Scientific Report*, **3**, 1792.
30. McLellan, J. M., Li, Z.-Y., Siekkinen, A. R., and Xia, Y. (2007). The SERS Activity of a Supported Ag Nanocube Strongly Depends on Its Orientation Relative to Laser Polarization, *Nano Letters*, **7**, pp. 1013–1017.
31. Coluccio, M. L., Das, G., Mecarini, F., Gentile, F., Pujia, A., Bava, L., Tallerico, R., Candeloro, P., Liberale, C., De Angelis, F., and Di Fabrizio, E. (2009). Silver-Based Surface Enhanced Raman Scattering (SERS) Substrate Fabrication Using Nanolithography and Site Selective Electroless Deposition, *Microelect. Eng*, **86**, pp. 1085–1088.
32. Jensen, L., and Schatz, G. C. (2006). Resonance Raman Scattering of Rhodamine 6G as Calculated Using Time-Dependent Density Functional Theory, *J. Phys. Chem. A*, **110**, pp. 5973–5977.
33. Vosgrhne, T., and Meixner, A. J. (2005). Surface- and Resonance-enhanced Micro-Raman Spectroscopy of Xanthene Dyes: From the Ensemble to Single Molecules, *Chem. Phys. Chem.*, **6**, pp. 154–163.
34. Balasubramanian, K., Burghard, M., Kern, K., Scolari, M., and Mews, A. (2005). Photocurrent Imaging of Charge Transport Barriers in Carbon Nanotube Devices. *Nano Letters*, **5**, pp. 507–510.
35. Balasubramanian, K., et al. (2004). Photoelectronic Transport Imaging of Individual Semiconducting Carbon Nanotubes, *App. Phys. Lett.*, **84**, pp. 2400–2402.
36. Ahn, Y., Dunning, J., and Park, J. (2005). Scanning Photocurrent Imaging and Electronic Band Studies in Silicon Nanowire Field Effect Transistors, *Nano Lett.*, **5**, pp. 1367–1370.
37. Gu, Y., et al. (2006). Quantitative Measurement of the Electron and Hole Mobility—Lifetime Products in Semiconductor Nanowires, *Nano Lett.*, **6**, pp. 948–952.
38. Chakraborty, R., et al. (2013). Broad Spectral Photocurrent Enhancement in Au-Decorated CdSe Nanowires, *Nanoscale*, **5**, pp. 5334–5340.
39. Kumar, P., et al. (2011). Photocurrent Spectroscopy of Single CdS Nanosheets: Valence Band Structure and Two Photon Absorption, *App. Phys. Lett.*, **98**, 143102-3.

40. Zhang, Y., et al. (2011). Spatially Resolved Photoconductivity of Thin Films Formed by Colloidal Octapod-Shaped CdSe/CdS Nanocrystals, *Nanoscale*, **3**, pp. 2964–2970.
41. Kudera, S., et al. (2012). Spatial Analysis of the Photocurrent Generation and Transport in Semiconductor Nanocrystal Films, *Phys. Rev. B*, **86**, 075307.
42. Huynh, W. U., Dittmer, J. J., and Alivisatos, A. P. (2002). Hybrid Nanorod-Polymer Solar Cells, *Science*, **295**, pp. 2425–2427.
43. Gur, I., et al. (2005). Air-Stable All-Inorganic Nanocrystal Solar Cells Processed from Solution, *Science*, **310**, pp. 462–465.
44. Chanyawadee, S., et al. (2009). Photocurrent Enhancement in Hybrid Nanocrystal Quantum-Dot p-i-n Photovoltaic Devices, *Phys. Rev. Lett.*, **102**, 077402.
45. Talapin, D. V., et al. (2010). Prospects of Colloidal Nanocrystals for Electronic and Optoelectronic Applications, *Chemical Reviews*, **110**, pp. 389–458.
46. Leatherdale, C. A., et al. (2000). Photoconductivity in CdSe Quantum Dot Solids, *Phys. Rev. B*, **62**, pp. 2669–2680.
47. Drndic, M., et al. (2002). Transport Properties of Annealed CdSe Colloidal Nanocrystal Solids, *J. App. Phys.*, **92**, pp. 7498–7503.
48. Romero, H. E., and Drndic, M. (2005). Coulomb Blockade and Hopping Conduction in PbSe Quantum Dots, *Phys. Rev. Lett.*, **95**, 156801.
49. Porter, V. J., et al. (2006). Temperature-, Gate-, and Photoinduced Conductance of Close-Packed CdTe Nanocrystal Films, *Phys. Rev. B*, **73**, 155303.
50. Franchini, I. R., et al. (2010). Phototransport in Networks of Tetrapod-Shaped Colloidal Semiconductor Nanocrystals, *Nanoscale*, **2**, pp. 2171–2179.
51. Mongin, D., et al. (2012). Ultrafast Photoinduced Charge Separation in Metal-Semiconductor Nanohybrids, *ACS Nano*, **6**, pp. 7034–7043.
52. Atwater, H. A., and Polman, A. (2010). Plasmonics for Improved Photovoltaic Devices, *Nat. Mater.*, **9**, pp. 865–865.

Chapter 5

Surface-Enhanced Raman Scattering: Fundamentals and Biosensing Applications

Cosmin Farcau and Simion Astilean

Nanobiophotonics and Laser Microspectroscopy Centre,
Institute for Interdisciplinary Research in Bio-Nano-Sciences and Faculty of Physics,
Babes-Bolyai University,
1, M. Kogalniceanu Str. 400084, Cluj-Napoca, Romania
simion.astilean@phys.ubbcluj.ro

5.1 Introduction

Raman spectroscopy is a powerful optical spectroscopic method of chemical analysis with a wide field of applications, ranging from pharmaceutical and biomedical studies to art and archaeology [1–3]. These are based on the fingerprint-like information that it provides by analyzing molecular vibrations. Its capabilities are however outperformed in several applications where a very high sensitivity is required, for example, for detecting molecular trace amounts, mainly due to the very low efficiency of the Raman scattering process. Surface-enhanced Raman spectroscopy (SERS)

Handbook of Enhanced Spectroscopy
Edited by Pietro Giuseppe Gucciardi, Marc Lamy de la Chapelle, and Nathalie Lidgi-Guigui

ISBN 978-981-4613-32-3 (Hardcover), 978-981-4613-33-0 (eBook)
www.panstanford.com

provides the means for improving the sensitivity by amplifying the original Raman scattering intensity for several up to ten orders of magnitude. It was more than 30 years ago that an enhanced Raman signal was observed from molecules adsorbed at roughened silver electrode surfaces [4, 5]. Since then a huge amount of work was done in order to understand its multiple-faceted aspects, and develop SERS applications, this work being still in progress. Although a big part of the fundamentals of the technique had been established in the 1980s, important contributions to the advancements have been achieved during recent years, in which we have witnessed an intense blooming in the development of SERS, mainly in direct relationship with the rising field of plasmonics. In turn, this latter was boosted by the rapid evolution of nanotechnology in general, which provided nanofabrication and characterization methods of continuously increasing performance. Current challenges addressed in the last years, and ongoing, are both on controlling the fabrication of SERS substrates with nanoscale precision and on furthering the current understanding of the details of SERS effects.

In this chapter, we give a brief introduction to the fundamentals of the SERS effect, then we highlight some of the contributions to the recent advancements in the understanding of the SERS phenomenon. Where appropriate we also briefly describe the nanofabrication techniques that were employed for fabricating those SERS substrates. Guidelines for the morphology and optical properties required for fabricating an efficient solid SERS substrate are thus indicated. The inter-dependence of optical properties, plasmon modes and SERS efficiency implies that it is imperative to use well-designed solid SERS substrates in order to get an optimum signal for a particular excitation laser and target molecule.

The number of SERS publications in the last years is overwhelming, with more than 2000 article-type publications found on the Web of Science database, only from 2009 to present. Considerably broad is also the palette of noble metal nano-architectures demonstrated to be SERS-active substrates. That is why we don't aim at being exhaustive here, we will certainly highlight only partially the recent advances.

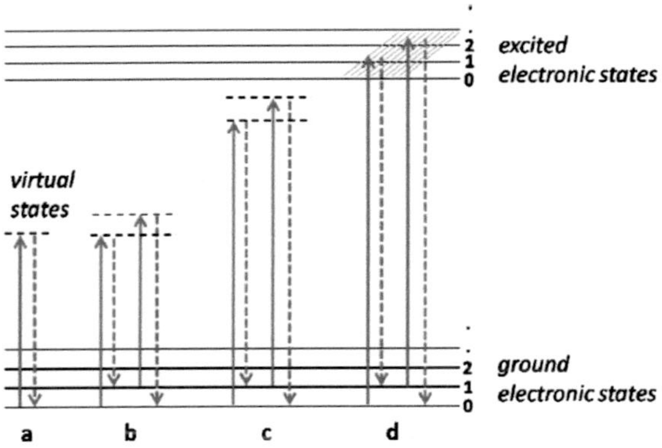

Figure 5.1 Various light-scattering processes: (a) Rayleigh; (b) non-resonance (normal) Raman; (c) pre-resonance Raman; (d) resonance Raman.

5.2 Fundamental Aspects of SERS

The Raman effect consists of an inelastic scattering process between a photon and a molecule, mediated by a fundamental vibrational or rotational mode of the molecule, as depicted in Fig. 5.1(b–d). The Raman-scattered photon can be of either lower energy (Stokes) or higher energy (anti-Stokes) than the incident laser photon. The efficiency of Raman scattering is much weaker (more than 10 orders of magnitude) than that of fluorescence transitions for example, with typical Raman scattering cross sections σ^{RS} having values of 10^{-31} cm^2/molecule $\leq \sigma_{RS} \leq 10^{-29}$ cm^2/molecule, depending on whether the scattering is non-resonant or resonant. Raman intensities roughly scale as the product of the incident electric field intensity and the molecule's polarizability derivative. This is an intuitive reason for acceptance of the two commonly considered mechanisms for increasing the Raman intensity (SERS): the first one involves enhancements of the electric field intensity, resulting from the excitation of surface plasmon resonances, while the other, the enhancement in polarizability, is attributed to chemical effects such as charge-transfer excited states.

The first of these is known as the *electromagnetic enhancement mechanism* (EEM), and in this mechanism the enhancement factor

EF at each molecule is generally approximated by

$$\mathrm{EF} = |\mathrm{EF}(\nu_{\mathrm{Laser}})|^2\, |\mathrm{EF}(\nu_{\mathrm{Stokes}})|^2$$

where $\mathrm{EF}(\nu_{\mathrm{Laser}})$ is the local electric-field enhancement factor at the incident frequency ν_{Laser} and $\mathrm{EF}(\nu_{\mathrm{Stokes}})$ is the corresponding factor at the Stokes-shifted frequency ν_{Stokes} of the Raman-scattered photons. The enhancement factor quantitatively represents the ratio between the enhanced electric fields at the metallic surface where the Raman scatterer molecule is located and the incident field is located away from the surface. Because the electric field intensity, amplified by surface plasmon excitations, is at maximum at the metallic surface and decrease exponentially away from the surface, the typical SERS experiment involves a Raman scatterer located at the metallic surface or very close to it; these metallic surfaces with electric field enhancement properties are referred to as *SERS substrates*. One can thus consider that the electromagnetic contribution to the total SERS enhancement is roughly proportional to the fourth power of the field enhancement factor, by assuming that the difference in frequency $\nu_{\mathrm{Laser}}-\nu_{\mathrm{Stokes}}$ is small enough to consider a similar enhancement at both the incoming and scattered photons frequency; this implies that the linewidth of the involved plasmon modes, generating the enhancement, is larger than the Raman Stokes shift, which is in most cases true. Note however that this fourth power law is only an approximation (also known as the E^4 approximation [6]), and that in some specific conditions this can be a pretty poor estimate.

In the chemical mechanism, which is now thought to contribute a maximum enhancement factor of 100, a charge-transfer state is created between the metal and adsorbate molecules. In a manner analogous to that observed in resonance Raman spectroscopy, the existence of this charge-transfer state increases the probability of a Raman transition by providing a pathway for resonant excitation. This mechanism is site-specific and analyte-dependent. The molecule must be directly adsorbed to the metallic nanostructured surface to experience the chemical enhancement.

In the last decade, many efforts have been concentrated in establishing connections between the morphology and optical properties of the metallic nanostructure on one side, and the SERS efficiency on the other side. In the following, we continue with a selection of literature results dealing with establishing such correlations between the SERS enhancements and the optical properties of those SERS substrate. The size and shape of metal nanostructures are crucial parameters determining their plasmon resonances, which in their turn can be observed and monitored by simple optical spectroscopic methods (such as UV–Vis extinction). In this section, by presenting a few experimental results of fundamental relevance, which clearly evidenced the role of SPs in SERS, we will indicate some guidelines for fabricating an efficient SERS substrate.

5.2.1 *Localized Surface Plasmons and SERS*

Light incident on a noble metal nanostructured surface can resonantly excite surface plasmons, that is, collective oscillations of the free conduction electrons. These excitations are localized in case of isolated nanoparticles (note that a single hole in a metal film behaves as a particle), or can propagate on extended surfaces. Among the important consequences of surface plasmons excitation we name: a significant increase of the absorption and scattering cross sections for electromagnetic waves, as well as a strong enhancement of the near-fields in the immediate vicinity of the metallic surface [7]. It is this resonantly enhanced near-fields that are exploited for the enhancement of Raman scattering in SERS.

The ongoing evolution of nanofabrication techniques allows preparing metallic nanoparticles of desired size and shape, arranged in a regular array on a solid support. Such substrates allowed studying the evolution of optical spectra, which are determined by the surface plasmon resonances excited, with the geometrical parameters of the particles. Also the dependence of SERS efficiency on the optical properties could be monitored in this way, either by using multiple laser lines relative to one given surface plasmon band or by designing substrates with different plasmon resonances relative to one fixed excitation line.

Haynes and Van Duyne reported on such a study in which they employed nanoparticle arrays obtained by nanosphere lithography (NSL) to perform a correlated fundamental study of optical properties and SERS enhancement efficacy [8]. NSL mainly consists of evaporating metal films through a mask made of close-packed arrays of self-organized polystyrene (or silica) spheres (typically hundreds of nanometer in diameter), followed my mask removal. The authors made use of the small morphological variations inherently found on NSL (more details on this technique will be given later) fabricated nanoparticles, variations which determine a different local optical response. Therefore, by comparing optical extinction and SERS spectra acquired from the same sample area, the conditions for optimized enhancement were determined. Engineering the SERS substrate in such a way that its surface plasmon resonance lies between the laser excitation and the Raman shifted photons is the guide line for maximizing the SERS EF of isolated nanoparticle substrates. This is because both the incident incoming photon and the Raman-scattered photon can interact with the plasmon resonances, providing a multiplicative enhancement effect.

Felidj and co-workers refined these results by employing regular gold nanoparticle arrays fabricated by electron-beam lithography [9]. This technique allows for a more precise control over the particle size, shape, and interpaticle distance, which are the parameters determining the position of the localized plasmon resonance. They designed elongated nanoparticles having two LSP resonances: a plasmon band located at 647 nm corresponding to excitation of LSP along the direction perpendicular to the particle long axis (transversal) and a band at 723 nm due to LSP excited along the long axis (longitudinal). Vibrational bands of trans-1,2-bi-(4-pyridyl) ethylene (BPE), located around 1600 cm^{-1}, were selected to monitor the SERS intensities for three different excitation wavelengths: 633 nm, 647 nm, and 676 nm, for both transversal and longitudinal polarizations. On a wavelength scale, the positions of the Raman bands are at about 705 nm, 723 nm, and 760 nm respectively for the three excitation lines. The maximum enhancement was obtained for 676 nm laser excitation for the longitudinal LSP, which demonstrates the optimization of SERS efficiency by positioning the LSP resonance midway between the energy of the exciting photons and that of the Raman-scattered photons.

Work in Van Duyne's group showed the same results by adopting a different approach: several laser lines were used for excitation of the SERS spectra on a given nanoparticles array sample

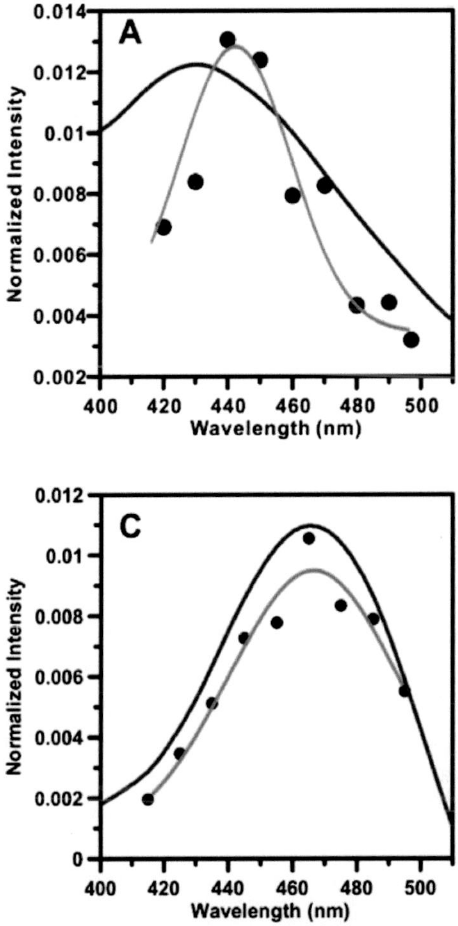

Figure 5.2 Correlation between localized surface plasmon resonance profile (blue line) and surface-enhanced resonance Raman enhancement profile (dots and red line fit) on Ag nanoparticles made by NSL. Reprinted with permission from Ref. [10], Copyright 2005 American Chemical Society.

with a well-defined plasmon response [10]. Figure 5.2 presents some examples of good correlation between the localized surface plasmon resonance profile (blue dark line) and surface-enhanced resonance Raman enhancement profile (dots and red line fit) on Ag nanoparticle arrays made by NSL.

However, it was demonstrated, by using regular arrays of nanoparticles of different shapes (fabricated by EBL), that the precise optimal spectral position of the LSP relative to that of incident and Raman-scattered photons is dependent on the particle geometry [11]. It was found that for particles having a large aspect ratio (length/width) such as ellipsoids, the best enhancement is observed for LSP close to the frequency of Raman-scattered photons. This was attributed to the increased scattering efficiency of elongated particles, which play the role of a nanoantenna for the Raman radiation. This is linked to the fact that the enhancement process has two contributions, both at the incident and Raman frequency.

A similar result was more recently demonstrated for colloidal gold nanorods in suspensions [12]. When SERS measurements are performed in suspensions, a competition takes place between the SERS enhancement and the extinction through the sample, that is, SERS signals can be absorbed and/or scattered by the colloids. Therefore, even if a strong enhancement takes place at the nanorod surface, the SERS signal can be strongly attenuated and doesn't reach the detector. The study employed a set of gold nanorods of different aspect ratios, with longitudinal surface plasmon resonances at wavelengths in the range 600–800 nm, and a 785 nm excitation wavelength. The results showed that the maximum SERS signal was obtained from that plasmon resonance that is blue-shifted from excitation.

Beside frequently encountered dipolar resonances, multipolar plasmon resonances can arise when the size of nanoparticles increases and the electrostatic approximation doesn't hold anymore. It was demonstrated experimentally that such multipolar resonances can produce SERS enhancements comparable to those determined by more usually met dipolar resonances [13]. This conclusion was reached by measuring the optical extinction and SERS spectra from arrays of long nanorods prepared by EBL, which exhibit high odd order (third, fifth) plasmon multipoles.

The strong increase in the local field EF that is obtained by bringing two particles very close to each other deserves some attention. When two particles are separated by distances typically lower than the size of the particles, strong coupling of

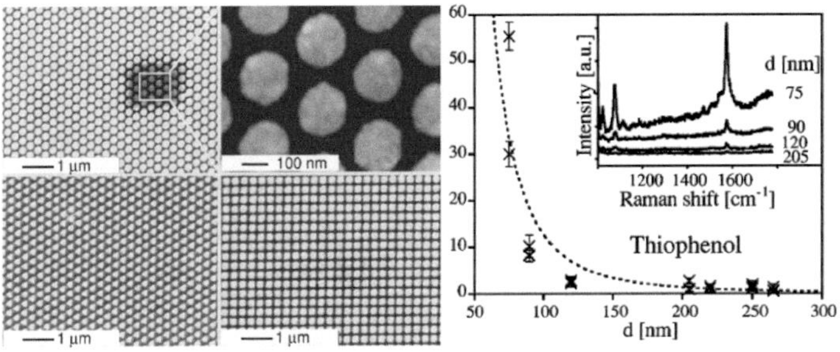

Figure 5.3 (left) SEM images of nanoparticle arrays fabricated by EBL. (right) Dependence of the SERS enhancement on the spacing between particles, with spectra shown in the inset. Reprinted with permission from Ref. [15], Copyright 2001, AIP Publishing LLC.

the dipolar resonances can occur, which produce an increase of the electromagnetic field in the gap between the particles (for light polarized along the interparticle axis) [14]. To test this and demonstrate the enormous benefits of SERS EF from this strong coupling regime, Gunnarsson et al. devised and performed an elegant experiment: they have fabricated regular arrays of Ag nanoparticles with increasingly smaller interparticle distances by EBL, while keeping the particle size constant [15]. A strong increase in the SERS EF was observed with the decrease of the distance between the particles (see Fig. 5.3), evidencing the great importance of electromagnetic coupling in small gaps for SERS.

A similar study showed an increase of the SERS EF by a factor of 200 when decreasing the interparticle distances from 35 to 10 nm [16]. However, in their approach, by growing Ag nanowires in highly regular porous aluminum oxide, the particle size was simultaneously increased, which could obscure the pure interparticle coupling effects.

Note that the nanoparticles size has itself an effect on the SERS efficiency. While both reports [15, 16] emphasized the role of small gaps between particles, it is a pity that they lack the correlation with the optical properties, that is, it was not shown which array is in better resonance condition with respect to the SP modes excited in the array.

After the examples discussed above, an approach to increase the efficiency of SERS substrates would be to produce substrates with interparticle distances as small as possible, approaching 1–2 nm. A question can arise: how does the size of the particles influence the SERS from such strongly interacting, close-packed particles? Lithographic techniques such as EBL, very useful in providing fundamental insights as those discussed above, fail in controlling such small distance. Instead, colloidal nanoparticles synthesized by wet chemistry can be subsequently assembled into close-packed arrays. Wei et al. prepared gold colloids of different size, which were self-organized into two-dimensional lattices on a solid quartz substrate [17]. They have observed a size-dependent SERS efficiency, the trend being that larger nanoparticles are more favorable. We have also attempted to look at this problem in our laboratory, by using gold colloids of different sizes obtained by standard citrate reduction [18]. Our approach was rather different, in the sense that colloidal solutions were allowed to age and aggregate, and these nanoparticle aggregates were then captured on functionalized glass substrates. The SERS efficiency was monitored for adsorbed aminothiphenol, and higher SERS intensities were generated by the nanoparticles of larger size. Interestingly, both in our study and that in Ref. [17], the SERS EF could not be well correlated to the optical response of the substrates. This is an indication that in strongly interacting nanoparticles systems the far-field monitored optical properties do not rigorously mirror the enhanced electromagnetic near-fields. Thus, a rational design and optimization of substrates with a complex, non-regular distribution of touching particles is not straightforward.

As if things were not already complicated, there are particular observations which show that the rules of thumb for obtaining an SERS substrate of optimized efficiency are not always strict rules. As for example work in Mirkin's group [19] demonstrated that the smallest gap between metal nanoparticles is not always optimal for SERS enhancement. However, they study a system of very particular geometry, consisting of nanodisks arranged parallel to each other, fabricated by an original method named On Wire Lithography. This arrangement is very different from nanoparticles arrays obtained by other techniques, in which, being cylinders, ellipsoid, rods, or disks, they are always coplanar. We would add that their result is valid for the single wavelength employed in their study, and it is well known that by decreasing the interparticle distance the coupled SP resonance redshifts. Thus, in such nanoparticle ensembles of known and controlled morphology, one should take into account the plasmonic response, and not only the distance between particles, in order to explain the SERS behavior.

5.2.2 *Propagative Surface Plasmons (PSP) and SERS*

Contrary to isolated metal nanoparticles, on which localized surface plasmons can be excited, propagative surface plasmon polaritons can be excited by incident light on extended metallic surfaces. In their original approach, Baumberg et al. fabricated ordered arrays of spherical nanovoids of controlled depth, by electrochemical deposition through a template of self-assembled polymer spheres. They showed that PSPs also play a role in the SERS process, and they observed a resonant enhancement by monitoring the angle dependence of SERS signals on periodically ordered gold substrates [20]. Results in Fig. 5.4 show a good correlation between the extinction spectra at different incidence angles and SERS spectra collected at different angles.

The angle dependence is due to the fact that SPP modes can be excited by light of a given wavelength and for a given incident angle, which are determined by wavevector matching conditions [21]. Also

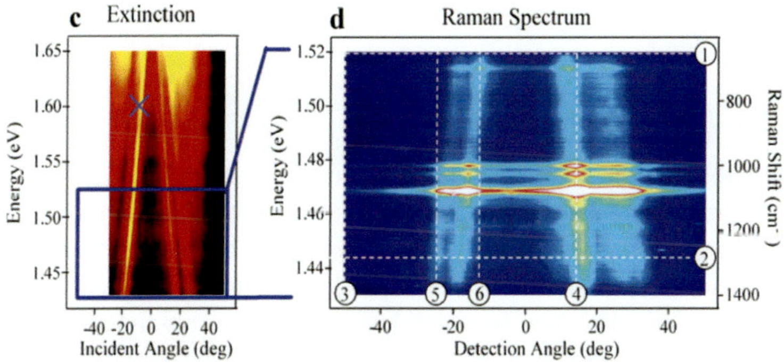

Figure 5.4 (left) Extinction spectra at different incident angles on ordered arrays of spherical nanovoids; excitation laser is marked by X. (right) SERS spectra recorded at different angles. Reprinted with permission from Ref. [20], Copyright 2005, American Chemical Society.

in the case of extended metallic surfaces, the engineering of PSP modes in resonance with the excitation and the Raman-scattered photons is required to obtain an efficient substrate.

A very interesting paper on this topic concerns the observation of directional SERS on periodically ordered silver nanowell array having well-defined propagative surface plasmon modes [22]. They employed a home-made setup in which they could rotate the sample in the plane parallel to the laser excitation, and modify the angle of detection relative to the sample surface normal axis, while keeping fixed the direction of excitation relative to the sample surface. The results show that the SERS emission is restricted to a certain direction in space, with a half-divergence angle of about 10°.

Another report deals with gold/silver coated parallel (biharmonic) gratings, on which by adjusting the grating strength its surface plasmon resonance was modified. By analyzing SERS of rhodamine 6G adsorbed on these different gratings, it was observed that the higher SERS enhancement occurs when the plasmon resonance wavelength is better overlapping the Raman Stokes signal. It is suggested therefore that a plasmon-coupled enhancement mechanism is the dominant factor for the SERS electromagnetic enhancement on this kind of substrates. [23]. In fact this behavior is very similar to what was observed in metal-enhanced fluorescence, where a plasmon-coupled emission is very often responsible for the intensity enhancement [24].

5.2.3 *Polarization Effects in SERS*

Note that in the case of ordered arrays the polarization of the incident laser field has to be taken into consideration. As mentioned, EM fields are amplified in the gap between the particles when the incident electric field is polarized parallel to the interparticle axis. To illustrate this, the simplest case of a one-dimensional nanoparticle array is very useful. Linear arrays of Ag nanoparticles were obtained by physical vapor deposition on highly oriented pyrolytic graphite substrate [25]. The authors report a 20-fold stronger SERS signal for incident light polarized parallel to the 1D array than for perpendicular polarization.

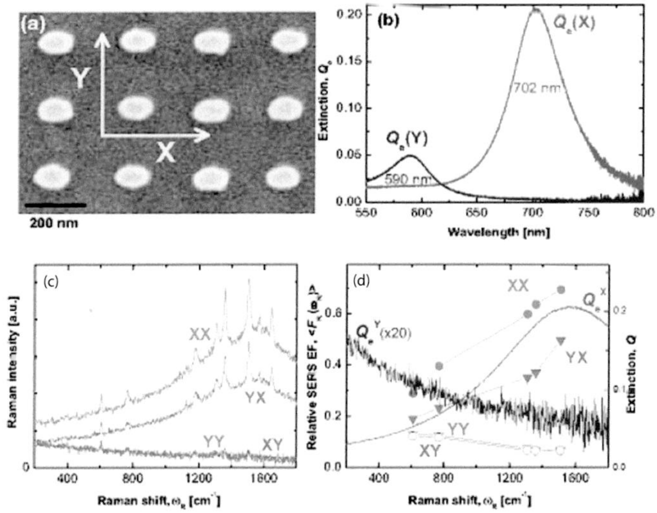

Figure 5.5 (a) Electron microscopy image of the array of gold NPs prepared by EBL. (b) Extinction spectra for *x* and *y* polarizations. (c) SERS spectra for the four excitation-detection configurations indicated. (d) Correlation between SERS EFs and extinction profiles. Reprinted with permission from Ref. [26], Copyright 2008, American Chemical Society.

A very interesting result with implications in the current understanding of SERS has been obtained by Le Ru et al. by employing polarized SERS analyses on arrays of elongated particles prepared by electron beam lithography [26]. The prepared arrays of elongated particles present two distinct surface plasmon resonances, which can be excited when light is incident parallel to the long axis (longitudinal) or the short axis (transversal) of the prolate nanoparticles (Fig. 5.5 (a, b)). Briefly, their experiments proved that the strongest plasmon resonance (in this case the longitudinal one) controls the polarization of the SERS signal: the SERS emission is polarized longitudinally not only when the incident laser is polarized longitudinally, but also when the excitation is transversal (Figs. 5.5 (c, d)). These results provide a clear experimental demonstration that the simple E^4 approximation is indeed not valid in all circumstances. Obviously, some similar polarization effects would also be expected on the metallic gratings in Ref. [22] discussed in Section 5.2.2.

5.2.4 *Single-Molecule SERS*

A very hot topic within the SERS field is that of single-molecule SERS, which has been observed in many instances since the first reports more than 10 years ago [27]. The possibility to observe and analyze single molecules is of high interest both for the fundamental understanding of SERS and for accessing the properties of single molecules. Most of the single-molecule SERS reports dealt with aggregates of nanoparticles, in which hot-spots created by short-range electromagnetic interactions have led to enormous Raman enhancement factors. These so-called hot-spots are usually highly localized spatial regions in which the electromagnetic fields are most intense, and they appear typically at gap junctions between metallic nanofeatures. The optical properties of aggregated colloidal structures are difficult to control, and even to interpret, being dominated by very broad resonances in the measured spectra, with SP modes galore being excited. Thus, attempts have been made to understand which is the local nanoscale morphology required for the irregular nanostructures to produce giant SERS enhancements.

Wang et al. [28] embarked on such a study and for this they elaborated a large variety of nanostructured silver SERS substrates: compacted nanoparticles films (high-density silver nanoparticles films covalently attached to silanized glass slides), films with cluster aggregates (colloidal aggregates attached to the glass slides), mirror reaction films, silver nanorod films, and evaporated Ag films. The topology of the Ag films was mapped by atomic force microscopy (AFM) and scanning electron microscopy (SEM). The SERS investigations revealed that only the films possessing a fractal character exhibited Raman hot-spots.

Another class of systems that has been predicted and demonstrated in the literature to produce super-enhancing locations is pairs of nanoparticles, in which the hot-spot position is located in the small gap between the two particles. Regular arrays of nanoparticle dimers were prepared by using a variant of NSL, in which metal evaporation was conducted in two sequences, each of them with a different angle of the evaporating beam relative to the substrate's normal [29]. In this way, two arrays of nanoparticles are formed, corresponding to the two depositions, and the gap between closest-neighbor particles can be controlled by the deposition angles. Although the small nanometric gap between the particles could not be resolved by AFM, an optimum was found for which the SERS signature from a low number of molecules could be recorded.

Although many claims of single-molecule observation exist, an important issue along the way was to prove that indeed single-molecule SERS is measured in a particular experiment. The efforts of Le Ru and Etchegoin have brought considerable contributions in this sense, especially by pointing out the limitations and drawbacks of currently employed low-concentration approach, and by the development of the bi-analyte technique [30, 31]. Very often single-molecule observation is claimed on the basis of "blinking" of the SERS signal and fluctuations observed in the spectral position of Raman bands (spectral wandering). However, other effects, such as very tiny modifications of the local metal surface morphology at the hot-spot, could be responsible for a blinking behavior. For a more complete overview of the topic of single-molecule SERS, Ref. [32] is recommended.

5.2.5 *The SERS Background*

The SERS signal of interest (peaks in a spectrum) is always accompanied by an underlying continuum, called the background. This SERS background is not yet well understood and probably for this reason is in many cases simply neglected. In many cases, it is even subtracted, researchers not presenting it in their papers. However, as this background is itself a part of SERS, its understanding would only benefit the development of the technique. Some recent efforts in this direction have been done by the Baumberg group [33, 34]. They employed a snow-jet technique in order to induce perturbations at the molecular layer on the Klarite SERS nanostructured surface, systematically monitored both SERS peaks and background as a function of time and laser exposure time. Their results indicate a different dynamics of the peak and background after the perturbation process, suggesting a different origin/mechanism of the two SERS components.

A contribution to this topic has also been presented by our group, by the approach of SERS confocal imaging on a metal film over nanosphere system [35]. Maps of the SERS intensity across the nanostructured surface were constructed both from Raman bands and from background signal. Our results proved that there was a clear correlation between the maps of the SERS background and that of the SERS signal of probe molecules. Both maps showed the same spatial locations of the most enhanced signals. Furthermore, it was demonstrated that the measured intensity and the spectral profile of the SERS background are strongly dependent on the matching between the laser excitation wavelength and the substrate plasmon resonance band.

5.3 Bio-Applications of SERS

As mentioned in the introduction, normal Raman spectroscopy has a wide range of applications, from pharmaceutical and biomedical studies to art and archaeology. Likewise, the applications of SERS span these fields, with numerous demonstrations reported in the literature. We will discuss here some examples related to biosensing SERS applications. SERS is a rather promising candidate for developing biosensing platforms due to the high sensitivity achievable through LSP and PSP design, as discussed above, and due to its specificity, that is, Raman spectra provide a fingerprint-like information. This makes possible a rapid, label-free detection of trace amount of small target analytes, such as biomarkers, drugs, to spores, or viruses.

An approach toward SERS glucose sensing was developed in Van Duyne's group, by employing silver film over nanosphere substrates, and a surface functionalization with decanethiol (DT) and 6-mercapto-1-hexanol (MH). It is assumed that the mixed layer creates a kind of pocket effects, which allowed glucose to be trapped near the metallic surface and detect it by SERS. With this system implanted in animal rat models, they have proved the feasibility of in vivo transcutaneous glucose sensing [36]. Other approaches toward glucose detection include SERS substrates functionalized with a Raman probe molecule, 3,3′-boronic benzyl viologen (BBV), designed to bind specifically to glucose. Then, detection of glucose is implied via monitoring the SERS signals of BBV [37].

The potential of SERS has also been demonstrated in the context of early detection of cancer, as for the detection of circulating tumor cells (CTCs). Gold nanoparticles about 60 nm in diameter were functionalized with a Raman reporter, a polyethylene glycol protection layer and bio-conjugated with epidermal growth factor peptide, in order to target CTCs in peripheral blood. By mixing both conjugated and non-conjugated nanoparticles with blood samples and monitoring the SERS signal of the pre-attached Raman reporter, a strong targeting effect was demonstrated [38].

In the same context of cancer studies, the so-called SERS were proved to be useful nanotools for in vitro/in vivo characterization of single cells or as diagnostic agents for the detection and imaging of cancer cells. For this purpose, noble metal nanoparticles are tagged with a Raman active molecule, coated with a biocompatible polymer or protein, and functionalized with a cell recognizing agent such as antibodies or peptides. Very often anisotropic, non-spherical nanoparticles are preferred due to their better field enhancement properties. Some efforts in this direction have also been made in our group. Flower-shaped gold nanoparticles were synthesized by using L-ascorbic acid, commonly known as vitamin C, as a mild reducing agent of metallic species, conjugated with a Raman active dye and then biocompatibilized with α-methoxy-ω-mercapto poly(ethylene glycol) (mPEG-SH) [39]. These SERS nanotags were tested on epithelial cells from human retina and exhibited a low level of toxicity, at the concentrations required for SERS measurements. The observation of their encoded SERS signal in cells (see Fig. 5.6), under 785 nm excitation, makes this kind of tags potential candidates for spectral imaging inside living cells.

In another study, we developed a novel class of SERS nanotags based on chitosan-coated silver nanotriangles (Chit-AgNTs) and para-aminothiophenol (p-ATP) Raman active molecules and demonstrated their utility as biocompatible, highly sensitive agents for noninvasive detection and imaging of lung cancer cells (A549) under a wide window of excitation wavelengths, ranging from visible to near infrared (NIR) [40].

Another study performed by our group suggests that the SERS method can be used as an alternative approach for a noninvasive and early diagnosis of renal diseases [41]. In this paper, an

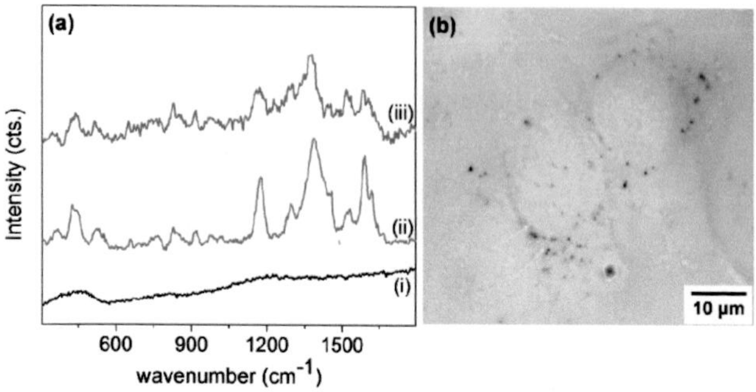

Figure 5.6 (A) SERS spectra of: (i) cells free of particles; (ii) inside a cell, at region marked by arrows; (iii) the tagged gold NPs in aqueous solution. (B) Optical image of cells loaded with Raman tagged gold nanoflowers.

ordered plasmonic substrate was simply produced by replicating the patterns from a recordable DVD and gold film coating. By SERS, we managed to not only identify Raman bands of urea molecules in different biological fluids such as urine, but also fingerprint sebum and tears, without using any processing of these biological fluids. These results prove that SERS could be employed to detect/analyze the trace of relevant biomolecules even in real biological fluids. Similar steps toward this approach were made for the detection of uric acid, which is an important diagnostic biomarker in urine and serum samples for a variety of health issues. Goodall et al. proved the potential for early diagnosis of preeclampsia (a sort of hypertensive disorder occurring during pregnancy) based on electrochemical SERS detection of uric acid in simulated urine [42].

Very often SERS is/should be combined with some other experimental techniques of separation, such as thin-layer chromatography (TLC). Such a report combining TLC and SERS presents results that are promising for the detection of some nicotine metabolites (cotinine and trans-3′-hydroxycotinine) in urine samples [43]. Easier to use dialysis or gel-filtration methods were also employed in conjunction with SERS for the detection of melamine in milk and infant formula [44]. An example of SERS spectra of melamine in milk is presented in Fig. 5.7.

A currently emerging field is that of developing SERS-based sensing schemes integrated into microfluidic platforms. Such

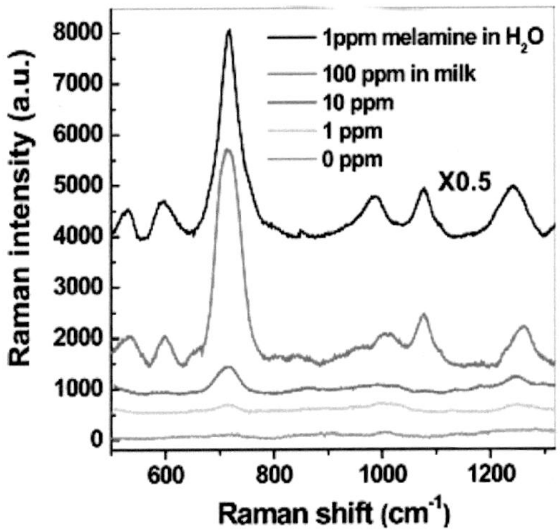

Figure 5.7 SERS spectra of melamine in milk after treatment by dialysis filtration, recorded on gold nanostructured platform. Figure reprinted with permission from Ref. [44], Copyright 2012, American Chemical Society.

Lab-on-a-Chip systems allow for online, real-time detection within a small and definite area, by handling of small sample volumes and a high throughput. By this kind of approach, the detection of a drug of abuse, methamphetamine, was demonstrated [45]. Colloidal silver nanoparticles, saliva samples, and a salt to induce aggregation were separately introduced in the microfluidic chamber, where they are mixed and allowed for SERS analysis. Another report concerns patient undergoing treatments with the immunosuppressive drug 6-mercaptopurine, which is used in the treatment of leukemia, inflammatory bowel disease, or after organ transplant. The authors demonstrate the SERS detection of the enzyme activity of thiopurine S-methyltransferase (TPMT) in lysed red blood cells, suggesting that this approach appears promising for developing a point-of-care application for a rapid test of the TPMT activity [46]. From the same Popp research group comes another report that demonstrates the potential of SERS in microfluidic devices for a highly specific and reliable detection of *E. coli* bacteria [47]. By employing a support vector machine kind of data analysis, the authors showed it is possible to discriminate between nine different strains of *E. coli*. Other examples concerning bacteria detection by SERS include the detection of *Bacillus anthracis* spores in orange juice, detection of *Cryptosporidium parvum* and *Giardia lamblia* [48], and even the identification of bacteria within bacterial mixtures [49]. For further reading on the topic of SERS for biosensing, the recent review by Bantz et al. is recommended [50].

5.4 Conclusions and Outlook

This chapter gave a brief introduction to the SERS effect. Several literature reports that made a considerable contribution, at least in authors' view, to the development of the field were highlighted. Specific examples discussing the impact of localized and/or propagative plasmon resonances on the SERS efficiency were briefly discussed. These point out some guidelines to follow for a rational design of an efficient SERS substrate/probe. To summarize some of these: surface plasmon resonances should be considered, and their separate contribution at both excitation and Raman scattering wavelength; in case of substrates with an ordered topography, the polarization of the incident laser must also be carefully aligned; however, very recent studies show that some "near-field type" plasmon modes can exist in parallel arrays of nanorods, which exhibit a very weak optical interaction [51], implying that the tuning of far-field properties does not necessarily affect the near-field in the same way. Then, when working with colloidal suspensions, extinction/attenuation effects should also be considered. Some reports on polarization effects, the background "problem" and the field of single-molecule SERS were also indicated.

A selection of SERS-based bio-applications was also presented, with examples as detection of bacteria, drugs, and glucose among others. The field of SERS tagging for imaging inside living cells was also briefly introduced. There are many promising results available nowadays, but we believe there are still some challenges for SERS, and there is a lot of room for improvements, at different levels: (i) on the synthesis/nanofabrication side, for a truly controlled fabrication of SERS substrates with a well-known morphology, down to the few nanometer scale, and a highly predictable optical/plasmon response; (ii) on the (bio)chemistry side, for controlling the interactions between SERS platform and analyte, for an efficient, specific trapping at the metal surface; (iii) on the collection and interpretation of SERS data, for example, the creation of reliable databases for identification of bacteria.

Acknowledgment

This work was supported by a grant of the Romanian National Authority for Scientific Research, CNCS—UEFISCDI, project number PN-II-RU-TE-2011-3-0134.

References

1. G. Turrell, and J. Corset (Ed.). *Raman Microscopy*. Elsevier 1996.
2. E. E. Lawson, B. W. Barry, A. C. Williams, and H. G. M. Edwards. Biomedical applications of Raman spectroscopy. *J. Raman Spectroscop.* 1997, 28, 111.
3. P. Vandenabeele. Raman spectroscopy in art and archaeology. *J. Raman Spectroscop.* 2004, 35, 607.
4. D. L. Jeanmaire, and R. P. Van Duyne. Surface Raman spectroelectrochemistry: Part I. Heterocyclic, aromatic, and aliphatic amines on the anodized silver electrode. *J. Electroanal. Chem.* 1977, 84, 1.
5. M. G. Albrecht, and J. A. Creighton. *Anomalously intense Raman spectra of pyridine at a silver electrode. J. Am. Chem. Soc.* 1977, 99, 5215.
6. E. C. Le Ru, and P. G. Etchegoin. Rigorous justification of the $|E|^4$ enhancement factor in surface enhanced Raman spectroscopy. *Chem. Phys. Lett.* 2006, 413, 63.
7. S. A. Maier. *Plasmonics: Fundamentals and Applications*. Springer, 2007.
8. C. L. Haynes, and R. P. Van Duyne. Plasmon-sampled surface-enhanced Raman excitation spectroscopy. *J. Phys. Chem. B.* 2003, 107, 7426.
9. N. Felidj, J. Aubard, G. Levi, J. R. Krenn, A. Hohenau, G. Schider, A. Leitner, and F. R. Aussenegg. Optimized surface-enhanced Raman scattering on gold nanoparticle arrays. *Appl. Phys. Lett.* 2003, 82, 3095.
10. A. D. McFarland, M. A. Young, J. A. Dieringer, and R. P. Van Duyne. Wavelength-scanned surface-enhanced Raman excitation spectroscopy. *J. Phys. Chem. B.* 2005, 109, 11279.

11. J. Grand, M. Lamy de la Chapelle, J.-L. Bijeon, P.-M. Adam, A. Vial, and P. Royer. Role of localized surface plasmons in surface-enhanced Raman scattering of shape-controlled metallic particles in regular arrays. *Phys. Rev. B* 2005, 72, 033407.
12. S. T. Sivapalan, B. M. DeVetter, T. K. Yang, T. van Dijk, M. V. Schulmerich, P. S. Carney, R. Bhargava, and C. J. Murphy. Off-resonance surface-enhanced Raman spectroscopy from gold nanorod suspensions as a function. *ACS Nano* 2013, 7, 2099.
13. G. Laurent, N. Félidj, J. Aubard, G. Lévi, J. R. Krenn, A. Hohenau, G. Schider, A. Leitner, and F. R. Aussenegg. Evidence of multipolar excitations in surface enhanced Raman scattering. *Phys. Rev. B* 2005, 71, 045430.
14. L. A. Sweatlock, S. A. Maier, H. A. Atwater, J. J. Penninkhoff, and A. Polman. Highly confined electromagnetic fields in arrays of strongly coupled Ag nanoparticles. *Phys. Rev. B* 2005, 71, 1.
15. L. Gunnarsson, E. J. Bjerneld, H. Xu, S. Petronis, B. Kasemo, and M. Kall. Interparticle coupling effects in nanofabricated substrates for surface-enhanced Raman scattering. *Appl. Phys. Lett.* 2001, 78, 802.
16. S. J. Lee, Z. Guan, H. Xu, and M. Moskovits. Surface-enhanced Raman spectroscopy and nanogeometry: The plasmonic origin of SERS. *J. Phys. Chem. C* 2007, 111, 17985.
17. A. Wei, B. Kim, B. Sadtler, and S. L. Tripp. Tunable surface-enhanced Raman scattering from large gold nanoparticle arrays. *ChemPhysChem* 2001, 12, 743.
18. S. C. Boca, C. Farcau, and S. Astilean. The study of Raman enhancement efficiency as function of nanoparticle size and shape. *Nucl. Inst. Meth. B* 2009, 267, 406.
19. L. Qin, S. Zou, C. Xue, A. Atkinson, G. C. Schatz, and C. A. Mirkin. Designing, fabricating, and imaging Raman hot spots. *Proc. Natl. Acad. Sci.* 2006, 103, 13300.
20. J. J. Baumberg, T. A. Kelf, Y. Sugawara, S. Cintra, M. E. Abdelsalam, P. N. Bartlett, and A. E. Russell. Angle-resolved surface-enhanced Raman scattering on metallic nanostructured plasmonic crystals. *Nano Lett.* 2005, 5, 2262.
21. W. Knoll. Interfaces and thin films as seen by bound electromagnetic waves. *Ann. Rev. Phys. Chem.* 1998, 49, 569.
22. H. Li, Y. Gu, H. Guo, X. Wang, Y. Liu, W. Xu, and S. Xu. Tunable plasmons in shallow silver nanowell arrays for directional surface-enhanced Raman scattering. *J. Phys. Chem. C* 2012, 116, 23608.

23. A. Kocabas, G. Ertas, S. S. Senlik, and A. Aydınli. Plasmonic band gap structures for surface-enhanced Raman scattering. *Opt. Express* 2008, 16, 12469.

24. Y. Zhang, K. Aslan, M. J. R. Previte, and C. D. Geddes. Metal-enhanced fluorescence: Surface plasmons can radiate a fluorophore's structured emission. *Appl. Phys. Lett.* 2007, 90, 053107.

25. W. Luo, W. van der Veer, P. Chu, D. L. Mills, R. M. Penner, and J. C. Hemminger. Polarization-dependent surface enhanced Raman scattering from silver 1D nanoparticle arrays. *J. Phys. Chem. C* 2008, 112, 11609.

26. E. C. Le Ru, J. Grand, N. Félidj, J. Aubard, G. Lévi, A. Hohenau, J. R. Krenn, E. Blackie, and P. G. Etchegoin. Experimental verification of the SERS electromagnetic model beyond the $|E|^4$ approximation: Polarization effects. *J. Phys. Chem. C* 2008, 112, 8117.

27. S. Nie, and S. R. Emory. Probing single molecules and single particles by surface-enhanced Raman scattering. *Science* 1998, 275, 1102.

28. Z. Wang, S. Pan, T. D. Krauss, H. Du, and L. J. Rothberg. The structural basis for giant enhancement enabling single-molecule Raman scattering. *Proc. Natl. Acad. Sci.* 2003, 100, 8638.

29. Y. Sawai, B. Takimoto, H. Nabika, K. Ajito, and K. Murakoshi. Observation of a small number of molecules at a metal nanogap arrayed on a solid surface using surface-enhanced Raman scattering. *J. Am. Chem. Soc.* 2007, 129, 1658.

30. E. C. Le Ru, M. Meyer, P. G. Etchegoin. Proof of single-molecule sensitivity in surface-enhanced Raman scattering (SERS) by means of a two-analyte technique. *J. Phys. Chem. B* 2006, 110, 1944.

31. P. G. Etchegoin, M. Meyer, E. Blackie, and E. C. Le Ru. Statistics of single-molecule surface-enhanced Raman scattering signals: Fluctuation analysis with multiple analyte techniques. *Anal. Chem.* 2007, 79, 8411.

32. E. C. Le Ru, and P. G. Etchegoin. Single-molecule surface-enhanced Raman spectroscopy. *Annu. Rev. Phys. Chem.* 2012, 63, 65.

33. J. T. Hugall, J. J. Baumberg, and S. Mahajan. Disentangling the peak and background signals in surface-enhanced Raman scattering. *J. Phys. Chem. C* 2012, 116, 6184.

34. S. Mahajan, R. M. Cole, J. D. Speed, S. H. Pelfrey, A. E. Russell, P. N. Bartlett, S. M. Barnett, and J. J. Baumberg. Understanding the surface-enhanced Raman spectroscopy "Background." *J. Phys. Chem. C* 2010, 114, 7242.

35. C. Farcau, and S. Astilean. Evidence of a surface plasmon-mediated mechanism in the generation of the SERS background. *Chem. Commun.* 2011, 47, 3861.
36. J. M. Yuen, N. C. Shah, J. T. Walsh, Jr., M. R. Glucksberg, and R. P. Van Duyne. Transcutaneous glucose sensing by surface-enhanced spatially offset Raman spectroscopy in a rat model. *Anal. Chem.* 2010, 82, 8382.
37. A. Zhang, D. Wheeler, Y. Ling, G. Wang, X. Yang, B. Vilozny, B. Singaram, C. Gu, and Y. Li. Novel molecular specific detection of glucose using a Raman probe molecule with surface enhanced Raman scattering. *Sci. Adv. Mater.* 2012, 4, 1047.
38. X. Wang, X. Qian, J. J. Beitler, G. (Z.) Chen, F. R. Khuri, M. M. Lewis, H. J. C. Shin, S. Nie, and D. M. Shin. Detection of circulating tumor cells in human peripheral blood using surface-enhanced Raman scattering. *Cancer Res.* 2011, 71, 1526.
39. S. Boca, D. Rugina, A. Pintea, L. Barbu-Tudoran, and S. Astilean. Flower-shaped gold nanoparticles: Synthesis, characterization and their application as SERS active tags inside living cells. *Nanotechnology* 2011, 22, 055702.
40. M. Potara, S. Boca, E. Licarete, A. Damert, M.-C. Alupei, M. T. Chiriac, O. Popescu, U. Schmidt, and S. Astilean. Chitosan-coated triangular silver nanoparticles as a novel class of biocompatible, highly sensitive plasmonic platforms for intracellular SERS sensing and imaging. *Nanoscale* 2013, 5, 6013.
41. C. Leordean, V. Canpean, and S. Astilean. Surface-enhanced Raman scattering (SERS) analysis of urea trace in urine, fingerprint, and tear samples. *Spectrosc. Lett.* 2012, 45, 550.
42. B. L. Goodall, A. M. Robinson, and C. L. Brosseau. Electrochemical-surface enhanced Raman spectroscopy (E-SERS) of uric acid: A potential rapid diagnostic method for early preeclampsia detection. *Phys. Chem. Chem. Phys.* 2013, 15, 1382.
43. R. Huang, S. Han, and X. (S.) Li. Detection of tobacco-related biomarkers in urine samples by surface-enhanced Raman spectroscopy coupled with thin-layer chromatography. *Anal. Bioanal. Chem.* 2013, 405, 6815.
44. A. Kim, S. J. Barcelo, R. S. Williams, and Z. Li, Melamine sensing in milk products by using surface enhanced Raman scattering. *Anal. Chem.* 2012, 84, 9303.
45. C. Andreou, M. R. Hoonejani, M. R. Barmi, M. Moskovits, and C. D. Meinhart. Rapid detection of drugs of abuse in saliva using surface

enhanced Raman spectroscopy and microfluidics. *ACS Nano* 2013, 7, 7157.

46. A. März, B. Mönch, P. Rösch, M. Kiehntopf, T. Henkel, and J. Popp. Detection of thiopurine methyltransferase activity in lysed red blood cells by means of lab-on-a-chip surface enhanced Raman spectroscopy (LOC-SERS). *Anal. Bioanal. Chem.* 2011, 400, 2755.
47. L. He, B. D. Deen, A. H. Pagel, F. Diez-Gonzalez, and T. P. Labuza. Concentration, detection and discrimination of Bacillus anthracis spores in orange juice using aptamer based surface enhanced Raman spectroscopy. *Analyst* 2013, 138, 1657.
48. K. L. Rule, and P. J. Vikesland. Surface-enhanced resonance Raman spectroscopy for the rapid detection of *Cryptosporidium parvum* and *Giardia lamblia. Environ. Sci. Technol.* 2009, 43, 1147.
49. D. Cam, K. Keseroglu, M. Kahraman, F. Sahin, and M. Culha. Multiplex identification of bacteria in bacterial mixtures with surface-enhanced Raman scattering. *J. Raman Spectrosc.* 2010, 41, 484.
50. K. C. Bantz, A. F. Meyer, N. J. Wittenberg, H. Im, O. Kurtulus, S. H. Lee, N. C. Lindquist, S.-H. Oh, and C. L. Haynes. Recent progress in SERS biosensing. *Phys. Chem. Chem. Phys.* 2011, 13, 11551.
51. M. D. Doherty, A. Murphy, R. J. Pollard, and P. Dawson. Surface-enhanced Raman scattering from metallic nanostructures: Bridging the gap between the near-field and far-field responses. *Phys. Rev. X* 2013, 3, 011001.

Chapter 6

Polarized Surface-Enhanced Raman Scattering

Barbara Fazio, Antonino Foti, Cristiano D'Andrea, Elena Messina, Alessia Irrera, Onofrio Maria Maragò, and Pietro Giuseppe Gucciardi

Consiglio Nazionale delle Ricerche, Istituto per i Processi Chimico-Fisici, Viale F. Stagno D'Alcontres 37, Messina, I-98158, Italy

gucciardi@me.cnr.it

When molecules are near-field coupled with optical nanoantennas, their scattering properties are strongly affected by the radiation properties of the nanoantenna itself. In surface-enhanced Raman spectroscopy (SERS) the scattering cross section is greatly enhanced. As well, the field orientation and polarization are modified by the nanoantenna radiation modes. This chapter discusses how the polarization of the Raman field, inelastically scattered by molecules in SERS, is modified by the interaction with optical nanoantennas in a model beyond the E^4 approximation. The polarized SERS intensity is discussed for different nanoantenna geometries and insight on how to recover information on the Raman polarizability tensor based on the SERS measurements is given.

Handbook of Enhanced Spectroscopy
Edited by Pietro Giuseppe Gucciardi, Marc Lamy de la Chapelle, and Nathalie Lidgi-Guigui

ISBN 978-981-4613-32-3 (Hardcover), 978-981-4613-33-0 (eBook)
www.panstanford.com

6.1 Introduction

Metal nanostructures supporting localized surface plasmon resonances (LSPRs) [1, 2] offer new ways to focus, amplify, and drive optical fields at the nanoscale, acting as antennas for optical photons [3, 4]. Arranged in either isolated or near-field coupled architectures, optical nanoantennas are nowadays used for surface-enhanced Raman scattering (SERS) [5–7], metal-enhanced fluorescence (MEF) [8–10], surface-enhanced infrared absorption-scattering (SEIRA-SEIRS) [11, 12] aimed at the development of high-sensitivity, multiband spectroscopic nanosensors for label-free detection of chemical and biomolecular compounds [13–20]. Optical nanoantennas convert freely propagating optical radiation into amplified optical fields, localized on nanometric regions, the so called *hot spots*. In surface-enhanced Raman spectroscopy [5, 6, 21], the plasmonic enhancement yields a Raman scattering signal magnification of factors ranging between 10^2 and 10^8 [22], allowing one to carry out vibrational analysis with excitation at optical frequencies down to the single molecule level [23, 24].

According to the E^4 model [25–28], the fundamental of the current understanding of the SERS electromagnetic contribution [29–31], the nanoantenna plays a twofold role: on one hand it amplifies the local field (excitation-field enhancement effect), on the other, it magnifies the Raman scattering (re-radiation effect) [6, 21]. Molecules lying in the *hot spots* (located at the edges of individual nanoantennas or in the nanocavities between near-field coupled nanoparticles (NPs) [17, 32–36]) experience an amplified local field and an enhanced re-radiation whenever both the wavelengths of the laser pump (λ_L) and of the induced Raman dipole (λ_R) are close to the LSPR one (λ_{LSPR}) [31]. The polarization of the electromagnetic fields plays a crucial role in anisotropic nanoantennas. It is well known that the enhanced local field is polarization sensitive, since the only component of the incident field exciting the LSPR and yielding the local field amplification is the one parallel to the nanoantenna axis [37–40]. On the other hand, in near-field coupled nanoantennas the re-radiation effect yields a selective enhancement of the Raman dipole component parallel to the nanocavity axis at the single molecule level, linearizing the polarization of the Raman field [41–45]. Several issues on the re-radiation effect are still unclear and remain the subject of active research. Question, such as the polarization state of SERS from individual nanorods with spectrally distinct LSPR [46], the possibility to use SERS to probe the Raman polarizability tensor of individual molecules, or the spatial arrangement of molecules on a substrate [26, 28, 47, 48], are still open.

In this chapter we discuss the polarization properties of the SERS intensity in near-field coupled nanostructures. We start with a description of the classical Raman effect (Section 6.2), followed by a discussion of the polarization properties of the Raman field scattered by individual and randomly oriented molecules (Section 6.3). We then discuss a phenomenological model to describe the polarization properties of the SERS field in the case of dimers, trimers, and high-symmetry nanoantennas (Section 6.4). We calculate the correct angular relations for the unpolarized, polarized, parallel-, and cross-polarized SERS intensities. Special attention is given to the SERS depolarization factor (Section 6.5), which concludes the chapter, analyzing how its expression is related to the molecular depolarization factor in the presence of near-field coupling with the nanoantennas and highlights how it can be used to gain information on the orientation-averaged non-diagonal components of the Raman tensor of the probe molecule.

6.2 Classical Description of the Raman Effect

The classical treatment of the Raman effect correctly predicts the appearance of discrete vibrational peaks in the scattering spectrum of molecules at wavelengths different than the incident one [49]. It accounts, as well, for those aspects related to the polarization dependence of the signal intensities and for the polarization-related selection rules. When a molecule interacts with an electromagnetic field $\vec{E}_{\text{exc}}$, an electric dipole moment $\vec{P}_{\text{rad}}$ is induced:

$$\vec{\mathrm{P}}_{\mathrm{rad}} = \overleftrightarrow{\alpha} \cdot \vec{\mathrm{E}}_{\mathrm{exc}} \qquad (6.1)$$

where $\overleftrightarrow{\alpha}$ is the molecular polarizability tensor. The polarizability is a function of the nuclei positions and hence of the molecule's vibrational state. Equation 6.1 is a first-order approximation of the polarizability, linear in $\vec{E}_{\text{exc}}$, associated to the Rayleigh and Raman scattering. Higher-order terms, quadratic or cubic in $\vec{E}_{\text{exc}}$, lead to the hyper-Rayleigh and hyper-Raman and to the second hyper-Rayleigh and second hyper-Raman scattering, describing all the four-wave-mixing processes, among which the coherent anti-stokes Raman scattering (CARS) is one of the most important. Limiting our description to linear term, the dependence of $\overleftrightarrow{\alpha}$ on the nuclei positions can be accounted for, in case of small vibrations, by expanding each component of the polarizability tensor, α_{ij}, in a Taylor series near the nuclei equilibrium positions as follows:

$$\alpha_{ij} = \alpha_{ij}^{0} + \sum_{k} \left(\frac{\partial \alpha_{ij}}{\partial \mathrm{q}_k} \right)_0 q_k + \cdots \tag{6.2}$$

where α_{ij}^{0} are the elements at the equilibrium positions and q_k the vibration normal coordinates components. The derivatives of the polarizability elements are calculated at the equilibrium position and summation is performed all over the possible vibrations k. In the following we will consider only one normal mode and indicate, for simplicity, with $\overleftrightarrow{\alpha}$ and $\vec{q}$ the respective polarizability tensor and normal coordinate. In light of Eq. 6.3, the polarizability tensor can be expressed as:

$$\overleftrightarrow{\alpha} = \overleftrightarrow{\alpha}_0 + \overleftrightarrow{\alpha}_{\mathrm{R}} q \tag{6.3}$$

where $\overleftrightarrow{\alpha}_{\mathrm{R}}$ is the Raman polarizability tensor, whose components are the derivatives of the polarizability tensor with respect to the k-th mode's coordinate. $\overleftrightarrow{\alpha}_{\mathrm{R}}$ is a symmetric tensor whose components are a function of the vibrational coordinates. Let's assume for q a simple harmonic behavior, $q = q_0 \cos \omega_{\mathrm{R}} t$, where q_0 and ω_{R} are amplitude and frequency of the vibration, respectively, and let's assume a monochromatic incident electromagnetic field $\vec{E}_{\text{exc}} = \vec{E}_0 \cos \omega_L t$ at frequency ω_L. The induced dipole moment due to the coupling of the electromagnetic wave with the k-th vibration mode turns out to be

$$\vec{P}_{\text{rad}} = \overleftrightarrow{\alpha}_0 \cdot \vec{E}_0 \cos(\omega_{\mathrm{L}} t) + \frac{1}{2} \overleftrightarrow{\alpha}_{\mathrm{R}} \cdot \vec{E}_0 q_0 [\cos(\omega_{\mathrm{L}} + \omega_{\mathrm{R}}) t + \cos(\omega_{\mathrm{L}} - \omega_{\mathrm{R}}) t] \tag{6.4}$$

The induced dipole $\vec{P}_{\text{rad}}$ has three components with different frequencies. The first term oscillates at the same frequency ω_{L}

of the incident field and gives rise to the elastic- or Rayleigh-scattering. Such radiation does not depend on the molecule vibrational states, but only on its polarizability at the equilibrium configuration. The second term in Eq. 6.4 describes the radiation of two oscillating dipoles emitting light at frequencies $(\omega_L + \omega_R)$ and $(\omega_L - \omega_R)$, corresponding to the inelastic Raman emission occurring, respectively, at energies higher (anti-Stokes) and lower (Stokes) than the excitation field. Two important observations can be drawn: (1) for the Raman scattering to take place, the Raman polarizability tensor $\overleftrightarrow{\alpha}_R$ must be different from zero, that is, the molecule must undergo a net change of polarizability during the vibration; (2) the polarization of the Raman dipolar field is related to the group-symmetry properties of the Raman polarizability tensor.

6.3 Polarization of the Raman Field

According to Eq. 6.4, the induced Raman dipolar field amplitude for a general vibrational transition is related to the excitation field amplitude by the relation:

$$\vec{P}_{\text{rad}} = \overleftrightarrow{\alpha}_R \cdot \vec{E}_{\text{exc}} \tag{6.5}$$

In the reference frame of the normal modes' eigenvectors, this can be made explicit as

$$\begin{pmatrix} P_x \\ P_y \\ P_z \end{pmatrix}_{\text{rad}} = \begin{pmatrix} \alpha_{xx} & \alpha_{xy} & \alpha_{xz} \\ \alpha_{yx} & \alpha_{yy} & \alpha_{yz} \\ \alpha_{zx} & \alpha_{zy} & \alpha_{zz} \end{pmatrix}_{\text{R}} \begin{pmatrix} E_x \\ E_y \\ E_z \end{pmatrix}_{\text{exc}} \tag{6.6}$$

where $\alpha_{ij} = \alpha_{ji}$. In the following we will drop the subscript "R" in the Raman polarizability tensor, which will be simply called $\overleftrightarrow{\alpha}$. The presence of non-diagonal components in the matrix of Eq. 6.6 shows that the orientation of the Raman dipole is in general non-coincident with the excitation field vector. In addition, the expression of the Raman polarizability tensor depends on the orientation of the molecule and on the symmetry group to which belongs the specific molecular vibration considered. As a result, in the most general case the Raman dipoles of molecules with different orientation will be differently oriented. Only for those symmetric vibrational modes in which $\overleftrightarrow{\alpha}$ is a multiple of the identity, the Raman dipole and the electric field vector will coincide for any molecular orientation. This has important consequences on the polarization of the Raman field. Assuming linearly polarized excitation, the Raman scattering of randomly oriented molecules will be generally unpolarized. For many molecules with a defined orientation, the polarization of the Raman field will be generally rotated with respect to the incident one. Only for total symmetric modes the polarization of the Raman field will coincide with that of the excitation field.

6.3.1 *Excitation-Detection Geometry. The Scattering Plane*

In order to probe the Raman polarizability tensor of a specific molecular vibration, or investigate the molecular orientation in a specific sample, polarization-sensitive Raman experiments are a powerful tool. Experiments are carried out using plane wave excitation (a laser beam) with specific polarization $\hat{e}_{\text{exc}}$ (linear, circular, etc.) and propagation direction $\hat{k}_{\text{exc}}$ (see Fig. 6.1a). The detection of the Raman signal is accomplished in the far-field, along a specific direction $\hat{k}_{\text{rad}}$, forming an angle η with respect to the excitation direction (Fig. 6.1a), and using a polarization analyzer oriented along $\hat{e}_{\text{det}}$ to investigate the polarization of the Raman field. The excitation-detection geometry is totally defined once we have identified the orientation of the unit vectors of the excitation field $\hat{k}_{\text{exc}}$ and of the scattered field $\hat{k}_{\text{rad}}$. Experimentally, two configurations are mostly used: the right angle detection, in which $\eta = 90°$, and the backscattering detection, in which $\eta = 180°$.

The plane defined by the two vectors $\hat{k}_{\text{exc}}$ and $\hat{k}_{\text{rad}}$ is called the scattering plane. It determines two mutually orthogonal polarization directions for the excitation and the scattered field: the parallel direction (also identified with the superscripts $||$ or p) in which the excitation and the scattered field unit vectors $\hat{e}^{||}_{\text{exc}}$ and $\hat{e}^{||}_{\text{det}}$ are in the scattering plane; and the orthogonal direction ($\perp$ or s) in which the fields unit vectors $\hat{e}^{\perp}_{\text{exc}}$ and $\hat{e}^{\perp}_{\text{det}}$ are orthogonal to the scattering plane. The scattering plane is univocally defined in the right-angle detection configuration. In Fig. 6.1b, without loss of generality we can identify $\hat{k}_{\text{exc}}$ with the z-axis and $\hat{k}_{\text{rad}}$ with the x-axis. In this case the scattering plane is identified by the (x, z) plane. The corresponding parallel- and orthogonal-polarization

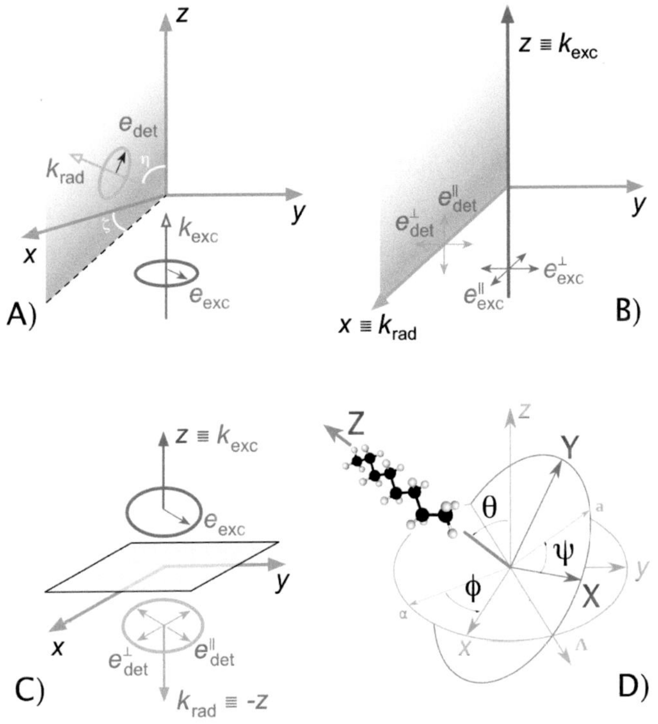

Figure 6.1 (a) Sketch of the excitation-detection geometry for a generic polarization-sensitive Raman experiment. (b) Scheme of the field's orientation for right-angle detection. The shaded area in a and b represents the scattering plane. (c) Fields orientation for the backscattering detection configuration (the shaded area is the sample plane). (d) Euler angles.

directions for the excitation and the Raman fields are, as well, univocally defined (see *arrows* in Fig. 6.1b). In the backscattering configuration, conversely, $\hat{k}_{exc}$ coincides with $\hat{k}_{rad}$ (Fig. 6.1c). The system becomes cylindrically symmetric and the scattering plane is no more univocally defined. Any plane containing $\hat{k}_{exc}$ can be considered as a scattering plane. Distinction between parallel and orthogonal polarization for the excitation field is meaningless since all the directions in the polarization plane are equivalent. The symmetry is broken when the polarization direction of the excitation field ($\hat{e}_{exc}$ in Fig. 6.1c) is chosen. The scattering plane is automatically defined by the unit vectors $\hat{k}_{exc}$ and $\hat{e}_{exc}$, and the two parallel- and orthogonal-polarization states of the Raman field are univocally defined with respect to the excitation field polarization ($\hat{e}_{\text{det}}^{||}$ and $\hat{e}_{\text{det}}^{\perp}$ in Fig. 6.1c).

6.3.2 *Intensity of the Scattered Radiation*

In experiments the measured quantity is the intensity of the Raman dipolar field I, time-averaged power per unit solid angle. The intensity radiated by an oscillating electric dipole passing along a polarizer oriented along $\hat{e}_{\text{det}}$ is given by:

$$\mathrm{I} = k\omega^4 \left| \vec{P}_{\text{rad}} \cdot \hat{e}_{\text{det}} \right|^2 \tag{6.7}$$

where $k=1/32\pi^2\varepsilon_0 c^3$ is a constant related to the vacuum dielectric constant, ε_0, and speed of light, c.

The intensity is dependent on the collection angle, η, the incident $\hat{e}_{\text{exc}}$, and the detection polarization $\hat{e}_{\text{det}}$. To take into account all these parameters, we define the intensity with the notation

$$I\ (\eta; \varepsilon_{\text{exc}}, \varepsilon_{\text{det}})$$

where ε_{exc}, ε_{det} represent the parameters describing the polarization state of the excitation and the Raman field, respectively, and will be identified in the following either with the polarization angles (angles formed by the fields with respect to some specific directions, to be defined each time) or with the symbols $||$ and $\perp$ when we refer to the scattering plane. As an example, $I\ (\pi/2; ||, \perp)$ will represent the intensity of the Raman field polarized orthogonally to the scattering plane when the excitation is polarized parallel to the scattering plane and the collection angle is $\pi/2$.

6.3.3 *Probing the Raman Polarizability Tensor*

The Raman polarizability tensor of the vibrational modes represents a basic information that, together with the vibration energy, and the oscillator strength, characterize the vibrational spectrum of a molecule. In the following we will see how different experimental geometries lead to the measurement of different components of the Raman polarizability tensor.

Let's start with a right-angle collection geometry. In the reference framework depicted in Fig. 6.1b, where $\hat{e}^{||}_{\text{exc}} \equiv \hat{x}$ and $\hat{e}^{\perp}_{\text{exc}} \equiv \hat{y}$, the excitation field can be written as

$$\vec{E}_{\text{exc}} = E_x\hat{x} + E_y\hat{y} = E_x\hat{e}^{||}_{\text{exc}} + E_y\hat{e}^{\perp}_{\text{exc}} \tag{6.8}$$

(the z-component being zero for a plane wave). Considering an excitation field polarized parallel to the scattering plane, that is, $E_x \neq 0$, $E_y = 0$, and observing that $\hat{e}^{||}_{\text{det}} \equiv \hat{z}$ and $\hat{e}^{\perp}_{\text{det}} \equiv \hat{y}$, the Raman dipole will be given by[a]

$$\vec{P}_{\text{rad}} = \left(\alpha_{yx}\hat{e}^{\perp}_{\text{det}} + \alpha_{zx}\hat{e}^{||}_{\text{det}}\right) E_x \tag{6.9}$$

The two intensities measured

$$I\left(\pi/2; \|, \|\right) = k\omega^4 \left(\alpha_{zx}\right)^2 E_x^2 \tag{6.10}$$

and

$$I\left(\pi/2; \|, \perp\right) = k\omega^4 \left(\alpha_{yx}\right)^2 E_x^2 \tag{6.11}$$

will bring information on different non-diagonal elements of the Raman polarization tensor.

If we instead consider orthogonally polarized excitation, that is, $E_x = 0$, $E_y \neq 0$, the Raman dipole will be given by

$$\vec{P}_{\text{rad}} = \left(\alpha_{yy}\hat{e}^{\perp}_{\text{det}} + \alpha_{yz}\hat{e}^{||}_{\text{det}}\right) E_y \tag{6.12}$$

and the two intensities measured

$$I\left(\pi/2; \perp, \perp\right) = k\omega^4 \left(\alpha_{yy}\right)^2 E_y^2 \tag{6.13}$$

and

$$I\left(\pi/2; \perp, \|\right) = k\omega^4 \left(\alpha_{yz}\right)^2 E_y^2 \tag{6.14}$$

will bring information on both the diagonal and the non-diagonal components of the Raman polarization tensor.

Let's now consider the backscattering configuration. Once set the excitation field polarization direction, the scattering plane is identified and the excitation will be, obviously, parallel polarized to the scattering plane. Without loss of generality we can assume $\hat{e}^{||}_{\text{exc}} \equiv \hat{x}$, and observe that $\hat{e}^{||}_{\text{det}} \equiv \hat{x}$ and $\hat{e}^{\perp}_{\text{det}} \equiv \hat{y}$. This leads to the expression

---

[a] In the right-angle configuration depicted in Fig. 6.1b we only have access to the $\hat{y}$ and $\hat{z}$ components of the scattered field.

$$\vec{P}_{\text{rad}} = \left(\alpha_{yx}\hat{e}_{\text{det}}^{\perp} + \alpha_{xx}\hat{e}_{\text{det}}^{\parallel}\right) E_x \tag{6.15}$$

for the Raman dipole. The two intensities measured

$$I\left(\pi; \parallel, \parallel\right) = k\omega^4 \left(\alpha_{xx}\right)^2 E_x^2 \tag{6.16}$$

and

$$I\left(\pi; \parallel, \perp\right) = k\omega^4 \left(\alpha_{yx}\right)^2 E_x^2 \tag{6.17}$$

will bring information on both the diagonal and the non-diagonal components of the Raman polarization tensor.

6.3.4 *Generalization to Randomly Oriented Molecules*

Equations 6.10–6.17 hold for a single molecule with fixed orientation in space. For N randomly oriented molecules the intensity of the Raman radiation can be calculated taking into account that the Raman dipoles associated to each molecule sum up incoherently. Equation 6.7 becomes:

$$I = k\omega^4 \left| \sum\nolimits_{i=1\ldots N} \vec{P}_{\text{rad}}^{(i)} \cdot \hat{e}_{\text{det}} \right|^2 = k\omega^4 \sum\nolimits_{i=1\ldots N} \left| \vec{P}_{\text{rad}}^{(i)} \cdot \hat{e}_{\text{det}} \right|^2 \tag{6.18}$$

This in turn gives an expression in which the intensity is N times the orientation averaged intensity scattered by a single molecule, that is

$$I = k\omega^4 N \left\langle \left| \vec{P}_{\text{rad}}^{(i)} \cdot \hat{e}_{\text{det}} \right|^2 \right\rangle_{\phi\theta\psi} \tag{6.19}$$

where ϕ, θ, and ψ are the Euler angles associated to the 3D spatial rotation of the molecule (Fig. 6.1d) in the reference frame considered.

The explicit intensity expressions for the polarized Raman intensities in the case of many randomly oriented molecules can be obtained by multiplying by N the intensity of the single molecule (Eqs. 6.10–6.17) and replacing the quantities $\left(\alpha_{ij}\right)^2$ with their isotropic averages $\left\langle \left(\alpha_{ij}\right)^2 \right\rangle_{\phi,\theta,\psi}$ given by

$$\left\langle \left(\alpha_{ij}\right)^2 \right\rangle_{\phi,\theta,\psi} = \frac{1}{8\pi^2} \int_0^{2\pi} d\psi \int_0^{\pi} \sin\theta d\theta \int_0^{2\pi} \left[\alpha_{ij}\left(\phi, \theta, \psi\right)\right]^2 d\phi \tag{6.20}$$

where

$$\alpha_{ij}(\phi, \theta, \psi) = \sum_{k,l} R_{ik}(\phi, \theta, \psi) R_{jl}(\phi, \theta, \psi) \alpha_{ij} \tag{6.21}$$

is the Raman tensor of the rotated molecule and $R_{ik}(\phi, \theta, \psi)$ are the elements of the rotation matrix

$$\overleftrightarrow{R}(\phi, \theta, \psi) = \begin{pmatrix} \cos\psi & \sin\psi & 0 \\ -\sin\psi & \cos\psi & 0 \\ 0 & 0 & 1 \end{pmatrix} \begin{pmatrix} 1 & 0 & 0 \\ 0 & \cos\theta & \sin\theta \\ 0 & -\sin\theta & \cos\theta \end{pmatrix} \times \begin{pmatrix} \cos\phi & \sin\phi & 0 \\ -\sin\phi & \cos\phi & 0 \\ 0 & 0 & 1 \end{pmatrix} \tag{6.22}$$

associated with the Euler angles in the reference frame in which the Raman polarizability tensor of the fixed molecule is α_{ij}.

As an example we report the Raman intensity for backscattering detection for the case of N randomly oriented molecules:

$$I(\pi; \|, \|) = Nk\omega^4 \left\langle (\alpha_{xx})^2 \right\rangle E_x^2 \tag{6.23}$$

and

$$I(\pi; \|, \perp) = Nk\omega^4 \left\langle \left(\alpha_{yx}\right)^2 \right\rangle E_x^2 \tag{6.24}$$

The right-angle and the backscattering detection configurations are examples in which the Raman dipole is simply related to one component of the Raman polarizability tensor. In the most general case the intensity will be given by a linear combination of the orientation averaged mixed products among the Raman polarizability elements, that is

$$\begin{aligned} I(\eta; \varepsilon_{\mathrm{exc}}, \varepsilon_{\mathrm{det}}) = & \sum_i A_i \left\langle (\alpha_{ii})^2 \right\rangle + \sum_{i \neq j} B_{i,j} \left\langle \left(\alpha_{ij}\right)^2 \right\rangle \\ & + \sum_{i \neq j} D_{i,j} \left\langle (\alpha_{ij}\ \alpha_{ji}) \right\rangle + \sum_{i \neq j} E_{i,j} \left\langle (\alpha_{ii}\ \alpha_{jj}) \right\rangle \\ & + \sum_{i \neq j} F_{i,j} \left\langle (\alpha_{ii}\ \alpha_{ij}) \right\rangle \end{aligned} \tag{6.25}$$

This expression can be greatly simplified using the rotation invariants, quantities that are invariant under rotations, as described below.

6.3.5 *Rotation Invariants*

Under rotation the elements of the Raman polarizability tensor do change. Some invariant quantities, however, can be found [49], the most simple studied in the basic linear algebra courses, being the trace of the tensor, that is $\mathrm{Tr}(\overleftrightarrow{\alpha}) = \alpha_{xx} + \alpha_{yy} + \alpha_{zz}$. Instead of the Placzek relations G [49] we will use the a^2, δ^2, γ^2 invariants that are, respectively:
the square of the mean polarizability

$$a^2 = \frac{1}{9}\left(\alpha_{xx} + \alpha_{yy} + \alpha_{zz}\right)^2 \tag{6.26}$$

the antisymmetric anisotropy

$$\delta^2 = \tfrac{3}{4}\left(\left|\alpha_{xy} - \alpha_{yx}\right|^2 + \left|\alpha_{yz} - \alpha_{zy}\right|^2 + \left|\alpha_{zx} - \alpha_{xz}\right|^2\right) \tag{6.27}$$

which is zero for symmetric Raman tensors, and the anisotropy factor

$$\gamma^2 = \tfrac{1}{2}\left(\left|\alpha_{xx} - \alpha_{yy}\right|^2 + \left|\alpha_{yy} - \alpha_{zz}\right|^2 + \left|\alpha_{zz} - \alpha_{xx}\right|^2\right) + \tfrac{3}{4}\left(\left|\alpha_{xy} + \alpha_{yx}\right|^2 + \left|\alpha_{yz} + \alpha_{zy}\right|^2 + \left|\alpha_{zx} + \alpha_{xz}\right|^2\right) \tag{6.28}$$

Calculation of the orientation-averaged of the tensor components gives us the following relations:

$$\left\langle\left(\alpha_{xx}\right)^2\right\rangle = \left\langle\left(\alpha_{yy}\right)^2\right\rangle = \left\langle\left(\alpha_{zz}\right)^2\right\rangle = a^2 + \frac{4}{45}\gamma^2 \tag{6.29}$$

$$\left\langle\left(\alpha_{xy}\right)^2\right\rangle = \left\langle\left(\alpha_{xz}\right)^2\right\rangle = \left\langle\left(\alpha_{yz}\right)^2\right\rangle = \frac{1}{15}\gamma^2 + \frac{1}{9}\delta^2 \tag{6.30}$$

$$\left\langle\left(\alpha_{xy}\,\alpha_{yx}\right)\right\rangle = \left\langle\left(\alpha_{yz}\,\alpha_{zy}\right)\right\rangle = \left\langle\left(\alpha_{zx}\,\alpha_{xz}\right)\right\rangle = \frac{1}{15}\gamma^2 - \frac{1}{9}\delta^2 \tag{6.31}$$

$$\left\langle\left(\alpha_{xx}\,\alpha_{yy}\right)\right\rangle = \left\langle\left(\alpha_{xx}\,\alpha_{zz}\right)\right\rangle = \left\langle\left(\alpha_{yy}\,\alpha_{zz}\right)\right\rangle = a^2 - \frac{2}{45}\gamma^2 \tag{6.32}$$

$$\begin{aligned}\left\langle\left(\alpha_{xx}\,\alpha_{xy}\right)\right\rangle &= \left\langle\left(\alpha_{xx}\,\alpha_{xz}\right)\right\rangle = \left\langle\left(\alpha_{yy}\,\alpha_{yz}\right)\right\rangle = \left\langle\left(\alpha_{yy}\,\alpha_{yx}\right)\right\rangle \\ &= \left\langle\left(\alpha_{zz}\,\alpha_{zx}\right)\right\rangle = \left\langle\left(\alpha_{zz}\,\alpha_{zy}\right)\right\rangle = 0\end{aligned} \tag{6.33}$$

The equivalence among the orientation-averaged diagonal terms of the Raman tensor (Eq. 6.29) is an obvious consequence of the fact that for randomly oriented molecules there is no preferential direction and the x, y, z laboratory directions are all equivalent. The same holds for the relations in Eqs. 6.30–6.33. A further relation among the orientation-averaged mixed products of the Raman comes from Eqs. 6.29–6.32.

$$\left\{\left\langle\left(\alpha_{ij}\alpha_{ji}\right)\right\rangle + \left\langle\left(\alpha_{ii}\alpha_{jj}\right)\right\rangle = \left\langle\left(\alpha_{ii}\right)^2\right\rangle - \left\langle\left(\alpha_{ij}\right)^2\right\rangle\right\}_{i\neq j} \tag{6.34}$$

6.3.6 *Probing the Orientation-Averaged Raman Polarizability Tensor Components*

The invariants defined in Eqs. 6.26–6.28 can be used to find simpler expressions of the Raman scattering intensities. Notably the Raman intensity will be the function of only two independent parameters (a^2 and γ^2, δ^2 being zero for symmetric Raman tensors), which reflects the fact that the orientation averages only distinguish between diagonal and off-diagonal components of $\overleftrightarrow{\alpha}$. The fact that we cannot access all the six independent components of the Raman polarizability but we are limited to only two parameters is the price we have to pay for the fact that we have no specific information on the spatial orientation of the molecule but we are averaging on all the possible orientations.

As for single molecules, different components of the Raman tensor can be gathered by switching the detection polarizer, although this is in general not always true and depends on the detection geometry and on the excitation polarization used. For example, the information gained in the right-angle detection with excitation orthogonal to the scattering plane (Eqs. 6.13 and 6.14) is equivalent to the one obtained in the backscattering configuration (Eqs. 6.16 and 6.17) and concerns both the diagonal and the off-diagonal components of the tensor, since

$$I\left(\pi/2;\perp,\perp\right) \propto \left\langle \left(\alpha_{yy}\right)^2 \right\rangle = a^2 + \frac{4}{45}\gamma^2 = \left\langle (\alpha_{xx})^2 \right\rangle \propto I\left(\pi;\parallel,\parallel\right) \tag{6.35}$$

and

$$I\left(\pi/2;\perp,\parallel\right) \propto \left\langle \left(\alpha_{yz}\right)^2 \right\rangle = \frac{1}{15}\gamma^2 = \left\langle \left(\alpha_{yx}\right)^2 \right\rangle \propto I\left(\pi;\parallel,\perp\right) \tag{6.36}$$

Conversely, in the right-angle detection, using excitation polarized parallel to the scattering plane, there is no new information switching the analyzer from parallel to orthogonal (Eqs. 6.10 and 6.11) since in both cases we probe the orientation averaged off-diagonal components of Raman tensor, which are all equivalent.

6.3.7 *The Raman Depolarization Ratio*

The polarized Raman intensities do depend on the excitation field intensity. In order to get rid of this and measure a parameter that only belongs to the molecular vibration, we can either retrieve the scattering cross-section or, more simply, take the ratio between some different Raman intensities. For linearly polarized excitation, two different parameters are usually considered: the depolarization ratio and the degree of depolarization. The depolarization ratio [49, 50] is defined as

$$\rho = \frac{I_{\perp}}{I_{\parallel}} \quad (6.37)$$

where $I_{\perp}$ and $I_{\parallel}$ are, respectively, the intensity of the Raman scattered fields polarized orthogonally and parallel to the laser field polarization. The degree of polarization, or depolarization factor [41, 42], in analogy to the visibility parameter, is instead defined as

$$\sigma = \frac{I_{\parallel} - I_{\perp}}{I_{\parallel} + I_{\perp}} \quad (6.38)$$

Together with the scattering energy and intensity the depolarization ratio (factor) is used for investigating the structure of liquids, mixtures, and solutions, and specifically provides information on the orientation averaged components of the Raman polarizability tensor. As such, it is dependent on the geometry of the experiment, that is, collection angle and excitation polarization direction.

As an example, we calculate the depolarization ratio for the right-angle and the backscattering configuration for the case of randomly oriented molecules. For the right angle configuration, using parallel polarized excitation we have

$$\rho = \frac{I\left(\pi/2; \parallel, \perp\right)}{I\left(\pi/2; \parallel, \parallel\right)} = 1 \quad (6.39)$$

Since the two experiments both probe the orientation-averaged off-diagonal components of the tensor, we do not expect to have some information from this ratio. Vice versa, for orthogonally polarized light we find (symmetric Raman tensors are assumed)

$$\rho = \frac{I\left(\pi/2; \parallel, \perp\right)}{I\left(\pi/2; \perp, \perp\right)} = \frac{\left\langle \left(\alpha_{yz}\right)^2 \right\rangle}{\left\langle \left(\alpha_{yy}\right)^2 \right\rangle} = \frac{3\gamma^2}{45a^2 + 4\gamma^2} \quad (6.40)$$

As expected (Eqs. 6.35 and 6.36) this value is the same as that can be retrieved using a backscattering configuration in which

$$\rho = \frac{I\left(\pi; \parallel, \perp\right)}{I\left(\pi; \parallel, \parallel\right)} = \frac{\left\langle \left(\alpha_{yx}\right)^2 \right\rangle}{\left\langle \left(\alpha_{xx}\right)^2 \right\rangle} = \frac{3\gamma^2}{45a^2 + 4\gamma^2} \quad (6.41)$$

Remarkably the values of the depolarization ratio in both these geometrical configurations are bounded, that is,

$$0 \leq \rho \leq 3/4 \tag{6.42}$$

Analogous considerations hold for the degree of polarization,

$$\sigma = \frac{I(\pi; \|, \|) - I(\pi; \|, \perp)}{I(\pi; \|, \|) + I(\pi; \|, \perp)} = \frac{45a^2 + \gamma^2}{45a^2 + 7\gamma^2} \tag{6.43}$$

being always greater than zero and bound between

$$\frac{1}{7} \leq \sigma \leq 1 \tag{6.44}$$

6.3.8 *Polarization Dependence of the Raman Scattering*

As we have discussed above, the Raman field radiated by randomly oriented molecules is unpolarized. Nevertheless if we analyze by means of a rotating linear polarizer the polarization dependence of the Raman scattering, we will measure some degree of polarization, that is, variation of the signal as a function of the polarizer orientation. The calculation is easily done for the backscattering configuration. We assume unit excitation field amplitude, $\hat{e}_{\text{exc}} \equiv \hat{x}$ and call ϕ the angle between the excitation field unit vector $\hat{e}_{\text{exc}}$ and the analyzer axis $\hat{e}_{\text{det}}$. In the $\{x, y, z\}$ base we will have $\hat{e}_{\text{det}} = (\cos\phi, \sin\phi)$. Recalling Eq. 6.15 for the Raman dipole, we obtain

$$I(\pi; \hat{e}_{\text{exc}}, \phi) = \left\langle \left| \hat{e}_{\text{det}} \cdot \vec{P}_{\text{rad}} \right| \right\rangle = \left\langle (\alpha_{xx})^2 \right\rangle \cos^2\phi + \left\langle (\alpha_{yx})^2 \right\rangle \sin^2\phi \tag{6.45}$$

or, in terms of the rotational invariants

$$I(\pi; \hat{e}_{\text{exc}}, \phi) = \frac{1}{45}\left[\left(45a^2 + 4\gamma^2\right)\cos^2\phi + 3\gamma^2\sin^2\phi\right] \tag{6.46}$$

As a result, the measured intensity will oscillate with sinusoidal behavior between a minimum, $I(\pi; ||, \perp)$ and a maximum $I(\pi; ||, ||)$.

6.4 Polarized Surface-Enhanced Raman Scattering

SERS differs from normal Raman spectroscopy because of the presence of a third element, the nanoantenna, that enters into play in the coupling between the electromagnetic field and the molecular vibration. Although the spectral dependence of the SERS gain is not yet fully understood [51–53], according to the E^4 model, if both the excitation and Raman photon energies are within the plasmonic resonance of the nanoantenna, the latter enhances both the local excitation and the re-radiated fields. We can model SERS as a three-step phenomenon [41, 45]:

(1) enhancement of the excitation field at wavelength λ_L and generation of the *hot spots*

$$\vec{E}^{\text{enh}}_{\text{exc}}(\lambda_L) = \overleftrightarrow{\Gamma}_{\text{exc}}(\lambda_L) \cdot \vec{E}_{\text{exc}}(\lambda_L) \tag{6.47}$$

(2) generation of a molecular Raman dipolar field at wavelength λ_R

$$\vec{P}_{\text{rad}}(\lambda_R) = \overleftrightarrow{a}_R \cdot \vec{E}^{\text{enh}}_{\text{exc}}(\lambda_L) \tag{6.48}$$

(3) amplification of the SERS field at wavelength λ_R for molecules located at the *hot spots*

$$\vec{P}^{\text{enh}}_{\text{rad}}(\lambda_R) = \overleftrightarrow{\Gamma}_{\text{rad}}(\lambda_R) \cdot \vec{P}_{\text{rad}}(\lambda_R) \tag{6.49}$$

where we have introduced the excitation field enhancement tensor $\overleftrightarrow{\Gamma}_{\text{exc}}(\lambda_L)$ and the re-radiation enhancement tensor $\overleftrightarrow{\Gamma}_{\text{rad}}(\lambda_R)$ to describe the amplification of the excitation and the Raman field, which are in principle wavelength dependent. Using tensors helps us to account for the different amplification of the three components of the electromagnetic fields. Nanosphere dimers, near-field coupled nanowires, trimers, or clusters, for example, feature hot spots at the nanocavities that are selectively excited when the field is parallel to the nanocavity axis and, in turn, selectively amplify the Raman field component parallel to the same axis [41, 45].

6.4.1 *Polarized SERS from Nanoantenna Dimers*

Let's consider a nanoantenna array made of two near-field coupled nanostructures. These could be nanospheres, nanorods arranged in a tip-to-tip configuration, or long nanowires arranged in a side-by-side geometry. The latter case will be considered in the following, since it is characterized by a single LSPR in the visible.

Let's assume (Fig. 6.2) a laboratory reference frame $\{x, y, z\}$ in which the laser beam propagates along $\hat{k} \equiv -\hat{z}$, and is linearly polarized along the x-axis, that is, $\hat{e}_{\text{exc}} \equiv \hat{x}$. Let's now define the nanoantenna reference frame defined by the three mutually

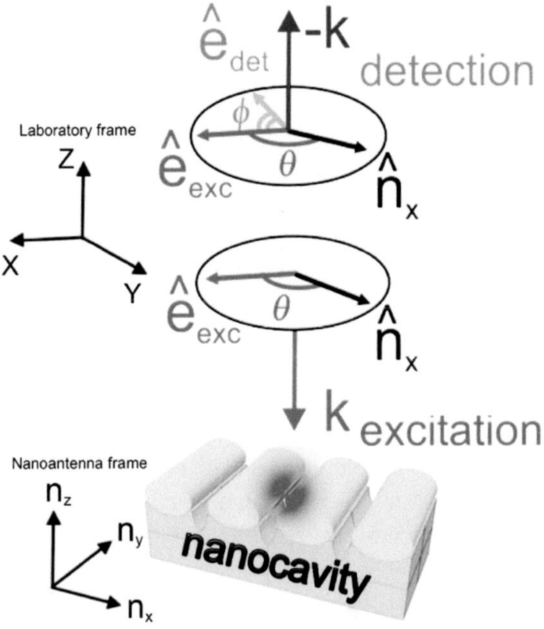

Figure 6.2 Sketch of the reference frameworks and fields polarizations in SERS from near-field coupled nanostructures generating hot spots at the nanocavities.

orthogonal axes $\{\hat{n}_x, \hat{n}_y, \hat{n}_z\}$ given, respectively, by the nanocavity axis (i.e., near-field coupling direction), the nanowires' long axis and the direction of propagation of the excitation electromagnetic field (-k), $\hat{n}_z \equiv \hat{z}$. The nanoantennas can be rotated with respect to the laboratory frame. We therefore call θ the angle between the nanocavity axis orientation $\hat{n}_x$ and the linear field-polarization direction, that is, $\hat{e}_{\text{exc}} \cdot \hat{n}_x = \cos\theta$. We assume a detection in backscattering, that is, the Raman field propagates in the $\hat{k} \equiv \hat{z}$ direction, and analyze the polarization state of the SERS radiation in the laboratory reference frame with a linear polarization analyzer whose optical axis forms and angle ϕ with respect to the excitation field polarization, that is, $\hat{e}_{\text{exc}} \cdot \hat{e}_{\text{det}} = \cos\phi$.

If both the excitation and the Raman photons wavelengths are in resonance with the LSPR of the sample, we can write the field enhancement tensor in the base $\{\hat{n}_x, \hat{n}_y, \hat{n}_z\}$ as:

$$\overleftrightarrow{\Gamma}_{\text{exc}}(\lambda_L) = \begin{pmatrix} \Gamma & 0 & 0 \\ 0 & 1 & 0 \\ 0 & 0 & 1 \end{pmatrix} \quad (6.50)$$

and

$$\overleftrightarrow{\Gamma}_{\text{rad}}(\lambda_R) = \begin{pmatrix} \Gamma(1+\varepsilon) & 0 & 0 \\ 0 & 1 & 0 \\ 0 & 0 & 1 \end{pmatrix} \quad (6.51)$$

in which we assume that the excitation and the re-radiated field components along the nanocavity axis $\hat{n}_x$ are amplified by the factors $\Gamma_{\text{exc}} = \Gamma$ and $\Gamma_{\text{rad}} = \Gamma(1+\varepsilon)$, and that the components along the other two axes are left unperturbed. This is a hypothesis with important physical consequences as will be discussed in Section 6.4.2. We use the factor $(1+\varepsilon)$ to distinguish the excitation field enhancement factor from the re-radiation one. The two are in principle different because of the wavelength dependence of the SERS amplification [54]. In analogy with Eqs. 6.7 and 6.19, the expressions for the SERS intensities for a single molecule and many randomly oriented molecules can be written, respectively, as:

$$I_{\text{SERS}} = k\omega^4 \left| \vec{P}^{\text{enh}}_{\text{rad}}(\lambda_R) \cdot \hat{e}_{\text{det}} \right|^2 \quad (6.52)$$

and

$$I_{\text{SERS}}(\pi;\theta,\phi) = k\omega^4 N \left\langle \left| \hat{e}_{\text{det}} \cdot \overleftrightarrow{R}_z(-\theta) \cdot \overleftrightarrow{\Gamma}_{\text{rad}}(\lambda_R) \cdot \overleftrightarrow{\alpha}(\phi,\theta,\psi) \cdot \overleftrightarrow{\Gamma}_{\text{exc}}(\lambda_L) \cdot \overleftrightarrow{R}_z(\theta) \cdot \hat{e}_{\text{exc}} \right|^2 \right\rangle_{\phi,\theta,\psi} \quad (6.53)$$

where the molecular orientation averages are carried out according to Eqs. 6.20 and 6.21, and we have introduced the matrix

$$\overleftrightarrow{R}_z(\theta) = \begin{pmatrix} \cos\theta & \sin\theta & 0 \\ -\sin\theta & \cos\theta & 0 \\ 0 & 0 & 1 \end{pmatrix} \quad (6.54)$$

to account for the rotation, along the z-axis, of the nanoantennas with respect to the laboratory frame (i.e., of the nanocavity axis with respect to the laser polarization). For a backscattering configuration the evaluation of the fields reduces to a calculation of the in-plane field components, neglecting the z-components of the fields, which is zero, and of the enhancement and Raman tensors. In the laboratory frame the excitation field is expressed as $\hat{e}_{\text{exc}} = (1, 0)$, the polarization analyzer as $\hat{e}_{\text{det}} = (\cos\phi, \sin\phi)$.

The polarized SERS intensity can be calculated according to Eq. 6.53 and gives:

$$\begin{aligned} & I_{\text{SERS}}(\pi;\theta,\phi) \propto \Gamma^4 (1+\varepsilon)^2 \left\langle (\alpha_{xx})^2 \right\rangle \cos^2\theta \cos^2(\theta-\phi) \\ & +\Gamma^2 (1+\varepsilon) \left\langle (\alpha_{xx})^2 \right\rangle \left[\frac{1}{2} \sin 2\theta \sin 2(\theta-\phi) \right] \\ & +\Gamma^2 \left\langle \left(\alpha_{xy}\right)^2 \right\rangle [\varepsilon \sin\theta \cos(\theta-\phi) + \sin\phi]^2 \\ & + \left\langle (\alpha_{xx})^2 \right\rangle \sin^2\theta \sin^2(\theta-\psi) \end{aligned} \tag{6.55}$$

The intensity is a sum of four terms, each one bringing information on the enhancement factors (excitation and re-radiation), the Raman polarizability tensor (orientation-averaged components), and the excitation–detection polarization dependence (θ, ϕ dependence). Notably the expression in Eq. 6.55 gives terms beyond the E^4 approximation. For high field enhancement factors, that is, $\Gamma^4 \gg \Gamma^2 \gg 1$ we can approximate Eq. 6.55 as

$$I_{\text{SERS}}(\pi;\theta,\phi) \propto \Gamma^4 (1+\varepsilon)^2 \left\langle (\alpha_{xx})^2 \right\rangle \cos^2\theta \cos^2(\theta-\phi) \tag{6.56}$$

The first term $\Gamma^4 (1+\varepsilon)^2 = \Gamma^2_{\text{exc}} \times \Gamma^2_{\text{rad}}$ tells us that the SERS intensity is the product of the square of the excitation field enhancement and the square of the re-radiation field enhancement, identical to the one derived from the E^4 model. In the case in which the two enhancement factors are equal ($\varepsilon = 0$) the SERS intensity goes as the Γ^4, that is, as the fourth power of the field enhancement factor. Equation 6.56 also tells us that SERS measurements, evaluated in the light of the E^4 approximation, only provide information on the orientation-averaged diagonal components of the Raman polarizability tensor, and therefore this information cannot be used to retrieve the molecular depolarization factor. Finally, the polarization dependence of the SERS intensity in a generic experiment is described by the geometrical factor $\cos^2\theta \cos^2(\theta-\phi)$. This correctly predicts the expressions of the SERS intensities to be expected in different typologies of experiments as [45, 47]: (i) the unpolarized SERS detection (excitation rotated, the analyzer is removed), given by the sum of parallel- and the cross-polarized

$I_{\text{SERS}}(\pi;\theta,\phi=0)+I_{\text{SERS}}(\pi;\theta,\phi=\pi/2)$ intensities

$$I_{\text{SERS}}^{\text{unpol}}(\pi;\theta)\propto\left\langle(\alpha_{xx})^2\right\rangle\cos^2\theta \tag{6.57}$$

(ii) the polarized SERS detection (excitation field forming an angle $\bar{\theta}$ to the nanocavity axis and rotating analyzer):

$$I_{\text{SERS}}^{\text{pol}}\left(\pi;\theta=\bar{\theta},\phi\right)\propto\left\langle(\alpha_{xx})^2\right\rangle\cos^2\left(\phi-\bar{\theta}\right) \tag{6.58}$$

(iii) the parallel-polarized detection (analyzer parallel to the excitation polarization and sample rotating)

$$I_{\text{SERS}}^{\parallel}(\pi;\theta,\phi=0)\propto\left\langle(\alpha_{xx})^2\right\rangle\cos^4\theta \tag{6.59}$$

(iv) the cross-polarized detection (analyzer orthogonal to the excitation polarization, sample rotating)

$$I_{\text{SERS}}^{\perp}(\pi;\theta,\phi=\pi/2)\propto\left\langle(\alpha_{xx})^2\right\rangle\sin^2\theta\cos^2\theta \tag{6.60}$$

Figure 6.3 (lines) shows the angular trends for the expressions 57–60, together with the experimental counterpart (hollow symbols) acquired on a sample of gold nanowires grown by ion beam scattering on which randomly oriented molecules of Methylene blue are absorbed [45]. Only the parallel-polarized SERS intensity is expected to vary as $\cos^4\theta$ (Fig. 6.3c).

Notably, the SERS radiation is always preferentially polarized along the nanocavity axis (Fig. 6.3b), no matter the direction of the excitation polarization. This is basically different from what occurs in normal Raman scattering and is a result of the strong coupling of the molecules with the nanoantenna that modifies the scattering properties of the molecules themselves.

6.4.2 *Terms Beyond the E^4 Model*

In addition to the most terms in Γ^4, Eq. 6.55 shows that the SERS intensity has two terms proportional to Γ^2, clearly beyond the E^4 approximation. These terms bring a weaker contribution to the SERS signal, of the order of $1/\Gamma^2$ with respect to the most intense one but are very important since they bring information on both the diagonal and non-diagonal components of the Raman polarizability tensor. The last element in Eq. 6.55 brings an even smaller contribution, of the order of $1/\Gamma^4$ with respect to the first term. However, the ability to probe experimentally this term

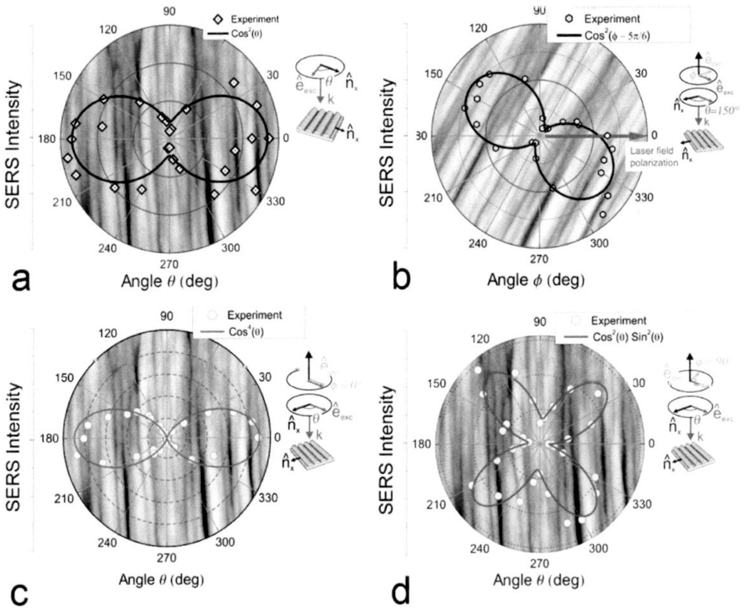

Figure 6.3 Polar plots of the SERS intensities for different excitation–detection polarization configurations: (a) unpolarized detection with excitation polarized parallel to the nanocavity, (b) polarized detection with excitation angle $\bar{\theta} = 150^\circ$ with respect to nanocavity axis, (c) parallel-polarized detection, (d) cross-polarized detection. The theoretical curves (lines) are superimposed to the experimental data (symbols). Reprinted with permission from Ref. [45], Copyright 2011 American Chemical Society.

is important since it is related to the Raman signal measured in absence of any enhancement and, therefore, can be used as a reference to precisely evaluate the SERS enhancement.

We can now write the complete expressions of the unpolarized SERS intensity

$$I_{\text{SERS}}^{\text{unpol}}(\pi;\theta) \propto \Gamma^4 (1+\varepsilon)^2 \left\langle (\alpha_{xx})^2 \right\rangle \cos^2\theta$$
$$+\Gamma^2 \left\langle \left(\alpha_{xy}\right)^2 \right\rangle \left[1+\varepsilon(2+\varepsilon)\sin^2\theta\right] + \left\langle (\alpha_{xx})^2 \right\rangle \sin^2\theta \qquad (6.61)$$

For the polarized SERS intensity when the excitation is polarized parallel and orthogonal to the nanoantenna (nanocavity) axis we, respectively, have

$$I_{\text{SERS}}^{\text{pol}}(\pi;0,\phi) \propto \Gamma^4(1+\varepsilon)^2\left\langle(\alpha_{xx})^2\right\rangle\cos^2\phi$$
$$+\Gamma^2\left\langle\left(\alpha_{xy}\right)^2\right\rangle\sin^2\phi \tag{6.62}$$

$$I_{\text{SERS}}^{\text{pol}}(\pi;\pi/2,\phi) \propto \Gamma^2(1+\varepsilon)^2\left\langle\left(\alpha_{xy}\right)^2\right\rangle\sin^2\phi$$
$$+\left\langle(\alpha_{xx})^2\right\rangle\cos^2\phi \tag{6.63}$$

for the parallel-polarized SERS intensity we have

$$I_{\text{SERS}}^{\parallel}(\pi;\theta,0) \propto \Gamma^4(1+\varepsilon)^2\left\langle(\alpha_{xx})^2\right\rangle\cos^4\theta$$
$$+\Gamma^2\left[\left\langle\left(\alpha_{xy}\right)^2\right\rangle\varepsilon^2+2\left\langle(\alpha_{xx})^2\right\rangle(1+\varepsilon)\right]\sin^2\theta\cos^2\theta$$
$$+\left\langle(\alpha_{xx})^2\right\rangle\sin^4\theta \tag{6.64}$$

and for cross- polarized SERS intensity we find

$$I_{\text{SERS}}^{\perp}(\pi;\theta,\pi/2) \propto \left[\Gamma^2(1+\varepsilon)-1\right]^2\left\langle\left(\alpha_{xx}\right)^2\right\rangle\sin^2\theta\cos^2\theta$$
$$+\Gamma^2\left\langle\left(\alpha_{xy}\right)^2\right\rangle\left(1+\varepsilon\sin^2\theta\right)^2 \tag{6.65}$$

Some observations can be drawn. First of all, we note from Eq. 6.63 that when the excitation polarization is orthogonal to the nanocavity axis we observe a 90° polarization rotation. The maximum SERS intensity is observed for $\phi = \pi/2$ (Fig. 6.4a), that is, with the analyzer set orthogonal to the laser field polarization, and parallel the nanoantenna (nanocavity) axis. This phenomenon is also experimentally verified [45] and, again, is a demonstration of how the SERS radiation follows the nanoantenna radiation properties and not the molecular radiation properties (Raman is always maximum in the parallel direction to the excitation field, as shown in Section 6.3.8). Such an alteration of the polarization properties of the Raman radiation makes non-trivial the recovery of the molecular Raman polarizability tensor component based on polarized SERS measurements, as will be discussed in Section 6.4.4. Secondly, in order to investigate the difference between the excitation and the re-radiation enhancement factor we must be able to measure the terms in Γ^2. This is particularly evident from Eq. 6.65 and can be done

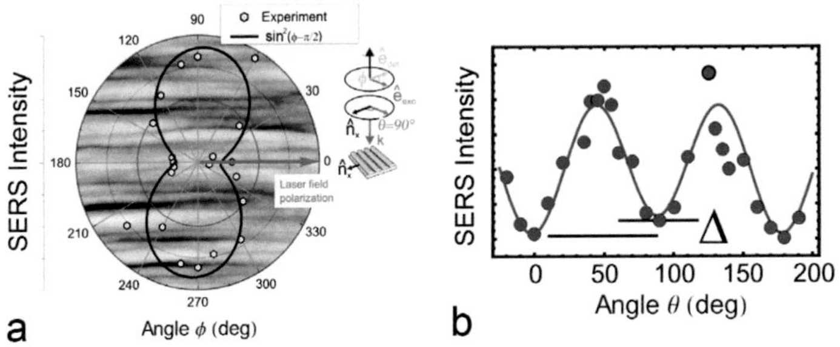

Figure 6.4 Plots of the SERS intensities for (a) polarized detection with excitation angle $\bar{\theta} = 90°$ with respect to nanocavity axis, (b) cross-polarized detection, highlighting the different intensity of the minima reflecting the different excitation and re-radiation enhancement values. Reprinted with permission from Ref. [45], Copyright 2011 American Chemical Society.

only for specific configurations in which the much more intense contributions proportional to Γ^4 are suppressed. Notably, in the cross-polarized configuration the terms in Γ^2 are singled out in the measured signals for $\theta = 0$ and $\theta = \pi/2$.

These are, respectively, proportional to the square of the excitation field enhancement factor, Γ^2, and to the square of the re-radiation enhancement, $\Gamma^2(1+\varepsilon)^2$. We therefore can retrieve the ratio between the two enhancement factors from the ratio

$$\Gamma^2_{\mathrm{rad}}(\lambda_\mathrm{R})/\Gamma^2_{\mathrm{exc}}(\lambda_\mathrm{L}) = I^{\perp}_{\mathrm{SERS}}(\pi;\pi/2,\pi/2)/I^{\perp}_{\mathrm{SERS}}(\pi;0,\pi/2) \quad (6.66)$$

Experimental evidence of the difference between the two enhancement factors has been given looking at the different minima values (see Fig. 6.4b) in the cross-polarized SERS signal from near-field coupled gold nanowires [45]. Thirdly, Eq. 6.62 reduces to the classical expression for the angular dependence of the polarized Raman intensity (Eq. 6.45) in absence of any enhancement, that is, for $\Gamma = 1$ and $\varepsilon = 0$, as expected. As well, in the absence of any field enhancement and of the consequent symmetry break given by the nanoantenna, all the excitation polarization directions are equivalent, so Eqs. 6.62 and 6.63 do coincide. An important final remark concerns the physical origin of the terms beyond the E^4 model. The calculations tell us that these are strongly related to the form of the field enhancement tensors (Eqs. 6.50, 6.51), and in particular to the assumptions made on the behavior of the field component orthogonal to the nanoantenna (nanocavity) axis at the *hot spot*. As soon as we assume that these are different from zero, that is, $\Gamma_{yy} \neq 0$, we find terms beyond the E^4 approximation. Conversely, only the E^4 terms are found if we assume that the nanoantenna has the capability to zero the amplitude of the electromagnetic field component orthogonal to its resonance axis, that is, $\Gamma_{yy} = 0$. Here we have assumed that these components are left unperturbed by the nanoantenna (i.e., $\Gamma_{yy} = 1$). Numerical calculations on a silver nanospheres dimer [47] show that the field component orthogonal to the dimer axis is strongly depleted at resonance, that is, $\Gamma_{yy} = 10^{-1} \div 10^{-2}$, but not zero. Assuming such values for the *yy*-component of the field-enhancement tensors in our model is physically equivalent to the case of unperturbed field. It still results in the appearance of terms beyond the E^4 approximation whose intensity, however, is further depleted by factors Γ_{yy}^2 and Γ_{yy}^4.

6.4.3 *Polarized SERS from Trimers and High-Symmetry Nanoantennas*

Metal nanoparticles trimers or tetramers have attracted the researchers interest because of the capability of behaving either as apolar plasmonic nanostructures (if symmetric) [55] or, if asymmetric, to rotate the polarization of the SERS radiation by changing the excitation light wavelength [41]. Theoretically it can be demonstrated that any complex nanoantenna structure with a C_n symmetry ($n \geq 3$) has an optical response insensitive to polarization when the wave vector is parallel to the symmetry axis. Moreover, if the nanoantennas has a C_{nv} or D_{nh} symmetry the polarization of the re-radiated field will be parallel to the Raman field [56, 57]. We therefore do not expect neither a dependence of the SERS intensity on the polarization vector, as experimentally verified, nor any polarization effect on the re-radiated photons. In terms of enhancement tensor, both tensors will be multiples of the identity. Individual nanocylinders, or nanospheres, as well as symmetric trimers or tetramers of near-field coupled nanospheres, nanocylinders, or nanorods satisfy this condition. Such nanoantennas provide an enhancement of the Raman intensity without altering the polarization properties of the molecular Raman emission.

The situation is totally different for asymmetric trimers. The formalism developed in the previous paragraphs can be used to model the SERS response from asymmetric structures, provided that the enhancement tensors for the specific geometry at the given excitation/Raman wavelengths are known (e.g., by generalized Mie Scattering Theory calculations [41]). In the case of trimers, for example, in the base $\left\{x', y'\right\}$ illustrated in Fig. 6.5, the tensors forms reported in Ref. [41] take the form

$$\overleftrightarrow{\Gamma}_{\text{exc}}(\lambda_{\text{L}}) = \begin{pmatrix} 0 & 0 & 0 \\ g_{yx}e^{-i\phi_x} & g_{yy}e^{-i\phi_y} & 0 \\ 0 & 0 & 1 \end{pmatrix} \tag{6.67}$$

and

$$\overleftrightarrow{\Gamma}_{\text{rad}}(\lambda_{\text{R}}) = \begin{pmatrix} 0 & g'_{xy}e^{-i\phi'_x} & 0 \\ 0 & g'_{yy}e^{-i\phi'_y} & 0 \\ 0 & 0 & 1 \end{pmatrix} \tag{6.68}$$

where g_{yx}, g_{yy}, g'_{xy}, and g'_{yy} are the real values and ϕ_x, ϕ_y, ϕ'_x, and ϕ'_y are the phases of the tensors elements. The trimer generates an enhanced excitation field polarized along the y'-direction in the 2–3 particles nanocavity (Fig. 6.5a), no matter how the laser field is polarized in the (x', y') plane (see, Fig. 6.5b,c). Considering a backscattering detection, the re-radiation tensor in Eq. 6.68 tells us that only the y'-component of the Raman dipolar field is enhanced and that the SERS field elliptically polarized in the x', y' plane. To model this situation we adopt a generic Raman tensor form for the molecular vibration considered. In the laboratory reference frame $\{x, y\}$ (Fig. 6.5a), which is rotated by an angle θ with respect to the trimer reference, we assume that the excitation is linearly polarized along the x-direction, that is, as $\hat{e}_{\text{exc}} = (1, 0)$, and the polarization analyzer oriented along $\hat{e}_{\text{det}} = (\cos\phi, \sin\phi)$. We now apply Eq. 6.53, taking care to send $\theta \rightarrow -\theta$ because we are rotating the field polarization and not the sample.

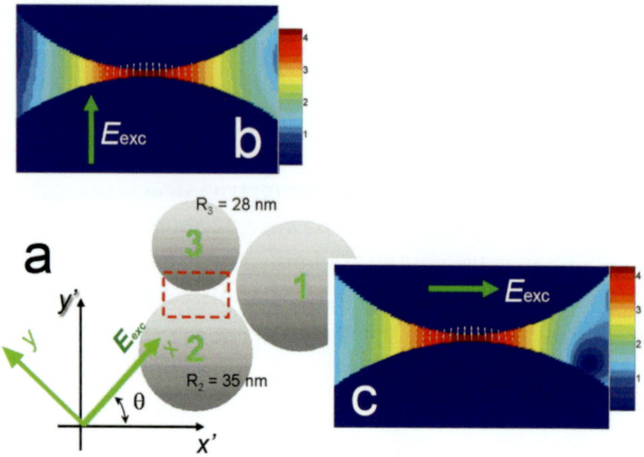

Figure 6.5 (a) Reference system and arrangement of the trimer. (b) The local field in one of the junctions of a nanoparticle trimer (indicated by a red dashed box in (a). The color plots show the absolute square magnitude of the calculated enhanced local electric field in the hot-spot area between particles 2 and 3 calculated at 532 nm for excitation polarization oriented along y' (b) and x' (c). Figure redrawn and reprinted from Ref. [41], Copyright 2008, National Academy of Sciences, U.S.A.

We find that the general form for the polarized SERS intensity:

$$I_{\text{SERS}}(\pi;\theta,\phi) \propto (\alpha_{yy})^2 \left[g_{yy}^2(1+r^2)\right]\left[g_{yy}'^2(1+r'^2)\right]$$

$$\times\left[1+\frac{(r^2-1)\cos 2\theta+2r\cos\Delta\sin 2\theta}{1+r^2}\right] \quad (6.69)$$

$$\times\left[1+\frac{(r'^2-1)\cos 2(\theta+\phi)+2r'\cos\Delta'\sin 2(\theta+\phi)}{1+r'^2}\right]$$

where $r = g_{yx}/g_{yy}$, $r' = g'_{yx}/g'_{yy}$, $\Delta = \phi_x - \phi_y$, and $\Delta' = \phi'_x - \phi'_y$. Notably, by assuming tensors in which the non-enhanced elements are strictly "0" (field totally depleted), instead of "1" (field unperturbed), yields only terms of the order $\left[g_{yy}^2(1+r^2)\right]\left[g_{yy}'^2(1+r'^2)\right]$, that is, proportional to the fourth power of the field enhancement, in the expressions of the polarized-SERS intensities. This puts forward the relevance of exactly knowing the amplitude of the electromagnetic field at the nanocavity between NPs (dimers or trimers) or between coupled nanowires (NWs) also when the polarization is orthogonal to the nanocavity axis.

We can now calculate the expressions for the most relevant experimental configurations. For the unpolarized SERS we have:

$$I_{\text{SERS}}^{\text{unpol}}(\pi;\theta) \propto \langle(\alpha_{yy})^2\rangle \left[g_{yy}^2\left(1+r^2\right)\right]\left[g_{yy}'^2\left(1+r'^2\right)\right] \times\left[1+\frac{\left(r^2-1\right)\cos 2\theta+2r\cos\Delta\sin 2\theta}{1+r^2}\right] \tag{6.70}$$

When the excitation is polarized parallel ($\theta = \pi/2$) or orthogonal ($\theta = 0$) to the nanocavity axis the polarized SERS intensity will be:

$$I_{\text{SERS}}^{\text{pol}}(\pi;\pi/2,\phi) \propto \langle(\alpha_{yy})^2\rangle g_{yy}^2\left[g_{yy}'^2\left(1+r'^2\right)\right] \times\left[1-\frac{\left(r'^2-1\right)\cos 2\phi+2r'\cos\Delta'\sin 2\phi}{1+r'^2}\right] \tag{6.71}$$

and

$$I_{\text{SERS}}^{\text{pol}}(\pi;0,\phi) \propto \langle(\alpha_{yy})^2\rangle g_{yy}^2 r^2\left[g_{yy}'^2\left(1+r'^2\right)\right] \times\left[1+\frac{\left(r'^2-1\right)\cos 2\phi+2r'\cos\Delta'\sin 2\phi}{1+r'^2}\right] \tag{6.72}$$

The parallel- and cross- polarized SERS intensities will be:

$$I_{\text{SERS}}^{\parallel}(\pi;\theta,0) \propto \left\langle(\alpha_{yy})^2\right\rangle \left[g_{yy}^2\left(1+r^2\right)\right]\left[g_{yy}'^2\left(1+r'^2\right)\right] \times\left[1+\frac{\left(r^2-1\right)\cos 2\theta+2r\cos\Delta\sin 2\theta}{1+r^2}\right] \times\left[1+\frac{\left(r'^2-1\right)\cos 2\theta+2r'\cos\Delta'\sin 2\theta}{1+r'^2}\right] \tag{6.73}$$

and

$$I_{\text{SERS}}^{\perp}(\pi;\theta,\pi/2) \propto \left\langle(\alpha_{yy})^2\right\rangle \left[g_{yy}^2\left(1+r^2\right)\right]\left[g_{yy}'^2\left(1+r'^2\right)\right] \times\left[1+\frac{\left(r^2-1\right)\cos 2\theta+2r\cos\Delta\sin 2\theta}{1+r^2}\right] \times\left[1-\frac{\left(r'^2-1\right)\cos 2\theta+2r'\cos\Delta'\sin 2\theta}{1+r'^2}\right] \tag{6.74}$$

Such expressions completely characterize the polarized SERS scattering of asymmetric trimers.

6.5 The SERS Depolarization Ratio

Since the early times [21], it was recognized that in SERS the relation between the depolarization ratio and the components of the Raman polarizability tensor was not straightforward. One striking example is given in Fig. 6.4a, where the polarized SERS intensity is found to be maximum in the direction orthogonal ($\phi = 90^\circ$) to the excitation field, whereas we have seen in Section 6.3.8 that the polarized Raman signal is maximum in the direction of the excitation field. Such a discrepancy, caused by nanoantenna radiation properties, suggests that a precise knowledge of the polarized SERS scattering is needed to understand how to retrieve the molecular information out of the signals measured experimentally. A detailed analysis of the SERS depolarization ratio is, therefore, useful in this context.

We remind that the depolarization ratio in Raman spectroscopy is defined as the ratio between the intensity of the Raman field polarized orthogonally and the intensity of the Raman field polarized parallel to the excitation field. We have seen in Section 6.3.7 that, once assigned the excitation–detection geometry, the depolarization ratio for a set of randomly oriented molecules is a parameter related to the intrinsic molecular Raman polarizability tensor. For these reasons in the following we will call ρ "molecular depolarization ratio." The latter, in the backscattering configuration, is given by

$$\rho_{\text{mol}} = \frac{\left\langle \left(\alpha_{yx}\right)^2 \right\rangle}{\left\langle \left(\alpha_{xx}\right)^2 \right\rangle} \tag{6.75}$$

In analogy, we can define the SERS depolarization ratio as the ratio between the cross- and the parallel-polarized SERS intensities, that is,

$$\rho_{\text{SERS}}\left(\theta\right) = \frac{I_{\text{SERS}}^{\perp}\left(\pi;\theta,\phi=\pi/2\right)}{I_{\text{SERS}}^{||}\left(\pi;\theta,\phi=0\right)} \tag{6.76}$$

ρ_{SERS} does not depend only on the Raman polarizability but also on θ, a parameter related to the geometrical coupling between the optical fields and the nanoantenna. An explicit model of such coupling is thus needed to relate ρ_{SERS} to ρ_{mol}, a model that will depend on the specific geometry of the nanoantenna we are dealing with. Analogous considerations can be drawn for the SERS degree of depolarization (or depolarization factor), which can be defined as

$$\sigma_{\text{SERS}}(\theta) = \frac{I_{\text{SERS}}^{||}(\pi;\theta,0) - I_{\text{SERS}}^{\perp}(\pi;\theta,\pi/2)}{I_{\text{SERS}}^{||}(\pi;\theta,0) + I_{\text{SERS}}^{\perp}(\pi;\theta,\pi/2)} \tag{6.77}$$

σ_{SERS} has been observed to assume values lower than zero in dimers and trimers [41–44], whereas only positive values, greater than or equal to 1/7 are expected in normal Raman scattering (Eq. 6.44). In the following we will discuss about the SERS depolarization ratio in near-field coupled dimers, high-symmetry nanoantennas, and in randomly distributed near-field coupled nanospheres.

6.5.1 *Nanoantenna Dimers*

For a nanoantenna dimer the SERS intensities are easily expressed in the E^4 approximation (Eqs. 6.59 and 6.60), yielding:

$$\rho_{\text{SERS}}(\theta) = \tan^2\theta \tag{6.78}$$

This expression only depends on the coupling between the light and the nanoantenna and does not bring any information on the Raman polarizability tensor. This proves that SERS experiments with strongly enhancing dimers cannot be used to probe the Raman polarizability tensor of molecules lying in the nanocavity [48], unless we are able to access the signal contribution given by the terms beyond the E^4 approximation, being the only that bring information on the non-diagonal components of the Raman tensor [45]. Equation 6.78 explains also why the SERS depolarization ratio, differently from the molecular depolarization ratio, can be higher than 3/4, showing that it even diverges for $\theta \rightarrow \pi/2$. A non-divergent expression of $\rho_{\text{SERS}}(\theta)$ can be obtained from the polarized SERS intensity formulas involving the terms beyond the E^4 approximation (Eqs. 6.64 and 65). Such an expression can be used to fit the experimental data and retrieve ρ_{mol}, although a detailed knowledge of the nanoantenna excitation and the re-radiation factors is required. Retrieving ρ_{mol} can be done more easily, by multiplying the SERS depolarization ratios measured at $\theta = 0, \pi/2$. It can be verified, in fact, that

$$\rho_{\text{mol}} = \sqrt{\rho_{\text{SERS}}(0) \cdot \rho_{\text{SERS}}(\pi/2)} \tag{6.79}$$

Experimental verification of the possibility to retrieve ρ_{mol} of Methylene blue molecules has been given using near-field coupled nanowires as underlying SERS-active substrate [45]. It is interesting to study $\rho_{\text{SERS}}(\theta)$ in the specific case when the excitation and the re-radiation enhancement factors are equal, $\varepsilon = 0$, in which the formula is much simpler, without loosing any physical insight.

$$\rho_{\text{SERS}}(\theta) = \frac{\Gamma^2 \rho_{\text{mol}} + \left(\Gamma^2 - 1\right)^2 \sin^2\theta\cos^2\theta}{\left(\Gamma^2\cos^2\theta + \sin^2\theta\right)^2} \tag{6.80}$$

This expression explicitly shows how the SERS depolarization ratio is linked to the molecular depolarization ratio through the light-nanoantenna coupling parameters θ and Γ in a non-straightforward way. Figure 6.6a compares the plots of $\rho_{\text{SERS}}(\theta)$ in the E^4 approximation (blue line) and including the terms beyond the E^4 approximation (red line). No divergences occur in the latter case, the maximum being limited. As well, the minimum is not zero, but assumes the value $\rho_{\text{mol}}/\Gamma^2$. Indeed the two models tend to coincide for $\Gamma^2 \gg 1$, the regime in which the E^4 approximation holds. A closer look at $\rho_{\text{SERS}}(\theta)$ shows that there are two possible behaviors. For molecular depolarization ratio $\rho_{\text{mol}} \geq 1/2$, $\rho_{\text{SERS}}(\theta)$ behaves as shown in Fig. 6.6a (red line) with maxima of intensity $\rho_{\text{mol}}\Gamma^2$ at $\theta = (n + 1/2)\pi$ (n is an integer), and minima of intensity $\rho_{\text{mol}}/\Gamma^2$ for $\theta = n\pi$. Vice versa, for $\rho_{\text{mol}} \leq 1/2$ there are two different regimes, depending on the specific value of ρ_{mol} and Γ. This is highlighted in Fig. 6.6b in which we compare the angular behavior of $\rho_{\text{SERS}}(\theta)$ for a nanoantenna featuring an SERS enhancement $\Gamma^4 \sim 10^2$ for three different molecules featuring vibrational modes with $\rho_{\text{mol}} = 0.47$ (red line), $\rho_{\text{mol}} = 0.3$ (yellow line), and $\rho_{\text{mol}} = 0$ (blue line). In the second and third case a local minimum is observed for $\theta = (n + 1/2)\pi$. The minimum goes down to zero for $\rho_{\text{mol}} = 0$.

It can be demonstrated that the local minimum occurs only when the condition

$$0 \leq \rho_{\text{mol}} \leq \frac{\Gamma^2 - 1}{2\Gamma^2} \tag{6.81}$$

is verified, corresponding to the couples of parameters $(\rho_{\text{mol}}, \Gamma)$ evidenced by the shaded area in Fig. 6.6c. Note that even if the SERS depolarization ratio has a local minimum in $\theta = (n + 1/2)\pi$,

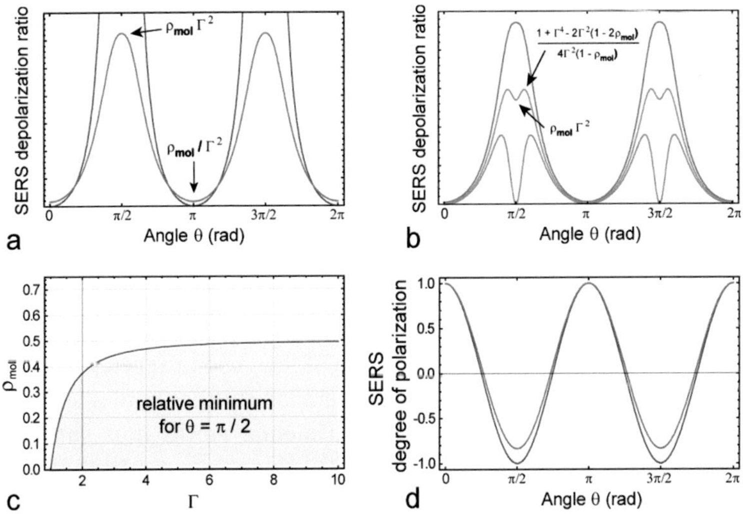

Figure 6.6 (a) Plot of the SERS depolarization ratio in the E^4 approximation (blue line) and accounting for the terms beyond the E^4 approximation (red line) for a vibrational mode featuring a depolarization ratio of 0.5. (b) Angular trend of the SERS depolarization ratios calculated for a nanoantenna featuring enhancement factor $\Gamma = 3$ on which are deposited molecules with different molecular depolarization ratios: $\rho_{\text{mol}} = 0.47$ (red line), $\rho_{\text{mol}} = 0.3$ (yellow line), and $\rho_{\text{mol}} = 0$ (blue line). (c) Plot of the function (blue line) delimiting the range of parameters (yellow solid area) for which an additional minimum is expected in the SERS depolarization ratio for $= (n + 1/2)\pi$, from that (hollow area) in which a maximum is expected for $\theta = (n + 1/2)\pi$. (d) Plot of the SERS degree of polarization in the E^4 approximation (blue line) and beyond the E^4 approximation (red line) for a vibrational mode featuring $\rho_{\text{mol}} = 0$. The nanoantenna is assumed to provide SERS signal enhancement ~ 100.

$\rho_{\text{SERS}}(\theta)$ is still a limited quantity and the relation given by Eq. 6.80 to retrieve the molecular depolarization ratio is always valid.

The degree of depolarization σ_{SERS} has been used in Refs. [41–43] as a physical quantity to prove the linear polarization induced by nanoantenna dimers to the SERS scattering of molecules in the cavities. In the E^4 approximation it can be easily found that

$$\sigma_{\text{SERS}}(\theta) = \cos 2\theta \tag{6.82}$$

which is expected to vary between -1 and 1 and, again, does not bring any information on the molecular Raman polarizability tensor.

The discrepancy to what expected in normal Raman spectroscopy ($\sigma_{\text{SERS}} > 0$) is, again caused by the polarization changes induced by the nanoantenna. If we include the terms beyond the E^4 approximation, assuming for simplicity equal enhancement factors ($\varepsilon = 0$), we find that the SERS degree of polarization reads as

$$\sigma_{\text{SERS}}(\theta) = \cos 2\theta - \left(\frac{1}{\Gamma^2}\right) \cdot \left[\frac{\rho_{\text{mol}}}{2} + (\rho_{\text{mol}} - 2)\sin^2\theta\right] \quad (6.83)$$

where we have neglected the terms proportional to $1/\Gamma^4$. Again, including terms such as Γ^2 we retrieve the information on the molecular depolarization ratio. Equation 6.82 has been experimentally verified [41, 43] in nanospheres, dimers, or clusters with a single hotspot. Nevertheless, in many experimental works it is shown that the minimum assumed by $\sigma_{\text{SERS}}(\theta)$ is higher than -1 [42–44]. Figure 6.6d displays a plot of $\sigma_{\text{SERS}}(\theta)$ and shows that when the full expression is taken in to account (red line), the minimum gets higher than -1 and the curve is slightly shifted with respect to the $\cos 2\theta$ trend expected in the E^4 approximation (blue line).

6.5.2 *Trimers and High-Symmetry Nanoantennas*

From the polarized SERS intensity calculations on asymmetric trimers (Eqs. 6.73 and 6.74) we can easily see that also in trimers the depolarization ratio, calculated in the E^4 approximation, does not bring any information on the molecular depolarization factor. $\rho_{\text{SERS}}(\theta)$ depends in a complex fashion only on the optical fields–nanoantenna coupling parameters (polarization angle, enhancement tensors). We therefore will not deserve particular attention to such systems.

Highly symmetric nanoantennas, both individual and near-field coupled, instead look extremely promising for high-sensitivity probing of the molecular Raman polarization tensor. As discussed in Section 6.4.3 nanoantennas with a C_{nv} or D_{nh} symmetry, $n \geq 3$, leave unperturbed the polarization of the Raman field [56]. As such, these nanoantennas are particularly important since they allow to amplify the Raman signal without altering the molecular vibrational geometric information, and so probe the orientation averaged diagonal and non-diagonal components of the Raman polarizability tensor of few molecules using simple polarized-cross- polarized measurements. Recent experiments carried out on gold nanocylinders, featuring a C_∞ symmetry, on which Methylene blue randomly oriented molecules has been absorbed confirm the possibility to retrieve the molecular depolarization factor in such a simple way [46].

6.5.3 *Randomly Distributed Nanoparticles*

Randomly distributed nanoparticles grown or cast on solid substrates represent one of the most easy to fabricate, reproducible, yet highly efficient class of SERS-active substrates. When densely packed with average distances of few nanometers, such systems behave as an ensemble of randomly oriented nanocavities. Such systems can be easy modeled averaging the polarized SERS response over all the possible nanocavities' orientations or, equivalently, averaging the response of a single dimer over all the possible incident polarizations in the plane. The SERS depolarization ratio can be therefore calculated as

$$\rho_{\mathrm{SERS}} = \frac{\int_0^{2\pi} I^{\perp}_{\mathrm{SERS}}\,(\pi;\theta,\,\pi/2)\,\mathrm{d}\theta}{\int_0^{2\pi} I^{\parallel}_{\mathrm{SERS}}\,(\pi;\theta,\,0)\,\mathrm{d}\theta} \tag{6.84}$$

Assuming equal excitation and re-radiation enhancements, we find

$$\rho_{\mathrm{SERS}} \sim \frac{1}{3} + (3\rho_{\mathrm{mol}} - 1)\,\frac{1}{\Gamma^2} \tag{6.85}$$

where we have neglected the terms in $1/\Gamma^4$. To the first order, the SERS depolarization ratio turns out to be $1/3$, constant for every molecule, as experimentally verified [28, 45]. Again, information on ρ_{mol} is brought by the terms beyond E^4 approximation, and can be obtained with such nanostructures, provided the SERS substrate is homogeneous enough and provides an enhancement factor such that the experimental error is below $1/\Gamma^2$. SERS enhancement factors $\Gamma^4 > 10^4$ are typical for such nanostructures, requiring an experimental accuracy of $\sim 0.1\%$ to measure molecular depolarization ratios of 0.5.

6.6 Conclusions

SERS is an extremely powerful technique capable to enhance the vibrational fingerprint of molecules deposited on optically resonant nanoantennas. One of the most promising applications of SERS, still under development, is to probe the Raman polarizability tensor of single or few molecules and, in a complementary fashion, to retrieve information on the spatial orientation of single molecules in nanocavities, for example, during a chemical interaction with another molecule. Polarization control is a keyword to accomplish such a task. As seen in this chapter, polarization-sensitive SERS experiments need an underlying model to extract the Raman polarizability tensor information encoded in the intensity signals. The field enhancement mechanisms and the influence of the nanoantenna radiation properties, in fact, play a crucial role in puzzling the polarization of the SERS fields so that any model developed for normal Raman spectroscopy has to be revisited in the light of the molecule-to-nanoantenna strong coupling. On the other hand, the synthesis of new nanoantenna structures, featuring high-symmetry and higher-field enhancement, looks extremely promising to accomplish the tasks cited above. An intense experimental effort is, however, required to study the re-radiation properties of such antennas and ascertain to what extent these structures can be considered apolar. In conclusion, polarization-sensitive SERS offers an exciting challenge from both the theoretical and the experimental point of view aimed both at a better understanding of the basic mechanisms of SERS and at allowing one to carry out molecular analysis of few or even single molecules.

Acknowledgments

We thank A. Iatì, M. G. Donato, G. Calogero, R. Saija, F. Borghese, M. Allegrini, F. Fuso, F. Tantussi, F. Buatier de Mongeot, A. Toma, E. Di Fabrizio, M. Lamy de La Chapelle, N. Felidj, and T. Toury for discussions. We wish also to thank A. Otto for critically reading the manuscript. Financial support from the European Union through the project Nanoantenna (FP7-Health-F5-2009-241818) and from MIUR through the projects PRIN 2008J858Y7 and Programma Operativo Nazionale Ricerca e Competitività 2007–2013, PON01 01322 PANREX.

References

1. Otto, A. (1968). Excitation of nonradiative surface plasma waves in silver by the method of frustrated total reflection, *Z. Phys.*, **216**, pp. 398–410.
2. Meier, S. A. (2007). *Plasmonics: fundamentals and applications* (Springer, New York).
3. Gramotnev, D. K. and Bozhevolnyi, S. I. (2010). Plasmonics beyond the diffraction limit, *Nat. Photonics,* **4**, pp. 83–91.
4. Schuller, J. A., Barnard, E. S., Cai, W. S., Jun, Y. C., White, J. S. and Brongersma, M. L. (2010). Plasmonics for extreme light concentration and manipulation, *Nat. Mater.,* **9**, pp. 193–204.
5. Kneipp, K., Moskovits, M. and Kneipp, H. (2006). *Surface enhanced Raman scattering* (Springer, New York).
6. Le Ru, E. C. and Etchegoin, P. (2009). *Principles of surface enhanced Raman spectroscopy* (Elsevier, Amsterdam).
7. Otto, A., Mrozek, I., Grabhorn, H. and Akemann, W. (1992). Surface-enhanced Raman scattering, *J. Phys. Condens. Mat.*, **4**, pp. 1143–1212.
8. Geddes, C. D. and Lakowicz, J. R. (2002). Metal-enhanced fluorescence, *J. Fluoresc.,* **12**, pp. 121–129.
9. Farahani, J. N., Pohl, D. W., Eisler, H. J. and Hecht, B. (2005). Single quantum dot coupled to a scanning optical antenna: a tunable superemitter, *Phys. Rev. Lett.,* **95**, p. 017402.
10. Kinkhabwala, A., Yu, Z. F., Fan, S. H., Avlasevich, Y., Mullen, K. and Moerner, W. E. (2005). Large single-molecule fluorescence enhancements produced by a bowtie nanoantenna, *Nat. Photonics,* **3**, pp. 654–657.
11. Hartstein, A., Kirtley, R. J. and Tsang, C. J. (1980). Enhancement of the infrared absorption from molecular monolayers with thin metal overlayers, *Phys. Rev. Lett.,* **45**, pp. 201–204.
12. Neubrech, F., Pucci, A., Cornelius, T. W., Karim, S., Garcia-Etxarri, A. and Aizpurua, J. (2008). Resonant plasmonic and vibrational coupling in a tailored nanoantenna for infrared detection, *Phys. Rev. Lett.,* **101**, art no. 157403.

13. Homola, J. (2008). Surface plasmon resonance sensors for detection of chemical and biological species, *Chem. Rev.*, **108**, pp. 462–493.
14. Aslan, K., Gryczynski, I., Malicka, J., Matveeva, E., Lakowicz, J. R. and Geddes, C. D. (2005). Metal-enhanced fluorescence: an emerging tool in biotechnology, *Curr. Opin. Biotech.*, **16**, pp. 55–62.
15. Xie, W. and Schlucker, S. (2013). Medical applications of surface-enhanced Raman scattering, *Phys. Chem. Chem. Phys.*, **15**, pp. 5329–5344.
16. Anker, J. N., Hall, W. P., Lyandres, O., Shah, N. C., Zhao, J. and Van Duyne, R. P. (2008). Biosensing with plasmonic nanosensors, *Nat. Mater.*, **7**, pp. 442–453.
17. D'Andrea, C., Bochterle, J., Toma, A., Huck, C., Neubrech, F., Messina, E., Fazio, B., Marago, O. M., Di Fabrizio, E., de la Chapelle, M. L., Gucciardi, P. G. and Pucci, A. (2013). Optical nanoantennas for multiband surface-enhanced infrared and Raman spectroscopy. *ACS Nano*, **7**, pp. 3522–3531.
18. Foti, A., D'Andrea, C., Bonaccorso, F., Lanza, M., Calogero, G., Messina, E., Maragò, O. M., Fazio, B. and Gucciardi, P. G. (2013). A shape-engineered surface-enhanced Raman scattering optical fiber sensor working from the visible to the near-infrared, *Plasmonics*, **8**, pp. 13–23.
19. Messina, E., Cavallaro, E., Cacciola, A., Saija, R., Borghese, F., Denti, P., Fazio, B., D'Andrea, C., Gucciardi, P. G., Iatì, M. A., Meneghetti, M., Compagnini, G., Amendola, V. and Maragò, O. M. (2011). Manipulation and Raman spectroscopy with optically trapped metal nanoparticles obtained by pulsed laser ablation in liquids, *J. Phys. Chem. C*, **115**, 5115–5122.
20. Maragò, O. M., Jones, P. H., Gucciardi, P. G., Volpe, G. and Ferrari, A. C. (2013). Optical trapping and manipulation of nanostructures, *Nat. Nanotech.*, **8**, 807–819.
21. Moskovitz, M. (1985). Surface-enhanced Raman spectroscopy, *Rev. Mod. Phys.*, **57**, pp. 783–826.
22. Le Ru, E. C., Blackie, E., Meyer, M. and Etchegoin, P. G. (2007). Surface enhanced Raman scattering enhancement factors: a comprehensive study, *J. Phys. Chem. C*, **111**, pp. 13794–13803
23. Ward, D. R., Halas, N. J., Ciszek, J. W., Tour, J. M., Wu, Y., Nordlander, P. and Natelson, D. (2008). Simultaneous measurements of electronic conduction and Raman response in molecular junctions, *Nano Lett.*, **8**, pp. 919–924.

24. Etchegoin, P. G., Le Ru, E. C. and Fainstein, A. (2011). Bi-analyte single molecule SERS technique with simultaneous spatial resolution, *Phys. Chem. Chem. Phys.*, **13**, pp. 4500–4506.

25. Le Ru, E. C. and Etchegoin, P. (2006). Rigorous justification of the E^4 enhancement factor in surface enhanced Raman spectroscopy, *Chem. Phys. Lett.*, **423**, pp. 63–66.

26. Nagasawa, F., Takase, M., Nabika, H. and Murakishi, K. (2011). Polarization characteristics of surface-enhanced Raman scattering from a small number of molecules at the gap of a metal nano-dimer, *Chem. Comm.*, **47**, pp. 4514–4516.

27. Le Ru, E. C., Grand, J., Félidj, N., Aubard, J., Lévi, G., Hohenau, A., Krenn, J. R., Blackie, E. and Etchegoin, P. G. (2008). Experimental verification of the SERS electromagnetic model beyond the E^4 approximation: polarization effects, *J. Phys. Chem. C.*, **112**, pp. 8117–8121.

28. Le Ru, E. C., Meyer, M., Blackie, E. and Etchegoin, P. G. (2008). Advanced aspects of electromagnetic SERS enhancement factors at a hot spot, *J. Raman Spectrosc.*, **39**, pp. 1127–1134.

29. Lee, S. J., Guan, Z., Xu, H. and Moskovits, M. (2007). Surface-enhanced Raman spectroscopy and nanogeometry: the plasmonic origin of SERS, *J. Phys. Chem. C*, **11**, pp. 17985–17988.

30. Schatz, G. C., Young, M. A. and Van Duyne, R. P. (2006). Electromagnetic mechanism of SERS, *Top. Appl. Phys.*, **103**, pp. 19–45.

31. Wokaun, A. (1984). Surface-enhanced electromagnetic processes. *Solid State Phys.*, **38**, pp. 223–294.

32. Hao, E. and Schatz, G. C. (2004). Electromagnetic fields around silver nanoparticles and dimers, *J. Chem Phys.*, **120**, pp. 357–366.

33. Schnell, M., Garcia-Etxarri, A., Huber A. J., Crozier, K., Alkorta, J., Aizpurua, J. and Hillenbrand, R. (2009). Controlling the near-field oscillations of loaded plasmonic nanoantennas, *Nat. Photonics.*, **3**, pp. 287–291.

34. Schnell, M., Garcia-Etxarri, A., Alkorta, J., Aizpurua, J. and Hillenbrand, R. (2010). Phase-resolved mapping of the near-field vector and polarization state in nanoscale antenna gaps, *Nano Lett.*, **10**, pp. 3524–3528.

35. Quin, L., Zou, S., Xue, C., Atkinson, A., Schatz, G. C. and Mirkin, C. A. (2006). Designing, fabricating, and imaging Raman hot spots, *Proc. Natl. Acad. Sci.*, **103**, pp. 13300–13303.

36. Garcia-Vidal, F. J. and Pendry, J. B. (1996). Collective theory for surface-enhanced Raman scattering, *Phys. Rev. Lett.*, **77**, pp. 1163–1166.

37. Nie, S. M. and Emory, S. R. (1997). Probing single molecules and single nanoparticles by surface-enhanced Raman scattering, *Science*, **275**, pp. 1102–1106.
38. Grand, J., De La Chapelle, M. L., Bijeon, J.-L., Adam, P.-M., Vial, A. and Royer, P. (2005). Role of localized surface plasmons in surface-enhanced Raman scattering of shape-controlled metallic particles in regular arrays, *Phys. Rev. B*, **73**, art. no. 033407.
39. Baik, J. M., Lee, S. J. and Moskovits, M. (2009). Polarized surface-enhanced Raman spectroscopy from molecules adsorbed in nano-gaps produced by electromigration in silver nanowires, *Nano Lett.*, **9**, pp. 672–679.
40. Wei, H., Hao, F., Huang, Y., Wang, W., Nordlander, P. and Xu, H. X. (2008). Polarization dependence of surface-enhanced Raman scattering in gold nanoparticle-nanowire systems, *Nano Lett.*, **8**, pp. 2497–2502.
41. Shegai, T., Li, Z., Dadosh, T., Zhang, Z., Xu, H. and Haran, G. (2008). Managing light polarization via plasmon-molecule interactions within an asymmetric metal nanoparticle trimer, *Proc. Natl. Acad. Sci. U.S.A.*, **105**, pp. 16448–16453.
42. Bosnick, K. A., Jiang, J. and Brus, L. E. (2002). Fluctuations and local symmetry in single-molecule rhodamine 6G Raman scattering on silver nanocrystal aggregates, *J. Phys. Chem. B*, **106**, pp. 8096–8099.
43. Jiang, J., Bosnick, K. A., Maillard, M. and Brus, L. E. (2003). Single molecule Raman spectroscopy at the junctions of large Ag nanocrystals, *J. Phys. Chem. B*, **107**, pp. 9964–9972.
44. Xu, H. X. and Käll, M. (2003). Polarization-dependent surface-enhanced Raman spectroscopy of isolated silver nanoaggregates, *ChemPhysChem*, **4**, pp. 1001–1005.
45. Fazio, B., D'Andrea, C., Bonaccorso, F., Irrera, A., Calogero, G., Vasi, C., Gucciardi, P. G., Allegrini, M., Toma, A., Chiappe, D., Martella, C. and de Mongeot, F. B. (2011). Re-radiation enhancement in polarized surface-enhanced resonant Raman scattering of randomly oriented molecules on self-organized gold nanowires, *ACS Nano*, **5**, pp. 5945–5956.
46. Foti, A. (2013). Proprietà di polarizzazione della Radiazione SERS diffusa da nanoantenne lineari singole, Laurea Thesis in Physics, University of Messina.
47. Etchegoin, P. G., Galloway, C. and Le Ru, E. C. (2006). Polarization-dependent effects in surface-enhanced Raman scattering (SERS), *Phys. Chem. Chem. Phys.*, **8**, pp. 2624–2628.

48. Shegai, T. O. and Haran, G. (2006). Probing the Raman scattering tensors of individual molecules, *J. Phys. Chem. B.*, **110**, pp. 2459–24561.
49. Long, D. A. (2002). *The Raman Effect: A Unified Treatment of the Theory of Raman Scattering by Molecules* (John Wiley: England).
50. Strommen, D. P. (1992). Specific values of the depolarization ratio in Raman spectroscopy: their origins and significance, *J. Chem. Educ.,* **69**, pp. 803–807.
51. McFarland, A. D., Young, M., Dieringer, J. and Van Duyne, R. P. (2005). Wavelength-scanned surface-enhanced Raman excitation spectroscopy, *J. Phys. Chem. B.*, **109**, pp. 11279–11285.
52. Guillot, N., Shen, H., Fremaux , B., Péron, O., Rinnert, E., Toury, T. and Lamy de la Chapelle, M. (2010). Surface enhanced Raman scattering optimization of gold nanocylinder arrays: influence of the localized surface plasmon resonance and excitation wavelength, *Appl. Phys. Lett.*, **97**, 023113.
53. D'Andrea, C., Fazio, B., Irrera, A., Artoni, P., Maragò O. M., Calogero, G. and Gucciardi, P. G. (2011). Spectral dependence of the amplification factor in surface enhanced Raman scattering, *Atti Accad. Pelorit. Pericol. Cl. Sci. Fis. Mat. Nat.*, **89**, C1V89S1P025.
54. Le Ru, E. C., Etchegoin, P. G., Grand, J., Felidj, N., Aubard, J., Levi, G., Hohenau, A. and Krenn, J. R. (2008). Surface enhanced Raman spectroscopy on nanolithography-prepared substrates, *Curr. Appl. Phys.*, **8**, pp. 467–470.
55. de la Chapelle, M. L., Guillot, N., Frémaux, B., Shen, H. and Toury, T. (2013). Novel apolar plasmonic nanostructures with extended optical tunability for sensing applications, *Plasmonics*, **8**, pp. 475–480.
56. Shen, H., Rouxel, J., Guillot, N., Lamy de la Chapelle, M. and Toury, T. (2012). Light polarization properties of three fold symmetry gold nanoparticles: model and experiments, *C. R. Phys.*, **13**, 830–836.
57. de la Chapelle, M. L., Guillot, N., Frémaux, B., Shen, H. and Toury, T. (2012). Novel apolar plasmonic nanostructures with extended optical tunability for sensing applications, *Plasmonics*, **8**, 475–480.

Chapter 7

Synthesis of SERS-Active Substrates by Pulsed Laser Ablation

Sebastiano Trusso,[a] Enza Fazio,[b] Fortunato Neri,[b] and Paolo Maria Ossi[c]

[a] *IPCF-CNR, Istituto per i Processi Chimico-Fisici, V.le F. Stagno d'Alcontres 37, 98158 Messina, Italy*

[b] *Dipartimento di Fisica ed Scienze della Terra, V.le F. Stagno d'Alcontres 67, 98166, Messina, Italy*

[c] *Dipartimento di Energia & Centre for NanoEneginered MAterials and Surfaces NEMAS, Politecnico di Milano, via Ponzio 34-3, 20133 Milano, Italy*

trusso@me.cnr.it

7.1 Introduction

Since the observation of the anomalous intensification of the Raman signal from pyridine molecules adsorbed on a rough metal surface [1] and the successive explanation of the phenomenon as the amplification of the Raman scattering cross section due to the presence of a nanostructured metal surface [2, 3], the research in this field grew rapidly and it is still growing as witnessed by about 2000 of papers published in this field just considering the year 2012[a]. Two enhancement mechanisms of different origin

[a] http://isiknowledge.com search string: "surface-enhanced Raman"

Handbook of Enhanced Spectroscopy
Edited by Pietro Giuseppe Gucciardi, Marc Lamy de la Chapelle, and Nathalie Lidgi-Guigui

ISBN 978-981-4613-32-3 (Hardcover), 978-981-4613-33-0 (eBook)
www.panstanford.com

were proposed to account for the observed giant amplification of the Raman signal: an electromagnetic and a chemical one [4, 5]. In the electromagnetic mechanism, the enhancement results from the coupling of the incident electromagnetic field with artificially metallic corrugated surface, in the chemical one, electronic states generated by the interaction between the metal and the molecules adsorbed on it are involved. In fact the order of magnitude of the resulting enhancements is very different: theoretical estimates for the enhancement factors point out for values up to 10^{10} [6, 7] for the electromagnetic and up to 10^3 for the chemical one [8]. Both mechanisms are still at the focus of intense research activity to clarify the physical basis underlying the phenomenon.

On the other hand, the role played by the enhancing substrate is of paramount importance [9]. A lot of efforts thus have been made in order to improve Raman enhancement and several production methods were explored, including chemical [10–15], electrochemical [16–19] and physical [20–24], electron beam lithography [25, 26], and metal film over nanosphere (MFON) methods [27–29]. The main goal is to optimize the SERS response through the control of the optical and morphological properties of the substrates.

In this chapter, we report about the production of SERS-active substrates produced by pulsed laser ablation. This technique allows for the production of noble metal nanoparticles (NPs) either deposited onto appropriate solid substrates or produced in colloidal form in a liquid medium, the latter resulting free of chemical contaminants that are present as residuals when chemical methods are employed. We will show that some experimental parameters are critical in the determination of the final optical and morphological properties of the substrates, so that their fine control can be used to tailor and to optimized their SERS activity.

7.2 Pulsed Laser Ablation: Technique and Instrumentation

Pulsed laser deposition developed as soon as high energy pulsed lasers were made available. In fact, it was soon clear that high energy pulsed lasers were able to vaporize any kind of material, even refractory ones, when properly focused on their surfaces. If laser fluence F (i.e., the energy delivered by a laser pulse per unit area (Jcm^{-2})) is above a critical value (F_{th}) that depends on the material and on the laser wavelength, a portion of the irradiated material is vaporized. Vaporized material is then collected on an appropriate substrate positioned in front of the target. Vaporization occurs on *ps* time scale, thus in case of a nanosecond laser the pulse lasts enough to interact with the vaporized species. Such an interaction leads to the formation of a plasma that, driven by its high internal pressure, starts to expand. Expansion takes place in the direction normal to the target surface, irrespective of the laser incidence direction. In the very first step of the interaction between the laser and the target, absorption of the laser pulse energy results in electronics excitations. Thereafter the system returns to equilibrium through electron-phonon coupling mechanisms that involve thermal conductivity and generation/recombination of lattice defects. The macroscopic result is a sudden increase of the local temperature. Such a process is very fast so that the boiling point of the target material can be easily reached with the formation of a large number of homogeneous nucleation sites. In the second step, the molten material on the target surface vaporizes, still absorbing energy in case of a nanosecond laser pulse. As a consequence, a thin layer of ionized vaporized material (plasma) develops. At the end of the laser pulse, plasma starts to expands. Most of the energy absorbed by the vaporized species is now converted into kinetic energy. Plasma species then can be collected on substrates where material growth proceeds.

With respect to other deposition techniques, pulsed laser deposition (PLD) presents some advantages:

- Any solid material can be used as target including refractories such as SiC, W, or TiC;
- Laser and deposition chamber are reciprocally independent so that different experimental arrangements are available (PLD can be easily coupled to other deposition techniques such as RF sputtering, or thermal, or e-beam evaporation systems);
- Energies of the ejected species can be very large, up to 10 eV;

- Complex stoichiometry can be transferred from the target to the material, with no, or little changes;
- Atomic species not present in the target can be included in the growing films by feeding reactive species in the chamber.

On the other hand, there are some well-known drawbacks:

- Besides atomic species, micrometric sized particulate is ejected from the target;
- Evaporated species are ejected normally to the surface of the target so that the thickness profile of the growing film results nonhomogeneous;
- Ablation efficiency can depend on the atomic mass of the removed atoms, so that films can results substoichiometric especially, concerning light, or volatile species.

Nevertheless, these issues can be limited, or eliminated. Particulate can be reduced by means of mechanical velocity filters or magnetic fields that prevent the arrival of slow and massive objects on the substrates [30, 31]. Homogeneous thicknesses can be obtained, even on large areas, by movimentation of the substrate holder, and correction of stoichiometry can be achieved by feeding gaseous species in the chamber [32] or by the use of multitarget holders [33, 34]. The above reported peculiarities made PLD the deposition technique of choice for superconductor, piezoelectric, or ferroelectric materials such as YBCO, PZT, or $Ba_xSr_{1-x}TiO_3$ whose complex stoichiometry made them difficult to produce as thin films using alternative deposition techniques [35–37].

A schematic of a typical pulsed laser deposition apparatus is shown in Fig. 7.1. The deposition process takes place in a vacuum chamber. The laser beam is focused onto the target surface by a lens, a beam homogenizer can be used to obtain an homogeneous energy distribution on the irradiated area. In order to avoid excessive target damaging, which can drastically alter the deposition conditions, target is usually mounted on a rotating holder; a better alternative is surface target rastering by moving the laser beam using mechanical or piezo movement. One or more gas inlets allow

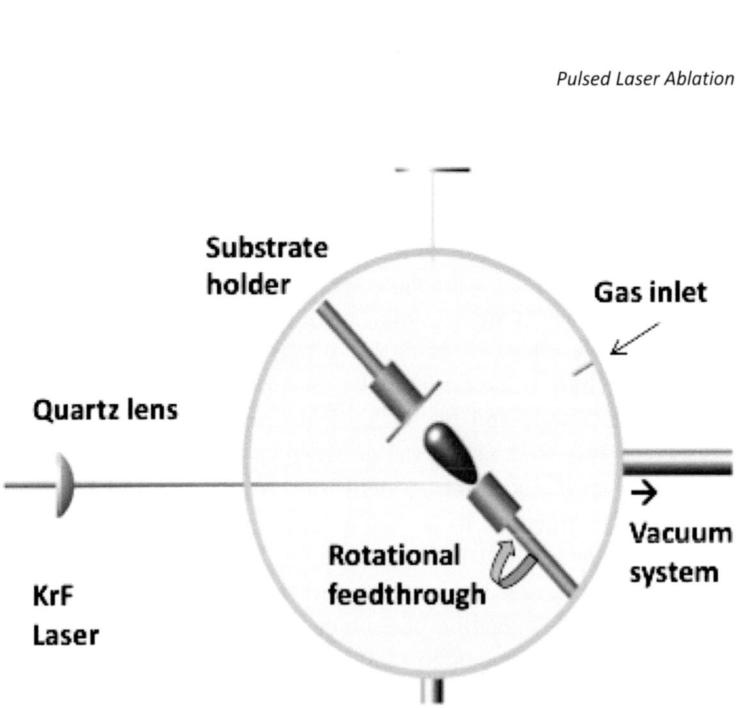

Figure 7.1 Setup of a typical pulsed laser deposition apparatus.

the introduction in the deposition chamber of gaseous species. Reactive gases interacting with the expanding plasma can produce chemical reactions, while inert gases are employed to modify plasma expansion kinetics. Typical deposition parameters that should be controlled in order to tailor optical, structural, and electronic properties of the growing material are as follows:

- laser wavelength and pulse duration;
- irradiated area;
- target to substrates distance;
- substrate temperature;
- density and nature of ambient gas;
- laser pulse number;
- fluence, F, i.e., the energy density released on a given area ($F = \mathrm{Jcm}^{-2}$).

All of these cooperate in very complex way to determine the final properties of the growing material.

7.3 Plasma Expansion

Thin film deposition by means of pulsed laser ablation can be sketched into four different stages: laser-target interaction, plasma formation, plasma expansion, and thin film growth on the substrate. Each one is the object of extensive research, nevertheless laser-target interaction is determined by the laser type (wavelength and pulse duration) and by the optical properties of the target material. In this first stage, the process can be controlled only through the laser pulse energy delivered on the irradiated target area, i.e., the fluence. It is worth to mention that different physical phenomena occur if nanosecond, or femtosecond lasers are employed. Nanosecond lasers, in fact, interact with the ablated species leading to ionization phenomena, while femtosecond laser pulses end before plasma formation. In the subsequent stages, the process can be more easily controlled by adjusting several experimental parameters. Plasma expansion dynamics, in fact, is strongly influenced whether deposition is performed in vacuum or in the presence of an ambient gas. Plasma kinetic energy can be controlled by the introduction in the chamber of inert gases, additional kinetic energy can be supplied to adsorbed species by keeping the substrate at a given temperature.

The laser generated plasma undergoes in vacuum an adiabatic expansion [38–40]. During the free expansion plasma assumes an elliptical shape, and the distribution of the species velocities is described by a shifted Maxwell–Boltzmann distribution:

$$f(v) = A\left(\frac{m}{2\pi k}\right)^{3/2} v^n \exp\left(\frac{-m(v-u)^2}{2kT}\right) dv \tag{7.1}$$

where m is the mass of the atoms, v the velocity, k the Boltzmann constant, T the temperature, and u the velocity of the center of mass. The model was also able to explain the thickness dependence of a deposited film $h(\theta)$ being $\theta = arctan\frac{R}{r}$, and r the distance of a point on the target surface from the normal straight line joining the laser spot and the substrate positioned at a distance R from the target:

$$h(\theta) = \frac{M}{2\pi\rho r^2}\cos^3\theta \tag{7.2}$$

here M is the deposited mass and ρ the density of the material. The presence of an ambient gas makes expansion a very complex process. A comprehensive model able to describe the phenomenon is still lacking, nevertheless phenomenological models able to give a description of one or more stages are present in the literature [41–43]. In the initial stage of expansion in the presence of a low pressure gas, a nearly free expansion is observed as it occurs in vacuum, after then a slowing down is observed: in such a case the drag model gives a good description of the expansion [44, 45]. The model describes the ejected species as an ensemble of particles subjected to a viscous force proportional to their velocity v:

$$R(t) = R_0(1 - e^{-\beta t}) \tag{7.3}$$

where R indicates the position of the front edge of the plasma, R_0 the distance at which the plasma propagation ceases (it is generally indicated as the stopping distance), and β a slowing down coefficient. Drag model holds only for low gas pressures; upon increasing gas pressure values, distances slightly shorter than those observed experimentally are predicted. At higher background pressures the blast model gives better results [46, 47]. According to this model, the plasma, expanding at a supersonic velocity, acts as a piston, compressing and accelerating the background gas leading to the formation of a shock wave ahead of the contact surface between the plasma and the ambient gas. A shock wave propagates away from the target and it is immediately followed by an expansion wave, which gradually reduces the shock strength. The presence of the background gas produces plasma confinement and as a consequence the increase of the pressure, temperature, and density of the ejected species in correspondence of the plasma-gas contact front [42]. The position of the front edge, as a function of the time t, is given by the following equation:

$$R(t) = \xi_0 \left(\frac{E}{\rho_0}\right)^{1/5} t^{2/5} \tag{7.4}$$

being ξ a geometrical factor nearly equal to unity, E the energy of the plasma, and ρ_0 the gas density [42]. The shock wave regime develops as long as the gas mass at the contact edge is larger than the plasma mass. In the final stage of expansion, a diffusive regime sets up:

$$R(t) = \sqrt{Dt} \tag{7.5}$$

where D is the diffusion coefficient given by $D = (\lambda v)/3$ with λ the mean free path and v the relative velocity of the colliding particles. None of the above reported phenomenological models is able to give a complete description of the process.

Recently the mixed-propagation model was introduced in an attempt to give a description of the expansion for larger pressure and temporal intervals [48, 49]. Mixed-propagation model describes plasma expansion by means of a two-step combination of modified version of drag and diffusion models. The drag model is assumed to hold in the initial stage of expansion but the parameters R_0 and β now assume a more physical significance being replaced by $R'_0 = v_0 D/\mu$ and $\beta' = \mu/D'$ respectively. v_0 is the initial plasma velocity, μ a slowing down coefficient and D' a modified diffusion coefficient. After the initial drag-like stage, plasma expansion is described by a diffusion-like equation, $R(t) = \sqrt{D't}$, $D' = kv_0(n_g\sigma)$ replaces the coefficient diffusion D, k is a constant that depends on the atomic mass of the plasma species, n_g the gas density, and σ the scattering cross section. Experimental data are fitted to the following equation:

$$R(t) = R_0^D(1 - e^{-\beta t}),\ \text{for } R < R_0^D$$

$$R(t) = R_0^D + \sqrt{D''(t - t_0^D)},\ \text{for } R > R_0^D \tag{7.6}$$

The R_0^D value represents, in this model, a point that *unphysically* separates the initial viscous motion and the subsequent diffusive regime. Such a position is determined by fitting experimental data with Eq. 7.6, the fitting procedure also provides a value for v_0 that represents the initial plasma velocity in the diffusive regime, or, equivalently, the residual velocity of the plasma species at the end of the viscous motion. Experimentally front edge positions of the expanding plasma can be obtained from pictures captured using an intensified nanosecond gated charge-coupled camera (iCCD), capturing shots of the expanding plasma at different time delays with respect to the arrival of the laser pulse. The investigated time interval ranges from the minimum accessible delay, which in our experimental setup depends on the electronics signal conditioning, and several μs. Luminous emission from the plasma originates from electronic relaxation processes of collisionally excited atomic species; as long as plasma expands the collision rate decreases, and

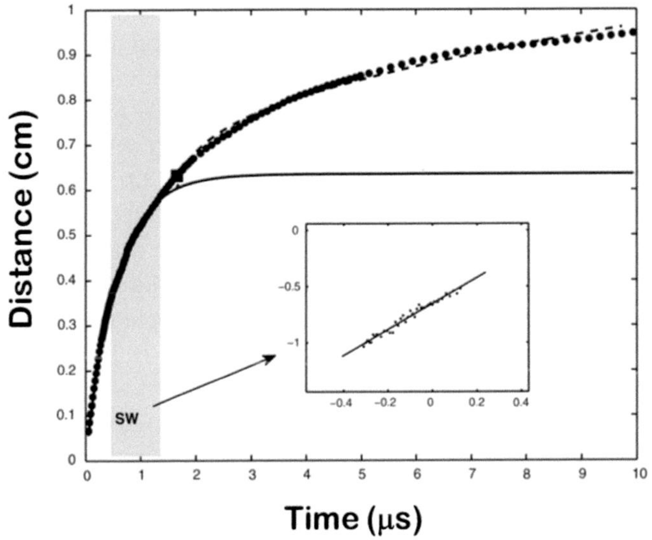

Figure 7.2 Position of the front edge of a silver laser generated plasma expanding in Ar at 70 Pa. Continuous and dashed lines refer to the fit performed adopting the mixed-propagation model (see Eq. 7.6). The full square indicates the position R_0^D and delay time t_0^D at which the plasma moves with the velocity v_0^D, used as the initial velocity for the diffusive regime. The gray area refers to the time interval where a shock wave regime holds, as can be seen from the data reported in the inset where are plotted as $\log(R(t))$ *vs.* $(2/5)\log(t)$.

so does the luminous emission. Integration times then are increased to maximize the signal-to-noise ratio. Typical integration times are of the order of few nanoseconds, in the initial stage of expansion, up to few μs in the late stage. In Fig. 7.2 (adapted from [50]), we report the position of the front edge of a silver plasma expanding in Ar at 70 Pa. Data are fitted to mixed-propagation model, and a satisfactory fit results over a large time interval, up to a time delay of 10 μs. From the fitting results some useful plasma parameters were extracted such as the initial velocity $v_0 = 11.5$ km·s^{-1}, the distance $R_0^D = 0.64$ cm, and the diffusion coefficient $D'' = 0.013$ cm^2 μs^{-1}. Such parameters can be useful for the description of nanoparticle (NP) formation in the gas phase and their subsequent deposition for the realization of SERS-active substrates [51].

First attempts to produce NPs by means of pulsed laser ablation were performed for silicon. The goal was the production of silicon thin films with emission in the visible due to quantum size effects. Electroluminescent diodes were produced by PKD [52], and the cluster growth in the gas phase was studied by time resolved optical spectroscopy [53] and theoretically [54]. PLD was used to obtain NPs of other materials: semiconductor, metals, oxides, and composite materials [55–58]. The formation of NP-assembled thin films can be described as a two step process: in the first step, the species ejected from the target are subject to intraplume collisions. When expansion takes place in an ambient gas, collisions occur at the contact surface between plasma and ambient gas. Collisions can lead to the formation of clusters whose dimensions depend on many parameters such as atomic masses of plasma and gas species, removed target mass per pulse, fluence, and ambient gas density. Cluster growth in an expanding plasma has been described by Wood et al. [59, 60] who put in evidence the role played by ambient gas. Mixed-propagation model gives a description of cluster growth owing to collisions between plasma species and ambient gas. In this model [48, 49], the steady-state number of atoms N in a cluster is given by the following equation:

$$N = \left(\sigma_{a-a}\langle n_a\rangle\langle v\rangle t_f\right)\left(\sigma_{a-g} n_g\langle v\rangle t_f\right) \tag{7.7}$$

where σ_{a-a} and σ_{a-g} are scattering cross sections for collisions between plasma species and plasma and gas ones respectively, $\langle n_a\rangle$ the average density of the plasma, n_g the ambient gas density, $\langle v\rangle$ the average velocity of the expanding plasma, and t_f the time at which the cluster reaches its steady dimension. An estimate of the cluster size growing in a laser-generated expanding plasma in a controlled atmosphere can be done using experimental data coming from fast photography experiments [50, 61]. Beside its position, the plasma mean velocity $\langle v\rangle$ can be evaluated as well, while $\langle n_a\rangle$ density is obtained from the estimation of the plasma volume and measuring the ablated mass per pulse removed by a single laser pulse by stylus profilometer measurements. Cluster formation time t_f is more difficult to evaluate, and in our analysis we estimated t_f as the time at which the plasma luminous emission decreased by three order of magnitude from its initial value [62]. We assume at t_f, collisions give no more contribution to the clusters growth.

As long as all these quantities are known, the number N of atoms in a cluster can be obtained from Eq. 7.7. The obtained values can be compared to TEM measurements from which NP size distributions can be extracted. The model correctly describes the increase of NP average size on rising the ambient gas density in case of silver samples deposited at laser fluence of 2.0 Jcm^{-2} and Ar pressure between 10 and 100 Pa. N values between 15 and 2344 for clusters growing in the presence of Ar pressure values between 10 and 100 Pa. The corresponding size d, calculated assuming sphere-like clusters with a packing efficiency of $\eta = 0.67$, results 0.7, 1.7, 2.8, and 4.1 nm. These values are compared to the results of TEM observation of samples deposited at Ar pressures of 70 and 100 Pa, using 10^4 laser shots. The average sizes of 2.9 ± 1.0 nm and 3.8 ± 2.0 nm were estimated from TEM pictures in good agreement with the prediction of the mixed-propagation model [63].

Under identical experimental conditions (fluence, laser spot area, Ar pressures values), gold plasma dynamics showed a different behavior with respect to the silver case. Gold plasma, in fact, was less affected by the presence of the ambient gas, expanding farther with respect to the silver case. Such a behavior is due to the larger atomic mass of gold ($m_{Au} = 196.97$ amu) with respect to the silver one ($m_{Ag} = 107.87$ amu). Taking into account that the plasma expansion, in the presence of a buffer gas, is governed by collisions, the mass ratio among plasma species and gas ones plays a critical role. For gold, mixed-propagation model predictions at 70 and 100 Pa of Ar gave average cluster size, 5.6 and 10.2 nm, while TEM observations gave average size of $5\div6$ and $7\div8$ nm, respectively [61]. Thus, surface morphology can be tuned either by changing the Ar pressure, keeping the laser pulses number fixed, or fixing the Ar pressure and performing the deposition at different laser pulse number. At fixed laser pulse number, ambient gas pressure affects NP size and the flux of ejected species that can reach the substrate. Lower pressure allows for higher substrate coverage than higher pressures. The same occurs when the laser pulses number is increased at a fixed ambient gas pressure. An exemplary case is report in Fig. 7.3 where TEM pictures of Au samples deposited at the fixed laser pulse number of 3×10^4 and at Ar pressure between 10 and 100 Pa are reported. As can be seen, the surface morphology

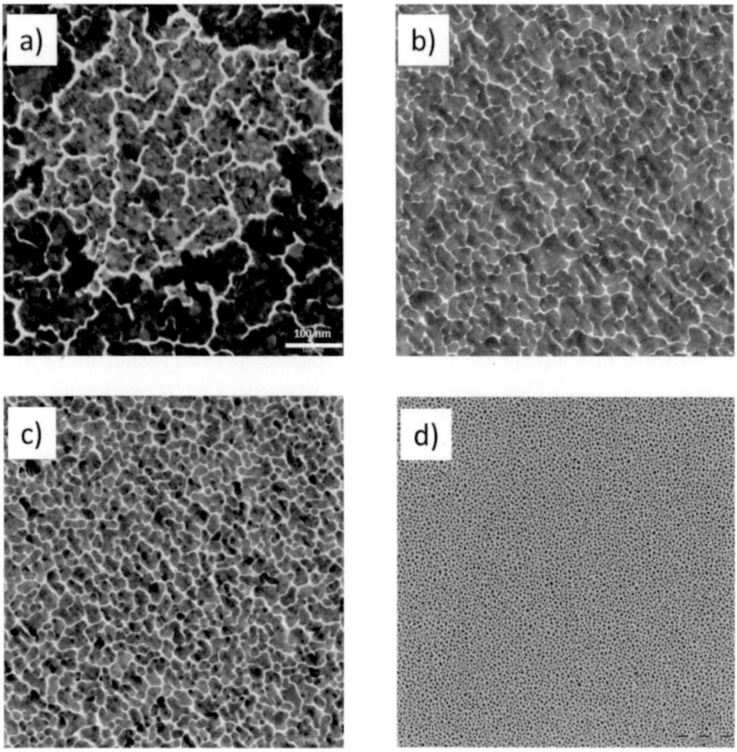

Figure 7.3 TEM pictures of Au samples deposited at Ar pressures of (a) 10 Pa, (b) 40 Pa, (c) 70 Pa, and (d) 100 Pa of Ar with 3×10^4 laser pulses number.

ranges from a nearly percolated structure at 10 Pa of Ar (Fig. 7.3a) to spherical isolated NPs at 100 Pa (Fig. 7.3d). Intermediate structures made by islands with smooth edges are obtained at 40 and 70 Pa (see Fig. 7.3b and c). Different morphologies result in different optical properties. Isolated silver or gold NPs show a surface plasmon resonance (SPR) absorption peak at about 400 and 520 nm. SPR position and linewidth strongly depend on size, shape, and aggregation properties. In Figs. 7.3a and b (adapted from [62]), we show the changes observed in the UV–Vis absorption spectra of silver nanostructured thin films deposited by PLD under different experimental conditions. As long as the surface morphology change

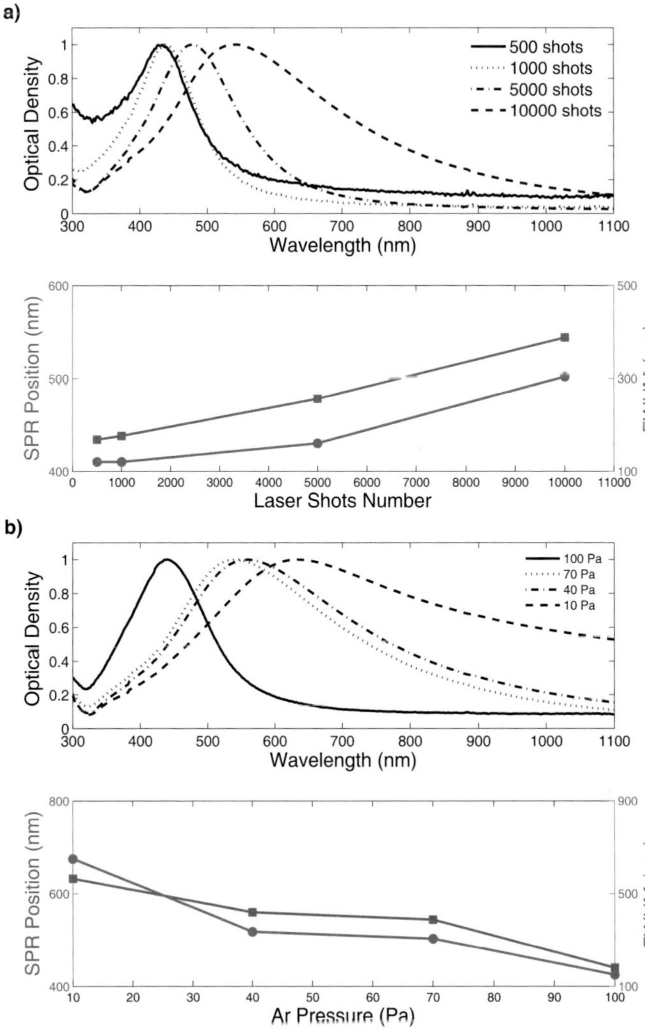

Figure 7.4 (a) In the top panel optical absorption spectra of silver films deposited at Ar pressure of 70 Pa at different laser pulse number, in the bottom panel the corresponding SPR positions and FWHM values. (b) In the top panel, optical absorption spectra of silver films deposited at different Ar pressures with fixed laser shot number of 10^3, in the bottom panel the corresponding SPR position and FWHM values; all spectra are normalized to their own maximum.

from isolated NPs toward a percolated structure, SPR position shifts to higher wavelength and its linewidth broadens, depending on the different deposition conditions adopted. Thus, controlling two easily accessible experimental parameters, i.e., Ar pressure and laser shot number morphologies and optical properties of nanostructured silver and gold thin films can be tailored and optimized.

7.4 SERS Activity of Pulsed Laser Ablated Thin Films

The amplitude of the SERS enhancements strongly depends on the size and shape of the NPs and also on their spatial density and interparticle spacing. Thus, the ability to control all these parameters is mandatory for the realization of highly SERS-active substrates. Silver and gold substrates grown under different experimental conditions, namely Ar pressure and laser shot number, being all the other relevant process parameters kept fixed, were tested against the presence of rhodamine 6G (R6G), which presents a very low Raman scattering cross section. Samples were soaked into R6G aqueous solution at concentration levels between 10^{-4} and 10^{-10} M, rinsed with deionized water, and then left to dry in air. Raman spectra were taken using a Jobin-Yvon HR800 Raman spectrometer equipped with different excitation lines, namely 632.8 nm (He-Ne) and 785.0 nm (solid-state diode laser). Integration times and power density at the sample surface were chosen so as to optimize the signal-to-noise ratio and to avoid R6G degradation. A set of silver samples was deposited at Ar pressures of 10, 40, and 70 Pa at the fixed laser pulse number of 1.5×10^4, a fourth sample was grown at 70 Pa of Ar with 3×10^4 laser pulses [65]. In Fig. 7.5, we report the Raman spectra of R6G 5×10^{-7} M acquired on the surface of silver samples deposited at 70 Pa of Ar and using (sample **A**) 1.5 $\times$ 10^4 and (sample **B**) 3×10^4 laser pulses. The two samples show a markedly different SERS activity. Such a difference depends on their surface morphology. The surface of sample **A** is characterized by the presence of isolated spherical NPs. A more dense network of NPs covers the surface of sample **B** owing to the adopted higher laser pulses number, even if the characteristics of coverage show that coalescence among NPs is still absent. The different morphology also

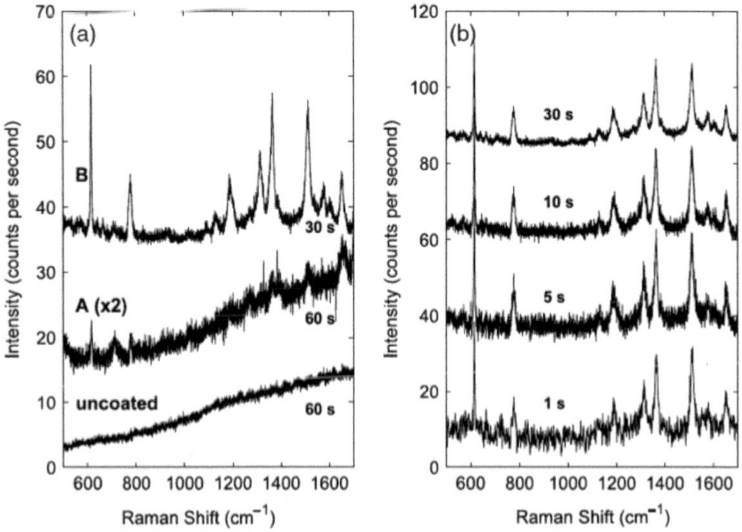

Figure 7.5 (a) Raman spectra of R6G 5×10^{-7} M adsorbed onto silver samples deposited at 70 Pa of Ar with 1.5×10^4 and 3×10^4 laser pulses number. The spectrum acquired on an uncoated substrate is also reported for comparison; (b) Raman spectra acquired using different integration times on the surface of the sample deposited at 70 Pa of Ar with 3×10^4 laser shots number.

affects SPR position observed at 430 and 560 nm for samples **A** and **B**, respectively. Such a large redshift of the SPR can be attributed to the strong coupling between closely spaced NPs [66–68].

The higher observed SERS activity of sample **B** can have a twofold origin; the redshift and broadening of the SPR that become resonant with the excitation wavelength (632.8 nm) and the higher spatial density of NPs. When two NPs are in proximity to each other, the electromagnetic field at the junction between the two NPs can be greatly enhanced [69–71] giving rise to the so-called *hot-spot*. The higher NP number density of sample **B**, thus, can lead to a higher density of hot-spots on its surface. The Raman enhancement factor (EF) is defined as the ratio between the observed SERS intensity (I_{SERS}) and the normal Raman intensity (I_R):

$$EF = \frac{I_{SERS}/N_{SERS}}{I_R/N_R} \tag{7.8}$$

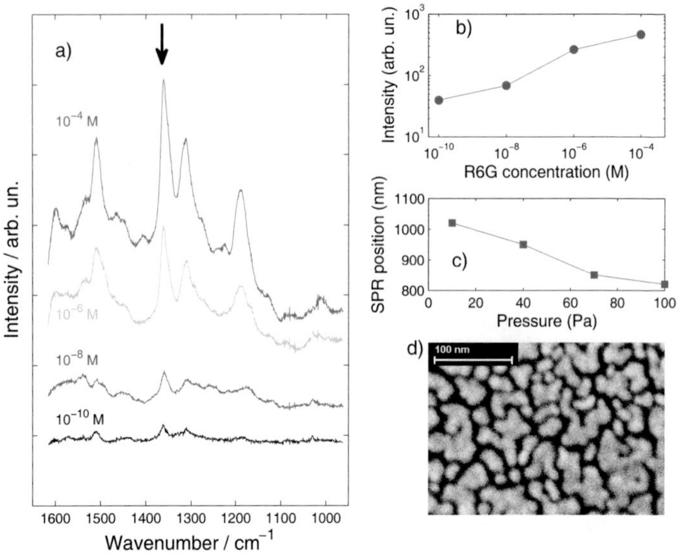

Figure 7.6 (a) Raman spectra acquired on the surface of a gold sample deposited at 100 Pa of Ar using 3×10^4 laser shots; (b) Intensities of 1262 cm^{-1} Raman peak as a function of R6G concentration. (c) SPR peak position for gold samples deposited using 3×10^4 laser shots at different Ar pressures; (d) SEM image of the surface of the gold sample deposited at 100 Pa with 3×10^4 laser shots.

where N_{SERS} and N_R are the number of molecules that contribute to SERS and normal Raman signals respectively. The spectrum acquired on bare glass substrate soaked into a rhodamine 6G solution at 10^{-4} M concentration showed no Raman signature. An EF estimation was made following the method of Gupta et al. [72] by comparison between Raman and fluorescence intensities [65]. An EF value of the order of 10^7 was obtained for the sample deposited at 70 Pa of Ar and using 3×10^4 laser shots.

The SERS activity of gold samples grown was investigated following the same procedure as for the silver ones. A set of samples was grown at Ar pressures between 10 and 100 Pa, keeping fixed fluence and laser shot number at 2.0 Jcm^{-2} and 3×10^4 respectively. SERS activity was tested for 10^{-4}, 10^{-6}, 10^{-8}, and 10^{-10} concentration of rhodamine 6G. The higher SERS activity was observed for the sample deposited at 100 Pa of Ar: in

Fig. 7.6a, Raman spectra acquired on its surface are shown for the different rhodamine concentration solutions. Spectra in this case were acquired using the 785 nm excitation line from a diode laser using a HR800 Jobin-Yvon spectrometer; a x50 microscope objective was used, spectra are normalized to their proper integration times. It can be seen that a good spectrum could be acquired even at concentration levels as low as 10^{-10} M. In Fig. 7.6b, we report the intensity of the Raman peak as a function of R6G concentration and in Fig. 7.6c, the position of the SPR peak of the gold samples as a function of the Ar pressure, a redshift of the SPR position, is observed as on decreasing the selected Ar pressure. It is worth to underline that the sample deposited at 100 Pa of Ar shows the SPR peak position in close resonance with the excitation line used in this experiment. We also notice that keeping fixed all the other deposition parameters, the best SERS activity for gold is observed at the pressure of 100 Pa; the latter is higher than the one at which the best SERS activity was observed for silver, i.e., 70 Pa, pointing out the crucial role played by collisions to determine surface morphology, as shown in Fig. 7.6c, and the SERS properties. The enhancement factor was obtained by comparing the Raman intensity of methylene blue at 10^{-5} M concentration adsorbed on a flat gold surface to the one acquired on a PLD deposited gold sample. In Fig. 7.7, the two spectra are shown: after a comparison between the intensities of the major observed peak, the resulting EF was of about 6×10^4.

7.5 Pulsed Laser Ablation in Liquid

In the past years, laser ablation synthesis (PLAL) in liquid solutions emerged as a reliable alternative to traditional chemical reduction methods to produce nanoparticles of various size, shape, material, or even to chemically model nanoparticles already produced, for example, by creating phases of oxidation in the nanoparticle body. PLAL is a green technique for the synthesis of stable noble metal nanoparticles in water or in organic solvents, which does not need stabilizing molecules or other chemicals [73]. Moreover, multiphase core–shell oxides, metal–semiconductor heterostructures, layered organometallic compounds and other complex nanostructures can

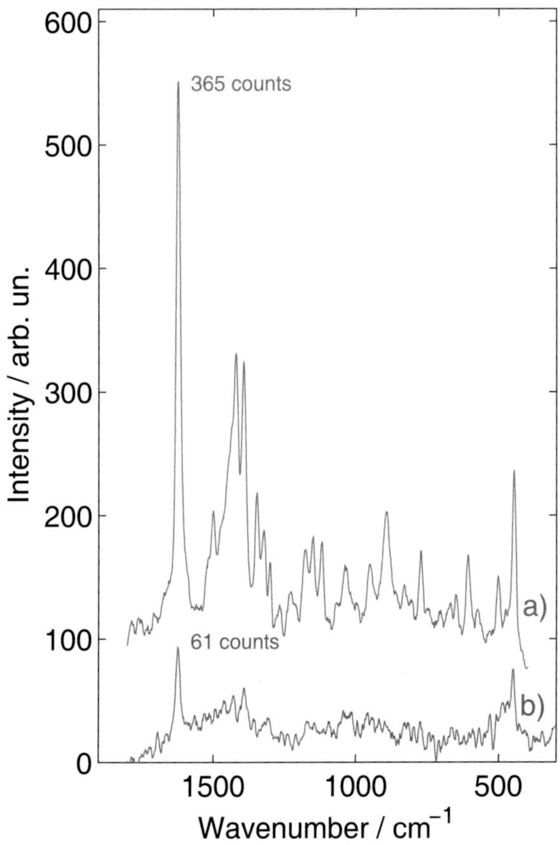

Figure 7.7 Raman spectra of methylene blue 10^{-5} M adsorbed onto (a) on nanostructured gold substrate, deposited at 70 Pa of Ar using 8000 laser shots and (b) a flat gold surface. Peak intensities used for the EF calculation are reported. Spectrum (a) was acquired using a neutral density filter with optical density 4 (10^{-4}).

be obtained with the same experimental setup, just by varying a few synthesis parameters. These parameters fall into two categories: material parameters (bulk target, solvent and solutes, system temperature, and pressure) and laser parameters (wavelength, duration, energy, repetition rate, number of laser pulses, and the spot area on the target). Nevertheless, the first and immediate PLAL interest factor lies in its simplicity: it is sufficient to have a pulsed laser source, with convenient emission frequency and power, and a system where the target to ablate is confined within an appropriate liquid. This technique can, in principle, produce moderate amounts of nanoparticulate, dispersed in colloidal phase, characterized by chemical and morphological properties centered in narrow ranges, and exhibiting high chemical stability for long time periods (months).

Compared to other preparation procedures used for the formation of nanocolloids, the pulsed laser ablation technique has the additional advantage that it can be done in a large number of different environments, not being dependent, such as its chemical counterpart processes, on the solubility of organic compounds used as precursors and/or reagents. At the same time, the presence of undesirable reaction products at the end of the colloid formation process is intrinsically avoided. Even when some surfactant molecules are introduced into liquid solutions, the chemical properties of NPs surface remain mostly unchanged. This advantage, as well as the possibility to control the nanoparticle (NP) size by changing laser parameters or liquid composition, makes PLAL a unique tool for NP synthesis. Generally, metastable phases can be found among the laser ablation products. The composition of the target is not necessarily found in the final products because of side reactions with the liquid solution. Especially with alloy targets, NPs stoichiometry can be very different from that of the bulk target. Frequently, a mixture of products with heterogeneous phase, structure, and morphology is obtained. In principle, the yield of desired products can be maximized by the choice of appropriate experimental parameters, once the connection between the two aspects is known. Therefore, the ability to control the phase and the structure of nanoparticles requires a thorough understanding of the physical and chemical processes involved in PLAL. The process of laser ablation in liquids is essentially based on the high degree of confinement of fast reactions in micrometric or nanometric size regions, contiguous to the focus area of the laser beam onto the solid target (see Fig. 7.8).

It was evidenced that laser ablation in liquids is accompanied by the formation of shock waves and cavitation bubbles. In addition, fragmentation and condensation and the roles of electrostatic

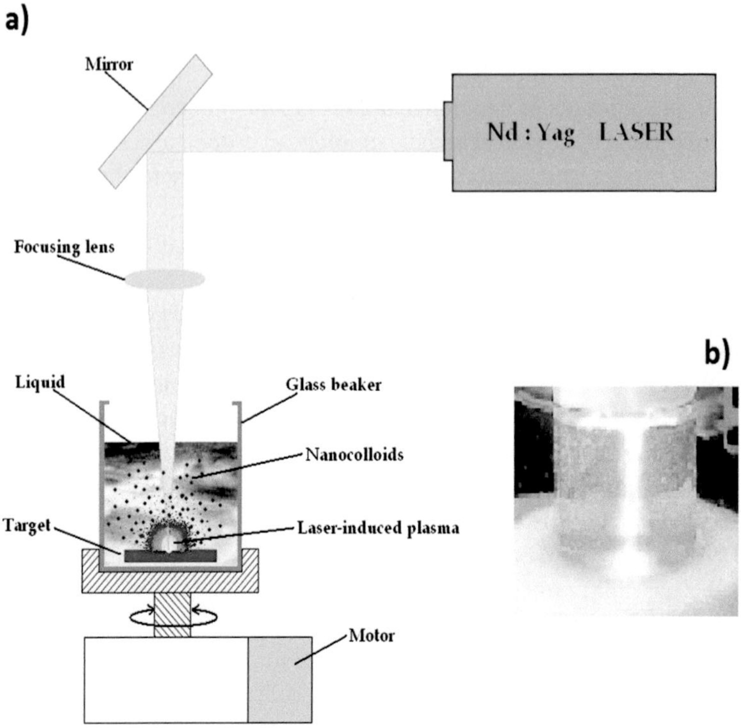

Figure 7.8 Scheme of the PLAL experimental setup and schematic illustration of the nanostructures formation in a PLAL process (a). Photograph during the ablation of a solid target in water (b).

interactions and chemical processes were envisaged. Especially, the presence of liquid solvent plays an important role in the evolution of nanoparticle formation processes, while providing to the system extreme environmental conditions that drive the same ablation mechanism. For instance, at the liquid–solid interface in correspondence of the laser spot, the high local temperature causes the formation of microbubbles whose vapor pressures are so high to inhibit the evaporation of the solid target species. The same solvent, subjected to very high thermal gradients in the vicinity of the laser spot, can suffer partial or total chemical dissociation, which translates into the ability to trigger solvent–nanoparticle chemical reactions, almost contextual to the same ablation process [74, 75].

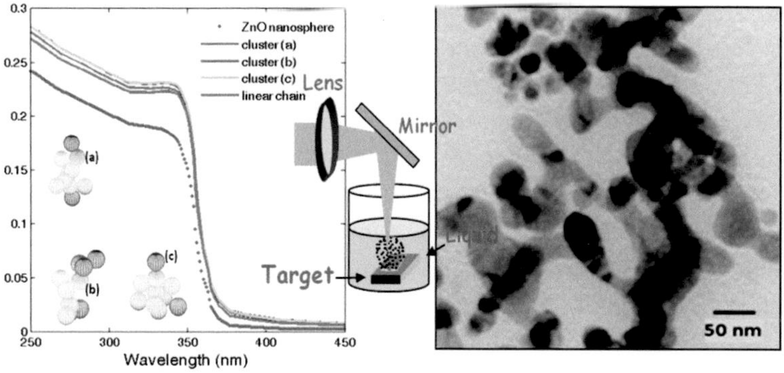

Figure 7.9 Analysis, through a theoretical approach based on the multipole expansion of the electromagnetic fields and the transition matrix method, of UV–Vis absorption spectra of ZnO water colloids prepared by PLAL (a) and TEM image of the ZnO colloids characterized by linear chains (b).

In synthesis, one of the main characteristics of laser-induced plasma inside liquids is fast quenching due to a dense medium. As observed for metal nanoparticles synthesis, when a nanosecond pulsed laser with sufficiently high energy density strikes the interface between liquid and the metal target, it forms a microplasma layer after a series of sequential events. At first, the laser beam is absorbed by the target to create a molten layer, and then further absorption leads to the local rise in temperature up to the boiling point of the metal, where several physical phenomena such as fragmentation, sublimation, and atomization occur simultaneously. The laser absorption takes place in the femtosecond-picosecond interval, with vaporization lasting less than a few nanoseconds, and the plasma formation taking place on a nanosecond time scale, followed by several events such as the plasma expansion and the expansion and collapse of the initial cavitation bubble, which end up within a microsecond. Finally, the condensation and the subsequent nanoparticle formation take place, usually within milliseconds [76]. It is well known that the primary stage of the ablation process in the presence of liquid is difficult to observe experimentally. Thus, to better understand the mechanisms of the NPs formation, Povarnitsyn et al. [77] carried out a theoretical modeling of femtosecond laser interactions with a gold target in the presence of liquid (water). Simulation of the ablation process over several nanoseconds shows that most of the primary NPs originate from the ablated metastable liquid layer, whereas only a minority is formed by condensation inside the cavitation bubble. These particles will further grow/evaporate and condensate during a much longer collision stage in the liquid colloid. Numerical simulations show that the large-sized particles are the result of liquid layer ejection and fragmentation. This regime is observed for low laser intensities when the thermodynamic trajectories enter the metastable liquid region far below the critical point. For higher laser intensities, the near-critical and supercritical trajectories appear for front target layers and the mass fraction of the liquid–gas mixture increases. Formation of small NPs can be observed for these trajectories and estimation of the NPs size matches quite well with the experimental findings [77]. Nevertheless, several issues require further clarification. In particular, the molecular species and the chemical reactions that participate in PLAL are largely unknown. The accurate quantitative determination of the space-time profiles of temperature, pressure, concentration of the ablated material, and concentration of the species in solution is crucial to understand what really happens during PLAL and to control the phase and structure of its products. In fact, temperature affects several physical properties of liquids such as viscosity, density, refractive index, surface tension, and compressibility, which in turn alter the efficiency of laser irradiation, plasma plume confinement, and the cavitation bubble dynamics [78]. In particular, the hydrodynamic size of Au NPs aggregates obtained by PLAL in H_2O was correlated with liquid temperature as the eject of the variation of liquid compressibility [79]. The eject of external pressure on PLAL can be investigated only by using a specific experimental setup, which is usually applied to laser ablation in supercritical fluids. In such cases, no dramatic differences have been reported by changing the liquid pressure, except that larger particles are obtained at larger pressures, probably because the ablation efficiency is improved by the more ejective plasma confinement just above the crater. Upon increasing the pressure, the cooling rate of ablated material increases, while the lifetime and maximum size of the cavitation bubble decrease. How these ejects influence the structure of the

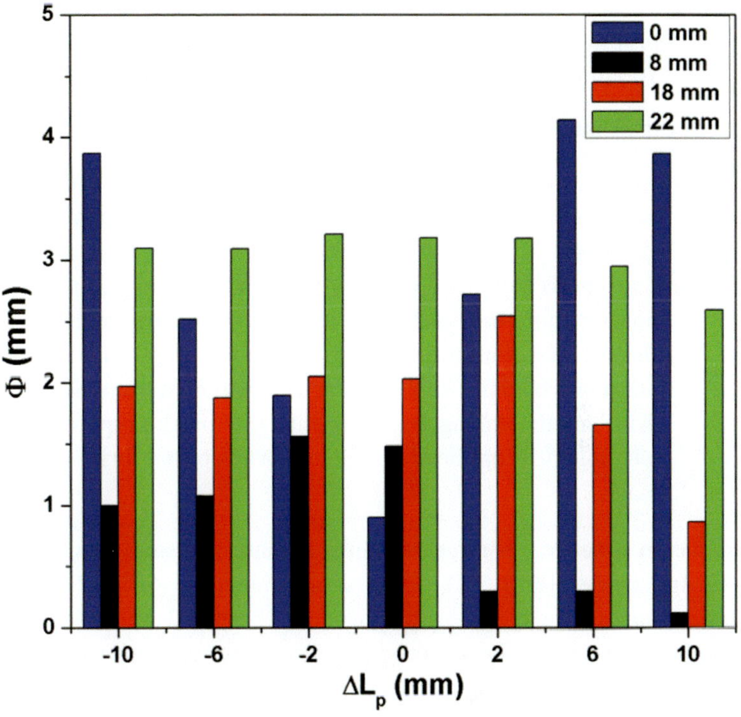

Figure 7.10 Effect of lens position and water layer thickness on the ablation spot diameter.

metal nanoparticles is not well understood at the moment, although it has been reported that Si NPs have more defects when obtained at high pressure, that is, when the cooling rate is higher [80, 81].

On the basis of the current knowledges, it can be summarized that, the main pros of PLAL are as follows:

- PLAL is a unique approach for the synthesis of a variety of nanomaterials in a variety of liquid solutions.
- PLAL does not necessarily require chemicals and does not necessarily produce waste.
- PLAL is a low-cost method: apart from the laser source, the experimental setup are minimal, chemical precursors are replaced by bulk materials, and other expensive chemicals are seldom used.

- PLAL is one of the fastest, cheapest ,and cleanest ways to produce colloidal nanoparticles with size dispersion below 100 nm on the milligram weight scale.

On the other hand, the cons of PLAL are the following:

- Large-scale productivity is still far from that achieved by other methods, such as spray pyrolysis or wet chemistry synthesis.
- The control of nanoparticles size distribution is usually poorer than that achieved by wet chemistry methods. Despite the limitations described above, there are many unique properties of PLAL-derived nanomaterials, which drive an increasing scientific interest for their potential technological applications. Below, we report a brief summary.

Submicron spheres: For various kinds of materials such as metals (Au, Ag), oxides (ZnO, TiO_2), and semiconductors (Si, GaP), the production of mostly spherical particles of several hundreds of nanometer in size can be achieved by pulsed laser melting, working with a low fluence (i.e., some hundreds of mJ cm^{-2} are sufficient) [82, 83] (see Fig. 7.9 (see Ref. [92]).

Bioconjugated nanoparticles: The use of laser-generated particles leads to significant improvements in medical technology since it allows the conjugation of gold nanoparticles with biomolecules. Moreover, laser-generated particles are more reactive, as well as less toxic, than chemically synthesized materials. The nanoparticle surface coverage with biofunctional ligands is five times higher than for chemically synthesized nanogold, resulting advantageous for applications that benefit from a high specificity, for example, in the case when antibody, DNA, or aptamer-targeting is used [84].

Catalysis: The large variety of nanoparticles that can be produced using PLAL is itself a great advantage for catalytic applications. Moreover, it is important to mention the significant increase of the particles' absorption efficiency as well as the fact that ligand-free heterogeneous catalysts do not require cleaning. Specifically, the deposition efficiency of laser-generated nanoparticles is 20 times higher in the ligand-free state compared to the citrate-stabilized nanoparticle surface in chemically synthesized nanoparticles (containing residual citrate).

Electrophoresis and electrodeposition: The combination of electrophoresis with PLAL greatly benefits from the plasma-based, liquid-confined process that releases charged species into the medium where an electric field is applied. In situ electric-field assisted PLAL aids the anisotropic assembly of charged precursors to form various nonspherical structures in the vicinity of the electrodes, possibly by orientation-directed ripening of precursors with a lower mass than the final nanostructure collected at the electrodes [85, 86].

Metastable phases: Quenching times in PLAL are so short that the metastable phases that form during the intermediate stage of the conversion can be frozen into this metastable state, forming the synthesized final products. Moreover, the use of a liquid medium confines the expansion of the plasma allowing the synthesis, in some special cases, of diamond and boron nitride particles [87, 88] as well as exotic nanostructure aggregations such as inorganic fullerene-like nanoparticles and metallic nanowires [89, 90].

Overall, it emerges that pulsed laser ablation in liquids and, eventually, subsequent pulsed laser treatment of colloidal nanoparticles allow the fabrication of advanced materials with integrated nanostructures that bear unique physical properties. Nevertheless, several issues, such as the limitations in size control and large volume productivity, still exist that deserve in-depth studies. A more refined knowledge of both temporal and spatial parameters involved in the procedure could allow for the process scale-up.

7.6 Growth and Characterization of Gold and Silver Nanostructured Colloids

In the case of metallic nanoparticles, the main goal is to obtain, by means of PLAL, colloidal solutions with tailored NPs concentration and distribution. Thus, it is vital to have: (1) an efficient control of the ablation process that determines the structural properties of the NPs; (2) a good knowledge of the surrounding environment (i.e., the solvent in which the nanoparticles are produced and dissolved) influence on NPs size and its distribution. As discussed largely in previous section, heterogeneous synthesis parameters are reported in the scientific literature and the comparison of results from different groups is sometimes difficult. Indeed, every laboratory equipped with a pulsed laser can produce its own NPs, without precisely defined procedures. Herewith we report about the preparation of colloidal water solutions of gold and silver NPs prepared by PLAL, with as much as possible specifications, together with a detailed investigation of their optical and structural properties. When laser processing is carried out in vacuum or in a gas, the energy density at the target is simply determined by the nominal deposited pulse energy and the ablation spot area. However, when the laser beam propagates from air into a liquid phase, several changes can take place. In particular, ultra-short, highly energetic laser pulses undergo self-focusing and filamentation. Thus, when NPs are generated during the ablation process, absorption and nonlinear interactions are enhanced due to the large linear and nonlinear absorption coefficients of the colloid. Such effects, although rarely considered, limit the ablation efficiency and require a careful analysis of the results [91]. It has been shown that the ablated mass increases with the energy density at the target, so that high intensities are desirable to increase the NPs production. Yet, when high energies are used, even determining the focal plane in the liquid via the ablation spot is not straightforward. For example, when gold NPs are prepared in water using nanosecond laser pulses, even a small change in the liquid thickness above the target requires refocusing with a new focal plane determination [93]. Moreover, both the liquid thickness above the target and the lens position affect the ablation spot diameter determining, in turn, the operative energy density, as shown in Fig. 7.10.

By ablating in air, the smallest ablation spot diameter is, rather obviously, obtained when the distance lens-target L_p nearly coincides with the lens focal length L_o. When the process takes place in water, defining $\Delta L_p = L_p - L_o$, the ablation spot remains almost unchanged, starting from negative values, until $\Delta L_p = +3$ mm,

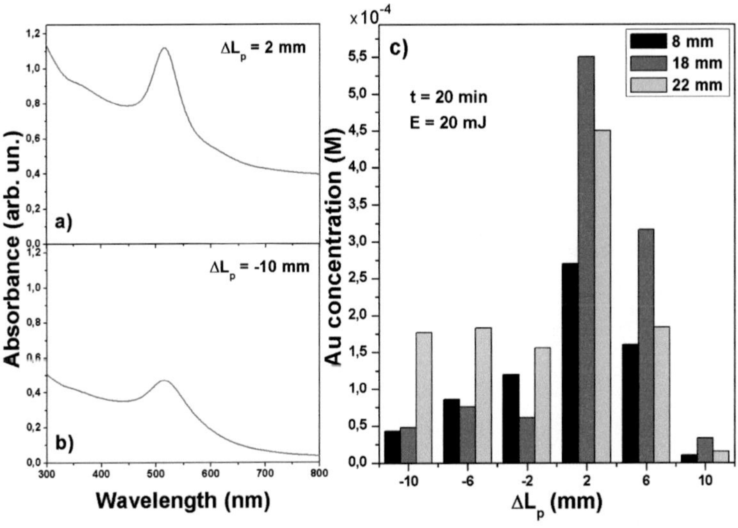

Figure 7.11 UV–Vis spectra (a, b) and the concentration of Au NPs (c).

then it decreases. This trend depends strongly on small water layer above the target thicknesses (8 and 18 mm) and becomes almost unappreciable at 22 mm. Another important parameter to look for is the NPs production yield. Generally, Au NPs concentration is deduced from the intensity of the optical absorption band, at about 520 nm, due to the surface plasmon resonance (SPR) of Au NPs in the UV–Vis spectra of the colloidal solutions [94]. In Fig. 7.11a, the SPR absorption bands of Au colloids prepared at two different ΔL_p values are shown. As shown in Fig. 7.11b, the largest amount of Au NPs is obtained for $\Delta L_p = 2.5$ mm, when the ablation process was performed with a 18 mm thick water layer above the target.

At variance with common observations in vacuum, or in gaseous atmosphere, the largest Au NP concentration is not obtained when the laser spot is smallest (in the experiment, here discussed, it was found at about $\Delta L_p = +10$ mm) and it sensitively depends on the water layer thickness. The result can be explained by considering that the laser-generated plasma strongly scatters and absorbs the incoming pulses; only a fraction of each energy pulse reaches the target: thus, the ablation efficiency is drastically reduced [91].

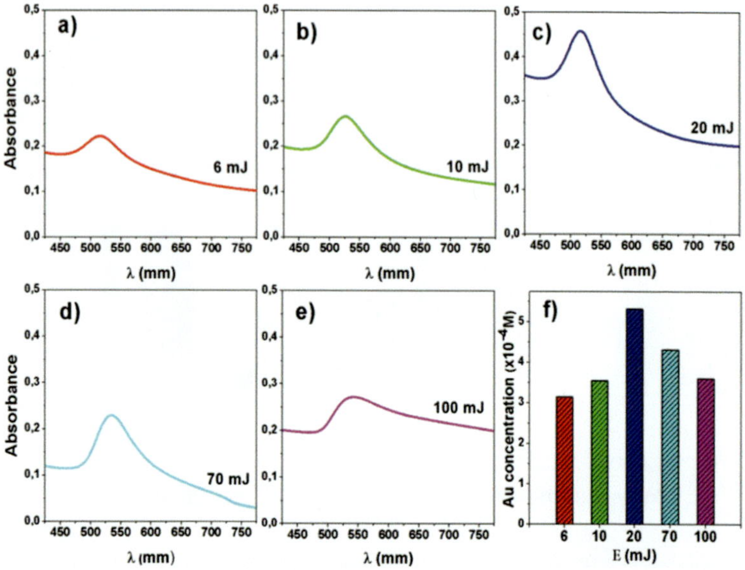

Figure 7.12 (a–e): Plasmon resonance of Au NP colloids prepared at different laser energies per pulse; (f): Au NP concentration versus laser pulse energy.

Finally, both the NPs concentration and morphology are strongly dependent on the laser pulse energy.

As shown in Fig. 7.12a–e, the already discussed SPR feature around 520 nm slightly redshifts when the pulse energy is increased; the overall absorbance decreases and a new contribution to absorption around 750 nm appears, presumably associated to a clustering process. From Fig. 7.12f, we see that the concentration of Au NPs initially increases up to the value of 5.5×10^{-4} M (corresponding to the laser energy of 20 mJ), then progressively decreases: indeed the extreme pressure and temperature conditions at the interface between the expanding plasma and the confining liquid may favor the fragmentation of existing NPs that will rearrange in clusters [95]. The observed considerable decrease of NPs concentration reflects such a structural rearrangement, as shown in Fig. 7.13. In fact, when the pulse energy is 10 mJ a uniform distribution of nearly spherical NPs smaller than 10 nm in diameter is observed, apart from some particles larger than 30 nm (Fig. 7.13a). By increasing the

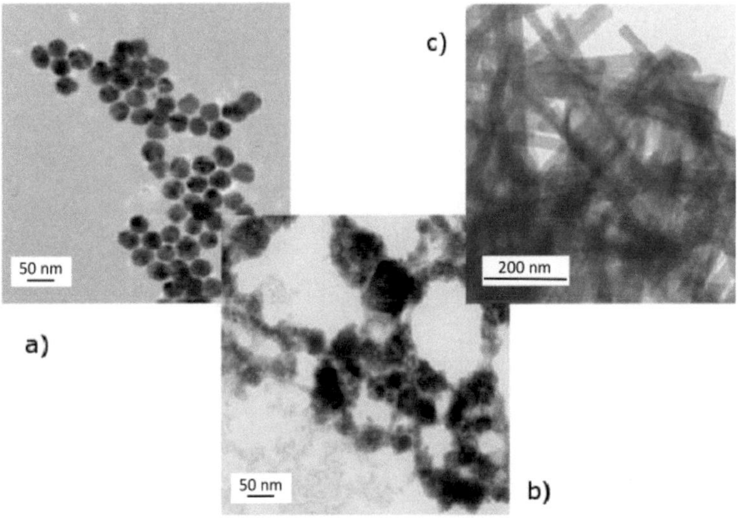

Figure 7.13 TEM images of samples prepared at different laser pulse energies: (a) 10 mJ, (b) 50 mJ, and (d) 100 mJ.

laser energy at 50 mJ (Fig. 7.13b), a restructuring of spherical NPs sized about 50 nm in chain-like agglomerations is evident. Finally, at the highest energy (100 mJ), some portions of the sample are characterized by columnar-shaped nanostructures (Fig. 7.13c).

Overall, the optical response of Au NPs can be tailored to a specific lineshape by changing the morphological properties of Au nanostructures that, in turn, strongly depend on the ablation parameters: mainly the water layer thickness above the target, the lens position, and the laser pulse energy. Similar considerations can be made for silver NPs colloids prepared with the same PLAL setup.

7.7 SERS Activity of Pulsed Laser Ablated Noble Metal Colloids

The SERS activity of Ag and Au NP colloids was generally tested into aqueous solutions of a reference dye such as rhodamine 6G (R6G), at different concentration level. Below, we describe the SERS characterization results obtained using gold and silver colloids

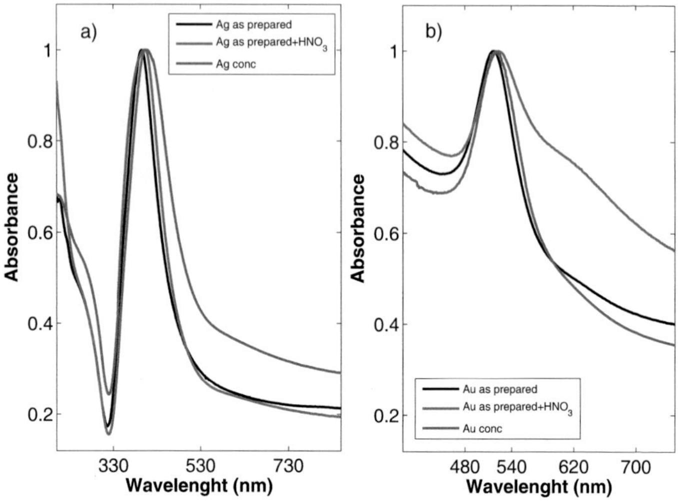

Figure 7.14 Surface plasmon resonance of (a) Ag and (b) Au colloids.

prepared by PLAL of high purity silver and gold rod targets. The ablation processes were carried out using the second harmonic (532 nm) of a Nd:YAG laser operating at 10 Hz repetition rate with a pulse width of 7 ns. The Ag and Au targets were irradiated in 4 ml of water at the laser pulse energy of 60 mJ and 20 mJ, for an ablation time of 20 and 5 min, respectively. The UV–Vis absorption spectra of PLAL prepared Ag and Au colloids are shown in Fig. 7.14.

The so produced solutions were subsequently reduced to about 1/4 in volume, by solvent evaporation. Moreover, a few milliliter of nitric acid (HNO_3 0.5%) was added to the metal colloids. The reduction in volume of the colloidal solutions and the addition of HNO_3 induces an increase of the NPs concentration and an agglomeration process of the nanoparticles themselves, as evidenced by the asymmetry of the SPR absorption spectra at larger wavelengths. The so prepared Ag and Au NPs solutions were used for the SERS characterization. To test the SERS activity, rhodamine 6G (R6G) aqueous solution at concentration level of 10^{-4} M was used. SERS measurements in the liquid phase were performed by adding equal volumes of metallic colloids and dye solutions. To avoid precipitation phenomena and to favor a uniform NPs

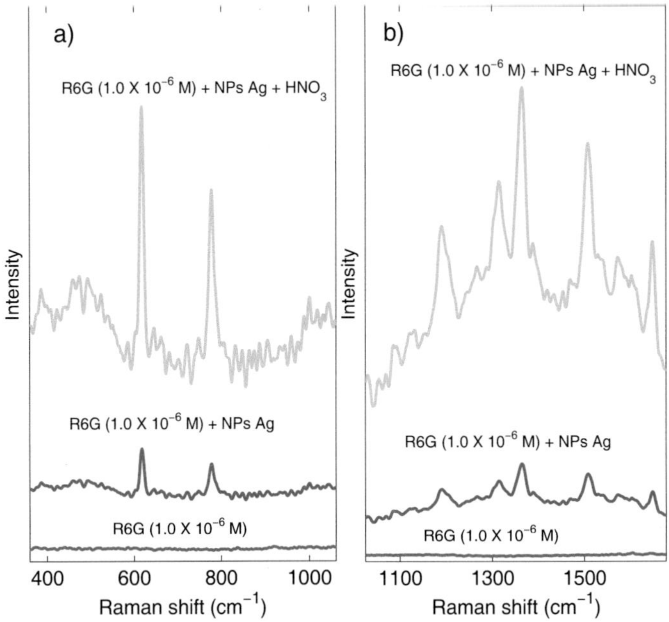

Figure 7.15 Raman spectra of R6G 10^{-4} M acquired in the presence of silver colloidal nanoparticles.

distribution, the mixed solutions were sonicated in an ultrasound bath for 15 min. SERS measurements were carried out immediately. As reported above, 0.5 ml of (HNO_3 0.5%) were added to the metallic nanocolloids in order to remove any oxide layer around the NPs and, in turn, to increase the SERS activity. In fact, in some cases metal nanoparticles aggregation could lead to higher SERS intensities when the investigated molecules lie in nanometric sized regions between NPs (the so-called hot-spots). Raman spectra were excited with the 638 nm laser line, keeping the laser power at the sample surface below 10 mW. Accumulation times were varied, depending on the signal-to-noise ratio, between 2 and 80 s; all the spectra shown here were normalized to their own integration times.

The Raman spectra of R6G 10^{-4} M acquired in the presence of silver and gold colloidal nanoparticles and the spectrum of the pure R6G solutions are shown in Figs. 7.15 and 7.16. All the SERS spectra show the peak at 615 cm^{-1} assigned to the C-C-C in-plane bending

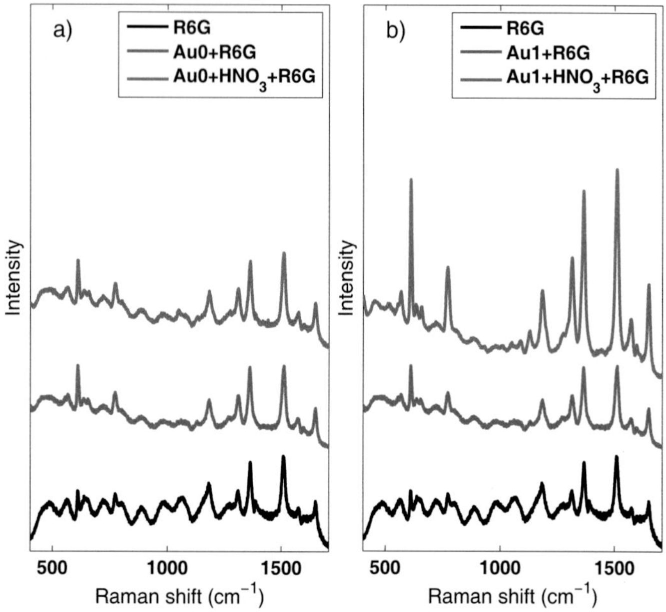

Figure 7.16 Raman spectra of R6G 10^{-4} M acquired in the presence of gold colloidal nanoparticles. The spectra shown in (a) refer to the as-prepared Au nanoparticles while the spectra shown in (b) refer to the Au NPs reduced in volume.

mode and some peaks in the 900–1600 cm^{-1} region assigned to stretching modes of aromatic C atoms [97]. The comparison of the three Raman spectra show that the peaks related to the vibrational modes at about 600 cm^{-1} and 800 cm^{-1} are clearly visible only in the solutions containing the metallic NPs.

For the R6G solutions in the presence of Ag nanoparticles, the Raman features below 900 cm^{-1} are more pronounced when few milliliter of nitric acid was added to the as-prepared solution. Furthermore, the SERS spectra (not shown) acquired using the Ag NPs colloid reduced in volume do not show significant differences with the as-prepared ones. Indeed, the induced NPs higher concentration and morphology's change (Ag NPs clustering) are not of particular relevance to the SERS effect. At variance, the SERS signal is significantly increased when the as-prepared Au NPs are reduced

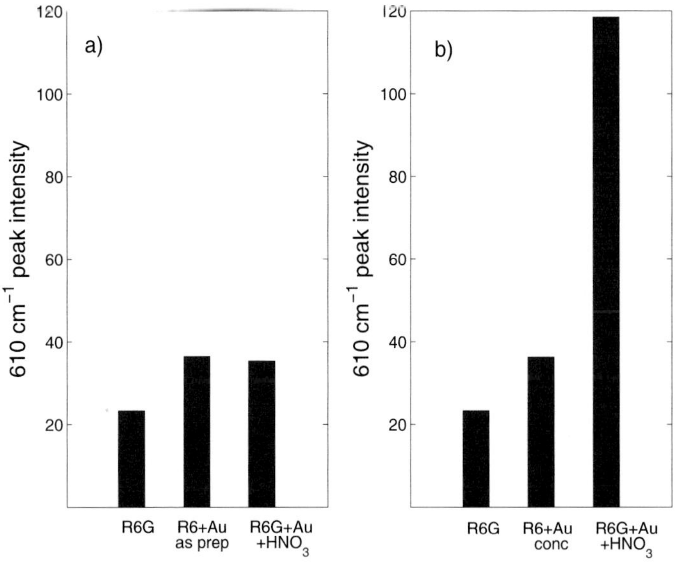

Figure 7.17 The intensity of the 610 cm^{-1} Raman peak for the collected SERS spectra shown in Fig. 7.16.

in volume and when few milliliter of HNO_3 is added to the colloids (see Fig. 7.17).

In the cultural heritage application field, SERS active colloids, produced by PLAL, can be effectively used to identify organic dyes in artworks for their dating, conservation, and restoration. When using other characterization techniques, several drawbacks concerning these dyes have been reported: X-ray fluorescence is missing elemental signature, Fourier Transform Infrared (FT-IR) absorption spectra also contain contributions from other substances present with the dyes, that is, binders and extenders, UV–Vis spectroscopy is often unable to selectively distinguish different chromophores. On the other hand, micro-Raman spectroscopy, due to its high selectivity, would be a good candidate to detect and provide the chromophore fingerprint. Yet, due to low Raman scattering cross-sections and possible fluorescence background signals, it is of vital importance to enhance Raman signals in order to obtain clear and well-resolved spectra. SERS offers a solution to the above difficulties. As an example, a brazilwood water solution at nominal

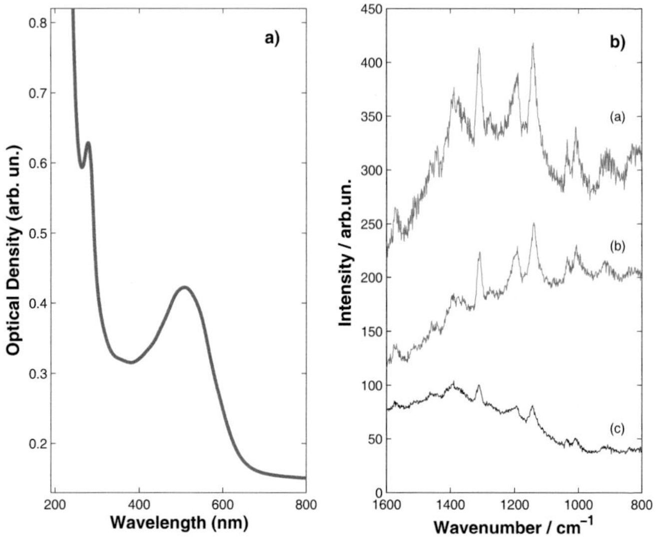

Figure 7.18 (a) UV–Vis absorption spectra of aqueous solutions of brazilwood, 5 × 10^{-3} M. (b) Raman spectra of brazilwood 5 × 10^{-3} M mixed with colloidal (a) gold and (b) aqueous solution compared to (c) pure brazilwood aqueous solution, multiplied by a factor of 10. Adapted from Ref. [96].

concentration of 5 × 10^{-5} M was considered. UV–Vis absorption spectrum of brazilwood lake water solutions is shown in Fig. 7.18a.

The optical absorption of brazilwood consists of a well-defined band at 280 nm and a broad band extending between 400 and 600 nm [98]. Attribution of spectral feature to brazilwood and/or to their different chromophores, usually referred to as brazilwood (brazilin and brazilein), is questionable. The situation is even worse in the case of dyes collected from, or analyzed on, original artworks due to the presence of other substances. Hence, UV–Vis absorption spectroscopy cannot discriminate among dyes used in artworks. Conversely, SERS measurements allow it, as shown below. Raman spectra of brazilwood 5 × 10^{-3} mixed with colloidal gold and silver aqueous solution are shown in Fig. 7.18b. The observation of the 1132, 1194, 1571, and 1412 cm^{-1} peaks attributed to brazilin and of the 1032 and 1600 cm^{-1} peaks (ring and C-C stretching respectively) attributed to brazilein seems to indicate that the analyzed brazilwood sample is composed of a mixture of the two molecules. This result is indicative of the Raman spectroscopy efficiency and selectivity when the SERS configuration can be settled in, by the appropriate use of noble metal NPs solutions.

7.8 Conclusions

In conclusion, we showed that pulsed laser ablation technique is a reliable method to prepare, following a bottom-up approach, SERS-active substrates. Despite its apparent experimental simplicity a detailed knowledge of the fundamental physical mechanisms, at the basis of the technique, is imperative to design the experimental parameters suitable to grow arrays of NPs with controlled structural and optical properties, and hence SERS activity. We showed that tailoring of morphological and optical properties can be done by ablating pure noble metal targets in the presence of a controlled inert atmosphere through the control of two accessible parameters, i.e., laser pulse number and ambient gas pressure, while keeping fixed all the other relevant process parameters. We showed that EF of the order of 10^7 for silver and of 10^5 for gold are achievable. We believe that even higher EF can be obtained with a deeper understanding of the physics underlying the process. Pulsed laser ablation in liquids (PLAL) can be regarded as the extension of the method to an ambient where pressure is fixed at one atmosphere with the difference that substrates are now obtained as colloidal solutions. Hence, laser ablation synthesis in solution has emerged as a reliable alternative to traditional chemical reduction methods to grow NPs as colloids. Moreover PLAL can be regarded as *green* technique for the synthesis of stable noble metal NPs in water or in organic solvents, without the need for stabilizing molecules or other chemicals. The so obtained NPs are highly available for further functionalization. Surface functionalization of metallic NPs can be monitored in real time by UV–Vis spectroscopy of the plasmon resonance. The exploitation of the prospects of NPs generation and excitation using lasers in liquids requires overcoming the limitations in size control and productivity. Although the gram scale productivity is already accessible using a few tens of watt output power during PLAL, now the present main technological challenge is to further define temporal and spatial parameters useful for a scale-up. The reason is that, above a certain threshold, the process does scale nonlinearly. The extent of nonlinearity depends on the material, the process and the laser beam parameters so that a general route for production at a larger scale should be drawn.

References

1. Fleishmann J., Hendra P. J., McQuillan A. J., *Chem. Phys. Lett.*, **26**, 163 (1974).
2. Jeanmarie D. L., Van Duyne R. P, *J. Electroanal. Chem.*, **84**, 1 (1977).
3. Albrecht M. G., Creighton J. A., *J. Am. Chem. Soc.*, **99**, 5215 (1977).
4. Otto A., Mrozek I., Grabhorn H., Akemann W., *J. Phys. Cond. Mater.*, **4**, 1143 (1992).
5. Moskovits M., *Rev. Mod. Phys.*, **57**, 783 (1986).
6. Michaels A. M., Jiang J., Brus L., *J. Phys. Chem. B*, **104** 11965 (2000).
7. Le Ru E. C., Blackie E., Meyer M., Etchegoint, P. G., *Journal of Physical Chemistry C*, **111**, 13794 (2007).
8. Doering W. E., Nie, S., *Journal of Physical Chemistry B*, **106**, 311 (2002).
9. Lin X. M., Cui Y., Xu Y. H., Ren B., Tian Z. Q., *Anal. Bioanal. Chem.*, **394**, 1729 (2009).
10. Creighton J. A., Blatchford C. G., Albrecht M. G., *J. Chem. Soc., Faraday Trans. 2*, **75** , 790 (1979).
11. Lee P. C., Meisel D., *J. Phys. Chem.*, **86**, 3391 (1986).
12. Ni F., Cotton M. T., *Anal. Chem.*, **58**, 3159 (1986).
13. R. Griffith Freeman, Grabar K. C., Allison K. J., Bright R. M., Davis J. A., Guthrie A. P., Hommer M. B., Jackson M. A., Smith P. C., Walter D. G., Natan M. J., *Science*, **267**, 1629 (1995).
14. Sun Y., Xia Y., *Science*, **298**, 2176 (2002).
15. Jana N. R., Pal T., *Advanced Materials*, **19**, 1761 (2007).
16. Miragliotta J., Furtak T. E., *Phys. Rev. B*, **35** , (1987).
17. Ren B., Lin X., Yan J., Mao B. Tian Z., *J. Phys. Chem. B*, **107**, 899 (2003).
18. Duan G., Cai W., Luo Y., Li Z., Li Y., *Appl. Phys. Lett.*, **89**, 211905 (2009).
19. Teslova T., Corredor C., Livingstone R., Spataru T., Birke R. L., Lombardi J. R., Cañamares M. V., Leona M., *J. Raman Spectrosc.*, **38**, 802 (2007).

20. Roark S. E., Rowlen K. L., *Anal. Chem.*, **66**, 261 (1994).
21. Semin D. J. and Rowlen K. L., *Anal Chem.*, **66**, 4324 (1994).
22. Van Duyne R. P., Hulteen J. C., Treichelb D. A., *J. Chem. Phys.*, **99**, 2101 (1993).
23. Yan B., Thubagere A., Premasiri W. R., Ziegler L. D., Dal Negro L., Reinhard B. M., *ACS Nano*, **3**, 1190 (2009).
24. Davies J. P., Pachuta S. J., Cooks R. G., Weaver M. J., *Anal. Chem.*, **58**, 1290 (1986).
25. Gunnarsson L., Bjerneld E. J., Xu H., Petronis S., Kasemo B., Käll M., *Applied Physics Letters*, **78**, 802 (2001).
26. Kahl M., Vogesa E., Kostrew S., Viets C., Hill W., *Sensors and Actuators B: Chemical*, **51**, 285 (1998).
27. Dick L. A., McFarland A. D., Haynes C. L., Van Duyne R. P., *J. Phys. Chem. B*, **106**, 853 (2002).
28. Yang W. H., Hulteen J., Schatz G. C., Van Duyne, R. P. *Journal of Chemical Physics*, **104** 4313 (1996).
29. Baia L., Baia M., Popp J., Astilean S., *J. Phys. Chem. B*, **110**, 23982 (2006).
30. Yoshitake T., Shiraishi G., Nagayama K., *Appl. Surf. Sci.*, **197-198**, 379 (2002);
31. Jordan, R., Cole D., Lunney J. G., *Appl. Surf. Sci.*, **109-110**, 403 (1997).
32. Koren G., Polturak E., Fisher B., Cohen D., Kimel G., *Appl. Phys. Lett.*, **53**, 2330 (1988).
33. Caricato A. P., Di Cristoforo A., Fernández M., Leggieri G., Luches A., Majni G., Martino M., Mengucci P., *Appl. Surf. Sci.*, **208-209**, 615 (2003).
34. Boughaba S., Islam M., McCaffrey J. P., Sproule G. I., Graham M. J., *Thin Solid Films*, **371**, 119 (2000).
35. Dat R., Lichtenwalner D. J., Auciello O., Kingon A. I., *Appl. Phys. Lett.*, **64**, 2673 (1993).
36. Gupta A., *J. Appl. Phys.*, **73** (1993).
37. Yoon S. G, Lee J. C., Safari A, *J. Appl. Phys.*, **76**, 2999 (1994).
38. Singh R. K., Narayan J., *Phys. Rev. B*, **41**, 8843 (1990).
39. Anisimov,S. I., BauerleD., Luk'yanchuk B. S., *Phys. Rev. B*, **16**, 48 (1998).
40. Dogget B., Lunney J. G., *J. Appl. Phys.*, **109**, 093304 (2011).
41. Geohegan D. B., *Appl. Phys. Lett.*, **60**, 2732 (1992).
42. Zel'dovich Y. B., and Raizer Y. P., *Physics of Shock Waves and High Temperature Phenomena*, New York, Academic Press, (1966).
43. Rode A. V., Gamaly E. G., Luther-Davies B., *Appl. Phys. A*, **70**, 135 (2000).

44. Dyer P. E., Issa A.,Key P. H., *Appl. Phys. Lett.*, **57**, 186 (1990).
45. Gonzalo J., Afonso C. N., Madariaga I., *J. Appl. Phys.*, **81**, 951 (1997).
46. Gehohegan D. B., *Thin Solid Films* , **230**, 138 (1992).
47. Trusso S., Fazio B., Fazio E., Neri F., Barreca F., *Thin Solid Films*, **518**, 5409 (2010).
48. Bailini A., Ossi P. M., Rivolta A., *Appl. Surf. Sci.*, **253**, 7682 (2007);
49. Bailini A., Ossi P. M., *EPL*, **79**, 35002 (2007).
50. Neri, F., Ossi, P. M., Trusso, S., *Laser Part. Beams*, **28**, 53 (2010).
51. Ossi P. M., Neri F., Santo N., Trusso S., *Appl. Phys. A*, **104**, 289 (2011).
52. Yoshida T., Takeyama, Yamada Y., Mutoh K., *Appl. Phys. Lett.*, **68**, 1772 (1996).
53. Geohegan D. B., Puretzky A. A., Duscher G., Pennycook S. J., *Appl. Phys. Lett.*, **72**, 2987 (1998).
54. Marine, W., Patrone, L., Luk'yanchuk, B., Sentis, M., *Appl. Surf. Sci.*, **154-155**, 345 (2000).
55. Jubert P. O., Fruchart O., Meyer C., *Surf. Sci.*, **522**, 8 (2003).
56. Ohresser P., Shen J., Barthel J., Zheng M., Mohan C. V., Klaua M., Kirschner, **59**, 3969 (1999).
57. Bolgiaghi D., Miotello A., Mosaner P., Ossi P. M., Radnoczi G., *Carbon*, **43**, 390 (2005).
58. Di Fonzo F., Bailini A., Russo V., Baserga A., Cattaneo C., Beghi M. G., Ossi P. M., Casari C. S., Li Bassi A., Bottani C. E., *Catal. Today*, **116**, 69 (2006).
59. Wood R. F., Chen K. R., Leboeuf J. N., Puretzky A. A., Geohegan D. B., *Phys. Rev. Lett.*, **79**, 1571 (1997).
60. Wood, R. F., Leboeuf, J. N., Chen, K. R., Geohegan, D. B., Puretzky, A. A., *Appl. Surf. Sci.*, **127–129**, 151 (1998).
61. Agarwal N. R., Neri F., Trusso S., Lucotti A., Ossi P. M., *Appl. Surf. Sci.*, **258**,9148 (2012).
62. Fazio E., Neri, F., Ossi, P. M., Santo N., Trusso S., *Laser Part. Beams*, **27**, 281 (2009).
63. Neri F., Ossi P. M., Trusso S., *La Rivista del Nuovo Cimento*, **34**, 103 (2011).
64. Fazio E., Neri F., Ossi P. M., Santo N.,Trusso S., *Appl. Surf. Sci.*, **255**, 9676 (2009).
65. D'Andrea C., Neri F., Ossi P. M., Santo N., Trusso S., *Nanotechnology*, **20**, 245606 (2009).
66. Micali N., Neri F., Ossi P. M., Trusso S., *J. Phys. Chem. C*, **117**, 3497 (2013).
67. Shalaev V., *Phys. Rep.*, **272**, 61, (1996).

68. Berciaud S., Cognet L., Tamarat P., Lounis B., *Nano Lett.*, **5**, 515 (2005).
69. Xu H., Aizpurua J., Käll M., Apell P., *Phys. Rev. E*, **62**, 4318 (2000).
70. Michels A. M., Jiang J., Brus L. E., *J. Phys. Chem. B*, **104**, 11965 (2008).
71. Kneipp J., Li X., Sherwood M., Panne U., Kneipp H., Stockmann M. I., Kneipp K., *Anal. Chem.*, **80**, 4247 (2008).
72. Gupta R., Weimer W. A., *Chem. Phys. Lett.*, **371**, 302 (2003).
73. V. Amendola, M. Meneghetti, *Phys. Chem. Chem. Phys.*, **11**, 3805 (2009)
74. V. Amendola, M. Meneghetti, *Phys. Chem. Chem. Phys.*, **15**, 3027 (2013).
75. S. Barcikowski, G. Compagnini, *Phys. Chem. Chem. Phys.*, **15**, 3022 (2013).
76. S. Z. Mortazavi, P. Parvin, A. Reyhani, R. Malekfar, S. Mirershadi, *RSC Adv.*, **3**, 1397 (2013).
77. M. E. Povarnitsyn, T. E. Itina, P. R. Levashov, K. V. Khishchenko, *Phys. Chem. Chem. Phys.*, **15**, 3108 (2013).
78. R. Zhao, R. Xu, Z. Shen, J. Lu and X. Ni, *Opt. Laser Technol.*, **39**, 968 (2007).
79. A. Menendez-Manjon, B. N. Chichkov, S. Barcikowski, *J. Phys. Chem. C*, **114**, 2499 (2010).
80. K. Saitow, *J. Phys. Chem. B*, **109**, 3731 (2005).
81. K. Sasaki, T. Nakano, W. Soliman and N. Takada, *Appl. Phys. Express*, **2**, 046501 (2009).
82. H. Wang, A. Pyatenko, K. Kawaguchi, X. Li, Z. Swiatkowska-Warkocka and N. Koshizaki, *Angew. Chem.*, **49**, 6361 (2010).
83. H. Wang, N. Koshizaki, L. Li, L. Jia, K. Kawaguchi, X. Li, A. Pyatenko, Z. Swiatkowska-Warkocka, Y. Bando and D. Golberg, *Adv. Mater.*, **23**, 1865 (2011).
84. J. G. Walter, S. Petersen, F. Stahl, T. Scheper and S. Barcikowski, *J. Nanobiotechnol.*, **8**, 21 (2010).
85. P. Liu, Y. Liang, X. Lin, C. Wang and G. Yang, *ACS Nano*, 2011, **5**, 4748.
86. G. Compagnini, M. Sinatra, P. Russo, G. C. Messina, O. Puglisi and S. Scalese, *Carbon*, **50**, 2362 (2012).
87. S. R. J. Pearce, S. J. Henley, F. Claeyssens, P. W. May, K. R. Hallam, J. A. Smith and K. N. Rosser, *Diamond Relat. Mater.*, **13**, 661 (2004).
88. Q. X. Liu, C. X. Wang and G. W. Yang, *Phys. Rev. B*, **71**, 15542 (2004).
89. E. T. Y. Lee, Y. Shimotsuma, M. Sakakura, M. Nishi, K. Miura and K. Hirao, *J. Nanosci. Nanotechnol.*, **9**, 918 (2009).
90. V. Lebedev, P. Moroshkin, B. Grobety, E. Gordon and A. Weis, *J. Low Temp. Phys.*, **165**, 166 (2011).

91. A. Menendez-Manjon, P. Wagener, S. Barcikowski, *J. Phys. Chem. C*, **115**, 5108 (2011).
92. E. Fazio, Cacciola A., Mezzasalma A. M., Mondio G., Neri F., Saija R., *J. Quant. Spectrosc. & Rad. Transf.*, **124**, 86 (2013).
93. E. Fazio, F. Neri, *Appl. Surf. Sci.*, **272**, 88 (2013).
94. S. Link, Z. L. Wang, M. A. El Sayed, *J. Phys. Chem.*, **103**, 3529 (1999).
95. U. Kreibig, M. Vollmer, in Optical Properties of Metal Clusters (Springer, Berlin, 1996) p. 265.
96. E. Fazio, F. Neri, A. Valenti, P. M. Ossi, S. Trusso, R. C. Ponterio, *Applied Surface Science* **278**, 259 (2013).
97. P. Hildebrandt, M. Stockenburger, *J. Phys. Chem.*, **88**, 5935 (1984).
98. H. N. Lioe, D. R. Adawiyah, R. Anggraeni, *Intern. Food Res. Journal*, **19**, 537 (2012).

Chapter 8

Surface-Enhanced Multipurpose Nanosensing with Microneedle-Shaped Fiber Optics

Andrea Lucotti,[a,b] Michele Casella,[a] and Matteo Tommasini[a,b]

[a] *Dipartimento di Chimica, Materiali e Ingegneria Chimica "G. Natta", Politecnico di Milano, Piazza Leonardo da Vinci 32, 20133 Milano, Italy*
[b] *Consorzio Interuniversitario per la Scienza e Tecnologia dei Materiali (INSTM), Unità di Ricerca del Politecnico di Milano (Dip. CMIC), Piazza Leonardo da Vinci 32, 20133 Milano, Italy*
matteo.tommasini@polimi.it

In a broad sense, nanodevices are the critical enablers that will allow exploiting the ultimate technological capabilities of chemical, pharmaceutical, and biological systems. In this scenario, surface-enhanced Raman scattering (SERS) spectroscopy is one of the most promising analytical techniques for identification of molecular species, able to detect several kinds of sample under different environmental conditions and a good potential with respect to low-concentration analytes. We illustrate here the possible uses of a surface-enhanced microneedle-shaped fiber optics acting as a multipurpose nanosensor. The optical probe takes advantage from SERS to improve its sensibility. It adopts an active substrate obtained by immobilization of nanoparticles on the manufactured

Handbook of Enhanced Spectroscopy
Edited by Pietro Giuseppe Gucciardi, Marc Lamy de la Chapelle, and Nathalie Lidgi-Guigui

ISBN 978-981-4613-32-3 (Hardcover), 978-981-4613-33-0 (eBook)
www.panstanford.com

fiber substrate. The multipurpose nanosensing capabilities of the optical probe will be highlighted by illustrating selected examples.

8.1 Introduction

Optical sensing is a powerful tool employed to investigate nature at different spatial scales [1]. Typically being noninvasive, though coupled to almost any kind of matter, optical waves have been largely exploited to quantitatively access the microscopic world. With the recent advent of nanoscience, the new frontier of optical sensing has become the nanoscale [2].

Nanotechnology opened doors to new ways of recognizing, characterizing, and quantifying several compounds through the use of nanosensors and nanoprobes [3]. The advantage of using nanotools capable to address several processes at the molecular level has received increasing interest for many diverse potential applications in, for example, environmental, industrial, scientific, and biomedical fields [4]. For instance, several nanosensors have been developed to improve environmental monitoring (air and water quality), food quality surveillance, pharmaceutical characterization, and to enable the reliable, real-time detection of chemical, biological, radiological, and nuclear hazards for military and antiterrorism applications [5–7]. The use of plasmonic nanoparticles has also significant potential as nanoscale optical tools in several molecular investigations [8, 9] like sensitive nanowire sensors allowing detection of biological molecules noninvasively [10].

The typically low scattering cross-section of normal Raman processes is boosted through the use of metallic nanoparticles (common choices are Ag and Au). The resulting SERS enhancement is thought to be the combined effect produced by electromagnetic and chemical causes [11–13]. Under laser irradiation of proper wavelength, noble metal nanoparticles can sustain plasmon oscillations. These produce intense and localized electromagnetic fields at the surface of metal nanostructures, which are at the basis of the strong electromagnetic enhancement of the SERS signal. Chemical effects upon adsorption of the analyte molecules at the metal surfaces (i.e., charge transfer or covalent bonding) can provide further signal enhancement and can affect the vibrational structure of the molecules compared to normal Raman. Hence, under these circumstances, peak shifts and spectral shape changes are commonly observed in SERS compared to normal Raman. For this reason, in this case, special care has to be taken to control environmental conditions (notably the kind of nanostructured substrate, pH, temperature, and laser power) so to ensure repeatability of the measurements and reliably correlate the observed SERS signal to the chemical species to be analyzed.

Nanosensors based on Raman spectroscopy offer a very good chemical specificity and can be applied for the characterization of drug and biological fluids [14, 15], recognition and dosimetry of vapors [16], or for the chemical/physical characterization of novel π-conjugated materials [17, 18].

8.2 Fiber Optics SERS Sensing

The concept of SERS sensing made through fiber-optic setup has been reported over the years by several authors [19–25]. We review here some possible applications of a fiber-optic SERS probe, whose sensitive tip has been fabricated with optimized double taper shape (see Fig. 8.1). We aim at illustrating the wide range of possible uses, which stem from the combination of the high molecular sensitivity of SERS with the versatility of the fiber-optic design that offers significant advantages with respect to nanostructured planar sensors. In fact the latter are usually supported by rigid and macroscopic substrates that are less prone to be used for time-dependent online sensing or remote sensing and offer a more limited spatial resolution with respect to the microscopic size of the fiber optics tip.

For developing the fiber-optic nanosensor we used a particular type of SERS-active substrate that employs a silver nanoparticle immobilization procedure [26–28] described in details in [29]. The whole fabrication process of the fiber-optic sensor, encompassing the fiber optics etching, the silver colloids synthesis, and immobilization on the fiber surface, is described elsewhere [14, 17, 29–31]. Nanosensing measurements with the fiber-optic setup

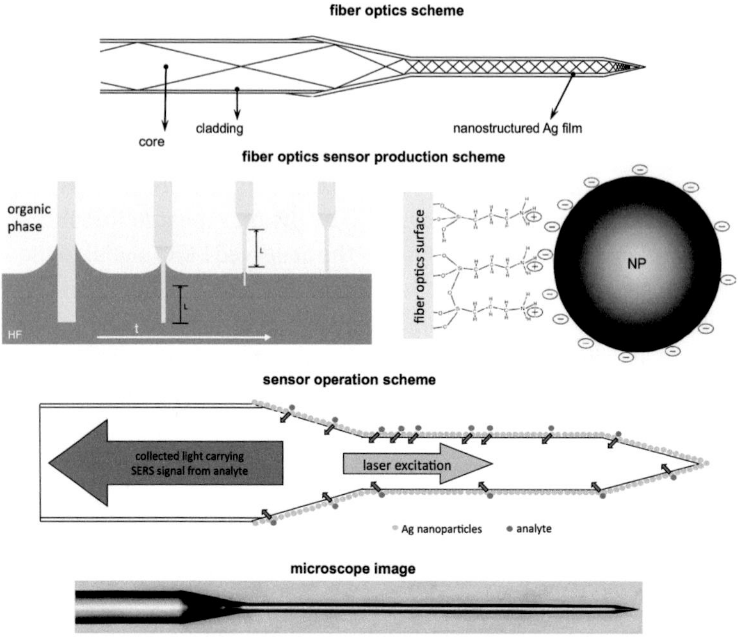

Figure 8.1 Schematic diagrams of the geometry, fabrication, and operation of the fiber optics–based SERS sensor described.

can be carried out very conveniently using the so-called *optrode* configuration [19, 20, 24]. This consists of a single optical fiber (with sensing tip at one end), which allows transportation of both the laser excitation at the fiber tip, where the sample is in contact with the SERS substrate. The resulting SERS signal travels back to the spectrometer through the fiber itself. The SERS process occurs at the fiber-optic nanofunctionalized surface and the resulting photons are collected by the fiber (see Fig. 8.1). The experimental setup as a whole (i.e., Raman spectrometer, fiber-optic sensor, and measurement chamber) is depicted in Fig. 8.2).

Compared with a traditional planar solid substrate, this kind of fiber optics setup allows to more closely bringing at the molecular sample both light excitation and light collection, thus in principle enabling a more effective and spatially resolved probing. Furthermore, the fiber-optic setup is nicely fitted for remote sensing and multiplexing, that is, probing from several tips located in

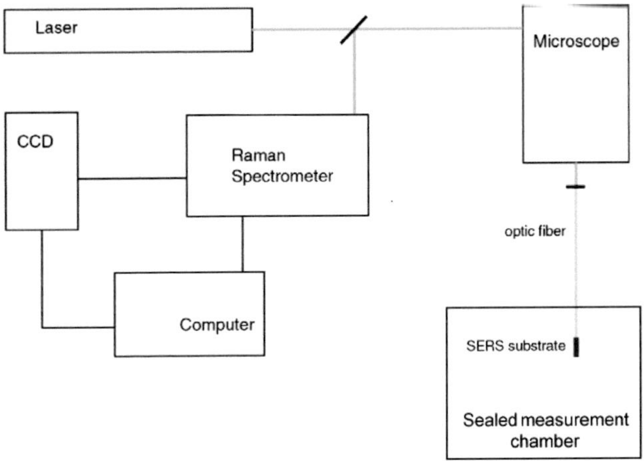

Figure 8.2 Diagram of the measurement setup for the fiber-optic SERS sensor in optrode configuration.

different locations to be monitored. A proper optical switch can then be used to read with the Raman spectrometer the signal from each individual tip, with no need of displacing the sensors. This scenario is particularly appealing in industrial applications where the environment could be particularly unfriendly to Raman spectrometers (even portable ones) for many reasons, such as encumbering or presence of harmful substances. The molecular sensitivity of SERS can then be potentially employed for remote sensing of several chemical species at once.

A commercial Raman equipment can be easily coupled with the fiber-optic sensor. In our case this is the Horiba Jobin Yvon LABRAM HR800 equipped with both 785 and 514 nm laser excitations and Olympus BX41 microscope. The latter allows injecting light into the fiber and collecting it back. The optical fiber used for the fabrication process is a common silica fiber with inner core diameter of 200 μm and a numerical aperture of 0.22.

Some worked-out examples can help to illustrate better the multipurpose capabilities of SERS sensing based on fiber-optic design in different fields: (i) healthcare (spectroscopic identification of apomorphine drug); (ii) materials science (in situ characterization of sp-carbon species during the synthesis process); and (iii) safety control (detection of hazardous volatile compounds).

(i) Drug identification

An interesting biomedical application for a fiber-optic SERS probe is the in vitro identification/sensing of drugs in biological fluids with minimal sample preparation. It is important to stress that this is not meant to replace existing well-established analytical methods in the clinical laboratory. Rather, the aim is to provide a complementary tool that is minimally invasive in the current clinical laboratory procedures and can add molecular information useful for cross-validation. With respect to standard clinical chemistry methods the benefits are the very limited sample volume required (few μL), the reduced analysis time (order of minutes), the sensitivity to molecular structure, and the little cost of the fiber-optic sensor (whose fabrication procedure is compatible with industrial scaling). Among the current drawbacks of this technique it has to be mentioned that each analyte requires a preliminary investigation of the SERS signal; furthermore, quantitative analysis cannot be as precise as traditional clinical chemistry methods, due to the inherent fluctuations of the SERS process and its dependence upon environmental conditions.

We illustrate here the characterization of the SERS signal collected from the fiber optics sensor on a low-concentration solution of apomorphine, a well-known potent short-acting dopamine agonist typically used in the treatment of patients with advanced Parkinson's disease. The commercial formulation of apomorphine (Apofin®) was obtained from Chiesi Farmaceutici (Parma, Italy). Apofin® is a clear and colorless water solution for subcutaneous injection (it contains 3.74×10^{-2} M apomorphine hydrochloride, with sodium metabisulfite and hydrochloric acid as excipients).

The SERS and fiber optics SERS (FO-SERS) spectra of apomorphine are reported in Fig. 8.3 together with the chemical structure of this drug. The main features in Raman and SERS of apomorphine have been assigned by means of density functional theory (DFT) calculations [14, 32]. We briefly review here the description of the main signals observed in Fig. 8.3. The strongest Raman lines are

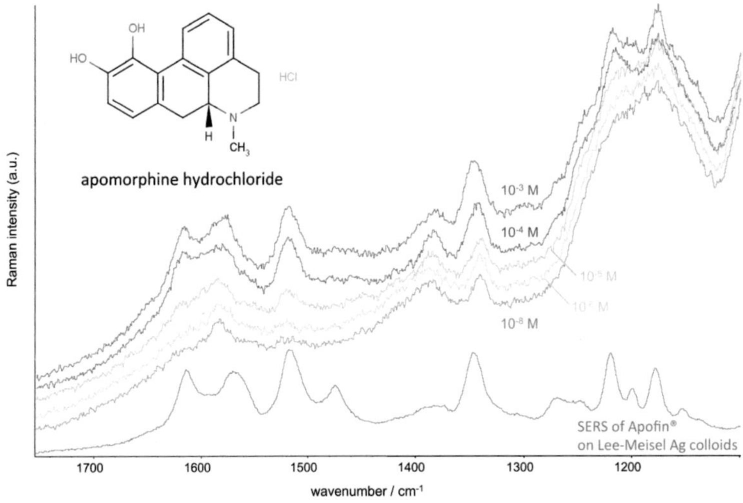

Figure 8.3 SERS and FO-SERS spectra of Apofin® and apomorphine hydrochloride at different concentrations in water solution (10^{-8} up to 10^{-3} M). Excitation line 785 nm. The spectrum recorded at 10^{-8} M concentration does not show any significant signal due to apomorphine and can be taken as the reference background spectrum of the fiber optics sensor.

observed in the 1200–1650 cm^{-1} region and are related to normal modes that involve C=C stretching or CH and OH in-plane bending vibrations [14]. In particular, in the C=C stretching region, we have two strong bands located at 1622 and 1594 cm^{-1}, both related to ring stretching modes of the two aromatic moieties of apomorphine. The peak at 1622 cm^{-1} is related to the stretching of catechol ring coupled with in-plane OH bending, while the peak at 1594 cm^{-1} is ascribed to the stretching of the aromatic ring coupled with in-plane CH bending. FO-SERS compares well with SERS signal of apomorphine; the broad feature of FO-SERS around 1050 cm^{-1} is a residual signal of the fiber optics substrate [30, 31]. It is evident that the SERS signals in the neighborhood of 1300 cm^{-1} differ from their Raman counterpart. The Raman peak at 1307 cm^{-1} is assigned to OH in-plane bending coupled with CH_2 bending [14]. The different SERS signal in this region has been interpreted with a chemisorption model of apomorphine, which involves the catechol ring and the OH groups [32]. This kind of interaction is required for a good SERS signal. In a more complex environment such as apomorphine in human plasma, the presence of transport proteins (i.e., albumin) able to bind apomorphine hinders the direct chemisorption of the drug on the nanosilver substrates, thus reducing the SERS activity [32]. This constitutes a potential problem for the detection of drugs in plasma at low concentration. Specific pretreatment of the sample aimed at weakening protein–drug interactions could improve the detection capability of the sensor.

(ii) Characterization of sp/sp^2 carbon allotropes in their synthesis environment

Many trends in current materials science ultimately depend on the chemistry and properties of the various forms of pure carbon [33], including diamond (sp^3-hybridized carbon), graphite/graphene (sp^2-hybridized carbon), as well as extended sp-hybridized carbon systems, which have been recently synthesized and characterized with different techniques [34–36].

These linear (i.e., one-dimensional) forms of carbon, referred to as polyynes or carbynes, represent interesting systems for fundamental aspects related to the science of carbon allotropes such as cluster [37] and nanotubes formation [38]. The molecular properties of polyynes are also of interest for nonlinear optics because of their remarkable second hyperpolarizability [39, 40].

The same fiber-optic SERS sensor illustrated above can be used also for the investigation of the formation of polyynes, directly in the synthesis environment. A very convenient and flexible route to the production of polyynes has been invented by Cataldo in 2004 [41–43]. It consists in creating an arc discharge between two graphite electrodes submerged in a selected solvent. By changing the operative conditions it is possible to produce polyynes with different terminations (e.g., nitrogen, in addition to hydrogen) [41–43]. Certainly the submerged arc discharge constitutes a typically unfriendly environment for most spectroscopic techniques. However, thanks to the fiber-optic design; it is possible to carry out real-time in situ SERS spectroscopy on the native sp-carbon species directly in the submerged arc-discharge synthesis environment. During the fiber-optic SERS measurements presented here the arc was turned off for practical reasons, but it is possible to carry out measurements while the arc is on, provided that the sensor is sufficiently far from the plasma region and spectral background from the light produced by the discharge is properly characterized and/or screened during the measurements. In principle, this setup allows studying production rates, cross-link interactions of the sp-chains in solution, and their stability with time.

Polyynes were produced in water solution by following Cataldo's submerged arc-discharge method. The native hydrogen-capped polyynes have been detected and analyzed by means of the fiber-optic SERS sensor illustrated in Fig. 8.1. SERS spectra (514 nm excitation wavelength) were recorded from the fiber-optic SERS sensor positioned in close proximity to the electric arc (approximately 2 cm). The spectra were collected at regular time intervals. After each arc discharge (duration about 10 s) the SERS spectrum was collected (integration time about 30 s) right after the arc discharge. By considering the obvious delay time required to switch the instrumentation, every discharge followed by the Raman measurement took about 1 minute, after which the subsequent discharge was initiated. We index the sequence of Raman spectra so recorded with the progressive number of "shots" (i.e., number of arc discharges).

The initial FO-SERS spectra (1–7 shots) reported in Fig. 8.4 show the generation of polyynes during the arcing process, as evidenced by the signal in the 1800–2300 cm^{-1} region (which has been assigned to polyynes [44]). Instead, the spectra at later times (7–10 shots, see Fig. 8.4) highlight the production of sp^2 hybridized carbon species (signals at 1300–1700 cm^{-1}). The signals from sp^2-hybridized carbon are generated by the cross-linking of polyynes and/or from the direct formation in the arc of condensed poly-aromatic species.

It is apparent that from shot 2 to 7 a progressive increase of the polyyne signals takes place, whereas from shot 7 to 10 we observe a progressive decrease of the sp signal. As mentioned above, the broad signals at 1055 and 1186 cm^{-1} are due to the fiber optics background and are present in all spectra (even recording in pure distilled water or air, see Fig. 8.5). Another strong background signal from the fiber is found at lower wavenumbers (100–600 cm^{-1}).

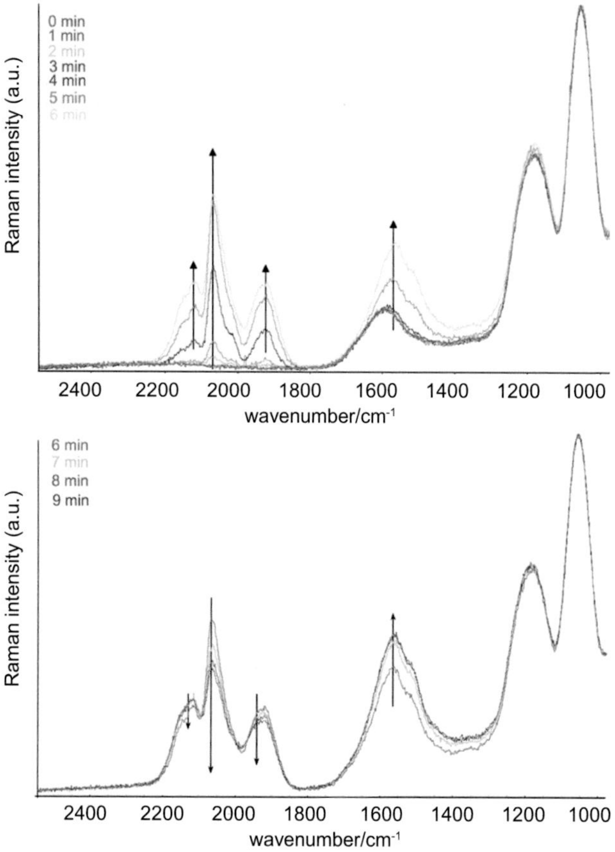

Figure 8.4 Evolution of SERS spectra recorded during the arcing process in water (see text for details). Excitation line 514 nm.

The sp-carbon signals (1800–2300 cm^{-1}) show a more definite series of peaks with smaller line widths. These are associated to sp-chains with different lengths, due to the characteristic frequency dispersion with molecular size observed in the Raman spectroscopy of polyynes [17, 36, 44]. Shorter chains produce Raman signals in the high wavenumber region (2100–2200 cm^{-1}), whereas longer chains produce Raman signals in the low wavenumber region (1900–2000 cm^{-1}). In polyynes the so-called я mode [17] is associated to the most intense Raman band; the associated nuclear displacement

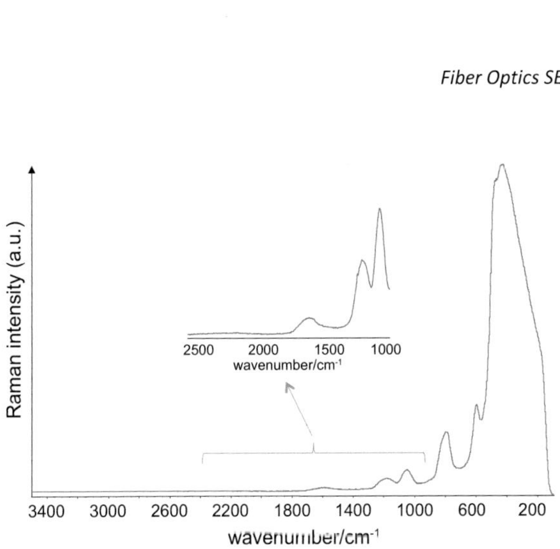

Figure 8.5 Fiber-optic background signal (excitation line 514 nm). The strong fiber-optic signal in the low-frequency region (100–600 cm^{-1}) hinders the possibility of using the SERS sensor in this region for low signal detection.

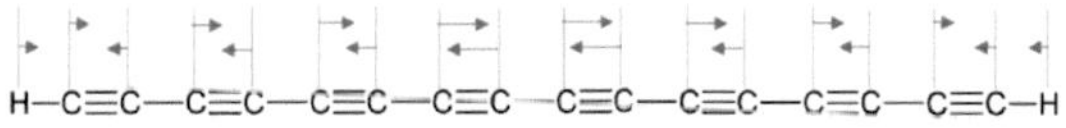

Figure 8.6 Nuclear displacements relative to the я mode of an exemplary polyyne (H–C_{16}–H).

consists of a collective stretching of the triple bonds and shrinking of the single bonds of the polyyne chain (...-C≡C-C≡C-C≡C- ...). See Fig. 8.6 for a sketch of the nuclear displacements of the я mode of polyyne H–C_{16}–H.

To better describe the evolution of SERS we have plotted the spectra in Fig. 8.4 in such a way to normalize them with respect to a stable reference, which is very conveniently provided by the 1055 cm^{-1} signal of the fiber optics itself.

The sp signal shows at first an increase with the number of shots and later a decrease. This phenomenon can be explained, respectively, by two mechanisms: (i) the combined polyyne production rate and adsorption kinetics on the nanostructured silver surface of the fiber-optic sensor and (ii) the saturation of the adsorbing sites and the cross-linking processes among polyynes, which produce shorter sp chains and amorphous sp^2-carbon.

Inspection of Fig. 8.4 reveals that in the initial stage (minute 1 to 4) longer polyynes are formed (1924 and 2065 cm^{-1} peaks) with a clear predominance of the 2065 cm^{-1} peak. From minute 4 to 6 the 2117 cm^{-1} peak increases faster than the other two peaks (1924 and 2065 cm^{-1}). It is also important to emphasize that the sp^2 signal (1560–1590 cm^{-1}) shows a behavior very similar to the 2117 cm^{-1} peak. A reasonable explanation of this trend can be the acceleration of the cross-linking process which takes place when the polyyne coverage of the silver surface approaches the monolayer. In this condition different polyynes get close to each other at the silver surface and are prone to cross-link. This is supported by the fact that the maximum value of the SERS signal is reached at minute 6 (maximum polyyne coverage of the silver surface) and then the signal starts to decrease. From minute 6 to 9 (see Fig. 8.4) one notices a progressive decrease of the whole sp-carbon signal accompanied by the increase of the sp^2-carbon signal: as cross-linking of polyynes proceeds more sp^2 carbon structures are produced and longer polyynes are shortened (the 2117 cm^{-1} signal decreases less than those associated to longer polyynes at 1924 and 2065 cm^{-1}).

Another interesting and straightforward use of the fiber optics allows to evaluate the stability of polyynes in air when they are interacting with the nanostructured silver surface of the sensor, which is known to stabilize sp-carbon species [45]. This can be done simply with a dip&dry procedure, where the sensor is first dipped into the polyyne solution, then removed from it and used for SERS measurements once the solvent completely evaporates (water in our case). A series of SERS spectra recorded from 2 to 8 minutes after the extraction from the water is reported in Fig. 8.7 (514 nm laser wavelength, 100 mW power injected into the fiber). The sp signal shows a progressive decrease with no appreciable differences among the three different sp-carbon peaks (i.e., 2117, 2065, and 1924 cm^{-1}) while the sp^2 signal shows a slight increase in the spectral region between 1250 and 1500 cm^{-1} (D-peak [46, 47]). This behavior is ascribed to a cross-linking process, which affects all polyynes, hence decreasing the sp signal, and brings upon the increase of the sp^2-carbon signal.

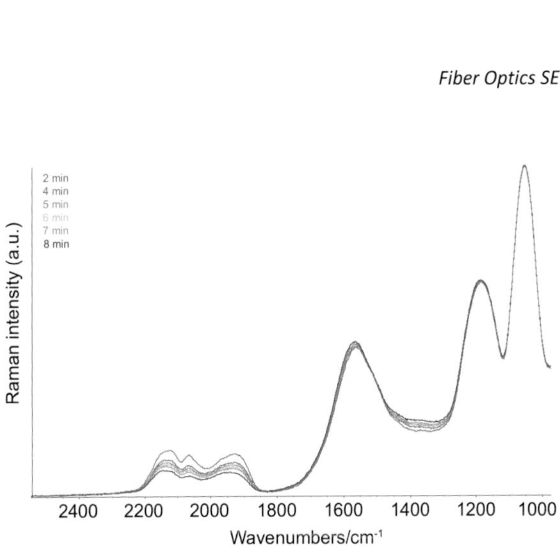

Figure 8.7 SERS spectra recorded from 2 to 8 minutes after the extraction of the fiber optics sensor from the water environment of the arc-discharge apparatus (see text for details).

(iii) Detection of volatile compounds

The estimation of health effects owing to the regular exposure in the workplace to chemicals in vapor form represents a goal for safety assessment [48]. An advantage of the fiber-optic SERS sensor is its straightforward use for gas-phase sensing. This can have many applications for air quality monitoring toward specific harmful pollutants able to be adsorbed on the nanostructured film that covers the fiber-optic tip. Sulfur-containing organic compounds, such as thiols, chemisorb on noble metals (Ag, Au) and are diffusely used in organic synthesis [49] and as surface-active agents [50]. The exposure to thiophenol may cause severe negative health effects [51–54]. Thiophenol was identified as a principal component to flavor-tainting found in fish and causes conversion of oxyhemoglobin to methemoglobin in human red blood cells [55, 56]. Within this scenario, the fiber-optic SERS sensor can be used as dosimeter for monitoring chronic exposure to toxic molecules. In this case, even at very low concentrations of the analyte, the time-integrated effect eventually becomes detectable since the number of chemisorbed analyte molecules steadily increases with time (with a given kinetics which is dependent upon the analyte).

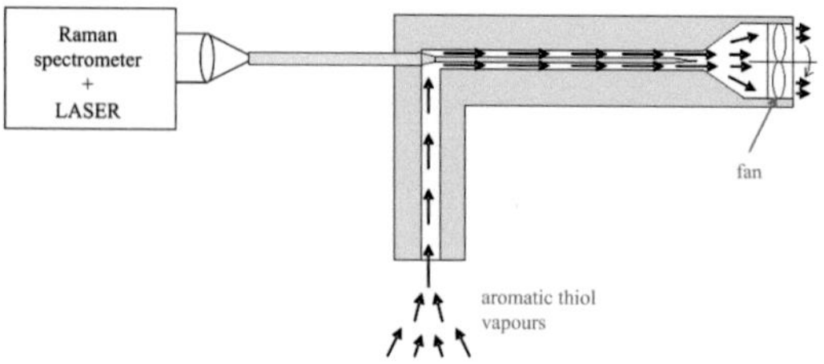

Figure 8.8 Scheme of operation of a fiber-optic SERS sensor for enhanced vapor detection under a forced flow.

It is also interesting to present a feasible technical improvement for easing adsorption of target molecules on the fiber-optic SERS sensor (see Fig. 8.8). It consists of a narrow channel where the fiber-optic SERS sensor is inserted and through which the gas to be analyzed can be circulated by means of a fan put on a distal end of the channel. This setup can enhance the volume of the gas probed by the sensor in a given time. This kind of measurement head could be very useful if combined with a portable Raman spectrometer to carry out on-field measurements as previously shown by Sylvia et al. with a similar set-up for on-field landmines detection [57]; however, in the cited case the SERS substrate is external to the optical fiber.

To test the FO-SERS sensor for applications in gas phase we have considered three different volatile thiols, namely thiophenol (**1**), 2-phenyl ethane thiol (**2**), and 2-naphthalene thiol (**3**). Low concentration ethanol solutions of the three compounds were prepared and a fixed small volume ($\approx$0.1 ml) of solution was left to completely evaporate within a sealed chamber of known volume (3 L). Depending on solution concentration this procedure allows reaching a controlled concentration of thiols in the sealed chamber (**1**: 10 ppb with 10^{-3} M, 0.01 ppb with 10^{-6} M; **2**: 1 ppb with 10^{-4} M; **3**: 10 ppb with 10^{-3} M, 0.1 ppb with 10^{-5} M). The fiber-optic sensor (previously placed in the sealed chamber) was excited with 785 nm laser to record the time evolution of the SERS signal of thiol molecules during chemisorption (spectra were recorded after about 30 minutes since the complete evaporation of the solution drops).

The background signal of the fiber is similar to that of the previous cases (i–ii) and it was determined at the beginning of the experiments with the fiber in free air (the main Raman features

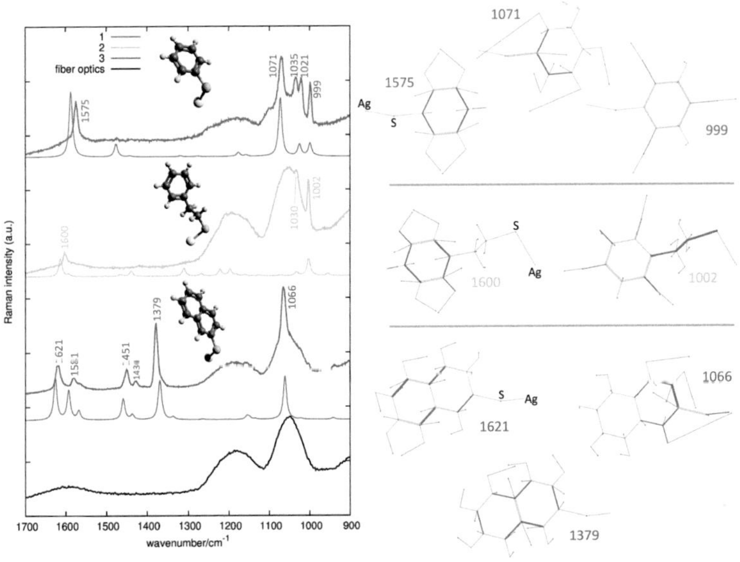

Figure 8.9 Fiber-optic SERS signals of the three thiols here considered (concentrations are 10, 1, and 10 ppb for molecules **1**, **2**, and **3**, respectively) compared with theoretical calculations. Results from DFT calculations (B3LYP/cc-pVTZ, frequency scaled by 0.98) are provided to guide the interpretation of the spectra through the nuclear displacement patterns of selected intense Raman modes (blue bonds shrink; red bonds stretch; nuclear displacements vectors are red lines). To account for chemisorption effects (within nearest-neighbor approximation) the –SH moiety of the three compounds has been replaced with the –SAg group.

are in the low wavenumber range, 100– 600 cm^{-1}, and at 1055–1186 cm^{-1}). The following signals (cm^{-1}) are observed in the three fiber-optic SERS spectra (see Fig. 8.9): (**1**): 1575, 1071, 1035, 1021, 999; (**2**): 1600, 1030, 1002; (**3**): 1621, 1581, 1451, 1430, 1379, 1066. These peaks can be assigned to specific nuclear displacement patterns with the help of DFT calculations, as shown in Fig. 8.9 for some selected modes (further details about the case of (**3**) can be found in [58]). In general, we can notice that the stronger SERS features are all due to vibrations involving the aromatic moieties of the three compounds. The associated nuclear displacements can be described as ring-stretching modes (falling in the higher wavenumber range, 1650–1550 cm^{-1}) or ring-breathing modes

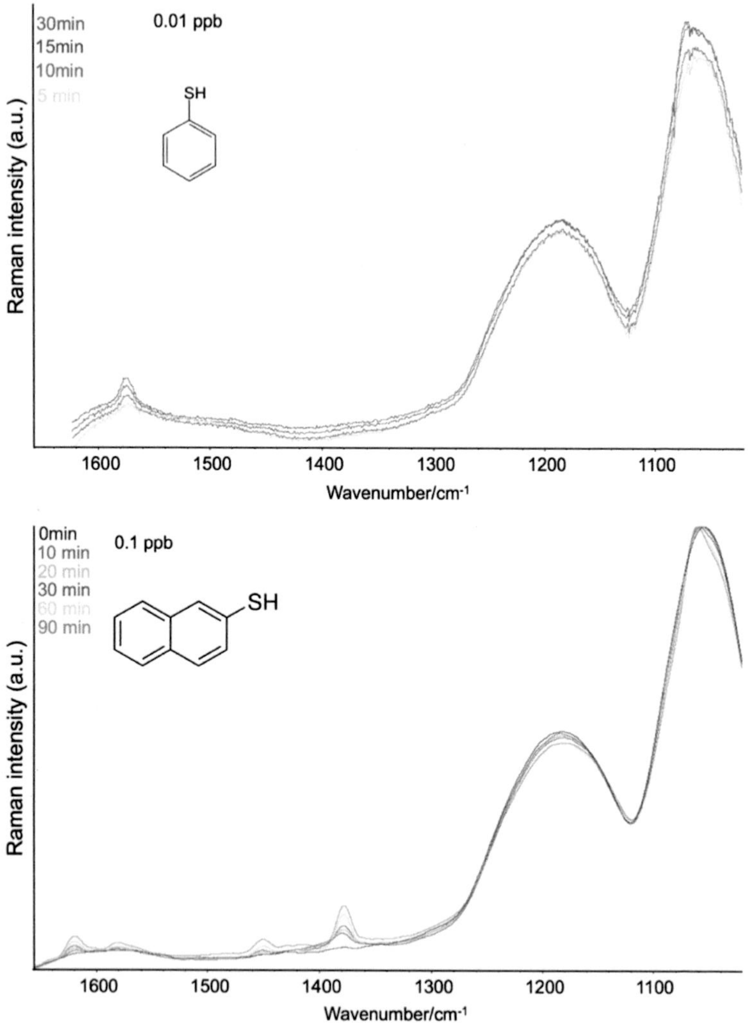

Figure 8.10 Time dependence of the fiber-optic SERS signal of thiols **1** (0.01 ppb) and **3** (0.1 ppb).

(falling in the neighborhood of 1000 cm^{-1}). For compound (**3**), because of its more complex chemical structure comprising two condensed rings, a series of peaks is found in the intermediate wavenumber range, between 1600 and 1000 cm^{-1}. These are all

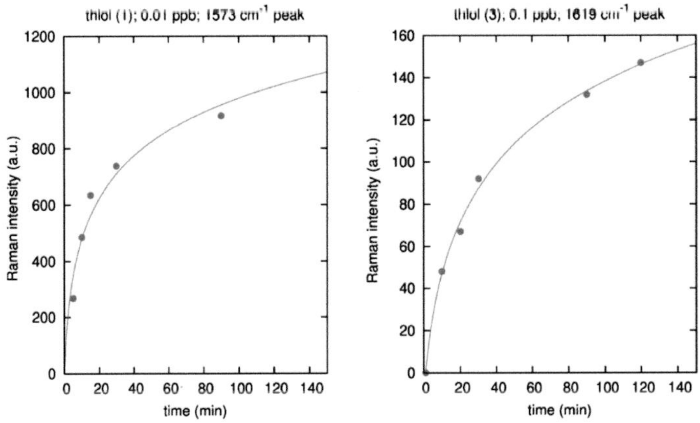

Figure 8.11 Elovich fits of the time dependence of the SERS signal of (**1**) and (**3**). The strongest peak of the two compounds was considered, which correspond to the ring-stretching mode. The time parameter $(a\alpha)^{-1}$ derived from the fits of (**1**) and (**3**) are respectively 1.4 and 4.9 minutes.

modes due to collective vibrations of CC bonds of the aromatic rings (details are given in [58]). Interestingly, for compound (**3**) the mode at 1066 cm^{-1} partially involves the breathing of one ring of the molecule and the stretching of the CS bond. The 1066 cm^{-1} mode of compound (**3**) is reminiscent of the 1071 cm^{-1} mode of (**1**)—see Fig. 8.9.

The fiber optics setup allows to straightforwardly carrying out time-dependent measurements in the sealed chamber with low starting concentration of the analyte thiol (**1** at 0.01 ppb; **3** at 0.1 ppb)- see Fig. 8.10. The analysis of the Raman intensity increase with time is summarized in Fig. 8.11 and clearly demonstrates the logarithmic time dependence of the signal, in the form of the Elovich law, well known in the context of chemisorption of volatile species [59]:

$$q(t) = \frac{\ln 10}{\alpha} \log\left[1 + a\alpha(t - t_0)\right] \tag{8.1}$$

where $q(t)$ is the quantity of chemisorbed molecules at time t, $q(t_0) = 0$ is the initial quantity chemisorbed (t_0 reflects the arbitrary origin of the time axis or the experimental error in time determination) and $\tau = (a\alpha)^{-1}$ is the typical time scale of the process. Since the intensity of the SERS signal is proportional to the number of molecules adsorbed at hot spots (which can be assumed to be randomly distributed all over the active nanostructured surface with a given surface density), it results that the SERS signal is proportional to $q(t)$, hence it also exhibits a logarithmic time dependence.

From the point of view of application as a sensor, a trend determined by the adsorption kinetics reported in Fig. 8.11 has two main features. The large increase of the signal in the first minute represents a fast response. Furthermore the sensor can be operated as a dosimeter for detecting the chronic exposure (even at very low concentrations) to a dangerous thiol substance.

8.3 Conclusions and Perspectives

Since the discovery of the enhancement of the Raman signal in proximity of noble metal nanostructures, back in 1974 [60], the field of SERS spectroscopy has very significantly grown, as it is testified by the steadily increasing number of works which have been published since then (see for instance the bibliometric data reported in [61]). Because of the nature of the processes involved, a fruitful application of SERS mainly requires a noble metal nanostructured substrate with suitable plasmonic properties (i.e., tuned for the chosen laser excitation) and a good understanding of the specific SERS signal of the analyte molecule. The latter point is particularly relevant for label-free detection of analytes, whose SERS signal can significantly differ from the normal Raman spectrum, due to interaction effects with the metal substrate (e.g., chemisorption, charge transfer, etc.). In other circumstances, chemical functionalization of the substrates lead to selective interaction for specific analytes, thus enhancing significantly the sensitivity of the SERS probe for a given application [65–67]. The use of reporter molecules with high Raman cross-section is a valid alternative which allows indirect detection of analytes, through their specific binding to a well-designed reporter (i.e., molecular tag) [68].

In any of the many possible scenarios for SERS applications the nanostructured substrate plays a central role, either in its pristine form for label-free detection or in some suitably functionalized form when adopting a setup that implies a selected reporter molecule. The fiber-optic SERS sensor described here can be operated straightforwardly for label-free direct detection of different analytes, as implied by the three examples described in this chapter, whose application field ranges from material science to pharmaceutics and environmental monitoring. We have shown that the fiber-optic sensor can be operated in fairly different setups and conditions: it can detect apomorphine drug molecule down to 10^{-5} M concentration in water solution; it can be profitably used to characterize the production of elusive carbon species (polyynes) directly in the submerged arc-discharge environment; it can be also employed as a dosimeter to detect volatile species (thiols).

Furthermore, the solid nature of the fiber-optic substrate is also well-suited for proper chemical functionalization which would significantly extend the application of this sensor to the biomedical field by imparting molecular recognition capabilities to the plasmonic surface of the fiber. Finally, we would like to mention that the presented fiber optics is also compatible with production techniques of the nanostructured surfaces other than chemical. Under this regard it is particularly appealing a physical method (pulsed laser deposition) which has shown a good potential for the production gold and silver arrays with controlled nanostructure and excellent SERS properties, both in terms of high response and reproducibility [62–64].

Acknowledgments

We thank Prof. Giuseppe Zerbi for his constant inspiration and illuminating scientific discussions on the many aspects of vibrational spectroscopy over the years.

We gratefully thank the Italian MIUR for financial support, under the auspices of the FIRB program RBFR08XH0H (Futuro in Ricerca 2008).

References

1. Zhang, L. M. and Uttamchandani, D. (1988). *Electron. Lett.* **24**, p. 1469.
2. Buckle, P. E., Davies, R. J., Kinning, T., Yeund, D., Edwards, P. R., Pollard-Knigh, D. and Lowe, C. R. (1993). *Biosens. Bioelectron.*, **8,** p. 355.

3. Vo-Dihn, T. et al. (2008). Nanosensors and nanoprobes for environmental health sensing and biomedical screening, IEEE/LEOS Summer Topical Meetings.
4. Saetz, G. et al. (2010). *NanoToday,* **5**, p. 251
5. Zhang, X. et al. (2005). *J. Am. Chem. Soc.,* **127**, pp. 4484–4489.
6. Shafer-Peltier, K. E. et al. (2003). *J. Am. Chem. Soc.,* **125**, pp. 588–593.
7. Bao, L. et al. (2003). *Anal. Chem.,* **75**, pp. 6614–6620.
8. Ozbay, E. (2006). Plasmonics: merging photonics and electronics at nanoscale dimension, *Science,* **311**, p. 189.
9. Atwater, H. (2007). The promise of plasmonics, *Sci. Am.,* **296**, p. 56.
10. Fahmy, T. et al. (2008). *Nanoletter,* **8**, p. 3310.
11. Van Duyne, R. P. et al. (2005). Surface enhanced Raman spectroscopy. *Anal. Chem.* **1**, p. 338A.
12. Schatz, G. C. and Van Duyne, R. P. (2002). *Handbook of vibrational spectroscopy,* eds. Chalmers, J. M. and Griffiths, P. R. (John Wiley & Sons, New York), Vol. 1, pp. 759–774.
13. Campion, A. and Kambhampati, P. (1998). *Chem. Soc. Rev.,* **27**, pp. 241–250.
14. Lucotti, A. et al. (2009). Experimental and theoretical investigation of the apomorphine Raman spectrum. *J. Raman Sectrosc.,* **40,** p. 2074.
15. Casella, M. et al. (2011). Raman and SERS recognition of β-carotene and haemoglobin fingerprints in human whole blood. *Spectrochim. Acta A,* **79,** p. 915.
16. Vo-Dihn, T. et al. (1999). SERS detection of chemical vapors with the use of personale dosimeters. *Field Anal. Chem. Tech.,* **3**, p. 346.
17. Lucotti, A., Tommasini, M., Fazzi, D., Del Zoppo, M., Chalifoux, W. A., Ferguson, M. J., Zerbi, G. and Tykwinski, R. R. (2009). Evidence for solution-state nonlinearity of sp-carbon chains based on IR and Raman spectroscopy: violation of mutual exclusion, *J. Am. Chem. Soc.,* **131**, pp. 4239–4244.
18. Milani, A., Lucotti, A., Russo, V., Tommasini, M., Cataldo, F., Li Bassi, A. and Casari, C. S. (2011). Charge transfer and vibrational structure of sp-hybridized carbon atomic wires probed by surface enhanced Raman spectroscopy. *J. Phys. Chem. C,* **115**, pp. 12836–12843.
19. MacDonald, H. L., Jorgeson, R. C., Schoen, C. L., Smith, B. F., Yee, S. S. and Mullen, K. I. (1994). *Proc. Int. Soc. Photo. Opt. Instrum. Eng. (SPIE),* **2293**, p. 198.

20. Stokes, D. L., Alarie, J. P. and Vo-Dinh, T. (1995). *Proc. Int. Soc. Photo-Opt. Instrum. Eng. (SPIE)*, **2504**, p. 552.
21. Kahl, M., Voges, E., Kostrewa, S., Viets, C. and Hill, W. (1998). *Sens. Actuators B*, **51**, p. 285.
22. Viets, C. and Hill, W. (2000). Single-fiber surface-enhanced Raman sensors with angled tips, *J. Raman Spectrosc.*, **31**, pp. 625–631.
23. Polwart, E., Keir, R. L., Davidson, C. M., Smith, W. E. and Sadler, D. A. (2000). *Appl. Spectrosc.*, **54**, p. 522.
24. Viets, C. and Hill, W. (2001). Fiber-optic SERS sensors with conically etched tips, *J. Mol. Str.*, **563** , pp. 163–166.
25. Foti, A., D'Andrea, C., Bonaccorso, F., Lanza, M., Calogero, G., Messina, E., Maragò, O. M., Fazio, B. and Gucciardi, P. G. (2013). A shape-engineered surface-enhanced Raman scattering optical fiber sensor working from the visible to the near-infrared, *Plasmonics*, **8**, pp. 13–23; DOI 10.1007/s11468-012-9371-3.
26. Shipway, A. N., Katz, E. and Willner, I. (2000). *ChemPhysChem*, **1**, p. 18.
27. Daniel, M. C. and Astruc, D. (2004). *Chem. Rev.* **104**, p. 293.
28. Grabar, K. C., Allison, K. J., Baker, B. E., Bright, R. M., Brown, K. R., Freeman, R. G. and Natan, M. J. (1996). *Langmuir*, **12**, p. 2353.
29. Lucotti, A. and Zerbi, G. (2007). *Sens. Actuators B*, **121**, p. 356.
30. Pesapane A., et al. (2010). Fiber-optic SERS sensor with optimized geometry: testing and optimization, *J. Raman Spectrosc.*, **41**, p. 256.
31. Lucotti, A., Pesapane, A. and Zerbi, G. (2007). *Appl. Spectrosc.*, **61**, pp. 260–268.
32. Lucotti, A. et al. (2012). TLC-surface enhanced Raman scattering of apomorphine in human plasma, *Vib. Spectrosc.*, **62**, p. 286.
33. Falcao, E. H. L. et al. (2007). *J. Chem. Technol. Biotechnol.*, **82**, p. 524.
34. Cataldo, F., ed. (2005). *Polyynes; synthesis, properties and applications* (CRC Press, Taylor & Franchising Publishing Group, Boca Raton).
35. Chalifoux, W. A. and Tykwinski, R. R. (2010). *Nat. Chem.*, **2**, p. 967.
36. Agarwal, N. R., Lucotti, A., Fazzi, D., Tommasini, M., Castiglioni, C., Chalifoux, W. A. and Tykwinski, R. R. (2013). Structure and chain polarization of long polyynes investigated with IR and Raman spectroscopy, *J. Raman Spectrosc.*, **44**, pp. 1398–1410.
37. Van Orden, A. (1998). *Chem. Rev.*, **98**, p. 2313.
38. Liu, Y., Jones, R. O., Zhao, X., Ando, Y. (2003). *Phys. Rev. B*, **68**, p. 125413.

39. Slepkov, A. D., Hegmann, F. A., Eisler, S., Elliott, E. and Tykwinski, R. R. (2004). The surprising nonlinear optical properties of conjugated polyyne oligomers, *J. Chem. Phys.,* **120**, p. 6807; DOI: 10.1063/1.1707011.
40. Lucotti, A., Tommasini, M., Fazzi, D., Del Zoppo, M., Chalifoux, W. A., Tykwinski, R. R. and Zerbi, G. (2012). Absolute Raman intensity measurements and determination of the vibrational second hyperpolarizability of adamantyl endcapped polyynes, *J Raman Spectrosc.*, **43**, pp. 1293–1298; DOI: 10.1002/jrs.3166.
41. Cataldo, F. (2004). *Carbon,* **42**, pp. 129–142.
42. Cataldo, F. (2004). *Tetrah. Lett.*, **45**, pp. 141–144.
43. Cataldo, F. (2004). Polyynes production in a solvent—submerged electric arc between graphite electrodes. I. Synthesis and spectroscopy, *Fullerenes Nanotubes Carbon Nanostruct.*, **12**, pp. 603–617.
44. Lucotti, A., Tommasini, M., Del Zoppo, M., Castiglioni, C., Zerbi, G., Cataldo, F., Casari, C. S., Li Bassi, A., Russo, V., Bogana, M. and Bottani, C. E. (2006). *Chem. Phys. Lett.*, **417**, p. 78.
45. Casari C. S., Russo, V., Li Bassi, A., Bottani, C. E., Cataldo, F., Lucotti, A., Tommasini, M., Del Zoppo, M., Castiglioni, C. and Zerbi, G. (2007). Stabilization of linear carbon structures in a solid Ag nanoparticle assembly, *Appl. Phys. Lett.*, **90**, p. 013111, http://dx.doi.org/10.1063/1.2430676.
46. Ferrari, A. C. (2007). Raman spectroscopy of graphene and graphite: disorder, electron–phonon coupling, doping and nonadiabatic effects, *Solid State Commun.,* **143,** pp. 47–57.
47. Castiglioni, C., Tommasini, M. and Zerbi, G. (2004). Raman spectroscopy of polyconjugated molecules and materials: confinement effect in one and two dimensions, *Phil. Trans. R. Soc. Lond. A*, **362**, DOI: 10.1098/rsta.2004.1448.
48. Gunningham, N. and Johnstone, R. (1999). *Regulating workplace safety* (Oxford University Press, New York).
49. Uchiro, H. and Kobayashi, S. (1999). *Tetrahedron Lett.*, **40**, p. 3179.
50. Tolaieb, B. and Aroca, R. (2003). *Can. J. Anal. Sci. Spect.*, **48**, p. 139.
51. National Institute for Occupational Safety and Health (1978). Publication No. (NIOSH) 78–213.
52. Parke, D. V. (1952). *Detoxification mechanism*, 2nd ed., (Thesis Univ. London: quoted by Williams, R. T., Chapman and Hall, London, 1959).
53. Parke, D. V. (1968). *The biochemistry of foreign compounds*, Volume 5. (Pergamon Press, Oxford, New York Toronto, Sydney).

54. Smith, J. E., Ross, D., Graham, A. B. and Skellern, G. G. (1992). *J. Pharm. Biomed. Anal.,* **10**, p. 461.
55. Heil, T. P. and Lindsay, R. C. (1988). *J. Env. Sci. Health B*, **23**, p. 489.
56. Amrolia, P., Sullivan, S. G., Stern, A. and Munday, R. (1989). *J. Appl. Toxicol.*, **9**, p. 113.
57. Sylvia, J. M., Janni, J. A., Klein, J. D. and Spencer, K. M. (2000). *Anal. Chem.*, **72**, pp. 5834–5840.
58. Agarwal, N. R. (2013). Controlled synthesis of noble metal nanoparticle arrays and their behavior as surface enhanced Raman scattering (SERS) substrates for application as biomedical sensors, PhD Thesis, Politecnico di Milano, Italy.
59. Low, M. J. D. (1960). Kinetics of chemisorption of gases on solids, *Chem. Rev.,* **60**, pp. 267–312; DOI: 10.1021/cr60205a003.
60. Fleischmann, M., Hendra, P. J. and McQuillan, A. J. (1974). Raman spectra of pyridine adsorbed at a silver electrode, *Chem. Phys. Lett.*, **26** (2), pp. 163–166.
61. Cialla, D., März, A., Böhme, R., Theil, F., Weber, K. Schmitt, M. and Popp, J. (2012). Surface-enhanced Raman spectroscopy (SERS): progress and trends, *Anal. Bioanal. Chem.*, **403**, pp. 27–54; DOI 10.1007/s00216-011-5631-x.
62. Fazio, E., Neri, F., Ossi, P. M., Santo, N. and Trusso, S. (2009). *Appl. Surf. Sci.*, **255**, p. 9676.
63. D'Andrea, C., Neri, F., Ossi, P. M., Santo, N. and Trusso, S. (2009). *Nanotechnology*, **20**, p. 245606.
64. Agarwal, N. R., Neri, F., Trusso, S., Lucotti, A. and Ossi, P. M. (2012). *Appl. Surf. Sci.*, **258**, p. 9148.
65. Bantz, K. C., Nelson, H. D. and Haynes, C. L. (2012). Plasmon-enabled study of self-assembled alkanethiol ordering on roughened Ag substrates, *J. Phys. Chem. C*, **116**, pp. 3585–3593.
66. Bantz, K. C. and Haynes, C. L. (2009). *Vib. Spectrosc.*, **50**, pp. 29–35.
67. Jones, C. L., Bantz, K. C. and Haynes, C. L. (2009). *Anal. Bioanal. Chem.*, **394**, pp. 303–311.
68. Hudson, S. D. and Chumanov, G. (2009). Bioanalytical applications of SERS (surface-enhanced Raman spectroscopy), *Anal. Bioanal. Chem.*, **394**, pp. 679–686; DOI 10.1007/s00216-009-2756-2.

Chapter 9

Isotope-Dilution Surface-Enhanced Raman Spectroscopy: Metrologically Traceable Reference Measurements at the Highest Precision Level and Their Application in Clinical Chemistry

Bernd Güttler, Sabine Zakel, Stefan Wundrack, and Rainer Stosch

Physikalisch-Technische Bundesanstalt, PTB, 38116 Braunschweig, Germany
bernd.guettler@ptb.de

In spite of the largely increasing use of Raman spectroscopy in research and development over the last few decades, it has, so far, rarely been applied for quantitative measurements in analytical chemistry and not at all as a reference measurement method on a metrological level. Especially when it comes to critical measurements in internationally regulated areas such as clinical chemistry other tools such as mass spectrometry are applied. The main factors limiting the application of Raman spectroscopy so far have been the lack of reproducibility and sensitivity of the method. We will describe here, how these limitations can be completely and simultaneously overcome by the application of isotope-dilution

Handbook of Enhanced Spectroscopy
Edited by Pietro Giuseppe Gucciardi, Marc Lamy de la Chapelle, and Nathalie Lidgi-Guigui

ISBN 978-981-4613-32-3 (Hardcover), 978-981-4613-33-0 (eBook)
www.panstanford.com

surface-enhanced Raman spectroscopy (ID-SERS). This chapter describes the critical demand for SI-traceable measurements of a high accuracy in clinical chemistry, the application of this new tool to solve the aforementioned problem and the analytical potential of ID-SERS beyond the limits of established analytical reference methods.

9.1 SI-Traceable Measurements in Clinical Chemistry

With the continuing globalization of trade and economy, the reliability of measurement results of any kind is of increasing importance. Traceability to recognized references, ultimately to the International System of Units (abbreviated as SI, from the French *Le Système International d'Unités*), is an indispensable prerequisite for measurement results to be comparable and trustworthy. This also holds for chemical measurements. Particularly in this field decisions and agreements on life and well being of each individual often critically depend on analytical results, be it in medical diagnostics or by the monitoring of limit values in foodstuffs and environmental analysis. It is the central aim of current activities in *Metrology in Chemistry* to build confidence in the reliability of chemical measurements and thereby afford both trade and society an internationally recognized metrological reference framework.

The connection between the SI and chemistry is established by the *mole*. The mole is the SI base unit for the base quantity "amount of substance" and corresponds to a number of identical, defined elementary entities according to its definition [1]:

(1) The mole is the amount of substance of a system which contains as many elementary entities as there are atoms in 0.012 kg of carbon 12; its symbol is "mol."
(2) When the mole is used, the elementary entities must be specified and may be atoms, molecules, ions, electrons, other particles, or specified groups of such particles.

With the aid of the mole, quantitative relations in chemistry can be traced back to the SI and, thus, internationally compared. In recent years, a large number of national standards for important measurands in chemistry were introduced on this basis.

Traceability to an SI unit is achieved by the practical realization of the unit in the form of primary standards with a specified quantity value and an associated measurement uncertainty [2]. Primary standards are standards which have been manufactured with the aid of a primary measurement procedure. This is a reference measurement procedure which is used to obtain a measurement result without relating to a measurement standard for a quantity of the same kind. Primary measurement procedures are, therefore, at the same time the first step in the dissemination chain from the SI to the end user [3].

Clinical chemistry may serve as an example to demonstrate how traceability is realized in practice. In Europe, the metrological traceability in clinical chemistry is a legal requirement. Values assigned to calibrators and/or control materials must be assured through reference measurement procedures and reference materials of a higher order. This requirement is laid down in national directives in Germany as well as European legislations [4, 5]. The legal demand has triggered international activities at the Meter Convention in order to establish an international reference framework for globally comparable measurements in clinical chemistry by providing primary reference measurement standards as described below as national standards of the highest order for the most important measurands in clinical chemistry.

In recent years, SI-traceable national standards were established for clinical markers of smaller molar masses (up to approx. 1000 g/mol) covering "classic" molecules such as cholesterol, creatinine, or uric acid in bodily fluids. These primary reference measurement standards are based on isotope dilution mass spectrometry (ID-MS) [3, 6, 7]. While this method is well established and provides excellent results, it puts some limits on SI-traceable reference methods. For example, more complex biomarkers having molar masses of up to several million g/mol exceed current instrumental limits. Over the last two decades, large biomarkers (proteins) have dramatically gained in importance as a result of the introduction of diagnostic tests based on immunochemical methods [8, 9]. Also, structural information (e.g., about different chemical conformations of a measurand) is indispensable and often extremely low concentration ranges have to be covered.

New approaches for primary methods are, therefore, urgently required and may either be strategies of how to circumvent the current limitations of the established methods [10] or the development of entirely new methods. In fact, the huge number, diversity, and complexity of potential biomarkers will most probably not be covered with a single universal analytical approach. Further to these fundamental limitations, alternative methodologies that are based on completely different physical principles are also strongly required even in case of "conventional" analytes because undisclosed and, hence, unknown contributions to the measurement uncertainty that were missed by a solitary measurement principle can never be excluded. Tackling particular problems in analytical chemistry with a complementary measurement technique improves, therefore, the reliability of the data. We will describe below how primary measurement methods based on optical spectroscopy can offer such a highly demanded new and complementary tool.

9.2 Realizing Primary Standards for Clinical Measurands

Establishing primary standards for the quantity *amount of substance* and the associated unit *mole* implies a measurement procedure with the aid of which a defined measurand can be *identified* and *quantified*.

The *identification* requires recognizing a molecule as a whole so that it can unambiguously be assigned to the original compound. In the case of mass spectrometry, this is, for example, done on the basis of the mass of the molecule. With the aid of optical (spectroscopic) methods, this identification is achieved from characteristic natural frequencies of the molecule. In the case of infrared or Raman spectrometry, these frequencies are observed as vibrational modes of which the entirety forms a molecule-specific spectral "fingerprint."

The *quantification* requires obtaining information on measurands, such as the substance concentration c_A (in mol/m^3) of an analyte A in a matrix or its amount fraction x_A (in mol/mol). Since quantification of these measurands by measuring the number of

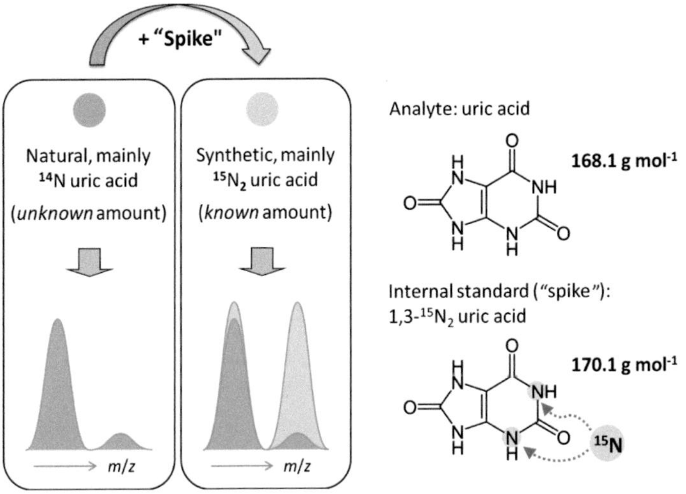

Figure 9.1 Principle of ID-MS as an example of a *ratio primary reference measurement procedure* for the quantification of biomarkers in human serum. Schematic mass spectra are shown (*m*: molecular mass number; z: charge number of the ion).

elementary entities is, in most cases, unrealistic, traceability to the mole is usually realized with the aid of so-called "ratio primary reference measurement procedures" most of which are based on the isotope dilution (ID) approach [3, 11].

The ID approach comprises the addition of an exactly known quantity of the analyte in a synthetic, isotope-modified form (isotopologue) to the sample to act as an internal, chemically identical standard, the so-called "spike." Then, signals of the spike and the pristine analyte can be distinguished employing a detection method which is sensitive to the molecular mass such as mass spectrometry (ID-MS, see Fig. 9.1).

The concentration of the analyte can be determined by comparing the signal intensities of the two isotopologues. This procedure allows measurement results to be determined with measurement uncertainties which are not influenced by recovery rates and analytical preparation steps such as purification for the removal of interfering components. A particularly high accuracy is achieved when the measuring operation is related to an exactly known, gravimetrically prepared reference mixture of the analyte in the same matrix and in a similar concentration. The same quantity of the spike is added to this reference mixture, and the results are compared ("exact matching") [12, 13]. In this case, the measurement accuracy is no longer influenced by the isotopically labeled standard.

It is the advantage of this approach originally developed for mass spectrometry that, at least in principle, it is transferable to infrared and Raman spectroscopy as these methods are also sensitive to mass differences and can, hence, distinguish different isotopologues of the same molecule. Such a tool does not only provide a much sought-after alternative route to the primary method of measurement for clinical markers, it would also offer new approaches to tackle fundamental limits of the available primary methods of measurement such as the mass resolution-dependent limits of mass spectrometry or the lack of direct information on molecular structure.

A major reason for the previous absence of this approach in optical spectrometry is certainly related to the limited sensitivity and reproducibility of these methods. This leaves the question as to whether such an approach is really capable of living up to the needs in clinical chemistry at a primary reference level and the expectations set by the well-established tools provided by mass spectrometry. We aim to demonstrate here by worked examples applied to the same problems how this gap between optical and mass spectrometry was closed in recent years.

9.3 Quantitative Measurements Based on Raman and Surface-Enhanced Raman Scattering: Potential and Limitations

Raman spectroscopy is most often associated with qualitative analytical problems such as the detection and identification of chemical compounds or the determination of molecular structure. Nevertheless, there is a direct relationship between the Raman scattering intensity $I(\nu)_R$ and the number of molecules N in the sample volume being probed [14]:

$$I(\nu)_{\mathrm{R}} = K \frac{h\, I_{\mathrm{L}} N (\nu_0 - \nu)^4}{\mu\, \nu \left(1 - e^{-h\nu/kT}\right)} \tag{9.1}$$

This relationship is the prerequisite that offers the opportunity for using Raman spectroscopy also as a means for quantitative analysis. To be more precise, the analyte concentration in a sample is proportional to the integrated intensity of an analyte Raman band. In practice, the integrated intensity is obtained as a summation of the measured intensities over the wavenumber range of the Raman band after baseline subtraction. The integrated intensity of a Raman band is proportional to its height as long as its band shape does not change. Therefore, quantitative Raman methods may also be based on band height measurements which could offer advantages whenever a partial overlap exists between analyte bands and bands from additional sample components [15].

Despite these seemingly straightforward relationships, several problems and pitfalls exist, which are likely to be encountered in practice. Most of the difficulties arise from the fact that Raman spectroscopy is a single-beam technique. For this reason absolute band intensities are usually influenced by numerous factors that are not associated with changes in the analyte concentration such as fluctuations in laser power or variations of the alignment of the sample with respect to the excitation laser and collection optics. If these experimental variables significantly change with time, it is difficult to reproduce absolute signal intensities, even when obtained from the same sample. However, quantitative evaluations are still possible by monitoring the variations with the aid of an external or internal standard and applying a correction to the measured absolute Raman band areas or heights [15, 16].

An external standard is a separate sample whose properties remain unchanged with time. The quantification of changes in laser power or Raman analyzer requires the continuous collection of separate Raman spectra for the corresponding standards and samples. The ratio of band intensities of analyte and standard is then used for quantitative evaluation of the analyte concentration. With the aid of an external standard, long-term drifts and trends can be successfully eliminated to ensure the reliability of a method over a long period of time. However, an external standard does not correct for changes related to the sample itself such as variations in the opacity or the refractive index. Generally, periodic switching between the sample and the external standard is not appropriate whenever a sample is to be measured continuously with time such as for analyzing chemical reaction kinetics or on-line/in-line monitoring of industrial processes.

The limitations of external standards can be overcome by applying an internal standard that is either mixed to the sample or, even better, is part of the sample itself. The latter option is often more appropriate as it does not affect the chemical and physical properties of the sample solution. For this reason, the solvent in general and water in particular have been frequently used as internal standards [17, 18].

For aqueous solutions, water is a suitable internal standard because of its high molar concentration that does not change as long as it is much larger than the changes in analyte concentration. Normally, the Raman bands of water (symmetric and anti-symmetric OH stretching modes ν_1 and ν_3 in the range 3000–3700 cm^{-1}, bending mode ν_2 around 1640 cm^{-1}) show negligible interference with most of the interesting analyte spectral regions and can be considered either individually [18] or together [17]. It is not applicable, however, in case of complex matrices such as serum when the analyte concentration is low and when the sample is exposed to elaborate extraction procedures as is usually the case in clinical chemistry and many other subject fields (e.g., environmental or food samples).

Regardless of whether external standard or internal standard was used for applying corrections to the measured Raman spectra, quantitative evaluation in either case requires the establishment of a mathematical relationship between the Raman intensity and the analyte concentration. Such a relationship is also referred to as an analytical model which is almost always created by measuring Raman spectra from a set of defined samples and relating the Raman intensity to the property of interest. The accuracy of the model is then validated using another set of defined samples in order to end up with an assessment of how well the model fits the test set and also of its robustness against variations that might occur in future samples or measurement conditions.

The most straightforward implementation of the direct proportionality between the Raman band intensity and the analyte concentration is represented by the linear model $y = a + bx$. After applying the internal standard correction, a plot of the area (or height) of a single Raman band, y, against the concentration, x, results in a straight line.

Often, variances in the analyte concentration are reflected by more than just one single characteristic Raman band. In order to extract the maximum information present in a set of spectra, the analytical model has to be generated with the aid of multivariate data analysis techniques such as partial least squares (PLS) regression or principal component regression (PCR) [19, 20].

Multivariate analytical models are particularly suitable for analyzing multicomponent systems or systems containing spectral interfering compounds. Nevertheless, the simultaneous analysis of several spectral segments or even of the entire Raman spectrum generally offers some fundamental advantages over univariate analytical models resulting from the benefit of signal averaging. This is clarified with the aid of a set of conventional Raman spectra obtained from aqueous creatinine solutions (Fig. 9.2) after background subtraction and intensity correction applying the

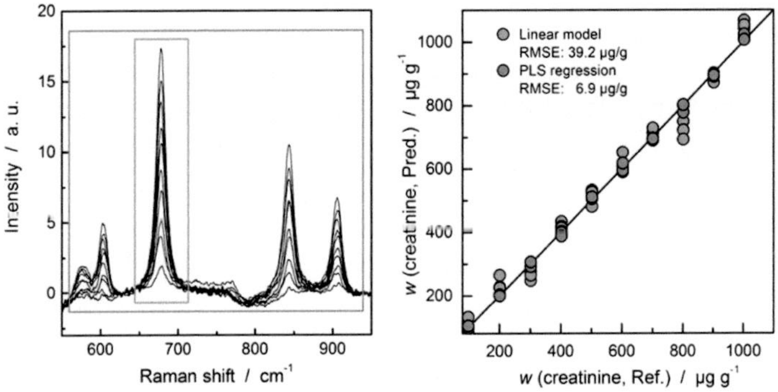

Figure 9.2 (Left) Conventional Raman spectra of creatinine in aqueous solutions ($w_x = 90$–1000 µg/g); the rectangles indicate the spectral ranges used for generation of a linear (orange) and a PLS (green) model; (right) internal cross validation results.

solvent as internal standard. While the linear model has been generated only from the signal intensities of a single (678 cm^{-1}) band, the PLS model is based on the full wavenumber range shown (550–950 cm^{-1}). The predictive ability of the two models were assessed by the root mean square error (RMSE) derived from internal cross-validation results [20]. Although the linear model results from a distinct Raman band without interference from other compounds, the PLS models yield a significantly smaller RMSE value indicating the superiority of the multivariate over the univariate approach.

If all the issues discussed so far are properly addressed, Raman scattering can be applied for measuring the concentrations of a couple of biomarkers such as creatinine, glucose, lactic acid, or urea dissolved in aqueous media [21–23]. However, due to the low Raman cross-sections of many biomarker molecules, relatively large analyte concentrations, typically present in human urine, were used in all these studies while the accurate and precise measurement of significantly lower serum concentrations is rarely possible [24, 25].

For a large number of analytes, the sensitivity range can be expanded toward lower concentrations or even trace amounts with the aid of SERS [26, 27]. The application of SERS requires that the molecule of interest is brought in close proximity to the surface of a metallic nanostructured substrate or a metallic nanoparticle. This is commonly achieved by spontaneous adsorption of the target molecules to the metallic surface which is supported by the presence of particular surface-seeking functional groups within the analyte molecule. In the case of the absence of such groups, strategies for capturing the analyte on the surface of the SERS substrate are a promising alternative [28].

On the other hand, there are only a limited number of surface sites being capable of both for interacting with analyte molecules and for contributing to the overall signal enhancement. In practice, this effect often manifests itself in producing a plateau in the intensity versus concentration plot at higher analyte concentrations, resulting from a gradual saturation of the available adsorption sites [29]. This leads to an unwanted narrowing of the linear dynamic range suitable for model generation and subsequent prediction. Consequently, reliable quantitative measurements are only possible on the basis of SERS *intensity ratios* obtained from analyte and internal standard rather than from absolute intensities.

When comparing the fundamental principles underlying conventional Raman spectroscopy and SERS, it becomes apparent that specific additional requirements are to be met by internal standards used in combination with SERS [30, 31]. Basically, the enhanced signal intensities (i.e., the enhancement factors) for both analyte and standard should be of the same order of magnitude while the spectra itself must be (at least partly) non-overlapping. As SERS signals are obtained exclusively from molecules adsorbed on the surface of the enhancing media, it is necessary that the proportion of the quantities (concentrations) of both compounds present in the initial sample solution is exactly reproduced in terms of surface coverage. This requirement is most readily met if the target analyte and the standard are physically and chemically similar which is true only for isotopologues of the analyte.

9.4 Overcoming the Limits: Combined Isotope Dilution and Surface-Enhanced Raman Scattering

Raman scattering also provides a suitable alternative measurement technique to discriminate between isotopologues of a chemical compound. Hence, truly quantitative measurements that overcome the limits of arbitrary internal standards and ensure SI traceability can be achieved when the principle of isotope dilution is applied. A procedure referred to as ID-SERS unites high sensitivity and molecular specificity with the advantages of reliability, accuracy, and repeatability during quantification. Based on molecular vibrations, a mass discrimination between analyte and spike is obtained from vibrational energies while the set of specific vibrations additionally provides a means for unambiguous identification of a compound including possible structural modifications of otherwise similar compounds.

An isotopologue of the analyte to be used as the spike is commonly enriched in either ^{13}C or ^{15}N (or both) at one or several positions within the backbone of the molecule. Such isotopic enrichment results in a slightly increased molar mass of the spike

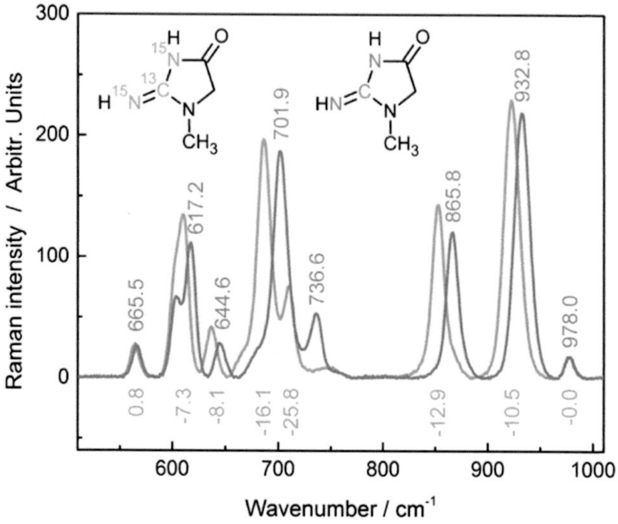

Figure 9.3 SERS spectra of pristine creatinine (green, with band positions above) and isotopically labeled ^{13}C-$^{15}N_2$-creatinine (orange, with isotopic peak shifts below) measured on colloidal Ag nanoparticles as SERS medium ($\lambda_{ex} = 532$ nm).

as compared to the pristine analyte, which causes a shift of several Raman bands toward lower wavenumbers. The SERS spectra of pristine creatinine and isotopically labeled ^{13}C-$^{15}N_2$-creatinine obtained using colloidal silver nanoparticles as signal-enhancing medium may serve as an example (Fig. 9.3).

Spectra were recorded within a limited wavenumber interval between ~500 and ~1000 cm^{-1}. In this region, the spectral variability caused by the isotope exchange effect is greatest and changes in the ratio of the two isotopologues would therefore result in multiple and most significant isotopic band shifts. Because of the mass increase at three positions within the molecular structure, a couple of Raman modes are shifted to lower wavenumbers, while the total number of bands remains unchanged. The individual values of the isotopic shifts depend on the degree to which the substituted atoms contribute to the particular normal modes of vibration. For the triple-substituted creatinine, a range between 0 cm^{-1} (978 cm^{-1} band) and ~26 cm^{-1} (736.6 cm^{-1} band) was observed.

Furthermore, SERS experiments on defined mixtures of analyte and spike show that the resulting spectrum contains the spectral features of both isotopologues with relative intensities proportional to the isotopologue abundance ratios present in the mixture [32, 33]. These findings imply that the slight difference of the molar masses neither affects the adsorption kinetics of the isotopologues nor their selectivity with respect to particular surface sites. Furthermore, the analyte-to-spike ratio, once adjusted in a sample solution, is exactly reproduced by spontaneous adsorption on the surface of the colloidal nanoparticles, thus providing a rational basis for applying the ID approach.

9.5 Applications in Clinical Chemistry

Based on these experimental facts, a complete analytical procedure for measuring biomarkers in bodily fluids such as human serum can be set up. The resulting working scheme comprises two main branches, whose essential working steps can be summarized in a process chart (Fig. 9.4).

The first branch starts by preparing a set of calibration mixtures, followed by the generation and validation of a PLS prediction model from the measured SERS spectra. The calibration mixtures are made from stock solutions of a high purity reference material of the target analyte (z) and its isotopically labeled analogue (y) with known amount-of-substance fractions ranging from 0 to 1 in the analyte.

The second branch comprises spiking, pretreatment, and measuring of the serum sample together with a reference specimen that contains the reference material and the spike in an equimolar composition. Adding the spike at the beginning of the procedure is essential to ensure that potential sample losses that might occur during pretreatment do not change the isotopologue ratio in the sample and the reference. The treatment of sample and reference ends with a measurement of the SERS spectra.

Such a parallel analysis of a spiked serum and a reference makes up the so-called "double ID approach." Sample treatment including spiking, purification, and SERS experiments is conducted in parallel with the sample itself (x, blood serum) but also with a

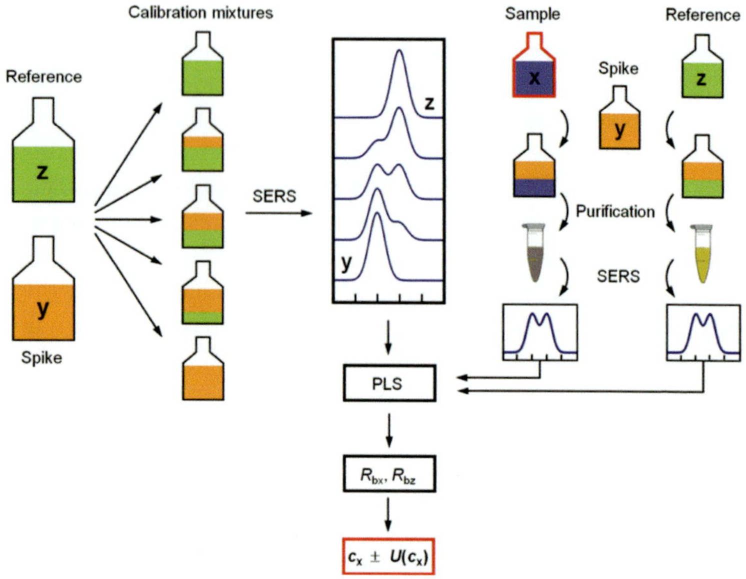

Figure 9.4 Process chart illustrating the sequence of a complete double ID-SERS analysis: A set of calibration mixtures is prepared from defined stock solutions of the two isotopologues (reference z and spike y) and from SERS measurements of these mixtures, a PLS model is generated. Sample (e.g., blood serum x) and reference (z) are both mixed with the spike and undergo further pretreatment to remove possibly interfering compounds. From SERS spectra of the purified solutions the isotopologue abundance ratios R_{bx} (sample) and R_{bz} (reference) are predicted and converted to the wanted analyte concentration and the associated measurement uncertainty, $c_x \pm U(c_x)$. Reprinted from Ref. [34], Copyright 2013, with permission from John Wiley and Sons.

reference specimen (z) containing a known amount of the analyte. The same amount of spike solution (y) is added to both specimens, and the amount of analyte in the reference is then adjusted in an iterative process to end up with isotopologue abundance ratios, R_{bx} and R_{bz} close to unity [12, 13, 33]. With this so-called "exact matching" procedure it is ensured that the relatively poorly known composition (chemical purity and isotopic enrichment) of the spike material does not appear in the measurement equation and its standard uncertainty does not significantly contribute to the overall uncertainty budget:

$$w_x = w_z \frac{m_{yx}}{m_x} \frac{m_z}{m_{yz}} \frac{R_{bx}}{R_{bz}} \tag{9.2}$$

In this equation, the wanted mass fraction w_x of the analyte in serum is expressed as the mass fraction w_z of the analyte in the reference solution, the certain masses of the reconstituted serum (m_x) and reference solution (m_z) that are blended with similar masses of the spike solution (m_{yx} and m_{yz}). The isotopologue abundance ratios $R_{bx} = \hat{x}_{bx}/(1 - \hat{x}_{bx})$ and $R_{bz} = \hat{x}_{bz}/(1 - \hat{x}_{bz})$ are predicted by the PLS model from the previously measured SERS spectra.

The measurement thereby remains traceable to the known amount of the analyte in the reference specimen, which is typically the gravimetrically known amount of the pure analyte with a known purity. In many cases, a certified reference material with known purity can be used for this purpose.

Considering the density ρ_x of the serum and the molar mass of the analyte M_x, the mass fraction w_x can easily be converted into the corresponding molar concentration c_x:

$$c_x = \frac{\rho_x}{M_x} w_x \tag{9.3}$$

In order to limit the number of cycles to reach the "exact matching" condition during the iterative process, it is helpful to have an estimate of the expected analyte concentration before starting the first cycle. In the case of clinical markers, this can be assessed from physiological serum levels including reduced or elevated levels caused by disease. A first estimate of the analyte concentration c_x is obtained from the isotopologue abundance ratios (R_{bx} and R_{bz}) predicted with the help of the PLS model. This concentration is then used as the starting point for the second cycle and the whole procedure is now repeated with this new concentration for the spike and the reference. Typically, a sufficiently close approximation to unity of the desired (R_{bx}/R_{bz}) value is already achieved in the second cycle.

Reliable data require quality assurance based on internationally agreed standards. This is achieved by the application of metrological principles such as SI traceability and associated measurement uncertainties accompanying the data. Unfortunately, these quality requirements are often missing in studies published on quantitative applications of Raman spectroscopy in general and SERS in particular. Analytical results without these qualifications can, however, hardly be interpreted or compared in terms of each other.

Traceability chains can only be built up if methods are available to evaluate the uncertainties of the links in the chain. Therefore, a full understanding of the uncertainty of the entire chemical measurement procedure is a key requirement. The *ISO Guide to the Expression of Uncertainty in Measurement* (GUM) provides a unified way to evaluate these uncertainties [2].

The uncertainty budget developed recently for the ID-SERS approach is in full accordance with the GUM approach. It shows that the main contributors to the standard uncertainty arise from the isotopologue abundance ratios R_{bx} and R_{bz}, as predicted from the PLS model [33]. This uncertainty, in turn, results from the statistical spread of the spectra. Those contributions, however, that result from imperfect knowledge of the isotopic enrichment of the spike are minimized by the double-ID principle. The individual uncertainty contributions were then assembled to form the combined standard uncertainty $u(c_x)$. For regulatory applications, particularly when health is concerned, a measure of uncertainty is required that defines an interval within which the value of the measurand can be expected to lie with a high (e.g., 95%) probability. This measure is termed expanded uncertainty, $U(c_x)$, and is obtained by multiplying $u(c_x)$ with a coverage factor $k = 2$.

9.6 Validation Through International Comparison Schemes

An integral part of setting up metrologically traceable measurement results is a validation procedure that includes the verification of the associated measurement uncertainty in international intercomparison schemes. Through regular participation in ring trials, laboratories demonstrate their competence as reference measurement laboratories in a particular analytical area and improve the reliability and the comparability of their results. International ring trials for reference laboratories (RELA) in clinical chemistry are organized and conducted annually by the German Society for Clinical Chemistry and Laboratory Medicine (DGKL) [35] on behalf of the Joint Committee for Traceability in Laboratory Medicine (JCTLM) [36]. The latter was established in response to the implementation of a European Community directive on in vitro medical devices [5]. Such comparisons on an international level are used as a tool for external quality control and are open to clinical reference laboratories and National Metrology Institutes (NMIs) that can provide a link to the SI.

The validation of ID-SERS has been achieved by simultaneous participation in two international comparisons. Out of a group of biomarkers having the highest medical priority, we focused our efforts on the determination of the creatinine concentration in human serum. Creatinine is a small product of protein metabolism and its serum concentration serves as a parameter for evaluating and monitoring the renal function. It was therefore selected as one of the key analytes in the RELA 2010/2011 ring trial and its concentration had to be measured in human serum provided by the study organizers [35]. The two serum samples contain different creatinine levels and were simultaneously treated as unknowns in another comparison conducted by the European Association of National Metrology Institutes (EURAMET). The study organizers particularly encouraged those NMIs assessing new approaches to participate in a parallel pilot study. The main outcome of the EURAMET.QM-K12 comparison is summarized in Fig. 9.5.

The five key participants are among the most experienced NMIs using their well-approved national standard procedures based on either GC- or LC-ID-MS [37]. As to the date of completion of this study, ID-MS represents the only accepted reference procedure for measuring creatinine in serum.

The results had to be reported as the mass fraction of creatinine in serum w_x (μg/g), together with the standard measurement uncertainty $u(w_x)$, which means that the serum density did not contribute to the final result. A so-called comparison reference value (CRV) and its associated uncertainty (uCRV) were reported as a mutually agreed outcome of the EURAMET study. These two values were calculated as the arithmetic mean and the standard deviation

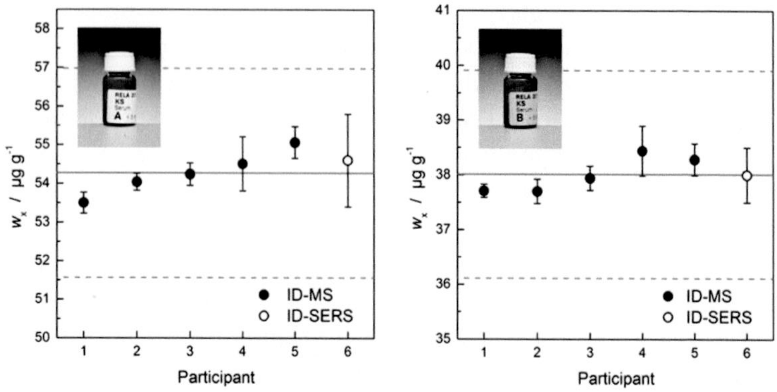

Figure 9.5 Individual participants' results of the creatinine mass fractions w_X measured within the EURAMET.QM-K12 comparison for sample A (left) and B (right) together with images of the vials containing lyophilized human serum samples provided as unknowns by the study organizer. The solid line represents the comparison reference value (CRV) while the dashed lines indicate the maximum allowed deviation (5%) from the reference value in ring trials for clinical reference laboratories in case of creatinine.

of the ID-MS data provided by the five key participants, where as the ID-SERS data were treated independently.

For both serum samples, an excellent agreement between ID-SERS and the CRV is obtained and the stated measurement uncertainties overlap with the interval specified by uCRV. Furthermore, all results are found well within a ±5% limit of equivalence around the reference values. This is the quality criterion for a laboratory to be accepted as a clinical reference laboratory in case of creatinine [35]. These limits are set by considerations based on medical diagnostics.

This validation at an international level confirms that a combination of SERS and ID can serve as a valuable tool for performing amount-of-substance measurements with the highest accuracy and reliability [34]. For the first time, a complete working procedure based on SERS for measuring biomarkers in human serum was successfully applied to unknown serum samples. Through the consistent application of basic metrological concepts, the method provides traceability to the SI and can compete even with established national standards based on ID-MS.

Isotope dilution surface enhanced raman scattering method for creatinine in blood serum	
▸ ID/SERS reference measurement method for serum creatinine	
Applicable matrice(s)	lyophilized, fresh, or frozen human serum
Full description of technique(s)	Isotope dilution surface enhanced Raman scattering (ID/SERS)
Quantity	Amount-of-substance concentration
Applicable range	10 µg/ml to 100 µg/ml
Expected uncertainty (level of confidence 95%)	2 % to 5 %
Reference(s)	Stosch et al., *Anal. Chem.*, 2005, **77**, 7386-7392 Zakel et. al., *Analyst*, 2011, **136**, 3956-3961
Comparability assessment study(ies)	IFCC-RELA 2010 EURAMET TC-MC Project 1176
JCTLM DB identification number	C9RMP1

Figure 9.6 Record generated in JCTLM database on laboratory medicine and in vitro diagnostic attesting the successful approval of ID-SERS as a reference measurement method for serum creatinine [36].

With the demonstrated potential to expand the existing pool of measurement techniques for metrology in chemistry, ID-SERS has been officially approved as a higher-order reference measurement method. As a consequence, a new record was created in the publicly accessible JCTLM database on laboratory medicine and in vitro diagnostics [36] maintained by the Bureau International des Poids et Mesures (BIPM) that attests the successful approval of ID-SERS as a reference measurement method for serum creatinine (Fig. 9.6).

References

1. Bureau International des Poids et Mesures, BIPM (2006), Le Système international d'unités (SI)—The International System of Units (SI). 8th ed., (Sèvres/Paris).
2. Joint Committee for Guides in Metrology (2008). *Evaluation of measurement data – guide to the expression of uncertainty in measurement*, 1st ed., JCGM, p. 100.
3. Milton, M. J. T. and Quinn, T. J. (2001). Primary methods for the measurement of amount of substance, *Metrologia*, **38**, pp. 289–296.
4. Bundesärztekammer (2011). Richtlinie der Bundesärztekammer zur Qualitätssicherung laboratoriumsmedizinischer Untersuchungen/ Guideline for quality assurance of quantitative medical laboratory tests, *Deutsches Ärzteblatt*, **108**, pp. 2298–2304.

5. European Commission (1998). Directive 98/79/EC of the European parliament and of the council of 27 October 1998 on in vitro diagnostic medical devices, *Off. J. Eur. Comm.*, **L331**, pp. 1–37.
6. De Bièvre, P. and Peiser, H. S. (1997). Basic equations and uncertainties in isotope-dilution mass spectrometry for traceability to SI of values obtained by this primary method, *Fresenius J. Anal. Chem.*, **359**, pp. 523–525.
7. Richter, W. (1997). Primary methods of measurement in chemical analysis, *Accred. Qual. Assur.*, **2**, pp. 354–359.
8. Thienpont, L. M. (2008). Accuracy in clinical chemistry – who will kiss sleeping beauty awake? *Clin. Chem. Lab. Med.*, **46**, pp. 1220–1222.
9. Vesper, H. W. and Thienpont, L. M. (2009). Traceability in laboratory medicine, *Clin. Chem.*, **55**, pp. 1067–1075.
10. Arsene, C. G., Ohlendorf, R., Burkitt, W., Pritchard, C., Henrion, A., O'Connor, G., Bunk, D. M. and Güttler, B. (2008). Protein quantification by isotope dilution mass spectrometry of proteolytic fragments: cleavage rate and accuracy, *Anal. Chem.*, **80**, pp. 4154–4160.
11. Joint Committee for Guides in Metrology (2008). *International vocabulary of metrology — basic and general concepts and associated terms (VIM)*, 3rd ed., JCGM, p. 200.
12. Henrion, A. (1994). Reduction of systematic errors in quantitative analysis by isotope dilution mass spectrometry (ID-MS): an iterative method, *Fresenius J. Anal. Chem.*, **350**, pp. 657—658.
13. Mackay, L. G., Taylor, C. P., Myors, R. B., Hearn, R. and King, B. (2003). High accuracy analysis by isotope dilution mass spectrometry using an iterative exact matching technique, *Accred. Qual. Assur.*, **8**, pp. 191–194.
14. Long, D. A. (2002). *The Raman effect* (John Wiley & Sons, Chichester), pp. 97–113.
15. Pelletier, M. J. (2003). Quantitative analysis using Raman spectrometry, *Appl. Spectrosc.*, **57**, pp. 20A–42A.
16. Favors, R. N., Jiang, Y., Loethen, Y. L. and Ben-Amotz, D. (2005). External Raman standard for absolute intensity and concentration measurements, *Rev. Sci. Instrum.*, **76**, 033108.
17. Moreno, T., Moran Lopez, M. A., Huerta Illera, I., Piqueras, C. M., Sanz Arranz, A., Garcia Serna, J. and Cocero, M. J. (2011). Quantitative Raman determination of hydrogen peroxide using the solvent as internal standard: online application in the direct synthesis of hydrogen peroxide, *Chem. Eng. J.*, **166**, pp. 1061–1065.

18. Sun, Q. and Qin, C. (2011). Raman OH stretching band of water as an internal standard to determine carbonate concentrations, *Chem. Geol.*, **283**, pp. 274–278.

19. Malinowski, E. R. (2002). *Factor analysis in chemistry*, 3rd ed., (John Wiley & Sons, New York).

20. Martens, H. and Naes, T. (1989). *Multivariate calibration* (John Wiley & Sons, New York).

21. Dou, X., Yamaguchi, Y., Yamamoto, H., Doi, S. and Ozaki, Y. (1996). Quantitative analysis of metabolites in urine using a highly precise, compact near-infrared Raman spectrometer, *Vib. Spec.*, **13**, pp. 83–89.

22. McMurdy, J. W. and Berger, A. J. (2003). Raman spectroscopy-based creatinine measurement in urine samples from a multipatient population, *Appl. Spectrosc.*, **57**, pp. 522–525.

23. Wickstedt, J. P., Erckens, R. J., Motamedi, M. and March, W. F. (1995). Raman spectroscopy studies of metabolic concentrations in aqueous solutions and aqueous humor specimens, *Appl. Spectrosc.*, **49**, pp. 987–993.

24. Berger, A. J., Koo, T.-W., Itzkan, I., Horowitz, G. and Feld, M. S. (1999). Multicomponent blood analysis by near-infrared Raman spectroscopy, *Appl. Optics*, **38**, pp. 2916–2926.

25. Qi, D. and Berger, A. J. (2007). Chemical concentration measurement in blood serum and urine samples using liquid-core optical fiber Raman spectroscopy, *Appl. Optics*, **46**, pp. 1726–1734.

26. Le Ru, E. and Etchegoin, P. (2009). *Principles of surface-enhanced Raman spectroscopy*, 1st ed., (Elsevier, Amsterdam).

27. Pieczonka, N. P. W., Moula, G., Skarbek, A. R. and Aroca, R. F. (2011). *Surface-enhanced Raman scattering*, ed. Schlücker, S., Chapter 4 "Single-molecule and trace detection by SERS", (Wiley-VCH, Weinheim), pp. 87–101.

28. Guerrini, L., Garcia-Ramos, J. V., Domingo, C. and Sanchez-Cortes, S. (2009). Sensing polycyclic aromatic hydrocarbons with dithiocarbamate-functionalized Ag nanoparticles by surface-enhanced Raman scattering, *Anal. Chem.*, **81**, pp. 953–960.

29. Bell, S. E. J. and Sirimuthu, N. M. S. (2008). Quantitative surface-enhanced Raman spectroscopy, *Chem. Soc. Rev.*, **37**, pp. 1012–1024.

30. Bell, S. E. J. and Stewart, A. (2011). *Surface-enhanced Raman scattering*, ed. Schlücker, S., Chapter 3 "Quantitative SERS methods", (Wiley-VCH, Weinheim), pp. 71–86.

31. Loren, A., Engelbrektsson, J., Eliasson, C., Josefson, M., Abrahamsson, J., Johansson, M. and Abrahamsson, K. (2004). Internal standard in surface-enhanced Raman spectroscopy, *Anal. Chem.*, **76**, pp. 7391–7395.
32. Stosch, R., Henrion, A., Schiel, D. and Güttler, B. (2005). Surface-enhanced Raman scattering-based approach for quantitative determination of creatinine in human serum, *Anal. Chem.*, **77**, pp. 7386–7392.
33. Zakel, S., Rienitz, O., Güttler, B. and Stosch, R. (2011). Double isotope dilution surface-enhanced Raman scattering as a reference procedure for the quantification of biomarkers in human serum, *Analyst*, **136**, pp. 3956–3961.
34. Zakel, S., Wundrack, S., O'Connor, G., Güttler, B. and Stosch, R. (2013). Validation of isotope dilution surface-enhanced Raman scattering (ID-SERS) as a higher order reference method for clinical measurands employing international comparison schemes, *J. Raman Spectrosc.*, **44**, pp. 1246–1256.
35. RELA – IFCC External quality assessment scheme for reference laboratories in laboratory medicine. http://www.dgkl-rfb.de.
36. Joint Committee for Traceability in Laboratory Medicine, BIPM (Sèvres), http://www.bipm.org/jctlm/
37. Siekmann, L. (1985). Determination of creatinine in human serum by isotope dilution mass spectrometry, *J. Clin. Chem. Clin. Biochem.*, **23**, pp. 137–144.

Chapter 10

Application of SERS to Chemicals Sensing

Inga Tijunelyte,[a] Nathalie Lidgi-Guigui,[a] and Emmanuel Rinnert[b]

[a] *CSPBAT, Sorbonne Paris Cité, Bobigny, France*
[b] *Laboratoire Détection, Capteurs et Mesures, Ifremer, Brest, France*
nathalie.lidgi-guigui@univ-paris13.fr

10.1 Introduction

During the last years, surface-enhanced Raman spectroscopy (SERS) has appeared as a relevant analytical technique. Thanks to its many advantages such as selectivity, sensitivity, and ease of operation; it is a promising analytical tool and is concerned with many fields of application, such as environmental sciences, clinical or food applications, etc.

The necessity of diagnosis tools in these domains and the progress in science—photonics, plasmonics, nanotechnologies, and molecular chemistry—ensure nowadays a research structuring around SERS technique for analysis and sensor applications for tomorrow.

The aim of this chapter is to present some recent investigations on the elaboration of SERS substrates in order to detect specific molecules in different field. Some studies are presented in detail

Handbook of Enhanced Spectroscopy
Edited by Pietro Giuseppe Gucciardi, Marc Lamy de la Chapelle, and Nathalie Lidgi-Guigui

ISBN 978-981-4613-32-3 (Hardcover), 978-981-4613-33-0 (eBook)
www.panstanford.com

to focus on the need of complementarities for the elaboration of SERS-active substrates. An exhaustive study is difficult to achieve in such a progressive scientific environment. We will first consider the nanostructures' production possibilities. Then a review of the strategies for specific detection will be given. Eventually, an overview of some SERS commercial solutions will also be described in order to confirm the potential market of the technique.

10.2 Nanostructures for SERS Detection

A good SERS substrate has to be sensitive, stable, easy to handle, and give reproducible results. Another very important point in the purpose of making commercial SERS surfaces is their price. As we will see in the following this is the determining parameter to choose the mean of fabrication.

Metallic nanoparticles chemically synthesized are possibly one of the simplest ways to produce SERS substrates. They are very easily synthesized; their production is well documented; and the literature is overflowing with strategies to control their shape, size, and surface chemistry. In the framework of SERS sensors, colloids can be used in several manners. An option is to form a single layer of nanoparticles on top of a functionalized glass substrates [1]. Another possibility is to take advantage of a fluidic system in order to aggregate the nanoparticles into a multitude of hot spots [2]. Eventually, fiber-optic SERS sensors are using colloids to functionalize standard fiber optics (see Chapter 8 of this book). While being cheap and easily handled, the use of colloids rarely make reproducible substrates. The reason for this is that the size and positions of the colloids cannot be finely controlled. The colloids assemblies often present defects such as a missing nanoparticle or a random number of aggregates forming hot spots. As it is not possible to control the number of defects, some substrates will exhibit extraordinary enhancement factors due to a high number of hot spots while other substrates will be less performant.

Gold and silver colloids are commercially available (see Section 10.5 below) for SERS, these two metals give the highest enhancement factors. Silver is known to give the best enhancement factor, whereas gold is highly stable, biocompatible, and its chemistry is well-known. However, a lot of work is in progress concerning metamaterials, which could concurrence the sensitivity of gold at lower cost.

The shape and organization of the nanostructures also contribute to the sensitivity of the SERS sensor. Tips and near-field coupling can lead to extraordinary enhancement factors. This is what is happening when colloids aggregates. But as these aggregates are difficult to control and thus reproduce, it can be a better option to choose lithography techniques to fabricate the nanoparticles. The resolution needed to get nanoparticles with tips (nanoantennas, triangles, stars, etc.) and spaced out of a few nanometers requires the use of electron-beam/e-beam lithography (EBL). In this technique an electron-sensitive resist is exposed to the electron beam of a scanning electron microscope (SEM) (see Fig. 10.1a).

At the same time, large nanostructured surfaces (at least 1 mm^2) have to be produced in order to be easily handled by a potential end-user. For example, in the case of *in situ* SERS experiments, it is not always possible to adjust the sample position. Fabrication of a 1×1 mm network of nanocylinders can take up to 72 hours with standards EBL. Such a long time for the production of a single substrate is in contradiction with any commercial view.

In order to overcome this issue other nanofabrication techniques can be used. Nanosphere lithography (NSL) and its derivatives can be cited [3] (Fig. 10.1c) as well as innovative techniques such as nanoimprint lithography (NIL) [4] (see Fig. 10.1b). NSL is well known to produce an assembly of nanotriangles. It is possible to tune the size and shape of the nanostructures by varying the way of stacking the nanospheres. Yet, it is not trivial to do so in a reproducible way. In the latter, a mother mold is made by EBL that is used to produce a stamp which in turn is used to pattern a substrate. Since the mother mold is made with EBL the nanoparticle assembly can be shaped for a specific requirement and then duplicated as many times as required.

As a conclusion, many different nanoparticles are available for SERS purposes, but they are more or less difficult to produce. Before choosing a substrate one has to finely define his need in term of price, reproducibility, and sensitivity. It is important to highlight

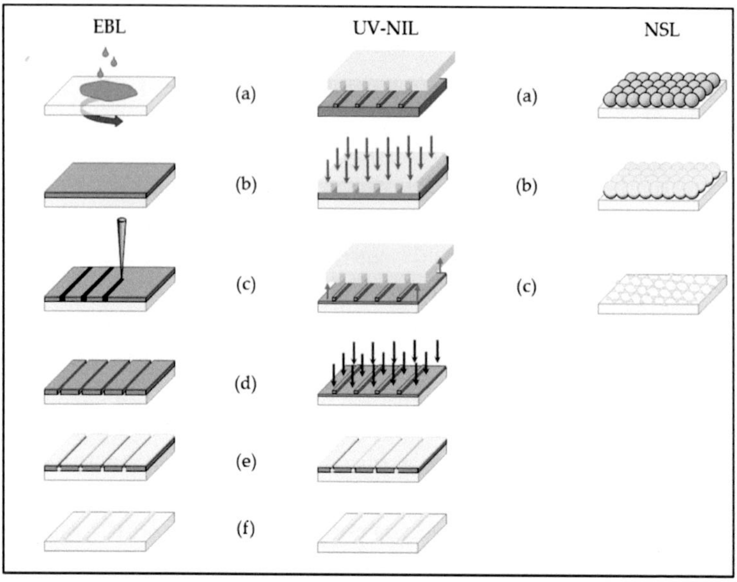

Figure 10.1 EBL: (a) spin coating, (b) metallization, (c) exposure to electron beam, (d) development, (e) gold evaporation, (f) lift-off. UV-NIL: (a) stamp fabrication from master mold, (b) the stamp is pressed against a resist and the system is exposed to UV, (c) separation, (d) etching, (e) gold evaporation, (f) lift off. NSL: (a) Nanospheres self-assembly, (b) gold deposition, this substrate can be used like this for SERS, it is then called film over nanospheres (FON), (c) nanospheres removal.

that the notion of sensitivity depends on the chemical aimed to be detected. For example, the European commission has fixed the maximum amount of naphthalene in sea water to 2.4 μg/L but only to 0.05 μg/L for benzo[a]pyrene.[a] The SERS industry is still in its young years and for the moment users still have to make their own test in order to validate a SERS substrate for a specific application.

10.3 Sensing Molecules Using SERS-Active Substrates

SERS is a powerful tool for toxic and hazard chemical compounds detection, identification, and quantification. For a molecular de-

[a]European Parliament Directive 2008/105/EC, published 24 December 2008

tection approach by SERS, optimization of the size and shape of plasmonic nanostructure is a very important parameter. However, in the matter of sensing, it is of prime importance to consider the issue of affinity between the analyte and the SERS surface. Some molecules, especially those containing oxygen, nitrogen, or sulfur atoms can be directly pre-concentrated near the surface due to strong interactions. These interactions are induced by charge-transfer complex formation, where molecule acts as electron donor and the metal surface as an acceptor [5]. In the literature, many examples have been reported on such molecules' detection and identification by SERS method. For instance trans-1,2-bis(4-pyridyl)ethylene (BPE) adsorbed on gold nanoparticles [6]. Other reported example is adsorption of azodicarbonamide (ADC) on as prepared gold nanoparticles of 50 nm in diameter, which have been monitored in low concentrations [7].

However, nonmodified metallic surface shows several disadvantages for a chemical sensing application. Interaction of some molecules with a surface is so strong that chemical change can be induced: molecular decomposition or polymerization has been observed [8–12]. Moreover, chemical sensor application for complex environment gives a crucial issue of nonselectivity against target compound. Metal surface functionalization is important to ensure affinity to a molecular group or even to a single molecule.

Another important disadvantage is related with detection of nonpolar substances such as polycyclic aromatic hydrocarbons (PAHs) and volatile toxic atmospheric pollutants that consist of fused aromatic rings and do not show affinity to metallic surfaces due to the absence of functional groups [13].

Finally, functionalization ensures metallic surface protection against environmental degradation especially by working with the silver-based substrates [14].

10.4 Surface Functionalization Strategies

10.4.1 *Self-Assembled Monolayers*

Self-assembly in general is the ability of organic molecules to construct different structures by noncovalent interactions [15].

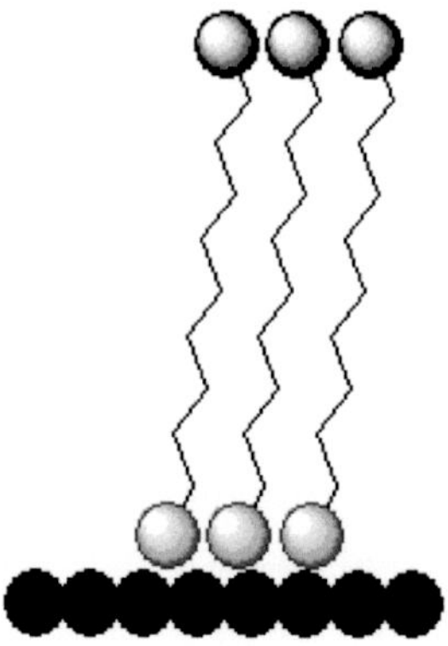

Figure 10.2 Graphical model for self-assembled monolayer of alkanethiols on metal surface.

Many types of compounds, mostly amphiphilic molecules can be involved in self-assembly processes. However, considering functionalization of solid surfaces for a steady and robust nanosensor design, molecules that strongly interact with metal are mostly appropriated.

Self-assembled monolayers (SAMs) on solid surfaces have structural similarities (Fig. 10.2). These molecules consist of three main parts, including head group, backbone or spacer unit, and a terminal or functional group [16]. The head group ensures strong interaction with the surfaces. Usually silane heads groups are devoted to oxide surfaces, whereas thiols or amine are more adapted to noble metals. In the framework of SERS it is thus no surprise that most of the functionalization procedures involve an -SH group. Considering noble metals (gold, silver, or platinum) surface functionalization, molecules bearing thiol groups are mostly used [17], because of the very strong bonding between metal and sulfur atoms (the true nature of the bond is still under debate, but most claims that it is irreversible).

The backbone of SAMs is used as a spacer between the head and the end group. It controls the orientated formation of homogeneous monolayer due to weak Van der Waals forces between the molecules forming the SAM.

In addition to easy functionalization procedure, a reason why SAMs are extremely attractive is the large range of possibilities offered by SAMs terminal groups. Depending on the purpose of the

SAMs, a vast range of end groups are available (commercially or *via* synthesis) and it can subsequently be used in various chemical reactions [18]. In this manner, hydrophobic decorated surface can be turned to hydrophilic [19], thanks to end groups modifications.

Nature of SAM-forming molecules is very important for a chemical nanosensor design. Considering small molecules detection, the hydrophobic nature of octadecylthiol, which was used to coat roughened silver surface, was found to preconcentrate nonpolar aromatic pollutants following the Langmuir isotherm [20]. The selectivity of this sensor was successfully tested with three isomers of xylene. The sensor sensitivity was measured by detecting benzene at 7.5 ppm and naphthalene at 2.3 ppm.

The length of carbon chain (spacer) in alkanethiol plays a crucial role in the formation of a well-packed monolayer [21]. However, nanosensor decorated by short alkanethiols, such as 1-propanethiol, have shown ability to preconcentrate small molecules at the similar concentration level via hydrophobic interactions [22]. Gold nanoparticles grafted to solid substrates were employed for more efficient detection of nonpolar molecules. Nanosensors were designed on quartz substrates, functionalized by SAMs of (3-mercaptopropyl)trimethoxysilane (MPTMS) or (3-aminopropyl)trimethoxysilane (APTMS), and gold nanoparticles were then grafted via S–Au or N–Au interactions [1]. With such SERS-active substrate, detection limit in artificial sea water was found to be 10 ppb for naphthalene and pyrene.

In addition to SAMs ability to preconcentrate chemicals by hydrophobic interactions they can also be applied for polar molecules detection via electrostatic interactions. Thus, for pentachlorophenol (PCP) sensing, cysteamine-decorated silver nanoparticle aggregates have been used by Jiang et al. [23]. Cysteamine SAM formed in this model has provided both rapid target molecule adsorption and internal reference of a quantitative sensing. In this manner, PCP has been monitored in a range from 0.5 to 100 μM.

10.4.2 *Polymers*

Polymer-based sensors are used for a wide range of chemical detection, a good review can be found in Ref. [24]. However, for a long time, polymer-decorated nanosensors were not applicable as a matrix for SERS detection due to the great thickness of the deposited layer. To overcome this disadvantage, sol–gel derived layers with encapsulated metal nanoparticles have been proposed. For instance, Murphy et al. [25] have reported a study, where SERS-active substrates were produced by using gold and silver nanoparticles along with tetraethoxysilane (TEOS) and methyltriethoxysilane (MTEOS) precursors. MTEOS with incorporated silver nanoparticles has shown greater detection limits than TEOS for chlorobenzene and phenylacetylene (100 and 10 ppb, respectively) [25]. In addition to the sensitivity, sol–gel-derived xerogels with encapsulated metallic nanoparticles are excellent SERS-based sensors for the *in situ* environmental conditions. The lifetime of at least 5 weeks in sea water has been estimated by Lucht et al. [26]. Moreover, the detection of nonpolar molecules has been improved by applying ethyltriethoxysilane (ETEOS) in the mixture with MTEOS. This has led to increased sensor hydrophobicity and reached detection limit as low as 9 ppb for phenylacetylene and 135 ppb for biphenyl [25].

Other strategies of nanostructures functionalized with polymers for the detection of organic molecules have been proposed. For instance, polystyrene (PS) beads coated with gold nanoparticles have been suggested as SERS-active substrate, by Peron et al. [27]. Prepared core–shell nanocomposites were immobilized on silanizated quartz surface. The successful preconcentration of naphthalene via hydrophobic interactions has been achieved in the range of 1 to 20 ppm (7.8–160 μM) in aqueous media. For the quantification, Raman signal of the PS has been used as an internal spectral reference. Another group proposed the synthesis of core–shell nanoparticles for melamine detection in the 100 nM range [28]. Therein *in situ* chemically reduced silver nanoparticles were immobilized on poly(styrene-co-acrylic acid) (PSA) nanospheres.

Recently, great improvement in sensor selectivity has been proposed by constructing a synthetic receptor for specific molecule detection employing molecularly imprinted polymers (MIPs). MIPs are made by imprinting molecules of interest into the polymeric matrix followed by washing out the template molecules to ensure permanent grooves [29] (Fig. 10.3). This technique led to create MIP-based system that is complementary to the molecule of interest in shape, size, and molecular interactions. Thus, MIPs can be applied for a complex environment condition for its excellent recognition of template as well as mechanical and chemical stability [30]. Coupling SERS with molecular imprinting technique has been applied to provide not only selectivity but also sensitivity to the nanosensor. In 2009, Kantarovich et al. have proposed a method to print a MIP on Klarite® substrate (see Section 10.5) for a direct monitoring of beta-blocking drug S-propranolol uptake and release [31]. MIP droplets were deposited on a surface using a pipette or a nanofountain pen. SERS signal of S-propranolol has been observed even if the MIP layer deposited on the surface was not homogeneous and thick. By minimizing distance between template molecules and metallic moieties, great enrichment of chemical detection has been achieved. The controlled thickness of MIP layer was applied by Bompart et al. choosing propranolol as a template molecule [32]. In this article, core–shell composite particles of submicrometer size were designed. For SERS signal amplification, gold nanoparticles were incorporated as interlayer between polymeric core and MIP shell. Detection limit of 100 nM for a template molecule has been reached on single MIP composite by a core–shell-based nanosensor [32]. Recently, similar system was applied for a spectroscopic monitoring of bisphenol A, known as endocrine disruptor [33]. In this work, gold nanoparticles were decorated with molecularly imprinted shell via sol–gel process. The imprinted polymeric layer of 2 nm has been achieved ensuring an effective recognition sites by complete removal of template molecules. This improvement of nanosensor has led to rapid binding kinetics and a great detection limit of 0.12 $mg{\cdot}l^{-1}$. MIPs concept has also been exploited for sensing explosive agent 2,4,6-trinitrotoluene (TNT). Detection of TNT at 3 μM has been reached, thanks to sol–gel-derived xerogels, adsorbed on an SERS-active surface [34]. Yang et al. have demonstrated a nanosensor capable of monitoring TNT molecules in ultratrace concentration as low as 10^{-12} M [35]. Silver molybdate nanowires, decorated by crosslinked silver nanoparticles, have been used as a chemical sensor. Chang et al. have reached a particularly high detection limit of 10^{-15} M for a 4-mercaptobenzoic acid (4-MBA) by employing core–shell Ag-MIP hybrid as SERS-active substrate [36]. In this case, the MIP had a thickness of 40 nm, which is 2.7 times larger than

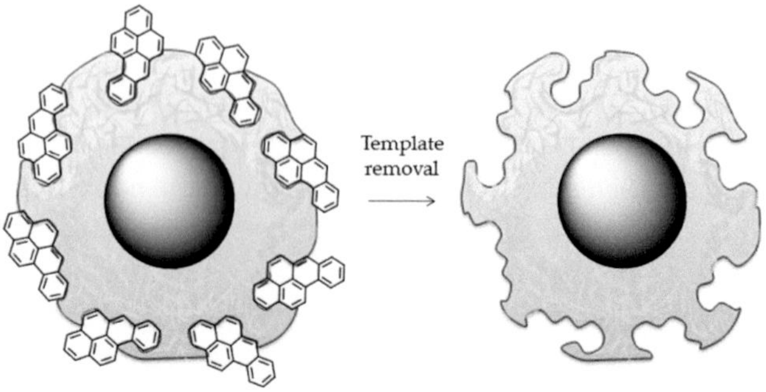

Figure 10.3 Graphical model of MIP preparation.

the maximum reported SERS distance [37]. A possible mechanism, dealing with a "gate effect" model [38] has been proposed to explain this great SERS enhancement.

10.4.3 *Host Materials*

Molecular host–guest systems have been studied for many years because of their great number of applications in different fields. Herein will be described some of the molecules which were applied for the design of nanosensor intended to detect trace concentrations of organic molecules such as PAHs, pesticides, or explosive agents by SERS. Nature of chemical interactions between host and guest molecules is noncovalent and provides the possibility of reusable sensor approach. Furthermore, organic molecule inside host material is protected from degradation by light or temperature, which can be induced by an electromagnetic field.

10.4.3.1 Cyclodextrins

Cyclodextrins (CDs) are cup-like molecules widely studied since the middle of 20th century. The enthusiasm was born after the demonstration of their ability to form host–guest complexes with hydrophobic molecules [39]. The most common CD forms are: α-CD, β-CD, and γ-CD. The difference between them is the number of glucopyranoside units (six for α-, seven for β-, and eight for γ)-attached by α-1,4-linkages [40]. CDs form specific 3D structure of

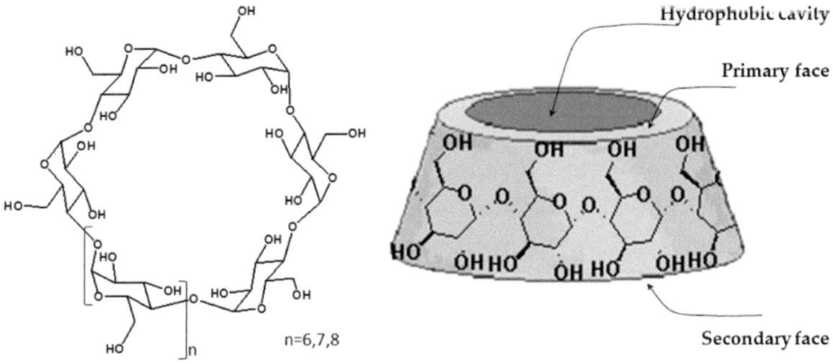

Figure 10.4 Chemical and 3D structure of cyclodextrins (CDs).

molecule, having inner hydrophobic cavity with a large entrance (Fig. 10.4). Cavity diameter depends on the number of glucosyl units. For this reason the smallest α-CD with the cavity size of 4.7 to 5.2 nm is known to form complexes with benzene, β-CD can capture molecules not bigger than 6.4 nm, and γ-CD having the cavity of 7.5 to 8.3 nm able to include two aromatic guest [40]. Since a wide variety of guest compounds can form complexes with CDs (aliphatic chains, aldehydes, ketones, alcohols, organic and fatty acids, PAHs, gases, and volatile compounds [41]), they can be applied for the design of nanosensor for pollutants detection based on SERS.

CDs have a number of primary hydroxyl groups on one edge called primary face as well as secondary hydroxyl groups on the other edge called secondary face. Those groups are orientated toward the outside of a cup-like structure, making the molecule soluble in water. Hydroxyl group's availability and reactivity are very important for CDs chemical modification in order to functionalize metallic surface. Many methods for the synthesis of controlled and selective CD derivatives have been studied and described [42, 43]. However, among all of them thiolated CDs remain the most used for surface decoration. Thiolated CDs can form packed monolayers as alkanethiols, but the chemisorption's kinetics strongly depends on the length between the SH function and the CD [44]. However, for an SERS-active surface functionalization, thiolated CDs without alkyl chain are preferred in order to increase the nanosensor sensitivity by preconcentrating chemicals closer to the metal surface.

Per-6-thio-β-cyclodextrins have been applied for decoration of rough silver surface to detect gaseous aromatic substances such as benzene, toluene, *m*-xylene, chlorobenzene, and *m*-dichlorobenzene, with the detection limits of 10000, 400, 1400, 250, and 15 ppm, respectively [45]. Detection limits have been greatly improved by applying metallic nanoparticles as SERS-active substrate. Xie et al. have described qualitative and quantitative detection of anthacene, pyrene, chrysene, and triphenylene with a per-6-deoxy-(6-thio)-β-cyclodextrin-modified gold and silver nanoparticles [46, 47]. The limit of detection was found to be 100, 10, 100, and 1000 nM for anthracene, pyrene, chrysene, and triphenylene, respectively. Triangular gold nanoprisms functionalized by mono-6-thio-beta-cyclodextrin have been applied for SERS detection of explosive agent 2,4-dinitrotoluene (DNT) at the concentration of 10 pM [48]. Methyl parathion insecticide (1 pM) was detected by investigating gold nanorods, decorated with mono-6-thio-beta-cyclodextrin [49].

10.4.3.2 Calixarenes

Calixarenes (CXs) are compounds, belonging to cavitands class as CDs. They are described as synthetic cyclo-oligomers whose structure forms cup-like shape (Fig. 10.5), which make them serve as host for various molecules [50]. Specific CXs structure ensures specificity in recognizing small species such as ions, anions, or neutral molecules [51]. Chemical modification of upper or lower rims ensures the molecular selectivity regarding functional affinity

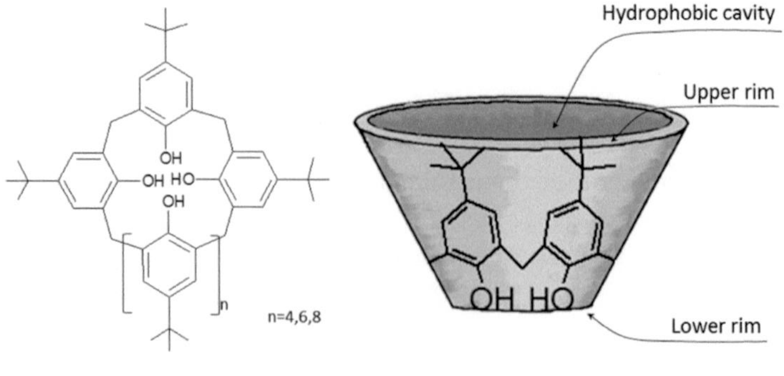

Figure 10.5 Chemical and 3D structure of calixarenes (CXs).

with the targeted chemicals. Because of the following reason this group of molecules is well applicable for SERS sensor design. As for the CD case, CXs offer different size of cavity depending on number of phenolic units (4, 6, 8) linked by methylene bridges. For the interest to detect pollutants, CXs propose hydrophobic cavity and π–π stacking between its structural unit of *p-tert*-butyl phenol and benzene ring–bearing compounds. Based on this interaction aromatic pollutants have been monitored by the mean of SERS. For instance, silver nanoparticles, covered by calix[4]arene and its derivatives having dicarboethoxy and dicyano substitutes in the lower rim have been studied by Leyton et al. [52, 53]. They have discovered that CXs show analytical selectivity to the guest systems bearing four benzene rings, mainly pyrene. Later, successful improvement of metallic surface functionalization have been proposed by Guerrini et al., using dithiocarbamate-modified calix[4]arenes [54]. In the mean of this substrate, PAHs such as benzo[c]phenanthrene, triphenylene, and coronene were monitored in the concentration interval between 10 and 100 ppb [55]. Ability of CXs-based nanosensor performance in seawater has also been demonstrated. Therein, SERS-active sensor was constructed by encapsulating Ag nanoparticles, previously decorated with 25,27-dimercaptoacetic acid-26,28-dihydroxy-4-tert-butylcalix[4]arene (DMCX) into sol–gel matrix [56]. A calibration procedure has approved sensitivity and stability in artificial sea water. The potential detection limits of 0.3 nM for pyrene and 13 nM for naphthalene have been calculated.

10.4.4 *Carbon Derivatives*

The interaction of chemical molecules with a single-walled carbon nanotube (SWNT) and two multiwalled carbon nanotubes (MWNTs) have been studied by Star et al. [57] and the adsorption affinity have been determined. Higher affinity has been found with nonpolar aromatics in comparison with nonpolar aliphatic molecules. This advantage took place in PAHs' preconcentration and detection by SERS. Experimental procedure to detect pyrene at the concentration as low 1 nM has been proposed by Leyton et al. Therein single-walled carbon nanotubes were deposited on colloidal silver nanoparticles, evaporated on glass slide. Pyrene then was preconcentrated near surface by a π–π electronic stacking with the nanotube [58].

10.4.5 *Biomolecules*

Aptamer are nucleic acid sequence [59], which can be generated against various molecules [60, 61]. Application of aptamers in nanosensor design deals with their high affinity and specificity toward the target molecule. A selective detection of ochratoxin-A (OTA) has been suggested by Galarreta et al. [62]. In this work two-dimensional gold nanostructures prepared by EBL were decorated with adsorbed oligonucleotides. The sensing ability of this system comes from the difference of the aptamer spectrum before and after it has been in contact with OTA. This suggests efficient aptamer–ochratoxin complex formation. Other model for indirect sensing of a melanine molecule has been proposed by Wen et al. [63], where silver nanoparticles of 10 nm, modified by oligonucleotide were used as SERS substrates. They have demonstrated that interaction of target molecules with aptamers induce nanoparticles release, which are then subsequently aggregating. This process has been monitored by SERS in the low frequency range. With this system melamine concentrations ranging from 6.3 to 403.6 μg/L have been analyzed.

10.5 Commercials Solutions for SERS Substrates

Concerning SERS commercial solutions, many providers appeared in the last ten years. Different kinds of companies share out the market: start-up, spin-out, multinational, etc. The commercialized solutions are various from colloids by chemical reduction to physical deposition of metals. Gold and silver are used as nanoparticles to enhance the Raman signal. The variety of substrates includes usually a chemical functionalization as described previously.

As far as we know, Mesophotonics was the first company to provide a SERS solution. Mesopthotonics is a spin-out from the University of Southampton. Their Klarite® substrates are made of gold deposited on nanostructured silicon through chemical etching. The sensitive pattern is a square of 4 × 4 mm available free or on a microscope slide. The sub-micron pattering allows an enhancement able to detect signal on concentrations as low as ppb. The manufacturing of these substrates ensures a standard deviation below 10% with both laser excitation of 633 and 785 nm. Klarite® substrates are now available at Renishaw since their investment in D3 Technologies Ltd., which holds different SERS patents. The application domains are various: analytical chemistry, pharmaceutical drug development, forensics, etc.

A new company based in Vilius, Lituania, and specialized in laser manufacturing supplies an SERS solution based on silver nanostructures named Randa. The possible range of excitation is foreshadowed from blue to near infrared. Standard areas are squares of 4 × 4 mm, but larger substrates of 10 × 10 mm are also available. The concept of manufacturing is 100% monolithic in order to avoid various risks during the elaboration or during the transportation. The samples are very reactive but consequently unstable so that they have to be stored under vacuum and are usable during 2 months after the manufacturing date. Integrated Optics is the only producer to mention an uncertainty principle: one has to choose either very high enhancement ratio or good repeatability of measurements. Randa samples are developed for trace detection but not for concentration determination.

Diagnostic anSERS Inc. sells original affordable substrates based on a production by ink-jet method. The patent-pending ink-jet printing process allows the company to produce the P-SERS™ substrates at very low cost without sacrificing sensitivity. Moreover, the substrates are very easy to use since only a pipette is needed to drop off the sample. Diagnostic anSERS Inc. also provides SERS solutions through their SERS-active gold or silver nanoparticles available in colloidal forms.

SILicon MEtal COating or Silmeco is a nanotechnology company specialized in SERS substrates. Designed for a wide range of applications and R&D-purposes, Silmeco provides very large area substrates on silicon wafers. The SERS signal is uniform on the covered surface that can reach up to 10 cm in diameter. Based on patented nanofabrication technologies without any lithography process, the fabrication method involves only two steps. Consequently, high-volume manufacturing processes and specific nanostructure patterns made of gold or silver are the know-how of Silmeco.

Nanova Inc., which is a high-tech company using cutting-edge technology provides also SERS solutions in their field of expertise in nanotechnology, sensors, medical science, and food safety. Thanks to its proprietary technology, the Q-SERS™ substrate is capable of detecting chemical and biochemical compounds at trace concentrations. It consists of nanoparticles from 15 to 60 nm of a metal (not defined) randomly organized on a silicon support glued on a microscope slide. The substrate is able to provide high enhancement factors, with reliable and reproducible results. The fields of application are analytical chemistry, food and water safety, pharmaceutical, medical, and forensic science.

In contrast to the the previous companies Enhanced Spectrometry, Inc. or EnSpectr develops SERS substrates in the framework of the promotion of the portable Raman spectrometry as a solution for advanced substances control, quality inspection, real-time testing, and innovative brand protection. This solution is based on a proprietary technology designed for substances detection and coding of information. EnSpectr SERS Substrates are able to identify various compounds at trace concentrations for chemical analysis or fundamental scientific research. The targeted applications are prohibited substances, cells and tissues, bioscience and pharmaceutical research, and authentication of products with spectral markers technology. Supplied substrates are unmounted squared chips of 5 × 5 mm. They are stored in an inert atmosphere until their utilization. The suggested wavelength is in the 450 to 550 nm range.

With the same market objectives, real-time analyzers provides SERS substrates with the analytical tools such as a portable Raman analyzer and with their process environment. On the opposite of all previous solutions, real-time analyzers has chosen to include silver nanoparticles in an organic sol–gel matrix. This choice allows to provide different substrates such as SERS Vials, SERS Microplates, or SERS Capillaries. These different presentations enlarge the field of application of liquid analyses. Furthermore, the presence of an organic matrix can exhibit a best sensitivity, thanks to the preconcentration effect for organic compounds. Coupled with the SERS Raman Analyzer, SERS substrates are applied to explosives, chemical weapons, narcotics, and hazardous materials.

The worldwide leader in Raman spectrometry Horiba JobinYvon proposes SERS substrates from ST Japan. This kind of substrates offers excellent reproducibility and significant enhancement. The substrates are available in different sizes, coated with pure gold or gold–silver alloy. Furthermore, the company has recently communicated about SERS through a study of nanoparticles themselves. It is well-known that metallic Au or Ag nanostructures provide ideal substrates for SERS measurements. Surface plasmon resonance (SPR) (see Chapter 2 of this book for details about SPR measurements) is used in order to optimize the SERS substrates. The use of SERS and SPR concomitantly promises great enthusiasm.

10.6 Conclusion

SERS has entered a few years ago in a new era. Several companies are selling SERS sensors. This evolution has been made possible, thanks to the evolution of the nanostructuration techniques (lithography, etching, metal deposition, etc.). However, surface chemistry has been another key factor to the selling of SERS sensors. As it was shown in the second part of this chapter, surface chemistry is essential to conceive a specific SERS sensor.

Numerous companies provide SERS solutions. Most of them use gold or silver nanoparticles made by the routes developed for the semiconductor or the nanotechnology industry. Chemical-based colloids are also used alone or coupled to organic matrix. A very high enhancement factor is always described. The reproducibility is often presented as good, with a standard deviation around 10%. But, in all likelihood, a lack of standards conducts to an impossibility to compare the SERS products: sensitivity, selectivity, and reproducibility are only mentioned but not really quantified.

Moreover, as other optical methods, SERS is suitable for miniaturization. Concomitantly to the advances in photonics and nanotechnologies, SERS will certainly be available in the future as on field sensors for many applications.

Acknowledgment

The authors want to acknowledge the ANR REMANTAS project (ANR-11-ECOT-0010) for financial support.

References

1. Péron, O., Rinnert, E., Lehaitre, M., Crassous, P. and Compère, C. (2009). Detection of polycyclic aromatic hydrocarbon (PAH) compounds in artificial sea-water using surface-enhanced Raman scattering (SERS), *Talanta,* **79**, pp. 199–204.
2. Delhaye, C., Bruneel, J.-L., Talaga, D., Guirardel, M., Lecomte, S. and Servant, L. (2012). Tailoring surface-enhanced Raman scattering effect using microfluidics, *J. Phys. Chem. C,* **116**, pp. 5327–5332.
3. Haynes, C. L. and Van Duyne, R. P. (2001). Nanosphere lithography: a versatile nanofabrication tool for studies of size-dependent nanoparticle optics, *J. Phys. Chem. B,* **105**, pp. 5599–5611.
4. Guo, L. J. (2007). Nanoimprint lithography: methods and material requirements, *Adv. Mater.,* **19**, pp. 495–513.
5. Guerrini, L., Jurasekova, Z., Domingo, C., Pérez-Méndez, M., Leyton, P., Campos-Vallette, M., Garcia-Ramos, J. V. and Sanchez-Cortes, S. (2007). Importance of metal–adsorbate interactions for the surface-enhanced Raman scattering of molecules adsorbed on plasmonic nanoparticles, *Plasmonics,* **2**, pp. 147–156.
6. Péron, O., Rinnert, E., Colas, F., Lehaitre, M. and Compère, C. (2010). First steps of in situ surface-enhanced Raman scattering during shipboard experiments, *Appl. Spectrosc.,* **64**, pp. 1086–1093.
7. Xie, Y., Li, P., Zhang, J., Wang, H., Qian, H. and Yao, W. (2013). Comparative studies by IR, Raman, and surface-enhanced Raman spectroscopy of azodicarbonamide, biurea and semicarbazide hydrochloride, *Spectrochim. Acta A,* **114**, pp. 80–84.
8. Alvarez-Ros, M. C., Sanchez-Cortes, S., Francioso, O. and Garcia-Ramos, J. V. (2001). Catalytic modification of gallic acid on a silver surface studied by surface-enhanced Raman spectroscopy, *J. Raman Spectrosc.,* **32**, pp. 143–145.
9. Lin, X. M., Cui, Y., Xu, Y. H., Ren, B. and Tian, Z. Q. (2009). Surface-enhanced Raman spectroscopy: substrate-related issues, *Anal. Bioanal. Chem.,* **394**, pp. 1729–1745.

10. Sanchez-Cortes, S., Francioso, O., Garcia-Ramos, J. V., Ciavatta, C. and Gessa, C. (2001). Catechol polymerization in the presence of silver surface, *Colloids Surf. A,* **176**, pp. 177–184.

11. Sanchez-Cortes, S. and Garcia-Ramos, J. V. (2000). FT surface-enhanced Raman evidence of the oxidative condensation reactions of caffeic acid in solution and on silver surface, *Appl. Spectrosc.,* **54**, pp. 230–238.

12. Sanchez-Cortes, S. and Garcia-Ramos, J. V. (2000). Adsorption and chemical modification of phenols on a silver surface, *J. Colloid Interface Sci.,* **231**, pp. 98–106.

13. Alvarez-Puebla, R. A. and Liz-Marzan, L. M. (2012). Traps and cages for universal SERS detection, *Chem. Soc. Rev.,* **41**, pp. 43–51.

14. Águas, H., Silva, R. J. C., Viegas, M., Pereira, L., Fortunato, E. and Martins, R. (2008). Study of environmental degradation of silver surface, *Phys. Stat. Sol. c,* **5**, pp. 1215–1218.

15. Discher, D. E. and Eisenberg, A. (2002). Polymer vesicles, *Science,* **297**, pp. 967–973.

16. Schreiber, F. (2000). Structure and growth of self-assembling monolayers, *Prog. Surf. Sci.,* **65**, pp. 151–256.

17. Smith, R. K., Lewis, P. A. and Weiss, P. S. (2004). Patterning self-assembled monolayers, *Prog. Surf. Sci.,* **75**, pp. 1–68.

18. Shon, Y. S. and Choo, H. (2003). Organic reactions of monolayer-protected metal nanoparticles, *C. R. Chim.,* **6**, pp. 1009–1018.

19. Schreiber, F. (2000). Structure and growth of self-assembling monolayers, *Prog. Surf. Sci.,* **65**, pp. 151–257.

20. Carron, K., Peitersen, L. and Lewis, M. (1992). Octadecylthiol-modified surface-enhanced Raman-spectroscopy substrates: a new method for the detection of aromatic-compounds, *Environ. Sci. Technol.,* **26**, pp. 1950–1954.

21. Love, J. C., Estroff, L. A., Kriebel, J. K., Nuzzo, R. G. and Whitesides, G. M. (2005). Self-assembled monolayers of thiolates on metals as a form of nanotechnology, *Chem. Rev.,* **105**, pp. 1103–1170.

22. Costa, J. C. S., Sant'Ana, A. C., Corio, P. and Temperini, M. L. A. (2006). Chemical analysis of polycyclic aromatic hydrocarbons by surface-enhanced Raman spectroscopy, *Talanta,* **70**, pp. 1011–1016.

23. Jiang, X., Yang, M., Meng, Y., Jiang, W. and Zhan, J. (2013). Cysteamine-modified silver nanoparticle aggregates for quantitative SERS sensing of pentachlorophenol with a portable Raman spectrometer, *ACS Appl. Mater. Interfaces,* **5**, pp. 6902–6908.

24. Hierlemann, A., Weimar, U., Kraus, G., Schweizer-Berberich, M. and Göpel, W. (1995). Polymer-based sensor arrays and multicomponent analysis for the detection of hazardous organic vapours in the environment, *Sens. Actuators, B,* **26**, pp. 126–134.
25. Murphy, T., Schmidt, H. and Kronfeldt, H. D. (1999). Use of sol-gel techniques in the development of surface-enhanced Raman scattering (SERS) substrates suitable for in situ detection of chemicals in sea-water, *Appl. Phys. B,* **69**, pp. 147–150.
26. Lucht, S., Murphy, T., Schmidt, H. and Kronfeldt, H. D. (2000). Optimized recipe for sol–gel-based SERS substrates, *J. Raman Spectrosc.,* **31**, pp. 1017–1022.
27. Peron, O., Rinnert, E., Toury, T., Lamy de la Chapelle, M. and Compere, C. (2011). Quantitative SERS sensors for environmental analysis of naphthalene, *Analyst,* **136**, pp. 1018–1022.
28. Li, J.-M., Ma, W.-F., Wei, C., You, L.-J., Guo, J., Hu, J. and Wang, C.-C. (2011). Detecting trace melamine in solution by SERS using Ag nanoparticle coated poly(styrene-co-acrylic acid) nanospheres as novel active substrates, *Langmuir,* **27**, pp. 14539–14544.
29. Cheong, W. J., Yang, S. H. and Ali, F. (2013). Molecular imprinted polymers for separation science: a review of reviews, *J. Sep. Sci.,* **36**, pp. 609–628.
30. Jung, B. M., Kim, M. S., Kim, W. J. and Chang, J. Y. (2010). Molecularly imprinted mesoporous silica particles showing a rapid kinetic binding, *Chem. Commun.,* **46**, pp. 3699–3701.
31. Kantarovich, K., Tsarfati, I., Gheber, L. A., Haupt, K. and Bar, I. (2009). Writing droplets of molecularly imprinted polymers by nano fountain pen and detecting their molecular interactions by surface-enhanced Raman scattering, *Anal. Chem.,* **81**, pp. 5686–5690.
32. Bompart, M., De Wilde, Y. and Haupt, K. (2010). Chemical nanosensors based on composite molecularly imprinted polymer particles and surface-enhanced Raman scattering, *Adv. Mater.,* **22**, pp. 2343–2348.
33. Xue, J.-Q., Li, D.-W., Qu, L.-L. and Long, Y.-T. (2013). Surface-imprinted core–shell Au nanoparticles for selective detection of bisphenol A based on surface-enhanced Raman scattering, *Anal. Chim. Acta,* **777**, pp. 57–62.
34. Holthoff, E. L., Stratis-Cullum, D. N. and Hankus, M. E. (2011). A nanosensor for TNT detection based on molecularly imprinted polymers and surface enhanced Raman scattering, *Sensors,* **11**, pp. 2700–2714.

35. Yang, L., Ma, L., Chen, G., Liu, J. and Tian, Z.-Q. (2010). Ultrasensitive SERS detection of TNT by imprinting molecular recognition using a new type of stable substrate, *Chem.-Eur. J.,* **16**, pp. 12683–12693.

36. Chang, L. M., Ding, Y. and Li, X. (2013). Surface molecular imprinting onto silver microspheres for surface enhanced Raman scattering applications, *Biosens. Bioelectron.,* **50**, pp. 106–110.

37. Singh, A. K., Khan, S. A., Fan, Z., Demeritte, T., Senapati, D., Kanchanapally, R. and Ray, P. C. (2012). Development of a long-range surface-enhanced Raman spectroscopy ruler, *J. Am. Chem. Soc.,* **134**, pp. 8662–8669.

38. Piletsky, S. A., Panasyuk, T. L., Piletskaya, E. V., Nicholls, I. A. and Ulbricht, M. (1999). Receptor and transport properties of imprinted polymer membranes—a review, *J. Membr. Sci.,* **157**, pp. 263–278.

39. Del Valle, E. M. M. (2004). Cyclodextrins and their uses: a review, *Process Biochem.,* **39**, pp. 1033–1046.

40. Ogoshi, T. and Harada, A. (2008). Chemical sensors based on cyclodextrin derivatives, *Sensors,* **8**, pp. 4961–4982.

41. Schmid, G. (1989). Cyclodextrin glycosyltransferase production—yield enhancement by overexpression of cloned genes, *Trends Biotechnol.,* **7**, pp. 244–248.

42. Boger, J., Corcoran, R. J. and Lehn, J.-M. (1978). Cyclodextrin chemistry. Selective modification of all primary hydroxyl groups of α- and β-cyclodextrins, *Helv. Chim. Acta,* **61**, pp. 2190–2218.

43. Khan, A. R., Forgo, P., Stine, K. J. and D'Souza, V. T. (1998). Methods for selective modifications of cyclodextrins, *Chem. Rev.,* **98**, pp. 1977–1996.

44. Weisser, M., Nelles, G., Wohlfart, P., Wenz, G. and Mittler-Neher, S. (1996). Immobilization kinetics of cyclodextrins at gold surfaces, *J. Phys. Chem.,* **100**, pp. 17893–17900.

45. Hill, W., Fallourd, V. and Klockow, D. (1999). Investigation of the adsorption of gaseous aromatic compounds at surfaces coated with heptakis(6-thio-6-deoxy)-beta-cyclodextrin by surface-enhanced Raman scattering, *J. Phys. Chem. B,* **103**, pp. 4707–4713.

46. Xie, Y., Wang, X., Han, X., Song, W., Ruan, W., Liu, J., Zhao, B. and Ozaki, Y. (2011). Selective SERS detection of each polycyclic aromatic hydrocarbon (PAH) in a mixture of five kinds of PAHs, *J. Raman Spectrosc.,* **42**, pp. 945–950.

47. Xie, Y., Wang, X., Han, X., Xue, X., Ji, W., Qi, Z., Liu, J., Zhao, B. and Ozaki, Y. (2010). Sensing of polycyclic aromatic hydrocarbons with cyclodextrin inclusion complexes on silver nanoparticles by surface-enhanced Raman scattering, *Analyst,* **135**, pp. 1389–1394.

48. Xu, J. Y., Wang, J., Kong, L. T., Zheng, G. C., Guo, Z. and Liu, J. H. (2011). SERS detection of explosive agent by macrocyclic compound functionalized triangular gold nanoprisms, *J. Raman Spectrosc.*, **42**, pp. 1728–1735.

49. Wang, J., Kong, L. T., Guo, Z., Xu, J. Y. and Liu, J. H. (2010). Synthesis of novel decorated one-dimensional gold nanoparticle and its application in ultrasensitive detection of insecticide, *J. Mater. Chem.*, **20**, pp. 5271–5279.

50. Gutsche, C. D. (1984) *Structural chemistry,* Chapter 1 "The calixarenes" (Springer, Berlin, Heidelberg), pp. 1–47.

51. Deligöz, H., Karakuş, Ö. Ö. and Çilgi, G. K. (2012). A brief review on the thermal behaviors of calixarene-azocalixarene derivatives and their complexes, *J. Macromol. Sci., Part A: Pure Appl.Chem.*, **49**, pp. 259–274.

52. Leyton, P., Sanchez-Cortes, S., Garcia-Ramos, J. V., Domingo, C., Campos-Vallette, M., Saitz, C. and Clavijo, R. E. (2004). Selective molecular recognition of polycyclic aromatic hydrocarbons (PAHs) on calix[4]arene-functionalized Ag nanoparticles by surface-enhanced Raman scattering, *J. Phys. Chem. B,* **108**, pp. 17484–17490.

53. Leyton, P., Domingo, C., Sanchez-Cortes, S., Campos-Vallette, M. and Garcia-Ramos, J. V. (2005). Surface enhanced vibrational (IR and Raman) spectroscopy in the design of chemosensors based on ester functionalized p-tert-butylcalix[4]arene hosts, *Langmuir,* **21**, pp. 11814–11820.

54. Guerrini, L., Garcia-Ramos, J. V., Domingo, C. and Sanchez-Cortes, S. (2006). Functionalization of Ag nanoparticles with dithiocarbamate calix[4]arene as an effective supramolecular host for the surface-enhanced Raman scattering detection of polycyclic aromatic hydrocarbons, *Langmuir,* **22**, pp. 10924–10926.

55. Guerrini, L., Garcia-Ramos, J. V., Domingo, C. and Sanchez-Cortes, S. (2009). Sensing polycyclic aromatic hydrocarbons with dithiocarbamate-functionalized Ag nanoparticles by surface-enhanced Raman scattering, *Anal. Chem.*, **81**, pp. 953–960.

56. Kwon, Y. H., Sowoidnich, K., Schmidt, H. and Kronfeldt, H. D. (2012). Application of calixarene to high active surface-enhanced Raman scattering (SERS) substrates suitable for in situ detection of polycyclic aromatic hydrocarbons (PAHs) in seawater, *J. Raman Spectrosc.*, **43**, pp. 1003–1009.

57. Star, A., Han, T.-R., Gabriel, J.-C. P., Bradley, K. and Grüner, G. (2003). Interaction of aromatic compounds with carbon nanotubes: correlation

to the Hammett parameter of the substituent and measured carbon nanotube FET response, *Nano Lett.,* **3**, pp. 1421–1423.

58. Leyton, P., Gómez-Jeria, J. S., Sanchez-Cortes, S., Domingo, C. and Campos-Vallette, M. (2006). Carbon nanotube bundles as molecular assemblies for the detection of polycyclic aromatic hydrocarbons: surface-enhanced resonance Raman spectroscopy and theoretical studies, *J. Phys. Chem. B,* **110**, pp. 6470–6474.
59. Sassolas, A., Blum, L. J. and Leca-Bouvier, B. D. (2011). Optical detection systems using immobilized aptamers, *Biosens. Bioelectron.,* **26**, pp. 3725–3736.
60. Osawa, Y., Takase, M., Sode, K. and Ikebukuro, K. (2009). DNA aptamers that bind to PQQGDH as an electrochemical labeling tool, *Electroanalysis,* **21**, pp. 1303–1308.
61. Subramanian, P., Lesniewski, A., Kaminska, I., Vlandas, A., Vasilescu, A., Niedziolka-Jonsson, J., Pichonat, E., Happy, H., Boukherroub, R. and Szunerits, S. (2013). Lysozyme detection on aptamer functionalized graphene-coated SPR interfaces, *Biosens. Bioelectron.,* **50**, pp. 239–243.
62. Galarreta, B. C., Tabatabaei, M., Guieu, V., Peyrin, E. and Lagugne-Labarthet, F. (2013). Microfluidic channel with embedded SERS 2D platform for the aptamer detection of ochratoxin A, *Anal. Bioanal. Chem.,* **405**, pp. 1613–1621.
63. Wen, G., Zhou, L., Li, T., Liang, A. and Jiang, Z. (2012). A sensitive surface-enhanced Raman scattering method for determination of melamine with aptamer-modified nanosilver probe, *Chin. J. Chem.,* **30**, pp. 869–874.

Chapter 11

Tip-Enhanced Raman Spectroscopy: Principle and Instrumentation

Jean-Christophe Valmalette
IM2NP Site de Toulon, Université Sud Toulon Var, BP 132, 83957 La Garde Cedex
jean-christophe.valmalette@im2np.fr

11.1 Introduction

The nanoscale identification of molecules is a driving force for the knowledge in nanoscience and a key parameter in the further development of nanotechnology such as nanoelectronics, biotechnology, or green technology. Vibration of atoms in molecules provides valuable information on chemical species and is considered as a molecular "finger-print." For example, vibration-based microscopy has recently shown the ability to track molecules in living tissues without the need for fluorescent labels [1, 2]. Furthermore, molecular vibrations or phonons in crystals are also sensitive to intrinsic perturbations (defects, plasmon–phonon coupling, etc.) and extrinsic perturbation (charge transfer, pressure, temperature, optical processes, etc.). Vibrations can be probed by only a limited number of techniques: infrared spectroscopy, inelastic neutron scattering, Raman scattering, and, more recently, inelastic scanning

Handbook of Enhanced Spectroscopy
Edited by Pietro Giuseppe Gucciardi, Marc Lamy de la Chapelle, and Nathalie Lidgi-Guigui

ISBN 978-981-4613-32-3 (Hardcover), 978-981-4613-33-0 (eBook)
www.panstanford.com

tunneling spectroscopy [3]. The last one "visualizes" the chemical nature of atoms and molecules on a surface but requires ultra-high vacuum (UHV) and very low temperature. The inelastic Raman scattering provides very rich information and does not require special sample preparation, but faces a low scattering cross section (10^{-31}–10^{-26} cm^2/molecule [4, 5]), and Abbe's diffraction limited spatial resolution.

These two obstacles can be overcome by localized electromagnetic field enhancement induced by a metallic nanostructure at the vicinity of point analysis combining surface-enhanced Raman spectroscopy (SERS) and near-field scanning probe techniques. Basically, tip-enhanced Raman spectroscopy (TERS) can be described as tip-localized SERS brought to the probe molecule deposited on a substrate. This combination allows more flexible configurations between light excitation, tip, probe molecule, substrate, and scattered light. The creation of a localized "hot spot" between a metallic tip and a substrate is directly related to the geometry of the experimental setup. The TERS is an emerging technique which is considered as the first label-free techniques able to provide the single-molecule detection spatially resolved in the near future.

The TERS technique has been conceptually predicted by Wessel [6] in 1985, the year after the first near-field scan using visible light [7, 8]. Considerable effort has been done to integrate chemical analysis into the ability of the scanning probe microscopy (SPM) to image at high spatial resolution, while SERS reached the single-molecule detection [9, 10]. The first experiments on near-field Raman imaging were obtained by scanning near-field optical microscopy (SNOM) on brilliant cresyl blue (BCB)-labeled DNA [11] and rhodamine 6G [12] by using near-field illumination of SERS substrate by optical fiber tip, which acts as a point light source with subwavelength. But this aperture SNOM Raman has been disused due to its low-intensity excitation and aperture size limitation.

Fifteen years later, the first TERS experimental was evidenced by several groups. The results were published the same year in 2000 by Zenobi's group from ETH Zuric (Fig. 11.1) [13], Anderson's group from Caltech [14] and Kawata's group form Osaka University [15]. The initial aim of these first TERS experiments was both to avoid the critical dependence of SERS enhancement on the substrate

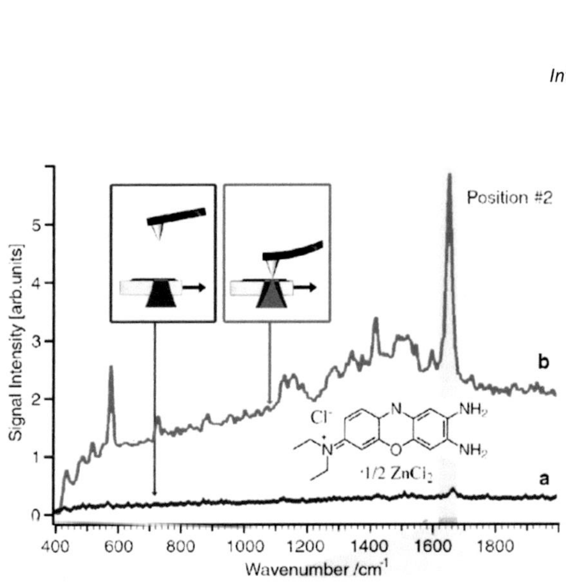

Figure 11.1 Tip-enhanced Raman spectra of brilliant cresyl blue (BCB) dispersed on a glass support measured with a silver-coated atomic force microscopy (AFM) probe under 488 nm excitation wavelength (1.5 mW aq. time 60s), tip retracted from the sample (a) and tip in contact with the sample (b) (from Stockle CPL 2000).

preparation across the sample and to reach the ultrahigh spatial resolution. TERS was also expected to eliminate the shortcomings of quantitative SERS without the need for depositing samples onto a special substrate.

Zenobi's group [16] used an inverted microscope with 488 nm excitation and immersion oil objective (NA = 1.4). The scattered light was fed into an avalanche photodiode (APD) (for integrated reflectivity measurements) and CCD (for spectrally resolved measurements). In the first experiment, they used a silver-coated atomic force microscopy (AFM) tip obtained by evaporation of 10 to 15 nm silver film and BCB deposited on glass substrate as probe molecule. An enhancement factor (EF) of 2000 was observed when the tip was in contact compared to withdrawn tip. In the second experiment, they analyzed C_{60} by using an electrochemical etched thin gold wire mounted on a tuning fork set-up. The C_{60} spectrum shown the two normal Raman modes and additionally four other modes assigned to a charge transfer between gold tip and C_{60} [15]. The same year, Anderson [17] used gold-coated silicon AFM tip and 785-nm laser excitation on a 10 μm sulfur film deposited on quartz

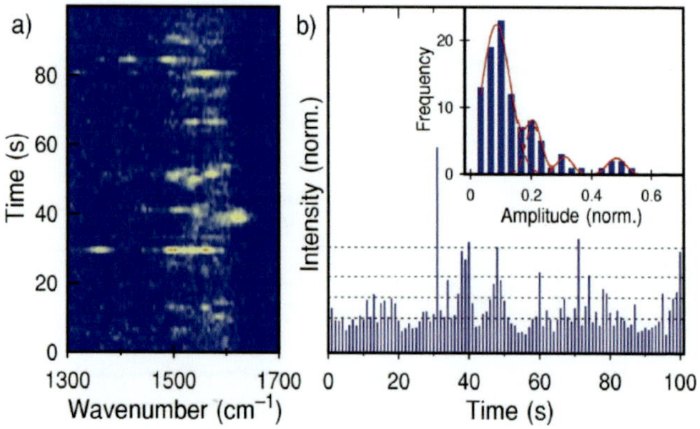

Figure 11.2 TERS signal from single malachite green (MG) molecules: time series (left) and temporal variation of the intensities of the integrated 1480 to 1630 cm^{-1} Raman band with 170 to 230 counts/mol s. Discretization of Raman intensities can be seen. Reprinted with permission from Ref. [19], Copyright 2006 by the American Physical Society.

glass substrate from acetone solution. Kawata's group [18] obtained near-field Raman scattering of rhodamine 6G by using apertureless near-field scanning optical microscope with 40-nm silver layer-coated AFM tip illuminated with 488 nm excitation wavelength and obtained a lateral resolution of about 50 nm.

A few years later, Neacsu et al. [19] reached the evidence of the single-molecule sensitivity in ambient atmosphere with shear force AFM and top illumination on submonolayer of triarylmethane dye malachite green on metallic substrate (see Fig. 11.2 and Gap-mode section below). This ultimate sensitivity has been confirmed more recently by different groups [20, 21].

The last major step on TERS history was the molecular inner structure resolution achieved in 2013 by Zhang et al. [22]. By the combination of high vacuum and gap mode, this group resolved a single molecule chemically by using TERS mapping of a single H_2TBPP molecule on Ag(111). The inner structure and surface configuration of the molecule is achieved by spectrally matching the resonance of the nanocavity plasmon to the molecular transition, particularly the downward transition responsible for the emission of Raman photons. The surface configuration of the single molecule is also achieved by TERS thanks to the orientation dependency of TERS spectra.

A few review papers [23–29] and book chapters on TERS [30–33] have been published these last years. In 2009, Le Ru and Etchegoin wrote "TERS is a technique that is (relatively speaking) still in its infancy but will surely see a wealth of activity in years to come." Indeed, since the last four years, major progresses have been already done, especially in UHV–TERS and in the use of new type of substrates for gap-mode (GM)-TERS.

11.2 Principle

The challenging task in TERS imaging consists in an efficient resonance enhancement for both laser excitation and Raman scattering at the vicinity of the molecule. The strong field enhancement can be achieved by using metallic nanostructures acting as optical nanoantennas. The concept developed from SERS experiments is that only a very few hotspots contribute to the total signal and most of the surface is inactive [33]. Only the few "active" molecules localized in these hotspots give rise to a huge enhancement, which makes the total SERS signal. Thus, TERS can be seen as a localized and ultimately controlled SERS to reach the single hotspot level at the apex of the tip. The EF $\geq 10^{14}$ observed in single-molecule SERS experiments make reasonable the possibility of single molecule mapping by TERS.

TERS differs from SNOM in the way that both excitation and collection are done under far field conditions (most of the case by using standard objectives). Thus, the contribution of the localized near-field intensity (I_{NF}) is the difference between two successive acquisitions: when the tip is approached (near-field + far-field contributions: $I_{appr} = I_{NF} + I_{FF}$) and withdrawn (far-field contribution only: $I_{withdr} = I_{FF}$). The net gain G_{net} due to the local field enhancement is then defined as the ratio:

$$G_{net} = \frac{I_{NF}}{I_{FF}} = \frac{I_{appr} - I_{withdr}}{I_{withdr}}.$$

In the early 2000s experiments, this net gain ranged from 1.5 to 40. The EF in TERS is not easy to define due to the uncertainty on the illuminated area, near-field volume, shadowing effect with tip position. Indeed, the increase of the signal comes from a very small volume compared to the far-field contribution. Only rough approximation can be done by the relation of EF [34]:

$$\mathrm{EF} = G_{\mathrm{net}} \left(\frac{r_{\mathrm{foc}}}{r_{\mathrm{tip}}} \right)^2$$

where r_{foc} and r_{tip} are respectively the focalization and tip radius. From this relation, we deduce EF of early 2000s experiments was about 10^2 to 10^4, which was still far from theoretical prediction.

The Raman scattering enhancement results from both excitation and emission processes induced by the tip apex, which acts as a nanoantenna [35, 36]. Scattering from the tip region is a two-stage process. First, the tip apex region is excited, then this region radiates. The overall enhancement of the scattered light intensity is given by:

$$I_{\mathrm{scat}} = E^2_{\mathrm{inc}}\left(\lambda_{\mathrm{inc}}\right) \cdot E^2_{\mathrm{rad}}\left(\lambda_{\mathrm{rad}}\right)$$

where $E^2_{\mathrm{inc}}(\omega_{\mathrm{inc}})$ and $E^2_{\mathrm{rad}}(\omega_{\mathrm{rad}})$ are respectively the incident and radiated electric field. In the elastic scattering, the incident and radiative processes are symmetric and occur at the same wavelength and electric field. Thus the intensity of elastically scattered light follows the famous "power of four" law. Theoretical predictions indicate field enhancement higher than 10^3 (see Section "Simulation and Modeling"), which should lead to more than 10^{12} in EF. But in the case of Raman scattering (inelastic) we must consider the red or blueshift of the radiation for few nm to about 100 nm. For a sharp resonance (e.g., with small silver nanoparticles), the difference between $E^2_{\mathrm{inc}}(\omega_{\mathrm{inc}})$ and $E^2_{\mathrm{rad}}(\omega_{\mathrm{rad}})$ occurring in a Raman process can be high as a consequence, the power four law might not be satisfied. This is one of the key parameters for achieving the optimal condition for TERS measurements. All the efforts consist in tailoring the tip resonance as the function of laser excitation and Raman shift.

Electromagnetic versus chemical enhancement: Both physical and chemical enhancements have been extensively investigated in SERS [37–46] and their relative contribution is still under discussion.

Similarly, these two effects can be expected in TERS. The insight into the chemical mechanism is still limited, but it is very important to reveal the nature of this chemical enhancement in TERS in a direct visual way. Sun et al. [47] calculated the charge density in four kinds of charge transfer: tip to molecule—surface to molecule—tip to surface to molecule simultaneously—tunneling charge transfer between tip and surface for different molecules and clusters tip. The static electronic polarizability components calculated in TERS are larger than those in SERS [47]. The EFs calculated for several pyrazine modes vary from 2 to 140.

Modification of Raman modes: Compared to normal Raman spectroscopy, spectra obtained by TERS show differences in band width and position, in relative intensities, time fluctuation, and additionally, non-Raman active modes can also be observed. The band narrowing is usually observed in TERS. It is generally assigned to the small number of molecules involved in the scattering process and the related lower density of state. The shift of Raman modes results from the molecule binding to the metallic structure (tip or conductive substrate) as it might be observed in SERS (or even in conventional Raman spectroscopy for chemically bounded molecules on surfaces), but this effect may be stronger because of the nondilution of the response in a large number of molecules. For a given molecule, the intensity of a Raman mode depends on the polarizability tensor of this mode, which gives the so-called "selection rules." As a consequence, the intensity of a Raman mode depends on the orientation of the incident electric field, the orientation of the molecule, and the orientation of the analyzer eventually used. The difference between conventional Raman and TER spectra comes from the polarization state [48, 49] of the confined and enhanced field underneath the tip [50]. For example, Blum et al. observed that contrary to the conventional Raman spectra of proteins, the amide I mode, which is widely used to identify secondary structure motifs of proteins, is not visible in GM-TERS [51]. The field lines orientation strongly depends on the morphology of the apex, which can vary drastically not only from tip to tip but also during lifetime of the tip. Molecules can also change in the orientation relative to the polarization with the tip interaction or be altered under the highly confined enhanced field.

Bleaching behavior: TERS generally also show a bleaching behavior. The signal decays under illumination at a rate that depends on the incident laser intensity, the Raman intensity. A complete description of different bleaching behaviors is given by Pettinger et al. [52]. The ratio of the bleaching time constants varies as the square of the field enhancement. As a consequence, this method can be used to evaluate the near-field volume from the measure of both EF and field enhancement.

Field gradient: In TERS configuration, the molecules are localized in a strong electric field gradient under the tip or in the gap between the tip and the metallic substrate. As a consequence, the gradient significantly alters the selection rules and activates IR-active modes [53, 54]. This mechanism has been described in details by Ayars et al. [55].

11.3 Instrumentation

A TERS setup consists in the creation of a hotspot highly localized on the probe molecule. The instrumentation design lies in the nanoscale control of the environment of the probe molecule: light excitation (positioning, polarization, incident angle, wavelength, etc.), tip (positioning, material, shape, etc.), and substrate (material, morphology, etc.). The configuration firstly depends on the exact application (transparent or nontransparent sample) and its environment (air, liquid, inert atmosphere, UHV [56, 57], etc.). The tip sample distance control is one of the key parameter for TERS detection. It can be achieved differently by using scanning tunneling microscopy (STM) or AFM.

11.3.1 *Configuration*

The first TERS experiments have been performed on AFM-based setup with inverted microscope used in laboratory environment

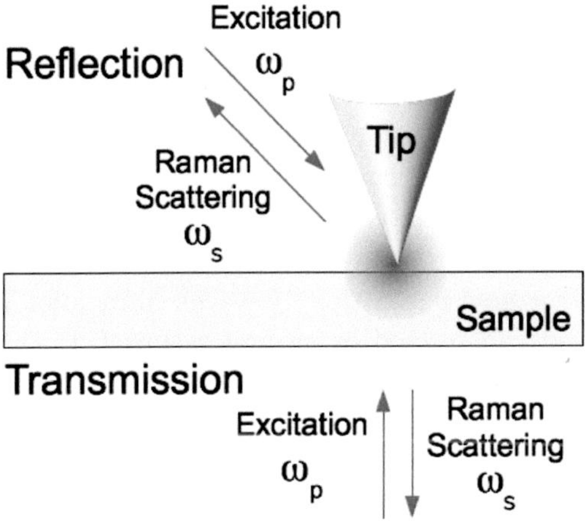

Figure 11.3 Reflection and transmission TERS configurations.

with nano-objects deposited on transparent substrates. Up to now, this is the most used configuration (transmission TERS configuration, Fig. 11.3). In the early 2000s, several apertureless optical configurations have been tested, for example, a total internal reflection (TIR) illumination/collection scheme through the same oil immersion objective was developed by Hayazawa et al. [58–60]. Bulgarevich and Futamata [61] reached net gain $G_{\text{net}} \sim 10^3$ by using an epi-illumination/collection (illumination and detection from one side of the sample) optics and semi-contact tapping mode on a diamond particle.

For opaque samples or samples deposited on nontransparent substrates that cannot be accessed by the laser from below, Sun and Shen [62] first attempted top illumination. Then Anderson and Pike [63] have developed an upright TERS configuration with side/top illumination and tested on optically trapped diamond particle. Since then, various molecular [64], polymeric [65], semiconductors [66–68], and carbon nanotubes (CNTs) [69] have been studied by side/top illumination. This configuration is now also used in "GM" excitation [70] (reflection TERS configuration, Fig. 11.3).

The STM makes possible the matching of the nanocavity plasmon resonance to the molecular transition with extremely precise tuning capability. Zhang et al. [22] suggested that ultrahigh resolution imaging results from the highly confined and broadband nature of the nanocavity plasmon field in the tunneling gap [22].

More recently, high-vacuum (HV-TERS) experiments have been performed for a better control of environment of molecular monolayers or sub-monolayers [29].

Most of the TERS setups use visible or near-infrared excitations on silver and/or gold-based tips. However, ultraviolet TERS (UV-TERS) has been constructed by several groups [71, 72] using an aluminum tip on silicon and CNTs film. A 20-nm spatial resolution can also be achieved by this technique.

For the alignment procedure [73] generally the focused laser beam is scanned over the tip, while spectrum is collected to give a spectral intensity map. In a second step, the laser is aligned on the higher enhancement and fixed at that position. Then, only the sample is moved by XYZ piezoelectric scanner.

11.3.2 *Tip Preparation*

The preparation of the tip depends on the mounting system. Metallic nanowires are used in STM and tuning fork–based AFM, whereas coated tip with various and complex geometry can be applied to AFM setup. AFM is more versatile and allows the use of different types of tip (metallic, semiconducting, or insulating) depending on the application. However, in STM, the tunneling current established through the sample implies a metallic tip and conductive substrates. The current allows studying a large variety of single-molecule photochemical phenomena.

Production of gold tips by electrochemical etching has been extensively described [74–76]. For example, the method proposed by Billot et al. [77] used a two-step etching of a gold wire 250 μm in diameter (anode) with a 500 μm diameter platinum wire cathode. A first step is operated in a mixed solution composed by chlorohydric and sulfuric acids followed by a second step with slow etching rate in a water/methanol mixture to induce a controlled roughness at the apex. A radius of curvature around 20 nm can be achieved by this method. By combination with the movement of gold wire while etching, it is also possible to produce different geometries: sharp smooth,

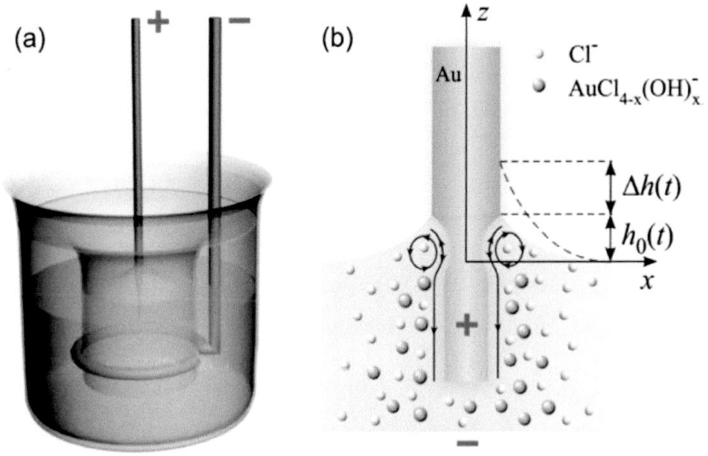

Figure 11.4 (a) Electrochemical cell, (b) an etching mechanism at an electric field–driven liquid meniscus. Reproduced with permission from Ref. [78]. Copyright © 2013 IOP Publishing. All rights reserved.

randomly rough (quasifractal) tapers, spheroids, edges, dimers, and patterned conical antennas with quasiuniformly spaced outgrowths (Figs. 11.4 and 11.5) [78]. Electrochemical etching is also used to produce a variety of morphologies of silver tips [79–85]. Deposition of Ag onto a tungsten tip is another approach to make Ag-coated tips [86]. Silver coating can be obtained by using a chemical mirror reaction leading to the formation of particles of 40 nm at the apex [87].

More recently, Valmalette et al. (2014) suggested that SWCNTs functionalized by silver NPs could be used as local probes to simultaneously provide nanoscale resolution for surface imaging and strong localized field enhancement [88]. Theoretical predictions show that the electromagnetic field strongly depends on how the NPs are organized on the CNTs (dimers, trimers, etc. of silver NPs). The predicted enhancement in the Raman signal of absorbed molecules could be as high as 106 on arrays of carbon nanotubes filled with silver cylinders [89]. The method proposed by Valmalette et al. is based on UV photo-reduction of a silver nitrate solution in the presence of sodium dodecyl sulfate (SDS)- functionalized SWCNT. The obtained nanostructure consists in 4-6 nm silver NPs chains or helix-wrapped SWCNT (Fig. 11.6).

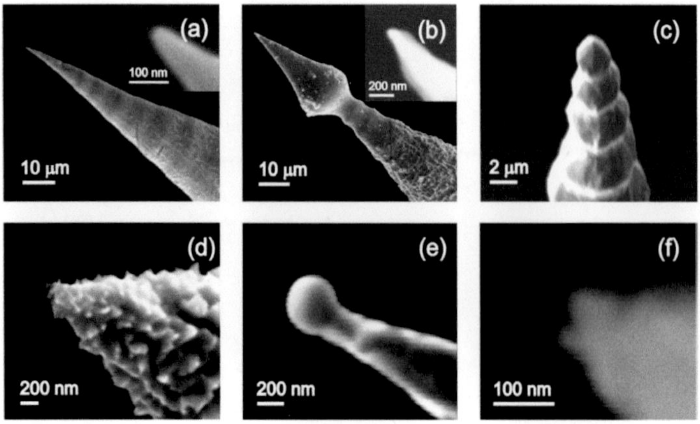

Figure 11.5 Electrochemically etched plasmonic optical antennas. Reproduced with permission from Ref. [78]. Copyright © 2013 IOP Publishing. All rights reserved.

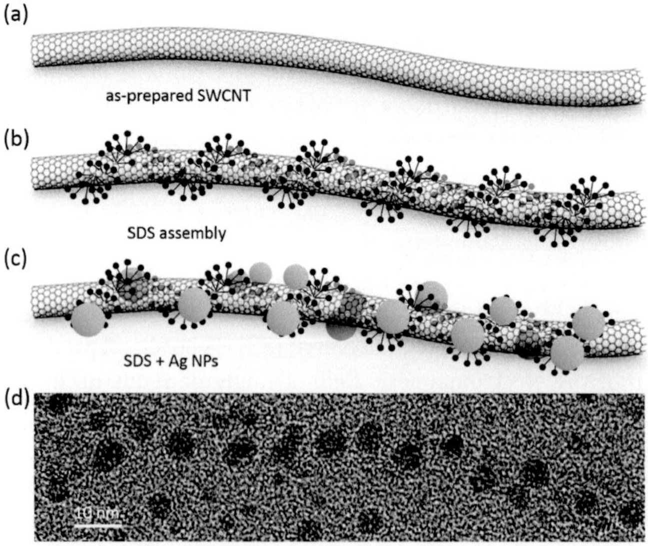

Figure 11.6 Schematic representation for Ag NPs created from a template on SDS-wrapped SWCNT bundles. Reprinted from Ref. [88], Copyright © 2014, Nature Publishing Group.

Dielectric coating: Loss of initial enhancement of the Raman signal is usually observed due to a change in the optical properties of the tip. A dielectric protective layer can be added on the tip to prevent contamination and/or oxidation of the tip. For example, Barrios et al. [90] studied the effect of a 1-nm thick layer of alumina over the surface of the metallic cone. The protection of the tip increases drastically the lifetime and reproducibility of TERS measurement. In addition, a dielectric coating layer is required to tune the plasmonic response of the tip. Indeed, discontinuity of a different material composition is necessary to increase the field enhancement [91] (see "Simulation and Modelling" section). Tuning of the discontinuity dimension is needed to maximize the resonance. A SiO_x layer on TERS tips was first investigated experimentally [92] and theoretically [93]. But from the work done by Murray et al. in 1980s showing the excellent protection properties of alumina coatings [94, 95] several groups started to increase the density of alumina films. Atomic layer deposition of an ultrathin layer of alumina has been show to improve the lifetime of silver SERS substrates [96], but the highest impermeability is achieved by a highly dense alumina coating obtained from atmospheric oxidation of aluminum layer [97]. However, the addition of a thin protective layer to the TERS tip to reduce oxidation and mechanical wear may decrease the intensity of the electric field by 50% [98] (Fig. 11.7).

Other routes are actively followed by different groups to improve the tip-induced conversion of photonic to plasmonic energy. Bao

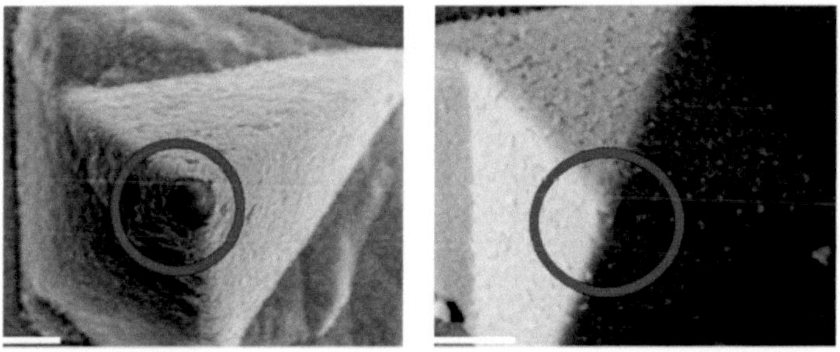

Figure 11.7 SEM images of unprotected (left) and Al_2O_3 unprotected (right) silver metallized silicon nitride tips after six scans on a standard silicon grid. Reprinted with permission from Ref. [96], Copyright 2010, American Chemical Society.

et al. [99] proposed a "campanile-probe" to efficiently merge a broadband field enhancement with bi-directional coupling of far- to near-field electromagnetic modes. They calculated the field confinement, enhancement, and polarization near the apex of the "campanile-probe" in both vacuum and tip-substrate GM. They demonstrated that more than 90% of the light radiated by the emitter is captured by this "campanile-probe". A multilayered nano-prism vertex tips for TERS has been also investigated recently [100].

The complete control of light manipulation requires a 3D control of the architecture which can be achieved only by fabrication methods deriving from microelectronics industry [101]. De Angelis et al. [102] have developed photonic–plasmonic device architectures based on the adiabatic generation of SPPs fully compatible with TERS setup, which could be another promising alternative for tip production.

In TERS experiments, the question of selection criterion for the tips is of major importance to guaranty a high sensitivity. The spectral response of the tip is achieved by the excitation of the optical response of the tip apex by using evanescent field illumination. Neacsu et al. [103] and Mehtani et al. [104] used a quartz prism for the excitation and a long working distance objective in the side-collection mode to collect the scattered light from tungsten and gold tips Figs. 11.8 and 11.9. The collection efficiency was improved by the use of total internal reflection fluorescence (TIRF) objective [105]. In this compact optical configuration, the backward scattered light is collected within the high opening angle, a circular diaphragm is inserted to block the directly reflected light and the scattered light is analyzed by a spectrometer.

11.3.3 *Light Polarization and Tip-Induced Depolarization*

One of the weak points of the initial TERS studies was the polarization of the laser light perpendicular to the tip axis (which is less efficient for surface plasmon excitations). A polarization component of the incident light along the tip axis induces a strong surface charge density at the sharp apex of the metallic tip which is a prerequisite for local enhancements [106]. In the later experiments, the polarization played a decisive role in the formation of a hotspot

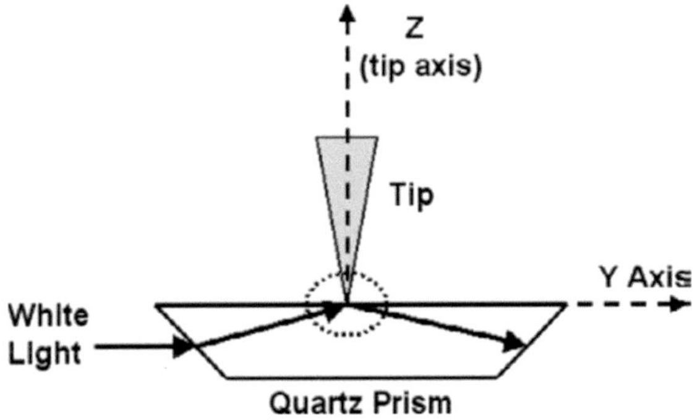

Figure 11.8 Schematic diagram of the total internal reflection setup used to measure optical properties of the tips. Reproduced with permission from Ref. [104]. Copyright © 2006 IOP Publishing. All rights reserved.

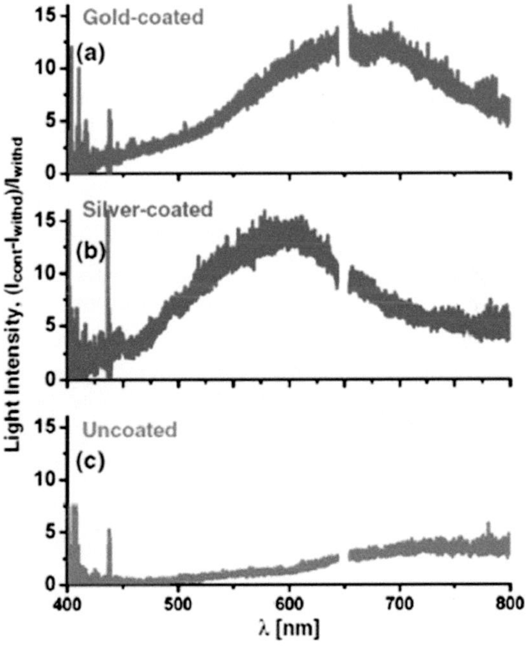

Figure 11.9 Optical scattering spectra for (a) gold-coated tungsten, (b) silver-coated tungsten, (c) tungsten tips. The gap in the range ~645–655 nm is due to the notch filter for the 647 nm laser. Reproduced with permission from Ref. [78]. Copyright © 2013 IOP Publishing. All rights reserved.

that can be used in TERS (see "Simulation and Modelling" section). For example, the simulations performed by Hayazawa et al. [107] on s- and p-polarized light transverse mode (TM) at $\lambda = 488$ nm emanating from the substrate in TIR configuration demonstrated an EF of 80 within 30 nm diameter of the area (diameter of tip apex) only by the excitation of p-polarization but not by that of s-polarization [108].

Linearly polarized source is commonly used for TERS; however, radially polarized modes are of great interest to conduct TERS experiments in axial illumination geometry [109–111]. Laser beams with radial polarization provide unique focusing properties including a strong longitudinal electric field component generated at the focal point [112]. While side illumination is preferable with linear input polarization orientated along tip axis, radial polarization promotes the strongest confinement on the electrical field. But this confinement strongly depends on the tip–substrate combination [113] (see "Simulation and Modeling" section).

The tip-induced depolarization on TERS experiment has been investigated on semiconductors single crystal [114, 115] Si(001) and GaAs(001). In GaAs, after far-field scattering minimization, with a proper choice of the incident polarization, the tip approach induces a different intensity increase for longitudinal optical (LO) and transverse optical (TO) modes. The tip plays a dual role in depolarizing the incident field, which boosts the Raman scattering and in deflecting a part of the laser beam. The measure of the differential LO–TO signal increase can be used as a fingerprint of tip-induced depolarization effect (Fig. 11.10).

11.3.4 *Substrate and Gap Mode*

The field enhancement in TERS does not only depend on the nature of the tip and the polarization of the excitation beam but also on the nature of the substrate. Raman enhancement can be increased by using a noble metal substrate. The excitations in the gap between the noble metal tip and the substrate are called "GM plasmons." In this configuration, molecules are localized in the gap between the metallic tip and the metallic nanoplate. A resonant optical cavity is thus created allowing the interaction between localized surface

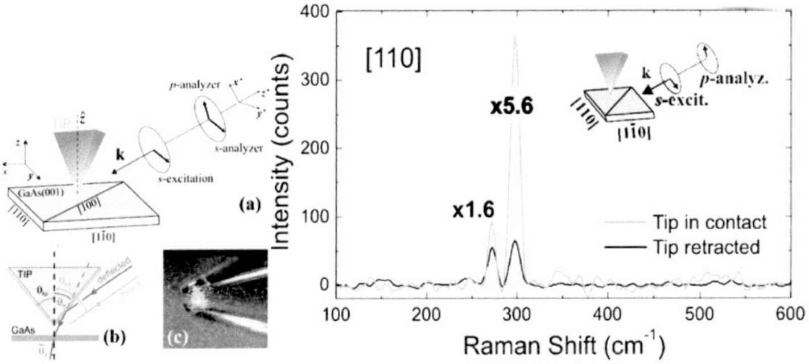

Figure 11.10 Sketch of the experimental setup with s-excitation and s- or p-analyzer (left) and TERS of GaAs: tip retracted (black) and tip in contact (green). Reprinted with permission from Ref. [114]. Copyright 2010, AIP Publishing LLC.

plasmons (LSPs) from the tip and surface plasmon polaritons (SPPs) from the substrate. A strong field enhancement is achieved leading to large enhancement of the Raman scattering [116, 117] up to 10^6–10^7. The substrates preferentially used are Au(111) substrates or an evaporated gold or silver films. For instance, immobilization of biomolecule (cystine) by absorption on thin crystalline gold plates provides evidence of local variation of cystine adsorption [118].

GM-TERS can be used to investigate the interaction between multilayer graphene and gap modes as well as localized emission of defects and inner unknown molecules enhanced by this coupling [119]. For back-illumination configuration (both excitation and scattered light are transmitted through the substrate), transparent gold of silver substrates may be used to achieve a considerable increase in sensitivity [120]. Triangular or hexagonal shaped noble metal nanoplates having Au(111) facets on their surfaces are now used since calculations demonstrate local field enhancement [121] strong enough to achieve the single-molecule detection level on biomolecules [122–125]. The net contribution of GM plasmons to the total enhancement spectrum can be achieved by the difference between the spectrum measured with tip located on a GM plasmon-active surface from the spectrum obtained with the tip on a GM plasmon inactive surface. Kazemi et al. [126] have demonstrated that heterometallic junction formed by a silver tip and a gold thin film appears to be very efficient (see "Simulation and Modeling" section).

Not only Au and Ag but also a wider variety of metallic substrates (Ni, Cu, etc.) could also be used in GM-TERS in the near future to increase the applicability of TERS and evolve this technique one step further toward use in standard [127].

11.3.5 *Ultrahigh Vacuum TERS (UHV-TERS)*

Van Duyne group [128] has developed a combination of molecular-resolution UHV-STM imaging with the detailed chemical information content of TERS. They demonstrated for the first time that the ability to probe the interaction between large polyatomic molecular adsorbates and specific binding sites on surface. Figure 11.11 shows copper phthalocyanine (CuPc) adlayers on Ag(111).

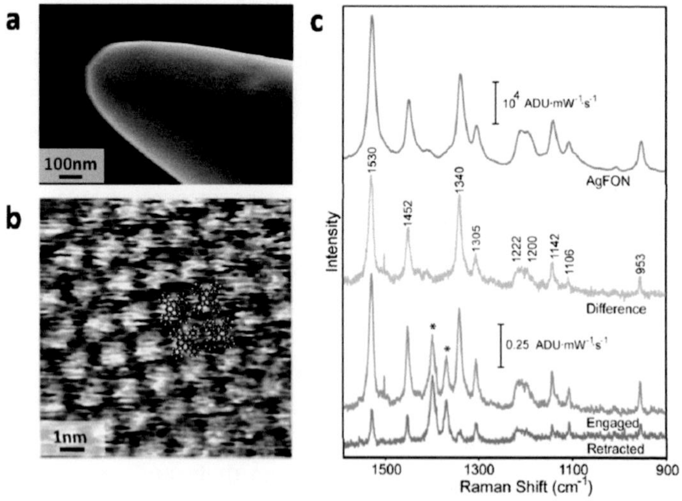

Figure 11.11 (a) SEM image of a silver tip showing a radius of curvature at the apex of 130 nm. (b) Molecular resolution topographic STM image (10 nm × 10 nm) of a CuPc adlayer on Ag(111) with the tip illuminated by the continuous wave (CW) laser. (c) UHV-TERS spectra of a CuPc adlayer on Ag(111) and SERS spectrum of a CuPc adlayer on AgFON. The asterisks (*) indicate peaks at 1374 and 1403 cm^{-1} that result from the sapphire windows. Reprinted with permission from Ref. [128], Copyright 2012, American Chemical Society.

More recently, Zhang et al. [129] have demonstrated unprecedented submolecular spatial and spectroscopic resolution with low-temperature UHV-TERS. They used the broadband nature of the nanocavity plasmon field between tip and substrate to obtained on-resonance very sharp Raman modes of a single H_2TBPP molecule on Ag(111) (Fig. 11.12).

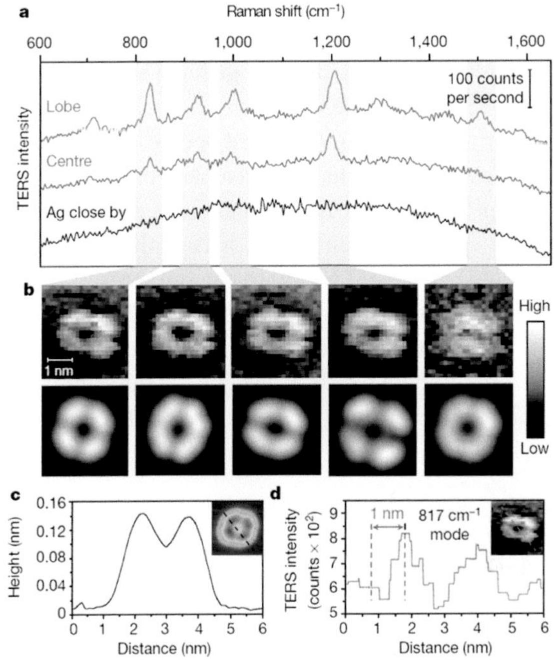

Figure 11.12 TERS mapping of a single H2TBPP molecule on Ag(111). (a) Representative single-molecule TERS spectra on the lobe (red) and center (blue) of a flat-lying molecule on Ag(111). The TERS spectrum on the bare Ag about 1 nm away from the molecule is also shown, in black (120 mV, 1 nA, 3 s). (b) The top panels show experimental TERS mapping of a single molecule for different Raman peaks (23323, 0.16 nm per pixel), processed from all individual TERS spectra acquired at each pixel (120 mV, 1 nA, 0.3 s; image size: 3.633.6 nm^2). The bottom panels show the theoretical simulation of the TERS mapping. (c) Height profile of a line trace in the inset STM topograph (1 V, 20 pA). (d) TERS intensity profile of the same line trace for the inset Raman map associated with the 817 cm^{-1} Raman peak, integrated over 800 to 852 cm^{-1}. Reprinted with permission from Ref. [128], Copyright 2012, American Chemical Society.

11.3.6 *Other Tip-Enhanced Spectroscopy*

TERS was one of the first tip-enhanced spectroscopy enabling to near-field image with extremely high spatial resolution [130], but other techniques have followed using other light-tip interactions (coherent anti-Stokes Raman scattering (CARS), infrared (IR), fluorescence): tip-enhanced CARS (TE-CARS) has been developed by Ichimura et al. [131] with a subwavelength resolution ($<\lambda/10$). Detailed review on the use of this technique for bioanalysis has been recently published by this group [132]. Keilmann and Hillenbrand [133] reported subwavelength microscopy in the mid-infrared region by tip-enhanced infrared spectroscopy (TEIRS) with a resolution of less than 20 nm, and Raschke et al. [134] obtained a resolution better than 10 nm on polymer films with this technique. Hartschuh et al. [135] demonstrated a two-photon tip-enhanced fluorescence (TEF) image of CNTs with the same resolution.

Recently, a first step for calibration procedure with internal standard has been proposed, based on the measure of the intensity of adenine and thiophenol bands. However, the determination of the absolute amount of molecule on the surface is not yet demonstrated [136].

11.4 Simulation and Modeling

The optical properties of different tip–substrate configurations have been simulated by 2D and 3D finite element models. After the 2000s' simulations [137–140], the number of papers published in theoretical prediction increased drastically the last few years and become essential tool to push-up the limit of the local field enhancement before testing new combinations.

The effective EF results from the convolution between the spatial distributions of the enhancement induced by the tip (or by the tip–substrate gap) and the spatial distribution of the molecules. A difference of 1 nm in position of the molecule underneath the apex can change the EF by several orders of magnitude [141, 142].

Several studies have modeled optical scattering from the isolated tip apex region. In 2000, Novotny [143] studied numerical solutions of the electromagnetic response of an isolated tip by using the discrete dipole approximation (DDA) approach. The local field in the apex region of a metallic tip strongly depends on the orientation of the polarization **E** and propagation vector **k** of the incident waves. Without substrate, the local field is maximum with polarization parallel to the tip, whereas only a very weak field enhancement is observed with polarization perpendicular to the tip axis. The next year, Martin et al. [144] calculated the EF of simple structures like cylinders, spheres, and ellipsoids by finite-element time-domain. These calculations revealed that phase retardation effects substantially reduce this factor for plasmon resonance when the size of the structure is larger than $\lambda/4$. They also demonstrated that discontinuity of different material composition on the tip apex can provide strong field enhancement.

The finite-difference time-domain (FDTD) modeling of the electromagnetic is a widely used method for solving Maxwell's equations in complex geometries [145–150]. The FDTD approach is applied to estimate the influence of parameters such as the material of the tip, the tip geometry [151, 152], the morphology of nanoplates [153], and the tip–sample distance for a given excitation wavelength [153–155]. The effect of the substrate on the EM enhancement have been extensively studied [157, 158] and references cited in, for example, Kazemi et al. demonstrated the strong field enhancement in GM induced by silver tip and gold substrate. Simulations show that not only gold but also other metals [159] are well suitable for TERS experiments. This result is of interest when the sample has unwanted interaction with gold or is difficult to transfer onto a gold substrate.

The impact of the incident laser polarization on the field enhancement has also been extensively studied by FDTD. The comparison of radially and linearly polarized source in the presence of a metallic substrate show strong differences on field intensity distribution [160] (Fig. 11.13).

In addition, simulation is a useful predictive tool for studying the influence of coating layer. This layer plays a major role in the protection to the tip (to prevent its oxidation and improve mechanical resistance, see section "Instrumentation") but also

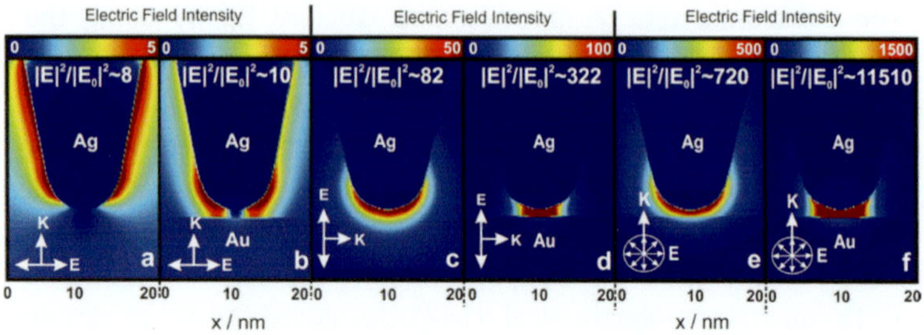

Figure 11.13 Electric field distribution at the 10-nm apex of a silver tip illuminated at 532 nm with 1 nm separation from a gold substrate [160].

allows the resonance tuning by changing the surrounding refractive index of metallic tip, layer, or nanoparticles.

Since the last years, the FDTD reliability to solve nano-optics problems is submitted to controversy. Kaminski et al. [161] even concluded that this method would not converge to the true solution because of the staircasing approximation in curved geometry even with negligible voxel size. For very small radius curvature, latest discontinuous Galerkin (DG) variant of the finite element methods has shown advantages with mesh generation with element sizes varying over several orders of magnitudes [162]. The ultralarge linear systems generated by this method, with several tens of millions of degrees of freedom, are a current challenge that should be overcome in a near future. At last, even if the charge transfer can be now visualized in clusters/molecules models [163], the calculation from the density functional theory (DFT) of the contribution of the so-called "chemical effect" will be the other step in the forthcoming progress in EF simulation in TERS.

11.5 Applications of TERS

TERS started in the early 2000s with dye molecules and CNTs due to their high Raman cross section under resonance. Since then it has been successfully used in a large variety of applications [164], nanoelectronics, biotechnology, materials, and environment and still has a great potential in many different fields. While a few configurations are more suitable for huge enhancement, TERS can also operate with molecules deposited on any substrate type (metal, semiconductor, and insulator) and requires no special sample preparation.

Applications of TERS in materials science:

- Carbon compounds: CNT bundles [165] or individual CNT [166–168], graphene [169–171], and 6H-SiC with graphene adlayers [172]. TERS has been successfully used for imaging the locally induced strain distribution in CNTs [173].
- Other inorganic compounds: semiconductors (InN [174], Si(001) [175], GaAs(001) [176]), oxides ($BaTiO_3$ [177], TiO_2 [178]), and silicone [179].
- Surfaces and interfaces: local interactions at the interface of styrene-butadiene rubber/multiwalled CNT nanocomposites [180], surface bonding on silicon [181], single surface states in inorganic semiconductors, such as Nb-doped TiO_2 [182]. Interaction between large polyatomic molecular adsorbates and specific binding sites on solid surfaces [183].
- Catalysts and heterogeneous catalytic reactions [184].
- Stress measurement [185].

Applications of TERS in molecular electronics:

- Molecular monolayers: copper porphyrin [186], copper phthalocyanine (CuPc) [187], thiol compounds [188, 189] (e.g., to distinguish the isomeric thiol regions).
- Singlemolecule TERS imaging [190].
- Electron transport processes in molecular junction, for example in the local measurement of single-molecule conductance by using "fishing mode" TERS [191].
- Electron transport from donor molecules to acceptor molecules in organic photovoltaic is deduced from TERS/photoluminescence (PL) measurements of correlation between polymer blends segregation and PL quenching efficiency [192–195].

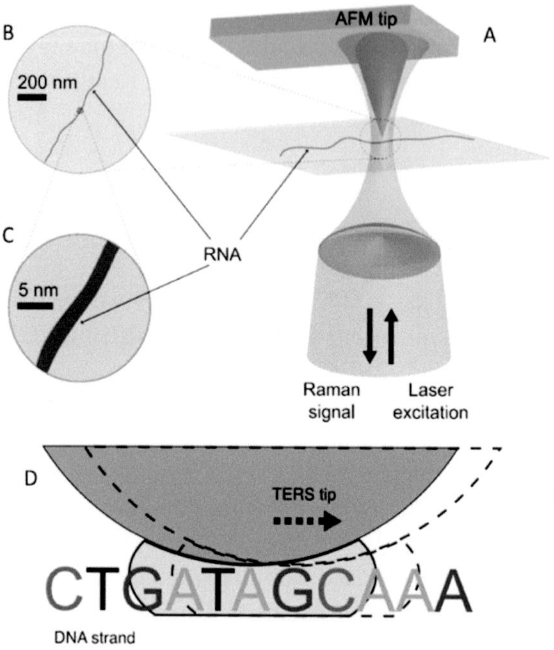

Figure 11.14 Direct base-sequencing (A) schematic and (D) procedure using TERS. (B) Magnified region roughly equal to the focal area of the laser beam spot. (C) Further magnified region approximating the enhancing region of the tip. (D) Sequence information can be obtained by laterally shifting the probe in intervals of one base-to-base distance. The yellow area refers to the enhancing region at the first tip position. Reprinted with permission from Ref. [201], Copyright 2013, American Chemical Society.

Application in the nanoscale control of chemical reaction

- In situ plasmon-driven chemical reactions under high vacuum [196].
- Monitoring and characterization of the catalytic processes [197, 198].

Applications of TERS in biotechnology

One of the first experiments on upright configuration was used to obtain a chemical variation in fatty acid esters and protein on and off the ridges of the ommatidial surfaces of dipteran compound eyes [199]. But since the recent development of environmental and highly sensitive TERS based on GM excitation by using dedicated new substrate (transparent nanoplates), nanoscale characterization biological system is now possible. TERS provides chemical and structural information on biological mechanisms leading to new fields of research in biotechnology Fig. 11.14 [200, 201].

- Biological units: proteins [202], peptides [203, 204], single RNA strands [205], cytochrome [206], lipid domains [207], alginate biofilms [208], insulin amyloid fibrils [209–211].
- Biological mechanisms: Interactions of proteins with biocrystals [212], protein–ligand binding [213], in vivo characterization of protein uptake by yeast cell envelope [214].

11.6 Conclusion

Since the first evidence of apertureless near-field Raman scattering, the 14 years of TERS has demonstrated its applicability to the characterization of a wide diversity of systems as single nano-objects (CNTs, single-molecules, molecular sheets, RNA strands, etc.), materials (organic, inorganic, hybrid), and biological systems. Although the theoretical description of the light scattered by a molecule localized in the gap between a substrate and a tip is not fully achieved, TERS allowed the description at the nanoscale of new mechanisms in the fields of physics, chemistry, and biology. This was not only due to the high spatial resolution but also to the extremely high sensitivity and to the ability to probe new interaction types thanks to the simultaneous contribution of other physical effects among them: tunneling current, PL, and light-induced electron transfers.

In a near future, TERS can be still possibly improved by new insight:

(1) Substrates: The use of new substrate types in GM excitation; among them, strong field enhancement has been demonstrated for graphene and this substrate is now used in the so-called grapheneenhanced Raman spectroscopy (GERS). If we are able to increase its manipulability, to take advantage of its doping ability and covalent surface modification, graphene could be a

good candidate for what is called graphene tipenhanced Raman spectroscopy (GTERS).

(2) Theoretical calculations: Field enhancement has been shown to depend drastically on the optical cavity response. Theoretical calculations will be soon more reliable for nano-optics, for example by using DG variant of the finite element methods. This improvement is required to provide a more accurate description of the λ- and space-dependent response of the nanocavity. The charge transfer and the shift of the resonance induced by the probed molecule must also be considered to improve the EF calculation reliability.

(3) Laser excitation tuning to fit the nanocavity plasmon resonance.

References

1. Saar, B. G., Freudiger, C. W., Reichman, Stanley, C. M., Holtom, G. R., Xie, X. S. (2010). Video-rate molecular imaging in vivo with stimulated Raman scattering, *Science*, **330** (6009), pp. 1368–1370.
2. Wei, L., Yu, Y., Shen, Y., Wang, M. C., Mina, W. (2013). Vibrational imaging of newly synthesized proteins in live cells by stimulated Raman scattering microscopy, *Proc. Natl. Acad. Sci.*, **110** (28), pp. 11226–11231.
3. Stipe, B. C., Rezai, M. A., Ho, W. (1998). *Science* **280** (1732), p. 217; Pascual, J. I., Gómez-Herrero, J., Sánchez-Portal, D. et al. (2002). *J. Chem. Phys.*, **117** (9531), p. 217.
4. Otto, A., Mrozek, I., Grabhorn, H. and Akemann, W. (1992). Surface enhanced Raman scattering, *J. Phys.: Condens. Matter*, **4** (5) p. 1143.
5. Pettinger, B., Picardi, G., Schuster, R. and Ertl, G. (2002). Surface-enhanced and STM-tip-enhanced Raman spectroscopy at metal surfaces, *Single Mol.*, **5** (5–6), pp. 285–294.
6. Wessel, J. (1985). *J. Opt. Soc. Am. B*, **2**, pp. 1538–1541.
7. Pohl, D. W., Denk, W. and Lanz, M. (1984). Optical stethoscopy: Image recording with resolution λ/20, *Appl. Phys. Lett.*, **44**(7), p. 651.
8. Lewis, A., Isaacson, M., Harootunian, A. and Murray, A. (1984). Development of a 500 Å spatial resolution light microscope. I. Light is efficiently transmitted through λ/16 diameter apertures, *Ultramicroscopy*, **13** (3), p. 227.

9. Nie, S. and Emory, S. R. (1997). *Science*, **275**, p. 21.
10. Kneipp, K., Wang, Y., Kneipp, H., Perelman, L. T., Itzkan, I., Dasari, R. R. and Feld, M. S. (1997). *Phys. Rev. Lett.*, **78**, pp. 1667–1670.
11. Deckert, V., Zeisel, D., Zenobi, R. and Vo-Dinh, T. (1998). Near-field surface-enhanced Raman imaging of dye-labeled DNA with 100-nm resolution, *Anal. Chem.*, **70**, pp. 2646–2650.
12. Zeisel, D., Deckert, V., Zenobi, R. and Vo-Dinh, T. (1998). Near-field surface enhanced Raman spectroscopy of dye molecules adsorbed on silver island films, *Chem. Phys. Lett.*, **283**, pp. 381–385.
13. Stöckle, R. M., Suh, Y. D., Deckert, V. and Zenobi, R. (2000). *Chem. Phys. Lett.*, **318**, p. 131.
14. Anderson, M. S. (2000). Locally enhanced Raman spectroscopy with an atomic force microscope, *Appl. Phys. Lett.*, **76**, pp. 3130–3132.
15. Hayazawa, N., Inouye, Y., Sekkat, Z. and Kawata, S. (2000). Metallized tip amplification of near-field Raman scattering, *Opt. Commun.*, **183**, pp. 333–336.
16. Stöckle, R. M., Suh, Y. D., Deckert, V. and Zenobi, R. (2000). *Chem. Phys. Lett.*, **318**, p. 131.
17. Anderson, M. S. (2000). Locally enhanced Raman spectroscopy with an atomic force microscope, *Appl. Phys. Lett.*, **76**, pp. 3130–3132.
18. Hayazawa, N., Inouye, Y., Sekkat, Z. and Kawata, S. (2000). Metallized tip amplification of near-field Raman scattering, *Opt. Commun.*, **183**, pp. 333–336.
19. Neacsu, C. C., Dreyer, J., Behr, N. and Raschke, M. B. (2006). Scanning-probe Raman spectroscopy with single-molecule sensitivity, *Phys. Rev. B*, **73**, p. 193406.
20. Zhang, W. H., Yeo, B. S., Schmid, T. and Zenobi, R. (2007). Single molecule tip-enhanced Raman spectroscopy with silver tips, *J. Phys. Chem. C*, **111** (4), p. 1733.
21. Sonntag, M. D. et al. (2012). Single-molecule tip-enhanced Raman spectroscopy, *J. Phys. Chem. C*, **116**, pp. 478–483.
22. Zhang, R., Zhang, Y., Dong, Z. C., Jiang, S., Zhang, C., Chen, L. G., Zhang, L., Liao, Y., Aizpurua, J., Luo, Y., Yang, J. L. and Hou, J. G. (2013). Chemical mapping of a single molecule by plasmon-enhanced Raman scattering, *Nature*, **498**, pp. 82–86.
23. Kawata, S., Inouye, Y. and Verma P. (2009). Plasmonics for near-field nano-imaging and superlensing, *Nat. Photon.*, **3** (7), pp. 388–394.

24. Opilik, L. et al (2013). Modern Raman imaging: Vibrational spectroscopy on the micrometer and nanometer scales, *Ann. Rev. Analyt. Chem.*, **6**, pp. 379–398.
25. Schmid, T., Opilik, L., Blum, C. and Zenobi, R. (2013). Nanoscale chemical imaging using tip-enhanced Raman spectroscopy: A critical review, *Angew. Chem. Intern. Ed.*, **52** (23), pp. 5940–5954.
26. Kawata, S. (2013). Plasmonics for nanoimaging and nanospectroscopy, *Appl. Spectrosc.*, **67** (2), pp. 117–125.
27. Lucas, M. and Riedo, E. (2012). Combining scanning probe microscopy with optical spectroscopy for applications in biology and materials science, *Rev. Sc. Instrum.*, **83** (6), p. 061101.
28. Pettinger, B. (2010). Single-molecule surface- and tip-enhanced Raman spectroscopy, *Molecular Phys.*, **108** (16), 2039–2059.
29. Zhang, Z., Chen, L., Sheng, S., Sun, M., Zheng, H., Chen, K. and Xu, H. (2014). High-vacuum tip enhanced Raman spectroscopy, *Front Phys.*, **9**(11), 17–24.
30. Kneipp, K., Moskovits, M. and Kneipp, H. (2006). *Surface enhanced Raman spectroscopy, physics and applications* (Springer, Berlin, Heidelbeg).
31. Hayazawa, N., Tarun, A., Taguchi, A. and Furusawa, K. (2012). *Raman spectroscopy for nanomaterials characterization* ed. Kumar, Challa S. S. R. (Springer, Berlin, Heidelbeg), pp. 445–476.
32. Ru, E. L. and Etchegoin, P. (2009). *Principles of surface enhanced Raman spectroscopy and related plasmonic effects* (Elsevier, Amsterdam).
33. Inouye, Y., Verma, P., Ichimura, T. and Kawata, S. (2007). *Tip enhancement*, ed. Kawata, S. and Shalaev, V. M. (Elsevier, Amsterdam).
34. Markel, V. A., Shalaev, V. M., Zhang, P. et al. (1999). *Phys. Rev. B* **59**, p. 10903; Shalaev, V. M. (2002). *Optical nonlinearities of fractal composites*, ed. Shalaev, V. M. "Optical properties of nanostructured random media" (Springer, Berlin, Heidelberg) pp. 93–112; Moskovits, M. and Jeong, D. H. (2004). *Chem. Phys. Lett.*, **397**, p. 91.
35. Kneipp, K., Moskovits, M. and Kneipp, H. (2006). *Surface enhanced Raman spectroscopy, physics and applications* (Springer, Berlin, Heidelberg).
36. Novotny, L. and van Hulst, N. (2011). Antennas for light, *Nat Photonics*, **5**, pp. 83–90.
37. Novotny, L. (2008). Optical antennas tuned to pitch, *Nature*, **455** (7215), p. 887.

38. Otto, A., Mrozek I., Grabhorn, H. and Akemann, W. (1992). *J. Phys. Condens. Matter*, **4**, p. 1143.
39. Xu, H. X., Bjerneld, E. J., Kall, M. and Borjesson, L. (1999). *Phys. Rev. Lett.*, **83**, p. 4357.
40. Stöckle, R. M., Suh, Y. D., Deckert, V. and Zenobi, R. (2000). *Chem. Phys. Lett.*, **318**, p. 131.
41. Pettinger, B., Ren, B., Picardi, G., Schuster, R. and Ertl, G. (2004). *Phys. Rev. Lett.*, **92**, p. 096101.
42. Watanabe, H., Ishida, Y., Hayazawa, N., Inouye, Y. and Kawata, S. (2004). *Phys. Rev. B: Condens. Matter Mater. Phys.*, **69**, p. 155418.
43. Domke, K. F., Zhang, D. and Pettinger, B. (2006). *J. Am. Chem. Soc.*, **128**, p. 14721.
44. Yong, M. A., Dieringer, J. A. and Van Duyne, R. P. (2007). *Tip enhancement*, ed. Kawata, S. and Shalaev, V. M. (Elsevier, Amsterdam), Chapter 1.
45. Zhang, W., Cui, X., Yeo, B., Schmid, T., Hafner, C. and Zenobi, R. (2007). *Nano Lett.*, **7**, p. 1401.
46. Chen, J. N., Yang, W. S., Dick, K., Deppert, K., Xu, H. Q., Samuelson, L. and Xu, H. X. (2008). *Appl. Phys. Lett.*, **92**, p. 093110.
47. Sun, M. T., Fang, Y. R., Yang, Z. L. and Xu, H. X. (2009). Chemical and electromagnetic mechanisms of tip-enhanced Raman scattering, *Phys. Chem. Chem. Phys.*, **11** (41), p. 9412.
48. Gucciardi, P. G. and Valmalette, J. C. (2010). Different longitudinal optical—transverse optical mode amplification in tip enhanced Raman spectroscopy of GaAs(001), *Appl. Phys. Lett.*, **97**, p. 263104.
49. Merlen, A., Valmalette, J. C., Gucciardi, P. G., Lamy de la Chapelle, M., Frigout, A. and Ossikovski, R. (2009). Depolarization effects in tip-enhanced Raman spectroscopy, *J. Raman Spectrosc.*, **40**, pp. 1361–1370.
50. Blum, C., Schmid, T., Opilik, L., Weidmann, S., Fagerer, S. R. and Zenobi, R. (2012). Understanding tip-enhanced Raman spectra of biological molecules: a combined Raman, SERS and TERS study, *J. Raman Spectrosc.*, **43** (12), pp. 1895–1904.
51. Blum, C., Schmid, T., Opilik, L., Metanis, N., Weidmann, S. and Zenobi, R. (2012). Missing amide I mode in gap-mode tip-enhanced Raman spectra of proteins, *J. Phys. Chem. C*, **116**, pp. 23061–23066.
52. Pettinger, B., Ren, B., Picardi, G., Schuster, R. and Ertl, G. (2005). Tip-enhanced Raman spectroscopy (TERS) of malachite green isothio-

cyanate at Au(111): bleaching behavior under the influence of high electromagnetic fields, *J. Raman Spectrosc.*, **36** (6–7) p. 541.

53. Sun, M., Fang, Y., Zhang, Z. and Xu, H. (2013). Activated vibrational modes and Fermi resonance in tip-enhanced Raman spectroscopy, *Phys. Rev. E*, **87**, p. 020401(R).
54. Zhang, Z., Sun, M., Ruan, P., Zheng, H. and Xu, H. (2013). Electric field gradient quadrupole Raman modes observed in plasmon-driven catalytic reactions revealed by HV-TERS, *Nanoscale*, **5**, p. 4151.
55. Ayars, E. J., Hallen, H. D. and Jahncke, C. L. (2000). Electric field gradient effects in Raman spectroscopy, *Phys. Rev. Lett.*, **85** (19), pp. 4180–4183.
56. Sun, M., Zhang, Z., Chen, L. and Xu, H. (2013). Tip-enhanced resonance couplings revealed by high vacuum tip-enhanced Raman spectroscopy, *Adv. Opt. Mater.*, **1** (6), pp. 449–455.
57. Zhang, Z., Tian, X., Zheng, H., Xu, H. and Sun, M. (2013). Tip-enhanced ultrasensitive stokes and anti-stokes Raman spectroscopy in high vacuum, *Plasmonics*, **8**, pp. 523–527.
58. Hayazawa, N., Inouye, Y., Sekkat, Z. and Kawata, S. *Opt. Commun.*, **183**, p. 333.
59. Hayazawa, N., Inouye, Y., Sekkat, Z. and Kawata, S. (2001). *Chem. Phys. Lett.*, **335**, p. 369.
60. Hayazawa, N., Yano, T., Watanabe, H., Inouye, Y. and Kawata, S. (2003). *Chem. Phys. Lett.*, **376**, p. 174.
61. Bulgarevich, D. S., Futamata, M. (2004). *Appl. Spectrosc.*, **58**, p. 757.
62. Sun, W. X. and Shen, Z. X. (2001). A practical nanoscopic Raman imaging technique realized by near-field enhancement, *Mater. Phys. Mech.*, **4**, p. 17.
63. Anderson, M. S. and Pike, W. T. (2002). A Raman-atomic force microscope for apertureless-near-field spectroscopy and optical trapping, *Rev. Sci. Instrum.*, **73**, pp. 1198–1203.
64. Mehtani, D., Lee, N., Hartschuh, R. D., Kisliuk, A., Foster, M. D., Sokolov, A. P. and Maguire, J. F. (2005). Nano-Raman spectroscopy with side-illumination optics, *J. Raman Spectrosc.*, **36**, p. 1068.
65. Mehtani, D., Lee, N., Hartschuh, R. D., Kisliuk, A., Foster, M. D., Sokolov, A. P. and J. F. Maguire (2005). Nano-Raman spectroscopy with side-illumination optics, *J. Raman Spectrosc.*, **36**, p. 1068.
66. Mehtani, D., Lee, N., Hartschuh, R. D., Kisliuk, A., Foster, M. D., Sokolov, A. P. and Maguire, J. F. (2005). Nano-Raman spectroscopy with side-illumination optics, *J. Raman Spectrosc.*, **36**, p. 1068.

67. Merlen, A., Valmalette, J. C., Gucciardi, P. G., Lamy de la Chapelle, M., Frigout, A. and Ossikovski, R. (2009). Depolarization effects in tip-enhanced Raman spectroscopy, *J. Raman Spectrosc.*, **40**, pp. 1361–1370.

68. Gucciardi, P. G. and Valmalette, J. C. (2010). Different longitudinal optical—transverse optical mode amplification in tip enhanced Raman spectroscopy of GaAs(001), *Appl. Phys. Lett.*, **97**, p. 263104.

69. Mehtani, D., Lee, N., Hartschuh, R. D., Kisliuk, A., Foster, M. D., Sokolov, A. P. and Maguire, J. F. (2005). Nano-Raman spectroscopy with side-illumination optics, *J. Raman Spectrosc.*, **36**, p. 1068.

70. Pettinger, B., Ren, B., Picardi, G., Schuster, R. and Ertl, G. (2004). Nanoscale probing of adsorbed species by tip-enhanced Raman spectroscopy, *Phys. Rev. Lett.*, **92**, p. 096101.

71. Poborchii, V., Tada, T., Kanayama, T. and Geshev, P. (2009). Optimization of tip material and shape for near-UV TERS in Si structures, *J. Raman Spectrosc.*, **40**, pp. 1377–1385.

72. Park, K. D., Kim, Y. H., Park, J. H., Park, J. S., Lee, H. S., Yim, S. Y., Lee, Y. H. and Jeong M. S. (2012). Ultraviolet tip-enhanced nanoscale Raman imaging, *J. Raman Spectrosc.*, **43** (12), pp. 1931–1934.

73. Stadler, J., Schmid, T. and Zenobi, R. (2010). Nanoscale chemical imaging using top-illumination tip-enhanced Raman spectroscopy, *Nano Lett.*, **10**, pp. 4514–4520.

74. Ren, B., Picardi, G. and Pettinger, B. (2004). Preparation of gold tips suitable for tip-enhanced Raman spectroscopy and light emission by electrochemical etching, *Rev. Sc. Instrum.*, **75** (4), pp. 837–841.

75. Billot, L., Berguiga, L., De La Chapelle, M. L., Gilbert, Y. and Bachelot, R. (2005). Production of gold tips for tip-enhanced near-field optical microscopy and spectroscopy: analysis of the etching parameters, *Eur. Phys. J. Appl. Phys.*, **31**, p. 139.

76. Lopes M., Toury, T., de La Chapelle, M. L., Bonaccorso, F. and Gucciardi, P. G. (2013). Fast and reliable fabrication of gold tips with sub-50 nm radius of curvature for tip-enhanced Raman spectroscopy, *Rev. Sc. Instrum.*, **84** (7), p. 073702.

77. Billot, L., Berguiga, L., De La Chapelle, M. L., Gilbert, Y. and Bachelot, R. (2005). Production of gold tips for tip-enhanced near-field optical microscopy and spectroscopy: analysis of the etching parameters, *Eur. Phys. J. Appl. Phys.*, **31**, p. 139.

78. Kharintsev, S. S., Hoffmann, G. G., Fishman, A. I. and Kh Salakhov, M. (2013). Plasmonic optical antenna design for performing tip-

enhanced Raman spectroscopy and microscopy, *J. Phys. D: Appl. Phys.*, **46**. p. 145501.

79. Gorbunov, A. A., Wolf, B. and Edelmann, J. (1993). *Rev. Sci. Instrum.*, **64**, p. 2393.
80. Dickmann, K., Demming, F. and Jersch, J. (1996). *Rev. Sci. Instrum.*, **67**, p. 845.
81. Iwami, M., Uehara, Y. and Ushioda, S. (1998). *Rev. Sci. Instrum.*, **69**, p. 4010.
82. Pettinger, B., Picardi, G., Schuster, R. and Ertl, G. (2002). *Single Mol.*, **3**, p. 285.
83. Zhang, W. H., Yeo, B. S., Schmid, T. and Zenobi, R. (2007). *J. Phys. Chem. C*, **111**, p. 1733.
84. Stadler, J., Schmid, T. and Zenobi, R. (2010). *Nano Lett.*, **10**, p. 4514.
85. Zhang, C., Gao, B., Chen, L. G., Meng, Q. S., Yang, H. Zhang, R., Tao, X., Gao, H. Y., Liao, Y., Dong, Z. C. (2011). Fabrication of silver tips for scanning tunneling microscope induced luminescence, *Rev. Sci. Instrum.*, **82**, p. 083101.
86. Yi, K. J., He, X. N., Zhou, Y. S., Xiong, W. and Lu, Y. F. (2008). *Rev. Sci. Instrum.*, **79**, p. 073706.
87. Saito, Y., Murakami, T., Inouye, Y. and Kawata, S. (2005). Fabrication of silver probes for localized plasmon excitation in near-field Raman spectroscopy, *Chem. Lett.*, **34**, p. 920.
88. Valmalette, J.-C., Tan, Z.Q., Abe, H. and Ohara, S. (2014). Raman scattering of linear chains of strongly coupled Ag nanoparticles on SWCNTs. *Sci. Rep.*, 4, article no. 5238; DOI:10.1038/srep05238.
89. Garcia-Vidal, F. J., Pitarke, J. M. and Pendry, J. B. (1998). Silver-filled carbon nanotubes used as spectroscopic enhancers. *Phys. Rev. B*, **58**, 6783–6786.
90. Barrios, C. A., Malkovskiy, A. V., Kisliuk, A. M., Sokolov, A. P. and Foster, M. D. (2009). Highly stable, protected plasmonic nanostructures for tip enhanced Raman spectroscopy, *J. Phys. Chem. C*, **113** (19), pp. 8158–8161.
91. Martin, Y., Hamann, H. and Wickramasinghe, H. (2001). Strength of the electric field in apertureless near-field optical microscopy, *J. Appl. Phys.*, **89**, p. 5774.
92. Barrios, C. A., Malkovskiy, A. V., Hartschuh, R. D., Kisliuk, A. M., Sokolov, A. P. and Foster, M. D. (2008). *Proc. of SPIE*, **6954**, p. 69540C.
93. Cui, X., Erni, D., Zhang, W. H. and Zenobi, R. (2008). *Chem. Phys. Lett.*, **453**, p. 262.

94. Murray, C. A., Allara, D. L. and Rhinewine, M. (1981). *Phys. Rev. Lett.*, **46**, p. 57.

95. Murray, C. A. and Allara, D. L. (1982). Atomic layer deposition of dielectric overlayers for enhancing the optical properties and chemical stability of plasmonic nanoholes, *J. Chem. Phys.*, **76**, p. 1290.

96. Im, H., Lindquist, N. C., Lesuffleur, A. and Oh, S. (2010). *ACS Nano.*, **4**, p. 947.

97. Agapov, R. L., Sokolov, A. P. and Foster, M. D. (2013). Protecting TERS probes from degradation: extending mechanical and chemical stability, *J. Raman Spectrosc.*, **44** (5), pp. 710–716.

98. Kazemi-Zanjani, N., Vedraine, S. and Lagugné-Labarthet, F. (2013). Localized enhancement of electric field in tip enhanced Raman spectroscopy using radially and linearly polarized light, *Optics Express*, **21**, p. 21.

99. Bao, W., Staffaroni, M., Bokor, J., Salmeron, M. B., Yablonovitch, E., Cabrini, S., Weber-Bargioni, A. and Schuck, P. J. (2013). Plasmonic near-field probes: a comparison of the campanile geometry with other sharp tips, *Optics Express*, **21** (7), pp. 8166–8176.

100. Kim, T., Jeon, K. S., Heo, K., Kim, H. M., Park, J., Suh, Y. D. and Hong, S. (2013). Multilayered nano-prism vertex tips for tip-enhanced Raman spectroscopy and imaging, *Analyst*, **138**, p. 5588.

101. Kino, G. (2008). Tuning into optical wavelengths, *Nature Photon.*, **2**, pp. 210–211.

102. De Angelis, F. et al. (2010). Nanoscale chemical mapping using three-dimensional adiabatic compression of surface plasmon polaritons, *Nat. Nanotechnol.*, **5**, pp. 67–72.

103. Neacsu, C. C., Steudle, G. A. and Raschke, M. B. (2005). Plasmonic light scattering from nanoscopic metal tips, *Appl. Phys. B*, **80** pp. 295–300.

104. Mehtani, D., Lee, N., Hartschuh, R. D., Kisliuk, A. M., Foster, D., Sokolov, A. P., Caiko, F. and Tsukerman, I. (2006). Optical properties and enhancement factors of the tips for apertureless near-field optics, *J. Opt. A: Pure Appl. Opt.*, **8**, pp. S183–S190.

105. Zhang, M., Wang, R., Zhu, Z., Wang, J. and Tian, Q. (2013). Spectral response of tips for tip-enhanced Raman spectroscopy, *J. Opt.*, **15**, p. 055006.

106. Novotny, L., Sánchez, E. J. and Xie, X. S. (1998). Near-field optical imaging using metal tips illuminated by higher order Hermite-Gaussian beams, *Ultramicroscopy*, **71** (1–4), pp. 21–29.

107. Hayazawa, N., Inouye, Y., Sekkat, Z. and Kawata, S. (2000). Metallized tip amplification of near-field Raman scattering, *Opt. Commun.*, **183**, pp. 333–336.

108. Furukawa, H. and Kawata, S. *Opt. Commun.*, **148**, p. 221.

109. Hayazawa, N., Saito, Y. and Kawata, S. (2004). Detection and characterization of longitudinal field for tip-enhanced Raman spectroscopy, *Appl. Phys. Lett.*, **85** (25), pp. 6239–6241.

110. Höppener, C., Beams, R. and Novotny, L. (2009). Background suppression in near-field optical imaging, *Nano Lett.*, **9** (2), pp. 903–908.

111. Steidtner, J. and Pettinger, B. (2008). Tip-enhanced Raman spectroscopy and microscopy on single dye molecules with 15 nm resolution, *Phys. Rev. Lett.*, **100** (23), p. 236101.

112. Dorn, R., Quabis, S. and Leuchs, G. (2003). Sharper focus for a radially polarized light beam, *Phys. Rev. Lett.*, **91** (23), p. 233901.

113. Kazemi-Zanjani, N., Vedraine, S. and Lagugné-Labarthet, F. Localized enhancement of electric field in tip enhanced Raman spectroscopy using radially and linearly polarized light, *Optics Express*, **21**, p. 21.

114. Gucciardi, P. G. and Valmalette, J. C. (2010). Different longitudinal optical—transverse optical mode amplification in tip enhanced Raman spectroscopy of GaAs(001), *Appl. Phys. Lett.*, **97**, p. 263104.

115. Merlen, A., Valmalette, J. C., Gucciardi, P. G., de la Chapelle, M. L., Frigout, A. and Ossikovski, R. (2009). Depolarization effects in tip-enhanced Raman spectroscopy, *J. Raman Spectrosc.*, **40**, pp. 1361–1370.

116. Pettinger, B., Ren, B., Picardi, G., Schuster, R. and Ertl, G. (2004). Nanoscale probing of adsorbed species by tip-enhanced Raman spectroscopy, *Phys. Rev. Lett.*, **92**, p. 096101.

117. Stadler, J., Oswald, B., Schmid, T. and Zenobi, R. Characterizing unusual metal substrates for gap-mode tip enhanced Raman spectroscopy, *J. Raman Spectrosc*, **44** (2), pp. 227–233.

118. Deckert-Gaudig, T. and Deckert, V. (2009). Ultraflat transparent gold nanoplates—ideal substrates for tip-enhanced Raman scattering experiments, *Small*, **5** (4), pp. 432–436.

119. Poliani, E., Nipper, F. and Maultzsch, J. (2013). Effect of gap modes on graphene and multilayer graphene in tip-enhanced Raman spectroscopy, *Phys. Status Solidi B*, **249** (12), pp. 2511–2514.

120. Pashaee, F., Hou, R., Gobbo, P., Workentin, M. S. and Lagugné-Labarthet, F. (2013). Tip-enhanced Raman spectroscopy of self-assembled thio-

lated monolayers on flat gold nanoplates using Gaussian-transverse and radially polarized excitations, *J. Phys. Chem. C*, **117** (30), pp. 15639–15646.

121. Hao, E. and Schatz, G. C. (2004). *J. Chem. Phys.*, **120**, p. 357.

122. Blum, C., Schmid, T., Opilik, L., Metanis, N., Weidmann, S. and Zenobi, R. (2012). Missing amide I mode in gap-mode tip-enhanced Raman spectra of proteins, *J. Phys. Chem. C*, **116**, pp. 23061–23066.

123. Deckert-Gaudig, T. and Deckert, V. (2009). *J. Raman Spectrosc.*, **40**, p. 1446.

124. Deckert-Gaudig, T. and Deckert, V. (2009). *Small*, **5**, p. 432.

125. Deckert-Gaudig, T., Erver, F. and Deckert, V. (2009). *Langmuir*, **25**, p. 6032.

126. Kazemi-Zanjani, N., Vedraine, S. and Lagugné-Labarthet, F. (2013). Localized enhancement of electric field in tip enhanced Raman spectroscopy using radially and linearly polarized light, *Optics Express*, **21**, p. 21.

127. Stadler, J., Oswald, B., Schmid, T. and Zenobi, R. (2013). Characterizing unusual metal substrates for gap-mode tip-enhanced Raman spectroscopy, *J. Raman Spectrosc.*, **44**, pp. 227–233.

128. Jiang, N., Foley, E. T., Klingsporn, J. M., Sonntag, M. D., Valley, N. A., Dieringer, J. A., Seideman, T., Schatz, G. C., Hersam, M. C. and Van Duyne, R. P. (2012). Observation of multiple vibrational modes in ultrahigh vacuum tip-enhanced Raman spectroscopy combined with molecular-resolution scanning tunneling microscopy, *Nano Lett.*, **12**, pp. 5061–5067.

129. Zhang, R., Zhang, Y., Dong, Z. C., Jiang, S., Zhang, C., Chen, L. G., Zhang, L., Liao, Y., Aizpurua, J., Luo, Y., Yang, J. L. and Hou, J. G. (2013). Chemical mapping of a single molecule by plasmon-enhanced Raman scattering, *Nature*, **498**, pp. 82–86.

130. Yu, J., Saito, Y., Ichimura, T., Kawata, S. and Verma, P. (2013). Far-field free tapping-mode tip-enhanced Raman microscopy, *Appl. Phys. Lett.*, **102**, p. 123110.

131. Ichimura, T., Hayazawa, N., Hashimoto, M., Inouye, Y. and Kawata, S. (2004). Tip-enhanced coherent anti-Stokes Raman scattering for vibrational nanoimaging, *Phys. Rev. Lett.*, **92**, p. 220801.

132. Ichimura, T. and Kawata, S. (2011). *Surface enhanced Raman spectroscopy: analytical, biophysical and life science applications*, ed. Schuckler, S. (WILEY-VCH, Weinheim).

133. Keilmann, F. and Hillenbrand, R. (2004). Near-field microscopy by elastic light scattering from a tip, *Philos. Trans. Roy. Soc. Lond. Ser. A Math. Phys. Eng. Sci.*, **362**, p. 787.
134. Raschke, M. B., Molina, L., Elsaesser, T., Kim, D. H., Knoll, W. and Hinrichs, K. (2005). Apertureless near-field vibrational imaging of block-copolymer nanostructures with ultrahigh spatial resolution, *Chem. Phys. Chem.*, **6**, p. 2197.
135. Hartschuh, A., Beversluis, M. R., Bouhelier, A. and Novotny, L. (2004). Tip-enhanced optical spectroscopy, *Philos. Trans. Roy. Soc. Lond. Ser. A Math. Phys. Eng. Sci.*, **362**, p. 807.
136. Bortchagovsky, E., Schmid, T. and Zenobi, R. (2013). Internal standard for tip-enhanced Raman spectroscopy, *Appl. Phys. Lett.*, **103**, p. 043111.
137. Hu, D., Micic, M., Klymyshyn, N. et al. (2003). *Rev. Sci. Instrum.*, **74**, p. 3347.
138. Festy, F., Demming, A. and Richards, D. (2004). *Ultramicroscopy*, **100**, p. 437.
139. Notingher, I. and Elfick, A. (2005). *J. Phys. Chem. B*, **109**, p. 15699.
140. Demming, A. L., Festy, F. and Richards, D. (2005). *J. Chem. Phys.*, **122**, p. 184716.
141. Demming, A. L., Festy, F. and Richards, D. (2005). Plasmon resonances on metal tips: understanding tip-enhanced Raman scattering, *J. Chem. Phys.*, **122**, p. 184716.
142. Cui, X., Zhang, W., Yeo, B., Zenobi, R., Hafner, C. and Erni, D. (2007). Tuning the resonance frequency of Ag-coated dielectric tips, *Opt. Express*, **15**, pp. 8309–8316.
143. Novotny, L. (2001). Forces in optical near-fields. In *Near-field optics and surface plasmon polaritons*, Kawata, S., Ohtsu, M., Irie, M. (eds.), **81**, pp. 123–141 (Springer-Verlag, Berlin).
144. Martin, Y. C., Hamann, H. and Wickramasinghe, H. (2001). Strength of the electric field in apertureless near-field optical microscopy, *J. Appl. Phys.*, **89**, p. 5774.
145. Furukawa, H. and Kawata, S. (1998). *Opt. Commun.*, **148**, p. 221.
146. Taflove, A. and Hagness, S. C. (2000). *Computational electrodynamics: the finite difference time domain method* (Artech House, London).
147. Demming, A. L., Festy, F. and Richards, D. (2005). *J. Chem. Phys.*, **122**, p. 184716.
148. Downes, A., Salter, D. and Elfick, A. (2006). Finite element simulations of tip enhanced Raman and fluorescence spectroscopy, *J. Phys. Chem. B*, **110**, pp. 6692–6698.

149. Yang, Z. L., Aizpurua, J. and Xu, H. X. (2009). *J. Raman Spectrosc.*, **40**, pp. 1343–1348.

150. Zhang, M., Wang, J. and Tian, Q. (2013). Tip-enhanced Raman spectroscopy based on plasmonic lens excitation and experimental detection, *Opt. Express*, **21** (8), pp. 9414–9421.

151. Wang, Y., Plouraboue, F. and Chang, H. C. (2013). Broadband converging plasmon resonance at a conical nanotip, *Opt. Express*, **21** (5), pp. 6609–6617.

152. Bao, W., Staffaroni, M., Bokor, J., Salmeron, M. B., Yablonovitch, E., Cabrini, S., Weber-Bargioni, A. and Schuck, P. J. (2013). Plasmonic near-field probes: a comparison of the campanile geometry with other sharp tips, *Opt. Express*, **21** (7), pp. 8166–8176.

153. Hao, E. and Schatz, G. C. (2004). *J. Chem. Phys.*, **120**, p. 357.

154. Demming, A. L., Festy, F. and Richards, D. (2005). Plasmon resonances on metal tips: understanding tip-enhanced Raman scattering, *J. Chem. Phys.*, **122** (18), p. 184716.

155. Sukharev, M. and Seideman, T. (2009). Optical properties of metal tips for tip-enhanced spectroscopies, *J. Phys. Chem. A*, **113** (26), pp. 7508–7513.

156. Treffer, R., Lin, X., Bailo, E., Deckert-Gaudig, T. and Deckert, V. (2011). Distinction of nucleobases—a tip-enhanced Raman approach, *Beilstein J. Nanotechnol.*, **2**, pp. 628–637.

157. Zhang, W. H., Cui, X. D. and Martin, O. J. F. (2009). *J. Raman Spectrosc.*, **40**, p. 1338.

158. Stadler, J., Oswald, B., Schmid, T. and Zenobi, R. (2013). Characterizing unusual metal substrates for gap-mode tip-enhanced Raman spectroscopy, *J. Raman Spectrosc.*, **44**, pp. 227–233.

159. Stadler, J., Oswald, B., Schmid, T. and Zenobi, R. (2013). Characterizing unusual metal substrates for gap-mode tip enhanced Raman spectroscopy, *J. Raman Spectrosc.*, **44** (2), pp. 227–233.

160. Kazemi-Zanjani, N., Vedraine, S. and Lagugnė-Labarthet, F. (2013). Localized enhancement of electric field in tip enhanced Raman spectroscopy using radially and linearly polarized light, *Optics Express*, **21**, p. 21.

161. Kaminski, F., Sandoghdar, V. and Agio, M. (2007). *J. Comput. Theor. Nanosci.*, **4**, p. 635.

162. Sarmany, D., Izsak, F. and van der Vegt, J. J. W. (2010). *J. Sci. Comput.*, **44**, p. 219.

163. Sun, M. T., Fang, Y. R., Yang, Z. L. and Xu, H. X. (2009). Chemical and electromagnetic mechanisms of tip-enhanced Raman scattering, *Phys. Chem. Chem. Phys.*, **11** (41), p. 9412.
164. Kawata, S. (2013). Plasmonics for nanoimaging and nanospectroscopy, *Appl. Spectroscop.*, **67** (2), pp. 117–125.
165. Hartschuh, A., Anderson, N. and Novotny, L. (2002). Near-Field Raman spectroscopy using a sharp metal tip, *J. Microsc.*, **210**, pp. 234–240.
166. Hartschuh, A. (2008). Tip-enhanced near-field optical microscopy, *Angew. Chem. Int. Ed. Engl.*, **47** (43), pp. 8178–8191.
167. Kawata, S. (2013). Plasmonics for nanoimaging and nanospectroscopy, *Appl. Spectrosc.*, **67** (2), pp. 117–125.
168. Nicklaus, M., Nauenheim, C., Krayev, A., Gavrilyuk, V., Belyaev, A. and Ruediger, A. (2012). Note: tip enhanced Raman spectroscopy with objective scanner on opaque samples, *Rev. Sci. Instrum.*, **83** (6), p. 066102.
169. Su, W. and Roy, D. (2013). Visualizing graphene edges using tip-enhanced Raman spectroscopy, *J. Vac. Sci. Technol. B*, **31** (4) p. 041808.
170. Poliani, E., Nipper, F. and Maultzsch, J. (2012). Effect of gap modes on graphene and multilayer graphene in tip-enhanced Raman spectroscopy, *Phys. Status Solidi B*, **249** (12), pp. 2511–2514.
171. Ghislandi, M., Hoffmann, G., Tkalya, E., Xue, L. and De With, G. (2012). Tip-enhanced Raman spectroscopy and mapping of graphene sheets, *Appl. Spectrosc. Rev.*, **47** (5), pp. 371–381.
172. Domke, K. F. and Pettinger, B. (2009). Tip-enhanced Raman spectroscopy of 6H-SiC with graphene adlayers: selective suppression of E1 modes, *J. Raman Spectrosc.* **40** (10) p. 1427.
173. Yano, T. (2013). Tip-enhanced nano-Raman analytical imaging of locally induced strain distribution in carbon nanotubes, *Nat Commun.*, **4**, p. 2592.
174. Poliani, E. , Wagner, M. R., Reparaz, J. S., Mandl, M., Strassburg, M., Kong, X., Trampert, A., Sotomayor Torres, C. M., Hoffmann, A. and Maultzsch, J. (2013). Nanoscale imaging of InN segregation and polymorphism in single vertically aligned InGaN/GaN multi quantum well nanorods by tip-enhanced Raman scattering, *Nano Lett.*, **13** (7), pp. 3205–3212.
175. Merlen, A., Valmalette, J. C., Gucciardi, P. G., Lamy de la Chapelle, M., Frigout, A. and Ossikovski, R. (2009). Depolarization effects in tip-enhanced Raman spectroscopy, *J. Raman Spectrosc.*, **40**, pp. 1361–1370.

176. Gucciardi, P. G. and Valmalette, J. C. (2010). Different longitudinal optical—transverse optical mode amplification in tip enhanced Raman spectroscopy of GaAs(001), *Appl. Phys. Lett.*, **97**, p. 263104.

177. Berweger, S. et al. (2009). *Nat Nanotechnol.*, **4**, pp. 496–499.

178. Sevinc, P. C., Wang, X., Wang, Y., Zhang, D., Meixner, A. J. and Lu, H. P. (2011). *Nano Lett.*, **11**, pp. 1490–1494.

179. Jose, D., Nijamudheen, A. and Datta, A. (2013). Tip enhanced Raman spectroscopy (TERS) as a probe for the buckling distortion in silicene, *Phys. Chem. Chem. Phys.*, **15**, p. 8700.

180. Suzuki, T., Yan, X., Kitahama, Y., Sato, H., Itoh, T., Miura, T. and Ozaki, Y. (2013). Tip-enhanced Raman spectroscopy study of local interactions at the interface of styrene-butadiene rubber/multiwalled carbon nanotube nanocomposites, *J. Phys. Chem. C*, **117** (3), pp. 1436–1440.

181. Zhuang, M., Liu, Z., Ren, B. and Tian, Z. (2010). Surface bonding on silicon surfaces as probed by tip-enhanced Raman spectroscopy, *Sc. China Chem.*, **53** (2), pp. 426–431.

182. Sevinc, P. C., Wang, X., Wang, Y., Zhang, D., Meixner, A. J. and Lu, H. P. (2011). *Nano Lett.*, **11**, pp. 1490–1494.

183. Jiang, N., Foley, E. T., Klingsporn, J. M., Sonntag, M. D., Valley, N. A., Dieringer, J. A., Seideman, T., Schatz, G. C., Hersam, M. C. and Van Duyne, R. P. (2012). Observation of multiple vibrational modes in ultrahigh vacuum tip-enhanced Raman spectroscopy combined with molecular-resolution scanning tunneling microscopy, *Nano Lett.*, **12**, pp. 5061–5067.

184. Kim, H., Kosuda, K. M., Van Duyne, R. P. and Stair, P. C. (2010). Resonance Raman and surface- and tip-enhanced Raman spectroscopy methods to study solid catalysts and heterogeneous catalytic reactions, *Chem. Soc. Rev.*, **39** (12), pp. 4820–4844.

185. Tarun, A., Hayazawa, N. and Kawata, S. (2009). Tip-enhanced Raman spectroscopy for nanoscale strain characterization, *Anal. Bioanal. Chem.* **394** (7), pp. 1775–1785.

186. Domke, K. F. and Pettinger, B. (2009). In situ discrimination between axially complexed and ligand-free Co porphyrin on Au(111) with tip-enhanced Raman spectroscopy, *ChemPhysChem*, **10** (11) p. 1794.

187. Jiang, N., Foley, E. T., Klingsporn, J. M., Sonntag, M. D., Valley, N. A., Dieringer, J. A., Seideman, T., Schatz, G. C., Hersam, M. C. and Van Duyne, R. P. (2012). Observation of multiple vibrational modes in ultrahigh

vacuum tip-enhanced Raman spectroscopy combined with molecular-resolution scanning tunneling microscopy, *Nano Lett.*, **12**, pp. 5061–5067.

188. Picardi, G., Chaigneau, M., Ossikovski, R., Licitra, C. and Delapierre, G. J. (2009). *Raman Spectrosc.*, **40**, pp. 1407–1412.
189. Stadler, J., Schmid, T., Opilik, L., Kuhn, P., Dittrich, P. S. and Zenobi, R. (2011). *Beilstein J. Nanotechnol.*, **2**, pp. 509–515.
190. Zhang, R., Zhang, Y., Dong, Z. C., Jiang, S., Zhang, C., Chen, L. G., Zhang, L., Liao, Y., Aizpurua, J., Luo, Y., Yang, J. L. and Hou, J. G. (2013). Chemical mapping of a single molecule by plasmon-enhanced Raman scattering, *Nature*, **498**, pp. 82–86.
191. Lui, Z. et al. (2011). Revealing the molecular structure of single-molecule junctions in different conductance states by fishing-mode tip-enhanced Raman spectroscopy, *Nature Comm.*, **2**, Article no. 305, DOI: 10.1038/ncomms1310.
192. Shaw, P. E., Ruseckas, A. and Samuel, I. D. W. (2008). *Adv. Mater.*, **20**, pp. 3516–3520.
193. Zhang, D., Wang, X., Braun, K., Egelhaaf, H. J., Fleischer, M., Hennemann, L., Hintz, H., Stanciu, C., Brabec, C. J. and Kern, D. P. (2009). *J. Raman Spectrosc.*, **40**, pp. 1371–1376.
194. Wang, X., Zhang, D., Braun, K., Egelhaaf, H. J., Brabec, C. J. and Meixner, A. J. (2010). *Adv. Funct. Mater.*, **20**, pp. 492–499.
195. Zhang, D., Heinemeyer, U., Stanciu, C., Sackrow, M., Braun, K., Hennemann, L. E., Wang, X., Scholz, R., Schreiber, F. and Meixner, A. J. (2010). *Phys. Rev. Lett.*, **104**, p. 056601.
196. Sun, M. (2012). In-situ plasmon-driven chemical reactions revealed by high vacuum tip-enhanced Raman spectroscopy, *Sci. Rep.*, **2**, p. 647.
197. van Schrojenstein Lantman, E. M. (2012). Catalytic processes monitored at the nanoscale with tip-enhanced Raman spectroscopy, *Nat. Nanotechnol.*, **7**, pp. 583–586.
198. Zhang, Z., Sun, M., Ruan, P., Zheng, H. and Xu, H. (2013). Electric field gradient quadrupole Raman modes observed in plasmon-driven catalytic reactions revealed by HV-TERS, *Nanoscale*, **5**, p. 4151.
199. Anderson, M. S. and Gaimari, S. D. (2003). Raman-atomic force microscopy of the ommatidial surfaces of Dipteran compound eyes, *J. Struct. Biol.*, **142**, pp. 364–368.
200. Pahlow, S., März, A., Seise, B., Hartmann, K., Freitag, I., Kämmer, E., Böhme, R., Deckert, V., Weber, K., Cialla, D. and Popp, J. (2012). Bioan-

alytical application of surface- and tip-enhanced Raman spectroscopy, *Eng. Life Sci.*, **12** (2) p. 131.

201. Pozzi, E. A., Sonntag, M. D., Jiang, N., Klingsporn, J. M., Hersam, M. C. and Van Duyne, R. P. (2013). Tip-enhanced Raman imaging: an emergent tool for probing biology at the nanoscale, *ACS Nano*, **7** (2), pp. 885–888.
202. Kurouski, D., Postiglione, T., Deckert-Gaudig, T., Deckert, V. and Lednev, I. K. (2013). Amide I vibrational mode suppression in surface (SERS) and tip (TERS) enhanced Raman spectra of protein specimens, *Analyst*, **138**, p. 1665.
203. Blum, C., Schmid, T., Opilik, L., Weidmann, S., Fagerer, S. R. and Zenobi, R. (2012). Understanding tip-enhanced Raman spectra of biological molecules: a combined Raman, SERS and TERS study, *J. Raman Spectrosc.*, **43** (12), pp. 1895–1904.
204. Paulite, M., Blum, C., Schmid, T., Opilik, L., Eyer, K., Walker, G. C. and Zenobi, R. (2013). Full spectroscopic tip-enhanced Raman imaging of single nanotapes formed from β-amyloid(1–40) peptide fragments, *ACS Nano*, **7** (2), pp. 911–920.
205. Bailo, E. and Deckert, V. (2008). Tip-enhanced Raman spectroscopy of single RNA strands: towards a novel direct sequencing method, *Angew. Chem. Int. Ed. Engl.*, **47** (9), pp. 1658–1661.
206. Böhme, R., Mkandawire, M., Krause-Buchholz, U., Rösch, P., Rödel, G., Popp, J. and Deckert, V. (2011). Characterizing cytochrome c states–TERS studies of whole mitochondria, *Chem. Commun. (Camb.)*, **47** (41), pp. 11453–11455.
207. Opilik, L., Bauer, T., Schmid, T., Stadler, J. and Zenobi, R. (2011). *Phys. Chem. Chem. Phys.*, **13**, pp. 9978–9981.
208. Schmid, T., Messmer, A., Yeo, B., Zhang, W. and Zenobi, R. (2008). Towards chemical analysis of nanostructures in biofilms II: tip-enhanced Raman spectroscopy of alginates, *Anal. Bioanal. Chem.*, **391**, pp. 1907–1916.
209. Moretti, M., Zaccaria, R. P., Descrovi, E., Das, G., Leoncini, M., Liberale, C., De Angelis, F. and Di Fabrizio, E. (2013). Reflection-mode TERS on insulin amyloid fibrils with top-visual AFM probes, *Plasmonics*, **8**, pp. 25–33.
210. Deckert-Gaudig, T. Kämmer, E. and Deckert, V. J. (2012). *Biophotonics*, **5**, pp. 215–219.
211. Kurouski, D., Deckert-Gaudig, T., Deckert, V. and Lednev, I. K. (2012). *J. Am. Chem. Soc.*, **134**, pp. 13323–13329.

212. Kazemi-Zanjani, N., Chen, H., Goldberg, H. A., Hunter, G. K., Grohe, B. and Lagugné-Labarthet, F. (2012). Label-free mapping of osteopontin adsorption to calcium oxalate monohydrate crystals by tip-enhanced Raman spectroscopy, *J. Am. Chem. Soc.*, **134** (41), pp. 17076–17082.

213. Wang, H. and Schultz, Z. D. (2013). The chemical origin of enhanced signals from tip-enhanced Raman detection of functionalized nanoparticles, *Analyst*, **138**, p. 3150.

214. Naumenko, D., Snitka, V., Serviene, E., Bruzaite, I. and Snopok, B. (2013). In vivo characterization of protein uptake by yeast cell envelope: single cell AFM imaging and mu-tip-enhanced Raman scattering study, *Analyst*, **138**, p. 5371.

215. Zhang, D., Domke, K. F. and Pettinger, B. (2010). *ChemPhysChem*, **11**, p. 1662.

Chapter 12

Tip-Enhanced Raman Spectroscopy: Applications to Carbon Nanomaterials

Nina Mauser and Achim Hartschuh
Department Chemie & CeNS, LMU Munich, 81377 Munich, Germany
achim.hartschuh@lmu.de

Sp^2-hybridized carbon nanomaterials such as graphene and carbon nanotubes feature unique electronic and vibronic characteristics. Consisting of a single type of element only, they can also help in understanding how structure and dimensionality determine solid state properties including electron and phonon dispersion. While Raman spectroscopy is very well suited for studying these phenomena, local structural variations, defects, and perturbations remain hidden in conventional microscopy studies due to the resolution limit imposed by diffraction. Tip-enhanced Raman spectroscopy (TERS) overcomes this limitation providing nanoscale spatial resolution and improved detection sensitivity. In this chapter we illustrate the main principles of TERS, possible experimental implementations, and review several recent applications to carbon nanotubes and graphene.

Handbook of Enhanced Spectroscopy
Edited by Pietro Giuseppe Gucciardi, Marc Lamy de la Chapelle, and Nathalie Lidgi-Guigui

ISBN 978-981-4613-32-3 (Hardcover), 978-981-4613-33-0 (eBook)
www.panstanford.com

12.1 Introduction to Carbon Nanomaterials

Carbon nanomaterials are discussed as prototypes for the development of nanoscience and nanometrology [1]. They exhibit unique electronic, optical, mechanical, and thermal properties making them an ideal reference/model material. Explicitly, their role as a model material is justified by the following three peculiarities: first, nanocarbons consist of only one type of atom arranged in a periodic hexagonal lattice—a system that can be easily studied by theory and experiments. Second, neighboring carbon atoms exhibit a covalent sp^2-bonding. In the sp^2 configuration the $2s$ orbital mixes with the $2p_x$ and $2p_y$ orbital to form a planar, hexagonal honeycomb lattice—the basis of all nanocarbons. These in-plane bonds are the strongest in nature making the nanocarbons extremely robust and stable. And third, sp^2 nanocarbons possess a very unique electronic structure in the form of a delocalized π- electron system formed by the $2p_z$ orbitals perpendicular to the planar lattice that are left unaffected by the hybridization.

sp^2 carbons can exist in different dimensionalities. From a structural point of view graphene, a two-dimensional (2D) crystal of carbon atoms tightly packed in a honeycomb lattice, can be seen as the mother of all sp^2 carbons [2]. Both fullerenes (0D) and single-walled carbon nanotubes (1D) can be thought of being formed by rolling up particular patterns cut in graphene (see Fig. 12.1), whereas graphite is composed of stacked graphene layers. Electron and phonon energies together with their dispersions of nanocarbons can thus be derived starting with those of graphene.

The electronic band structure of graphene is unique featuring linear dispersion around its K-point. In the visible the optical absorption of undoped graphene is spectrally constant, while it increases toward the UV. The electron dispersion of single-walled carbon nanotubes (SWCNTs) can be derived from the graphene band structure using the zone-folding. The quasi-1D structure of SWCNTs allows continuous k-vectors only in axial direction. In orthogonal direction the k-spectrum is quantized. SWCNTs thus sample the electron dispersion of graphene in parallel lines whose direction and spacing are fixed by the SWCNT's structure. The minima in the band structure of SWCNTs in which the dispersion has zero slope are the

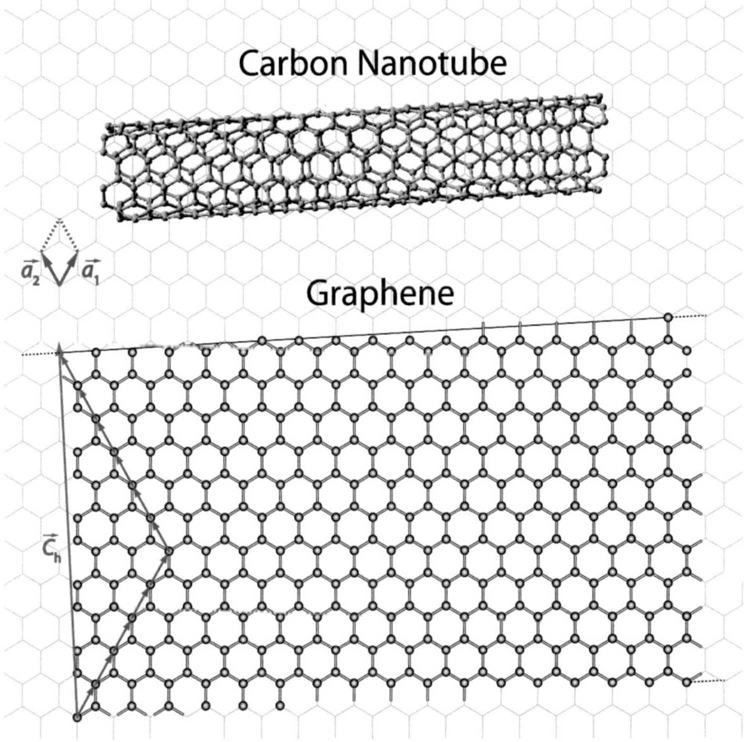

Figure 12.1 Schematic structure of graphene (below) and a single-walled carbon nanotube (top) illustrating the chiral vector $\mathbf{C}_h$ as a linear combination of the graphene lattice vectors $\mathbf{a}_1$ and $\mathbf{a}_2$.

origin of van Hove singularities, that is, energy levels with extremely high density of states.

The structure of a SWCNT is determined by the chiral vector $\mathbf{C}_h = n\mathbf{a}_1 + m\mathbf{a}_2$, that is, the vector around the nanotube circumference in terms of the unit cell vectors of graphene $\mathbf{a}_1$ and $\mathbf{a}_2$ (see Fig. 12.1). SWCNTs are therefore uniquely defined by the tuple (n,m). The chiral indices n and m determine if a nanotube is semiconducting or metallic. According to tight binding calculations [3] nanotubes with $(n - m)mod3 = 0$ are metallic or semimetallic. All others are semiconducting and exhibit a band gap that varies with the diameter. While tight-binding calculations were found to describe the optical properties of SWCNTs surprisingly well, more detailed theoretical considerations showed that multiparticle phenomena, that become particularly important in these materials due to their quasi-1D structure, need to be considered. The optical resonances of semiconducting nanotubes arise from excitons, electron–hole pairs with very high binding energies.

A very powerful and widespread method to study nanocarbons is Raman spectroscopy [4, 5]. A Raman spectrum of a sp^2 carbon exhibits several peaks coming from first-order and higher-order scattering processes. In general, only resonance Raman processes can be observed, that is, the incident laser photon or the scattered photon needs to be in resonance with a real electronic transition, that is, connecting van Hove singularities with extremely high densities of states. The signal intensity is then increased by orders of magnitude compared to the nonresonant case making it possible to detect the Raman scattering of single SWCNTs [6].

The main Raman peaks appearing for graphene and SWCNTs are [5, 7]: The **G-band**, the main signature of all carbon nanomaterials. It is a first-order process related to the C–C stretching mode and appears at around 1600 cm^{-1}. Uniaxial stretching as present in SWCNTs splits the band into a G^+ and G^- component. The **G'** or **2D-band** is a second-order process appearing in the range 2500–2800 cm^{-1}. The band is dispersive and can be used in graphene to determine the number of layers [8]. The D-band at around 1300 cm^{-1} results from a phonon mode with large momentum. D-band scattering is a second-order process that can occur only in the presence of defects in the carbon lattice. In a microscopic model a local defect is required for momentum conservation [9]. The **radial breathing mode (RBM)** exists only for CNTs. It is a tube-breathing-like motion transverse to the nanotube axis appearing at small Raman energies (50–500 cm^{-1}) that are inversely proportional to the nanotube's diameter.

12.2 Principles of TERS

Micro-Raman spectroscopy using a confocal microscope is a standard tool for the investigation of a variety of sample materials.

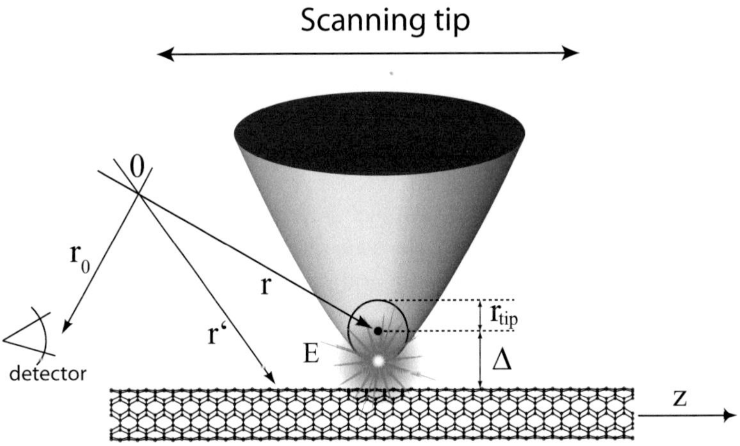

Figure 12.2 Schematics for the spatially resolved Raman scattering of carbon nanotubes. The electric field **E** confined to the apex of a laser-irradiated tip interacts locally with a carbon nanotube characterized by the Raman polarizability . The position vectors **r**, **r**′, and $\mathbf{r_0}$ denote the location of the center of the tip apex, the Raman dipole moment **p**, and the detector, respectively (see text). r_{tip} is the radius of the tip apex (dashed circumference), and Δ is the distance between the tip and the carbon nanotube. Reprinted with permission from Ref. [15], Copyright (2009), by the American Physical Society.

However, when it comes to single nano-objects it faces two major limitations: first, the spatial resolution is limited due to diffraction of light, making it impossible to study local properties of the nano-object; second, Raman scattering is a very rare event precluding the observation of nanoscale sample volumes for most materials. TERS is an approach that addresses both limitations using the local field enhancement at the apex of a scanning metal tip (see Fig. 12.2). In this approach the tip has the function of an optical antenna [10–12], being able to convert propagating radiation into localized energy and vice versa [13]. In other words, the evanescent near-field at the tip apex is used to interact with the sample and the tip provides the coupling between its near-field and the far-field. As a result, signal enhancement for objects close to the tip can be achieved using a laser source and a detector in the far-field. This signal enhancement and the correspondingly improved detection sensitivity are essential since high spatial resolution is connected to a shrinking sample volume and weaker signals [14].

12.3 Local Signal Formation

In the following the Raman-active object is assumed to be a point scatterer (0D) with dipole transition matrix element $\boldsymbol{\mu}$. The excitation and emission rates Γ_{exc} and Γ_{rad} are then given by

$$\Gamma_{exc} \propto |\boldsymbol{\mu} \cdot \mathbf{E}(\omega_{\mathbf{l}})|^2 \tag{12.1}$$

$$\Gamma_{rad} \propto |\boldsymbol{\mu} \cdot \mathbf{E}(\omega_{\mathbf{s}})|^2 \tag{12.2}$$

where $\mathbf{E}(\omega_{\mathbf{l}})$ and $\mathbf{E}(\omega_{\mathbf{s}})$ are the incident electric field at the laser frequency and the vacuum field at the frequency of the Raman scattered light in presence of the tip, respectively. Local field enhancement can then be described by the local field enhancement factor f through $\mathbf{E} = f\mathbf{E_0}$ with $\mathbf{E_0}$ being the field in the absence of the tip. In general, f is a frequency-dependent matrix that spatially varies with respect to the tip position.

In a classical description, the material response is represented by a Raman polarizability tensor $\overleftrightarrow{\boldsymbol{\alpha}}_R$ that determines the local dipole moment $\mathbf{p}(\omega_s) = \overleftrightarrow{\boldsymbol{\alpha}}_R\ \mathbf{E}(\omega_l) = \overleftrightarrow{\boldsymbol{\alpha}}_R\ f\mathbf{E_0}(\omega_l)$ at the Raman frequency ω_s as a response to the local electrical field at the laser frequency ω_l. The field radiated by the Raman dipole and seen by the detector depends on the environment characterized by the Green's function for which the effect of the tip can again be described by the local field enhancement factor f. Further details of this classical description are illustrated below in the discussion of 1D and 2D structures that also include the 0D case.

As can be seen from Eqs. 12.1 and 12.2 and the discussion made above the total signal enhancement for Raman scattering approximately scales with the fourth power of the enhancement factor f:

$$M_{Raman} \approx \Gamma_{exc}/\Gamma^0_{exc}\Gamma_{rad}/\Gamma^0_{rad} \approx f^4. \tag{12.3}$$

Note that this is a simplified description where four main approximations have been made: First, the same enhancement factor for the excitation and emission is assumed. This is justified in case of small shifts between the energy of the laser photon and the scattered photon for which the plasmon response of the tip remains spectrally uniform. Second, the vectorial character of the electric field and third, the angular dependence of the antenna gain and directivity are neglected [14]. Therefore, when designing a microscope setup, one has to take carefully into account that the antenna changes the angular emission [16–18]. This can have a strong impact on the collection efficiency [19]. And fourth, it is valid only for a 0D structure that can be described by a Raman point dipole or point dipole transition matrix element. Changing the dimensionality of the sample material to 1D or 2D requires the extension of the theory for Raman enhancement.

12.3.1 *1D Materials*

The signal formation for 1D nanostructures such as carbon nanotubes has to be revised as compared to the 0D case. Two major differences compared to a 0D system need to be considered: first, the enhanced electric field at the tip apex is modified by the presence of a SWCNT due to screening effects in transverse direction to the tube axis. Second, the Raman polarization dipole of the CNT cannot be described by a single vector. Instead a Raman dipole distribution along the CNT can be introduced. The following description is based on the work by Cancado and co-authors presented in Ref. [15].

The induced Raman dipole per unit length **p** is given by

$$\mathbf{p}(\mathbf{r}', \omega_s) = \overset{\leftrightarrow}{\boldsymbol{\alpha}}_R \, \mathbf{E}_{tot}(\mathbf{r}' - \mathbf{r}; \omega) \tag{12.4}$$

where $\overset{\leftrightarrow}{\boldsymbol{\alpha}}_R$ is the Raman polarizability tensor at position $\mathbf{r}'$ and Raman frequency ω_s and $\mathbf{E}_{tot}$ the Raman excitation field at the laser frequency ω depending on the tip position $\mathbf{r}$. The assignments used refer to Fig. 12.2.

For a 1D system $\overset{\leftrightarrow}{\boldsymbol{\alpha}}_R$ takes the form

$$\overset{\leftrightarrow}{\boldsymbol{\alpha}}_R = \begin{bmatrix} \alpha_{R,\perp} & 0 & 0 \\ 0 & \alpha_{R,\perp} & 0 \\ 0 & 0 & \alpha_{R,\parallel} \end{bmatrix} \tag{12.5}$$

where $\alpha_{R,\perp}$ is the polarizability transverse to the tube axis and $\alpha_{R,\parallel}$ is the polarizability parallel to the tube axis. Note that $\alpha_{R,\perp}$

is reduced due to screening effects caused by the finite diameter of the nanotube, whereas there is no screening along the axis that is treated as infinitely long. This polarization tensor holds for the *RBM* and the G^+ and the G^- phonon mode belonging to the same symmetry group.

The electric field at position of the detector $\mathbf{r}_0$ is obtained by spatial integration along the nanotube

$$\mathbf{E}(\mathbf{r}_0, \omega_s) \propto \int_{-\infty}^{+\infty} dz' \overleftrightarrow{\boldsymbol{\alpha}}_{tip}(\omega_s) \overleftrightarrow{\boldsymbol{G}}_0(\mathbf{r}, z'; \omega_s)\mathbf{p}(\mathbf{z}', \omega_s) \tag{12.6}$$

where $\mathbf{p}(\mathbf{z}', \omega_s)$ is the Raman dipole distribution along the z-axis (Eq. 12.4), $\overleftrightarrow{\boldsymbol{G}}_0(\mathbf{r}, z'; \omega_s)$ is the free-space dyadic Green's function in the absence of the tip and $\overleftrightarrow{\boldsymbol{\alpha}}_{tip}$ is the polarizability of the tip that is treated as a sphere of radius r_{tip}.

The near-field intensity of the detector $I_{NF}(\mathbf{r}_0, \omega_s)$ can then be calculated from the modulus square of the field derived in Eq. 12.6. In this case phase correlations between the Raman dipoles are considered, and Raman scattering is treated as spatially coherent. Incoherent scattering can be described by taking the modulus square of the integrand of Eq. 12.6.

Evaluating these integrals (see Ref. [15]) yields the following results: In the case of coherent scattering the near-field Raman intensity is inversely proportional to the 10th power of the separation between the tip and the nanotube (Δ in fig. 12.2), whereas for the incoherent case it is inversely proportional to the 11th power. As for a 0D structure, the enhancement scales with the fourth power of the enhancement factor for both coherent and incoherent scattering. By comparing theory with experimental data the authors found that the Raman enhancement process is not significantly influenced by the specific Raman mode, and is mainly defined by the tip properties.

Uryu and co-authors illustrate in Ref. [20] that the near-field generated by surface plasmons in gold films partially covering carbon nanotubes can lead to the efficient excitation of excitons with finite momentum in axial direction. These excitations, that are not-accessible in far-field experiments, could in principle influence resonance Raman scattering.

12.3.2 *2D Materials*

The theory for the Raman enhancement mechanism for two-dimensional systems is described in detail in Ref. [21]. Similar to the 1D case one can define a Raman dipole distribution per square length corresponding to Eq. 12.4.

The Raman tensor in 2D takes the general form

$$\overleftrightarrow{\boldsymbol{\alpha}}_R = \begin{bmatrix} \alpha_{R,11} & \alpha_{R,12} \\ \alpha_{R,21} & \alpha_{R,22} \end{bmatrix} \tag{12.7}$$

where only in-plane Raman modes are considered.

Similar to Eq. 12.6 the electric field at the detector is obtained by integrating the Raman fields radiated via the tip over the xy-plane.

$$\mathbf{E}(\mathbf{r}_0, \omega_s) \propto \int_{-\infty}^{+\infty} \int_{-\infty}^{+\infty} dx'dy' \, \overleftrightarrow{\boldsymbol{\alpha}}_{tip}(\omega_s) \, \overleftrightarrow{\boldsymbol{G}}_0(\mathbf{r}, \mathbf{r}'; \omega_s)\mathbf{p}(\mathbf{r}', \omega_s) \tag{12.8}$$

As described in the 1D case in order to calculate the near-field intensity at the detector $I_{NF}(\mathbf{r}_0, \omega_s)$ one has to distinguish between the coherent and the incoherent case and set the modulus square either outside or inside the integral, respectively. Evaluating these integrals (see Ref. [21]) with the polarizability tensor for the Raman G mode of graphene (E_{2g}) yields the following results:

The intensity is proportional to the 8th power of the tip-sample distance for coherent scattering and to the 10th power for incoherent scattering. Again, in both cases the near-field signal is proportional to the 4th power of the field enhancement factor. Moreover, the authors derive guidelines for choosing the optimal conditions for the tip inclination.

12.3.3 *Contrast Mechanism Based on Tip-Sample Contact*

In the discussion made so far the origin of the signal enhancement was electromagnetic in nature and resulted from an increase in the local electric field and its coupling to the far-field. The electronic and vibronic states of the sample material were considered to be unmodified.

Additional image contrast and information can be obtained using the tip to mechanically perturb the sample as has been demonstrated for SWCNTs by different groups [22–24]. This type of interaction requires direct contact and is thus very short-ranged. Experiments in which the Raman response was recorded while exerting pressure using the signal-enhancing tip indeed achieved sub-nanometer spatial resolution, substantially below the purely optical resolution (see also below Section 12.5.1).

12.4 Instrumentation

The instrumental setup of a TERS system consists of a confocal microscope combined with a tip-sample distance control mechanism [14]. The excitation laser usually is a CW-laser in the visible range focused by an objective with a high numerical aperture onto the sample. Depending on the substrate material different illumination schemes exist as presented in Fig. 12.3. Opaque substrates can be investigated with the side or top illumination scheme (Fig. 12.3(c)) having the advantage that the polarization requirement (polarization parallel to tip axis) is easily fulfilled. The drawback is the limitation to objectives with a rather low numerical aperture ($0.5 < \text{NA} < 0.7$) due to larger working distances. Transparent samples allow the illumination from below with an oil-immersion objective with a high numerical aperture ($NA > 1.49$) increasing the collection efficiency of the system and decreasing the confocal background, that is, the detected signal in the absence of the tip. In this configuration, a radially polarized laser beam with an electric field component in the direction of propagation can be used in order to efficiently couple to the antenna [25] (Fig. 12.3(a)). A configuration in order to investigate opaque samples at a high numerical aperture employs a parabolic mirror as focusing element [26, 27] (Fig. 12.3(b)).

The tip–sample distance control can be based on normal or shear-force detection used in atomic force microscopy or current detection as in scanning tunneling microscopy. Within the last few years a large variety of tip designs have been introduced aiming at strongest field enhancement and localization [14]. Standard tips are commercial cantilever tips coated with gold or silver and etched wire tips made of the same materials.

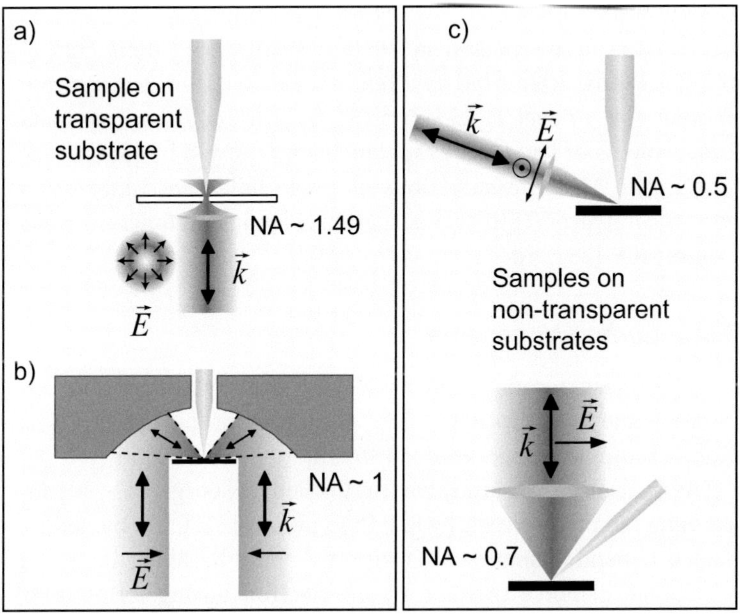

Figure 12.3 Illumination schemes used for TERS. (a) Schematic of an experimental setup employing on-axis illumination of a transparent sample. (b) Focusing of light using a parabolic mirror with high numerical aperture close to 1. (c) Side and top illumination of the tip on top of a nontransparent substrate.

For simultaneous optical and topographic image acquisition the sample is raster scanned. Therefore, the tip must be centered at the laser focus. The signal is collected by the same objective and can be detected by an avalanche photodiode in combination with a suitable filter set to suppress the laser and to filter out a certain Raman band. Alternatively, in order to achieve full spectroscopic information, a spectrometer equipped with a CCD camera can be used.

12.5 Applications to Carbon Nanotubes and Graphene

TERS has been demonstrated to be very well suited for the investigation of graphitic materials due to its rich phonon and electron–phonon coupling physics and the possibility to correlated topographical and optical properties. The application of TERS to SWCNTs was first reported in 2003 [28–30]. Since then, carbon nanotubes were often used as a reference material to demonstrate the quality of a novel TERS setup [31–33]. In 2009 the first TERS measurements of few-layer graphene were reported by [34]. So far there are no TERS investigations of single fullerenes reported in literature; however, TERS was successfully applied to fullerene films or aggregates [23, 35].

12.5.1 *Carbon Nanotubes*

Within the last decade TERS developed into a standard tool for the nanoscale investigation of the vibrational properties of CNTs. Several examples are reported in [36–38].

A more recent TERS study was performed by Georgi et al. [39] determining the phase-breaking length in metallic SWCNTs by defect-induced D-band Raman scattering. The phase-breaking length is the average distance traveled by an electron during its lifetime, that is, before being inelastically scattered by a lattice phonon. Figure 12.4(a–c) shows D-band and G-band intensity maps of a single metallic SWCNT with a spatial resolution of 15 nm. In this experiment a local defect was introduced on purpose by

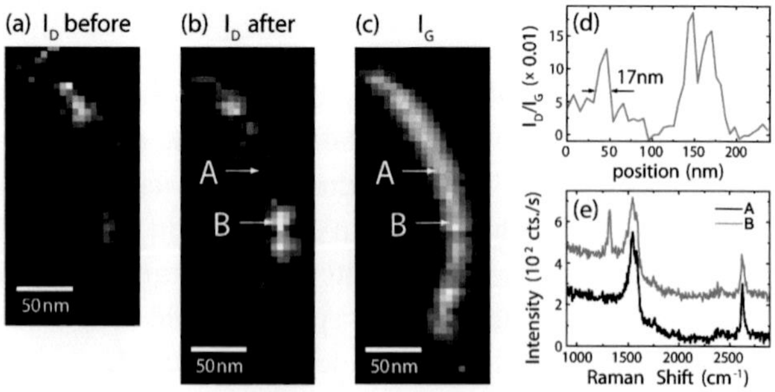

Figure 12.4 (a)–(c) Spectroscopic images of a single SWCNT before and after generation of an additional defect site (pixel size is 5.7 nm). (d) Distribution of the I_D/I_G-ratio along the SWCNT, showing strong local variations on a length scale of 17 nm. (e) Spectra from positions A and B, representing defect-free and highly defective positions within only 50 nm. Reprinted with permission from Ref. [39], Copyright 2010, American Chemical Society.

positioning the tip at position B and illuminating the SWCNT at high laser intensity for several minutes. The resulting photochemical modification gives rise to increased D-band scattering as seen in the scan image in Fig. 12.4(b) and the corresponding Raman spectra in Fig. 12.4(e), respectively. The ratio of D- and G-band intensities I_D/I_G that is a measure of the defect concentration taken along the SWCNT shows that the D-band scattering is localized within 17 nm (Fig. 12.4(d)). Considering the spatial resolution of the experiment the authors estimate the phase-breaking length to be equal to or smaller than 2 nm.

Yano et al. [40] used TERS for imaging of strain distributions along a single SWCNT. Local strain in form of tension and torsion was induced by bending the SWCNT with a cantilever tip leading to a modified G-band signature in the bended region. Okuno et al. [41] applied the same principle to investigate the electronic structure at the crossing point of two SWCNTs.

As described in Section 12.3.3 in TERS a limit for the spatial resolution exists that is determined by the tip size. Yano et al. [22] demonstrate an approach circumventing this limit by tip-pressurized Raman microscopy. By running the TERS system in a contact mode configuration that enables to exert a controlled amount of pressure onto a CNT the authors achieve a spatial resolution of 4 nm. This low value can be explained by the small number of molecules that are in physical contact with the tip. Only these molecules show a peak shift of the Raman G-band due to the induced pressure. Scanning a CNT under a constant force and mapping the peak shift reveals a spatial resolution that is higher by a factor of 5 than the topographic resolution (see Fig. 12.5).

In addition to TERS there are many reports on tip-enhanced photoluminescence studies of excitonic phenomena in SWCNTs such as energy transfer [42, 43], exciton localization and propagation [39, 44].

12.5.2 *Graphene*

So far, there have been only few tip-enhanced Raman measurements reported for graphene. Saito et al. [34] identified edge boundaries and estimated the number of layers of few-layer graphene using

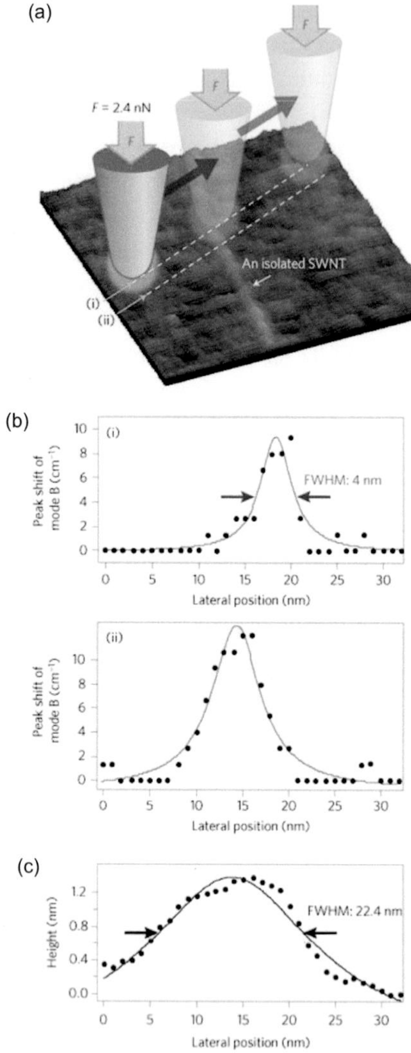

Figure 12.5 One-dimensional optical image of an SWCNT at extremely high spatial resolution. (a) An isolated SWCNT was scanned in steps of 1 nm under a constant tip-applied force of 2.4 nN. (b) The line profiles (i) and (ii) show the peak shifts (black dots) as a function of the tip position measured from the SWCNT, and corresponding best fits (red curves) for the scans are shown by lines (i) and (ii) in (a), respectively. The full width at half maximum (FWHM) of the line profile is 4 nm, which is more than five times smaller than that of the topographic profile shown in (c). Reprinted by permission from Macmillan Publishers Ltd: Ref. [22], copyright 2009.

TERS. The first TERS measurements on single-layer graphene are reported by Stadler et al. [45] mapping defects and contaminants with nanoscale resolution. Similar to the determination of the phase-breaking length in metallic SWCNTs described above, TERS was recently used to determine the phase-breaking length at the edges of a single graphene flake [46]. The edges are located by mapping the Raman D and G band scattering intensity (see Fig. 12.6). According to the authors, the width of the D band at the edge reveals a phase-breaking length of about 4 nm. This number is smaller by about a factor of 10 compared to confocal measurements of the same flake and is in general agreement with other literature reports [9].

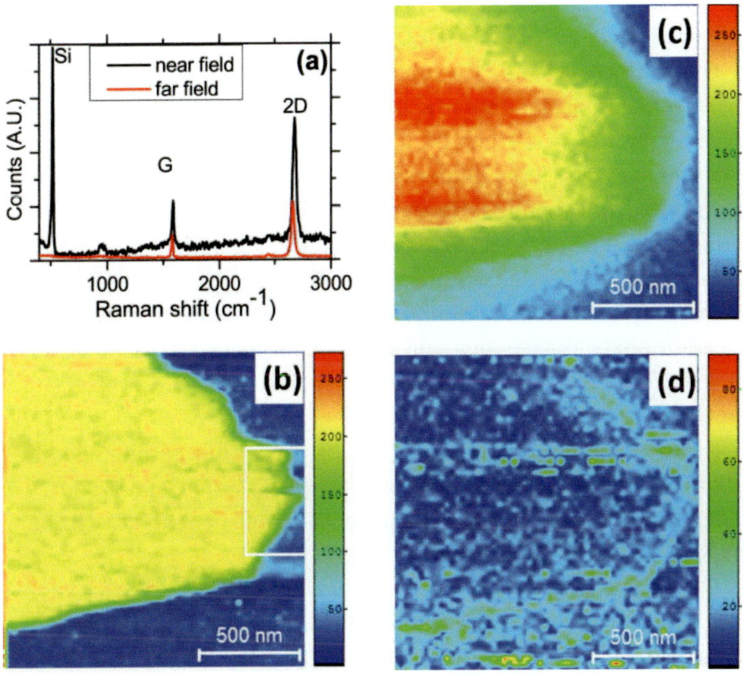

Figure 12.6 (a) Tip-enhanced and confocal Raman spectra of the monolayer graphene flake, (b) AFM lateral force image of the monolayer graphene recorded during TERS mapping, (c) TERS image formed using the 2D-peak height, and (d) TERS image formed using the D-peak height. (image size: 1.5 μm × 1.5 μm, 60 × 60 pixels). Reproduced with permission from Ref. [46]. Copyright 2013, American Vacuum Society.

12.6 Conclusions

In this chapter we reviewed the general principles of TERS and its application to carbon nanotubes and graphene. Both materials are photochemically stable and very efficient Raman scatterers that can also be detected topographically making them ideally suited for TERS studies. As quasi-one- and two-dimensional materials composed of sp^2 hybridized carbon they can serve as model systems for studying the role of the sample dimensionality in signal formation. In addition their electronic and vibronic properties are directly determined by the local structure. Structural variations and heterogeneities such as diameter changes, local defects, and the effect of extrinsic perturbations can thus be visualized on the nanoscale by taking Raman spectra.

On the other hand, the optical resolution achieved in TERS measurements so far remains insufficient to reveal the details of these structural variations on a scale of few atoms. For example, a detailed microscopic understanding of what types of defects actually lead to D-band scattering and with which yield. Here studies of the same samples with complementary methods, for example, high-resolution transmission electron microscopy (HRTEM), could provide essential information. It remains to be seen to what extent the additional enhancement and localization achieved for molecules on metal substrates [47] can be further improved and utilized for nanocarbons.

As a scanning probe method TERS can be readily combined with other signal detection schemes such as Kelvin probe, electric force, and photocurrent thus leading to unprecedented insight into a broad range of sample characteristics [14, 48, 49].

References

1. Jorio, A. (2012). *ISRN Nanotechnol.*, **2012**, article ID 234216, 16 pages, doi:10.5402/2012/234216.
2. Novoselov, K. S., Jiang, Z., Zhang, Y., Morozov, S. V., Stoermer, H. L., Zeitler, U., Maan, J.C., Boebinger, G. S., Kim, P. and Geim, A. K. (2007) *Science*, **315**, p. 1379.

3. Saito, R., Dresselhaus, G. and Dresselhaus, M. S. (1998). *Physical properties of carbon nanotubes*, (Imperial College Press, London).
4. Ferrari, A. C. and Robertson, J. (2004). *Phil. Trans. Soc. Lond. A*, **362**, pp. 2477–2512.
5. Dresselhaus, M. S., Dresselhaus, G., Saito, R. and Jorio, A. (2005). *Phys. Rep.*, **409**, pp. 47–99.
6. Jorio, A., Saito, R., Hafner, J. H., Lieber, C. M., Hunter, M., McClure, T., Dresselhaus, G. and Dresselhaus, M. S. (2001). *Phys. Rev. Lett.*, **86**, pp. 1118–1121.
7. Maultzsch, J., Pomraenke, R., Reich, S., Chang, E., Prezzi, D., Ruini, A., Molinari, E., Strano, M. S., Thomsen, C. and Lienau, C. (2005). *Phys. Rev. B*, **72**, pp. 241402(R)–241405 (R).
8. Ferrari, A. C., Meyer, J. C., Scardaci, V., Casiraghi, C., Lazzeri, M., Mauri, F., Piscanec, S., Jiang, D., Novoselov, K. S., Roth, S. and Geim, A. K. (2006). *Phys. Rev. Lett.*, **97**, p. 187401.
9. Casiraghi, C., Hartschuh, A., Qian, H., Georgi, C., Fasoli, A., Novoselov, K. S., Basko, D. M. and Ferrari, A. (2009). *Nano Lett.*, **9**, pp. 1433–1441.
10. Novotny, L. and van Hulst, N. (2011). *Nature Photon.*, **5**, pp. 83–90.
11. Mühlschlegel, P., Eisler, H.-J., Hecht, B. and Pohl, D. W. (2005). *Science*, **308**, pp. 1607–1609.
12. Hecht, B., Sick, B., Wild, U. P., Deckert, V., Zenobi, R., Martin, O. J. F. and Pohl, D. E. (2000). *J. Chem. Phys.*, **112**, pp. 7761–7774.
13. Bharadwaj, P., Deutsch, B. and Novotny, L. (2009). *Adv. Opt. Photon.*, **1**, pp. 438–483.
14. Mauser, N. and Hartschuh, A. (2014). *Chem. Soc. Rev.*, **43**, 1248–1262.
15. Cancado, L. G., Jorio, A., Ismach, A., Joselevich, E., Hartschuh, A. and Novotny, L. (2009). *Phys. Rev. Lett.*, **103**, pp. 186101–186104.
16. Kühn, S. and Sandoghdar, V. (2006). *Appl. Phys. B*, **84**, pp. 211–217.
17. Taminiau, T. H., Stefani, F. D., Segerink, F. B. and van Hulst, N. F. (2008). *Nature Photonics*, **2**, pp. 234–237.
18. Böhmler, M., Hartmann, N., Georgi, C., Hennrich, F., Hersam, M. C. and Hartschuh, A. (2010). *Opt. Express*, **18**, pp. 16443–16451.
19. Berweger, S. and Raschke, M. B. (2010). *Anal. Bioanal. Chem.*, **396**, pp. 115–123.
20. Uryu, S., Ajiki, H. and Ishihara, H. (2013). *Phys. Rev. Lett.*, **110**, p. 257401.
21. Maximiano, R. V., Beams, R., Novotny, L., Jorio, A. and Cancado, L. G. (2012). *Phys. Rev. B*, **85**, p. 235434.

22. Yano, T., Verma, P., Saito, Y., Ichimura, T., Kawata, S. (2009). *Nat. Photon*, 3, pp. 473–477.
23. Verma, P., Yamada, K., Watanabe, H., Inouye, Y. and Kawata, S. (2006). *Phys. Rev. B*, **73**, p. 045416.
24. Araujo, P. T., Barbosa Neto, N. M., Chacham, H., Carara, S. S., Soares, J. S., Souza, A. D., Canado, L. G., de Oliveira, A. B., Batista, R. J. C., Joselevich, E., Dresselhaus, M. S. and Jorio, A. (2012). *Nano Lett.*, **12**, p. 4110.
25. Novotny, L., Sánchez, E. J. and Xie, X. S. (1998). *Ultramicroscopy*, **71**, pp. 21–29.
26. Lieb, A. and Meixner, A. J. (2001). *Optics Express*, **8**, pp. 458–474.
27. Steidtner, J. and Pettinger, B. (2007). *Rev. Sci. Instrum.*, **78**, pp. 103104–103108.
28. Hartschuh, A., Sánchez, E. J., Xie, X. S. and Novotny, L. (2003). *Phys. Rev. Lett.*, **90**, pp. 095503–095504.
29. Hartschuh, A., Anderson, N. and Novotny, L. (2003). *J. Microscopy*, **210**, pp. 234–240.
30. Hayazawa, N., Yano, T., Watanabe, H., Inouye, Y. and Kawata, S. (2003). *Chem. Phys. Lett.*, **376**, pp. 174–180.
31. Roy, D., Wang, J. and Williams, C. (2009). *J. Appl. Phys.*, **105**, p. 013530.
32. Nicklaus, M., Nauenheim, C., Krayev, A., Gavrilyuk, V., Belyaev, A. and Rüdiger, A. (2012). *Rev. Sci. Instrum.*, **83**, p. 066102.
33. Oron-Carl, M. and Krupke, R. (2013). *Nanotechnology*, **24**, p. 415701.
34. Saito, Y., Verma, P., Masui, K., Inouyeb, Y. and Kawata, S. (2009). *J. Raman Spectrosc.*, **40**, pp. 14341440.
35. Stöckle, S. M., Suh, Y. D., Deckert, V. and Zenobi, R. (2000). *Chem. Phys. Lett.*, **318**, pp. 131–136.
36. Maciel, I. O., Anderson, N., Pimenta, M. A., Hartschuh, A., Qian, H., Terrones, M., Terrones, H., Campos-Delgado, J., Rao, A. M., Novotny, L. and Jorio, A. (2008). *Nature Mat.*, **7**, pp. 878–883.
37. Anderson, N., Hartschuh, A. and Novotny, L. (2005). *J. Am. Chem. Soc.*, **127**, pp. 2533–2537.
38. Anderson, N., Hartschuh, A. and Novotny, L. (2007). *Nano Lett.*, **7**, pp. 577–582.
39. Georgi, C., Green, A. A., Hersam, M. C. and Hartschuh, A. (2010). *ACS Nano*, **4**, pp. 5914–5920.
40. Yano, T., Ichimura, T., Kuwahara, S., H'Dhili, F., Uetsuki, K., Okuno, Y., Verma, P. and Kawata, S. (2013). *Nat. Comm.*, **4**, p. 2592.

41. Okuno, Y., Saito, Y., Kawata, S. and Verma, P. (2013). *Phys. Rev. Lett.*, **111**, p. 216101.
42. Qian, H., Georgi, C., Anderson, N., Green, A. A., Hersam, M. C., Novotny, L. and Hartschuh, A. (2008). *Nano Lett.*, **8**, pp. 1363–1367.
43. Qian, H., Georgi, C., Anderson, N., Green, A. A., Hersam, M. C., Novotny, L. and Hartschuh, A. (2008). *Phys. Stat. Sol. B*, **245**, pp. 2243–2246.
44. Georgi, C., Bøhmler, M., Qian, H., Novotny, L. and Hartschuh, A. (2009). *Phys. Stat. Sol. B*, **246**, pp. 2683–2687.
45. Stadler, J., Schmid, T. and Zenobi, R. (2011). *ACS Nano*, **5**, pp. 8442–8448.
46. Su, W. and Roy, D. (2013). *J. Vac. Sci. Technol. B*, **31**, p. 041808.
47. Zhang, R., Zhang, Y., Dong, Z. C., Jiang, S., Zhang, C., Chen, L. G., Zhang, L. Liao, Y., Aizpurua, J., Luo, Y., Yang, J. L. and Hou, J. G. (2013). *Nature*, **498**, p. 82.
48. Regler, M. E., Krenner, H. J., Green, A. A., Hersam, M. C., Wixforth, A. and Hartschuh, A. (2012). *Chem. Phys., http://dx.doi.org/10.1016/j.chemphys.2012.10.014*.
49. Burghard, M. and Mews, A. (2012). *ACS Nano*, **6**(7), pp. 5752–5756.

Chapter 13

20 nm–Resolved Stress Profile in SiGe Nano-Stripes Obtained by Tip-Enhanced Raman Spectroscopy

Marc Chaigneau,[a] Giovanni Maria Vanacore,[b*] Monica Bollani,[c] Gennaro Picardi,[a] Alberto Tagliaferri,[b] and Razvigor Ossikovski[a]

[a]*LPICM, Ecole Polytechnique, CNRS, F-91128 Palaiseau, France*
[b]*CNISM, Dipartimento di Fisica, Politecnico di Milano, I-20133 Milan, Italy*
[c]*CNR-IFN e L-NESS I-22100 Como, Italy*
razvigor.ossikovski@polytechnique.edu

In this chapter, we describe the determination of the stress profile in 150 nm–wide SiGe nano-stripes embedded into a Si matrix by using oblique incidence tip-enhanced Raman spectroscopy (TERS) with a spatial resolution of ~20 nm. The TERS spectra of the stripes exhibit a number of locally enhanced phonon modes that are absent when the tip is positioned out of the stripes. The hydrostatic stress component across the nano-stripe width is evaluated from the strain-induced frequency shift of the Si-Ge mode at ~380 cm^{-1}. The stress magnitude is found to be largest in the nano-stripe

*Present address: Physical Biology Center for Ultrafast Science and Technology, Arthur Amos Noyes Laboratory of Chemical Physics, California Institute of Technology, Pasadena (CA), USA

Handbook of Enhanced Spectroscopy
Edited by Pietro Giuseppe Gucciardi, Marc Lamy de la Chapelle, and Nathalie Lidgi-Guigui

ISBN 978-981-4613-32-3 (Hardcover), 978-981-4613-33-0 (eBook)
www.panstanford.com

center and decreases monotonously on each side down to zero at the boundaries. This behavior is quantitatively described by a classic continuous medium model. These results demonstrate the applicability of the TERS technique to stress determination in novel semiconductor structures at the nanometer scale.

13.1 Introduction

Advanced structures and devices based on the attractive idea of integrating silicon with other materials such as germanium [1] and carbon [2] have been recently developed due to their compatibility with the mainstream Si technology. Among these new engineered or "smart" materials, epitaxial SiGe/Si heterostructures have attracted considerable interest because of their ability to enhance the performance of metal oxide semiconductor (MOS) devices *via* band structure engineering [3]. Indeed, for a long time the improvement in the performance has been achieved through miniaturizing individual devices. Unfortunately, this so-called scaling-down approach has been facing significant physical and economical limitations. As a consequence, since the introduction of the 90-nm technology node, an alternative approach called "strain engineering" [4] has been adopted to further maintain the Moore's law trend.

For instance, introducing strain in the channel of a field-effect transistor (FET) results in large carrier mobility enhancements [5]. The material under stress can be strained Si (*s*Si), strained Ge (*s*Ge), or strained SiGe alloy (*s*SiGe) depending on the final application: MOSFETs, SiGe-based heterojunction bipolar transistors (HBT), or optoelectronic devices such as quantum well infrared detectors and quantum cascade emitters. The ability to control the growth process and to monitor the strain and composition of a SiGe alloy is therefore crucial for the engineering of the electrical and optical properties of the final heterostructure.

Most generally, Raman spectroscopy is a versatile nondestructive optical characterization technique whose numerous applications cover entire branches of modern electronic and optoelectronic device technology, from semiconductor growth to final device inspection and quality control [6]. Concisely, Raman spectroscopy probes the inelastic scattering of monochromatic light induced by the lattice vibrations in a crystalline solid. The strain, associated to internal (residual) or external (applied) stress, changes the equilibrium position of the atoms in the lattice. This modifies the phonon properties and the peak frequencies of the phonon lines originating from the strained crystal are shifted. By monitoring the shift of the Raman phonon lines, one can evaluate the strain by using perturbation theory; compressive and tensile strain result in (approximately) linearly proportional increases and decreases in the line frequency [7, 8]. More specifically, Raman spectroscopy has been widely used as a straightforward tool to determine the strain and the composition of SiGe alloys [9–12]. In Si-based structures, frequency shifts corresponding to a minimum stress value of several tens of MPa can be currently detected [13–15].

Nonetheless, the strain information over a given sample area provided by Raman spectroscopy at visible laser wavelengths is generally restricted to a spatial resolution of few hundreds of nanometers because of the diffraction limit. Strain and mobility optimization on semiconductor structures narrower than $\sim$0.3 μm is typically achieved by using spatially averaged responses and model-based interpretations [13]. As a consequence, the investigation of details within a nanoscale structure such as a single transistor channel is nowadays a challenging task since both the characterization and control of the strain in the channel down to the nanometer level are of fundamental importance for the design of new generation devices. As a matter of fact, existing techniques do not have the ability to map the distribution of strain in a transistor from the actual 45 nm technology.

TERS provides an interesting alternative for going beyond the diffraction limit by performing localized Raman spectroscopic scanning or mapping with high spatial resolution on a wide range of nanomaterials such as carbon nanotubes and graphene [16, 17], self-assembled organic monolayers [18, 19], and DNA molecules [20]. This is possible because of the local amplification and the spatial confinement of the electromagnetic radiation by the apex of a sharp metallic tip. TERS can thus offer unprecedented detail in the spatially resolved strain mapping of single nanostructures, without requiring destructive procedures for the sample preparation. This technique has already demonstrated its capability for the characterization of *s*Si and *s*SiGe structures [21–26], but the recorded TERS scans or maps do not always exploit the full spectroscopic information available from the Raman scattering experiment. In fact, it should be noted that, in order to obtain the strain distribution, the frequency shifts and not the intensities of the (locally enhanced) Raman phonon lines have to be monitored. This is a point of primary importance since the TERS line intensities can be affected to different extent by parasitic light interference [27] and tip depolarization effects [28, 29]. Recently, Ogawa et al. succeeded to measure the line shift attributed to the strain induced in a Si substrate by SiGe quantum dots, but depth averaging of the signal over the strongly nonuniform strain distribution prevented them from obtaining strain values in agreement with those previously reported on similar structures [30]. Hermann et al. reported on the TERS imaging and strain analysis of nanoscale SiGe structures [31]. However, the stress values obtained were not spatially resolved within a single nanoscale structure. At present, the spatial resolution required to monitor the strain in patterned semiconductor structures at the nanometer scale is not routinely achieved.

In this chapter, we report the nanoscale mapping of the strain profile of SiGe embedded nano-stripes with $\sim$20 nm spatial lateral resolution obtained by TERS in oblique backscattering configuration [32]. The TERS spectra, recorded each 20 nm across a single 150-nm-wide nano-stripe, exhibit Ge-Ge and Si-Ge phonon mode features that are absent in the far-field Raman response of the structure and of the inter-stripe area. The strain information was retrieved by monitoring the Si-Ge mode frequency as a function of the distance across the stripe while taking into account the modified Raman selection rules due to the presence of the tip. The stress magnitude was found to be largest at the center and to decrease monotonously toward the boundaries. A concentrated force model quantitatively reproduces the experimental results. These results represent a further advancement for the routine application of the TERS technique as an innovative tool for strain characterization of semiconducting nanostructures.

13.2 Experimental

13.2.1 *TERS Setup in Oblique Backscattering Configuration*

The TERS setup consists of a high resolution HORIBA Jobin Yvon *Labram HR800* Raman spectrometer optically coupled in an oblique (off-axis) backscattering geometry (at 60° with respect to the sample normal) to a Park Systems *XE-100* atomic force microscope (AFM) operated in scanning tunneling mode (STM) through a long-working-distance objective ($50\times$, NA $= 0.45$) from Olympus. The optomechanical coupling provided by Horiba Jobin Yvon is motorized along the x, y, and z axes and allows for an accurate alignment of the exciting laser spot with respect to the tip apex at the sample. The entire setup is operated under an acoustic enclosure providing the stability needed during the TERS mapping. The motorized optomechanical coupling is controlled with a joystick located outside the enclosure. The excitation wavelength used, 633 nm, was provided by the built-in HeNe laser of the spectrometer. The polarization state of the incident laser was set at p-polarization (electric field parallel to the scattering plane defined by the laser beam and the sample normal) by using a half-wave plate inserted in a motorized rotating mount with a $\pm 5^\circ$ accuracy. The sample is mounted on a manual rotating stage allowing for a full 360° rotation and placed so that its [110] crystallographic axis (corresponding to the nano-stripe direction) was aligned with the scattering plane, see Fig. 13.3(a). A schematics of the scattering configuration is represented in Fig. 13.1.

13.2.2 *TERS Tips Manufacturing Process*

The STM tips were prepared from a 0.25-mm-thick gold wire (from Goodfellow) by electrochemical etching in a concentrated HCl/ethanol 1:1 mixture [33]. Before being plugged deep into the HCl solution, a 2 cm piece of gold wire was positioned in the center of a 1-mm-thick gold counter-electrode (Fig. 13.2(a)). An electronic circuit was used to automatically cut off the etching process as soon as the current drops below a set-point value, a relay switch being used as a circuit breaker. The automatic cut off avoids over-

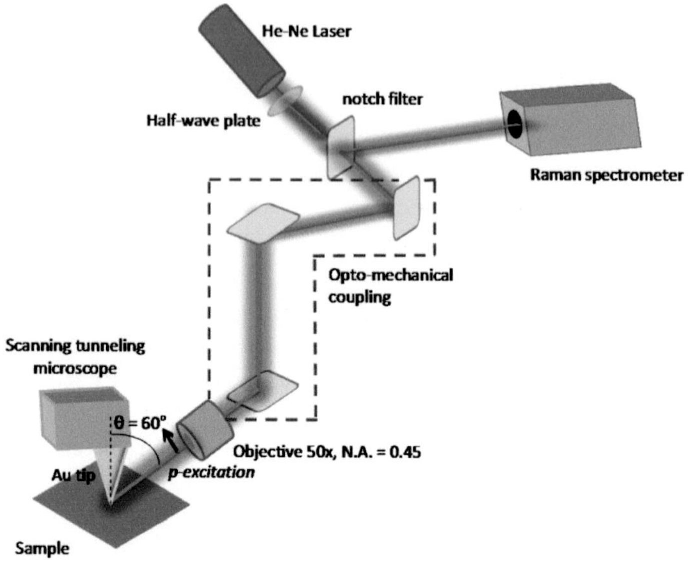

Figure 13.1 Schematics of the oblique backscattering-mode TERS experiment.

etching as the lower part of the wire separates. The system allows for the etching of five wires, one by one, in the same acid bath carried out in a glass beaker. The typical time at room temperature to create a conical tip by this electrochemical etching method was 8 min in HCl at 37% concentration. The tips were then washed in water. Tips with apex radii smaller than 30 nm can be reproducibly fabricated by using this technique (Fig. 13.2(b)), which furthermore provides minimized surface roughness avoiding spurious spectral contributions due to carbonaceous contaminants in the tip apex vicinity [34].

Tunneling experiments took place in ambient conditions by using a 1 V tip-positive bias voltage and a 0.1 nA tunneling current. Prior to each TERS measurement sequence, native silicon and germanium oxides naturally present on the sample surface were removed by means of diluted hydrofluoric acid (DHF) (10%; 30 s dip at room temperature). It should be noted that, despite the requirement for a conducting sample for tunneling, TERS experiments performed in

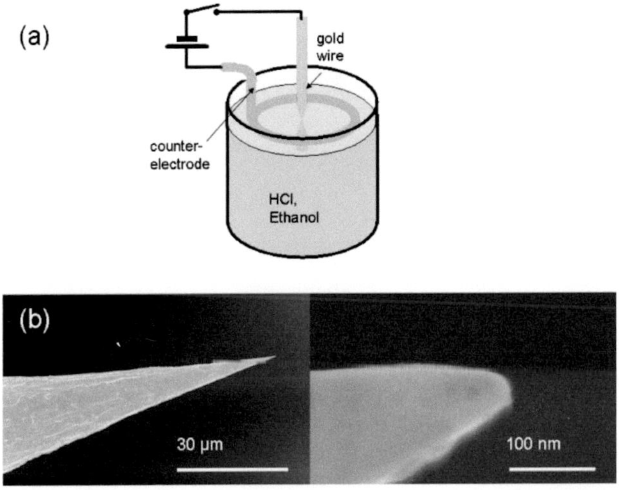

Figure 13.2 (a) Schematics of the etching setup for preparing the gold tip. (b) SEM images of an etched tip typically obtained.

STM mode generally have better reproducibility and exhibit larger enhancements than those performed in AFM mode [35].

13.2.3 *Strained SiGe Nano-Stripes*

The SiGe embedded nano-stripes were elaborated by coupling electron beam lithography (EBL) and low-energy plasma-enhanced chemical vapor deposition (LEPECVD) techniques [36]. A n^+(As)-doped Si(001) substrate was patterned with a series of trenches (nominal 110 nm depth and 150 nm width) aligned along the [110] direction by means of EBL. The epitaxial deposition of pure Ge by LEPECVD allowed for the filling of the trenches and the formation of the embedded nano-stripes. The SiGe epilayer possibly formed in-between the patterned structures after the Ge growth was completely etched away by a gentle mechanical polishing [32]. Figure 13.3(b) shows an SEM image of the periodic array of the nano-stripes. They exhibit lateral widths of about 150 nm and thicknesses of 110 nm (determined from an SEM image of a sample cross-section; not shown), and are evenly spaced by about 1 μm.

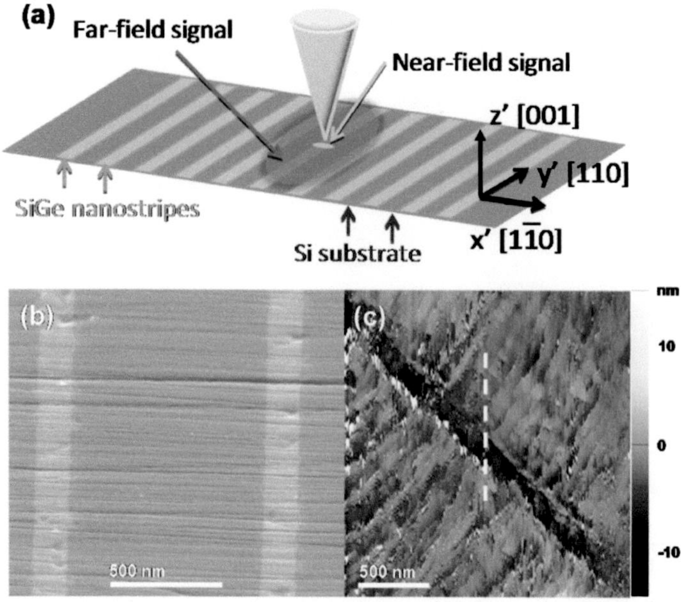

Figure 13.3 (a) Schematic representation of the tip in tunneling position on the sample together with reference frame used. The far-field and the near-field regions within the sample are represented in red and green, respectively. The scattering plane is the $y'z'$ plane; the nano-stripe array is parallel to it. (b) SEM image of the periodic array of SiGe nano-stripes [32]. The stripe width is about 150 nm. (c) STM image of a nano-stripe (−1 V sample bias, 0.1 nA tunneling current).

13.3 Results and Discussion

13.3.1 *TERS Measurement Procedure*

Prior to the TERS experiment, an area containing a single SiGe nano-stripe was selected by imaging the sample surface in STM mode (see Fig. 13.3(c)). The tunneling tip was then moved point by point along the dotted line as shown in Fig. 13.3(c) with a 20 nm step. The Raman spectra were recorded at each tip position with an integration time of 20 s and a 1 s delay between two consecutive measurements.

The Raman spectra recorded with the tip tunneling and the tip retracted on the Si substrate region between two adjacent

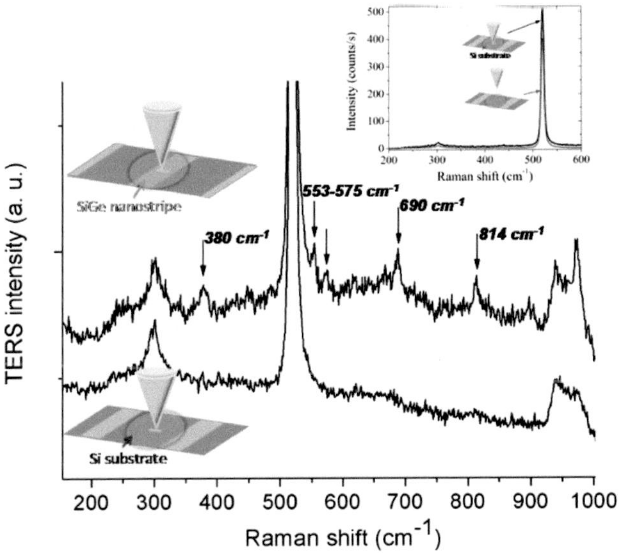

Figure 13.4 TERS spectra obtained with the tip in tunneling position on the Si substrate between two stripes (lower trace) and on a SiGe nano-stripe (upper trace). The enhanced Raman peaks originating from the SiGe nano-stripe are identified by arrows. Inset: Raman spectra measured on the Si substrate in between adjacent nano-stripes with the tip tunneling (near-field spectrum, black line) and the tip retracted (far-field spectrum, blue line).

nano-stripes are presented in the inset of Fig. 13.4. A 1.3-fold increase of the Raman intensity is observed for the first-order triply degenerate Si-Si optical phonon peak at ~521 cm^{-1}, accompanied by about a twofold increase for the 2TA (transverse acoustical) overtone at ~300 cm^{-1}. The “tip up–tip down” differences in intensities (contrasts) of these two pairs of Raman peaks are not only due to a near-field enhancement mechanism but can also be affected by the partial depolarization of the incident light induced by the tip apex, and by the different symmetry of the Raman polarizability tensors for the degenerate optical modes and the 2TA mode, as evidenced in previous works [28, 29]. This situation illustrates the extreme care that must be taken in interpreting the nature of the apparent “enhancements” observed in TERS experiments on bulk crystalline materials.

The lower trace in the main panel of Fig. 13.4 represents a baseline-corrected TERS spectrum recorded with the tip tunneling on the Si substrate between two stripes. As mentioned, the signal is mostly related to the Raman scattering from the entirely illuminated volume. The oblique backscattering geometry produces an elliptical laser spot with approximate long and short axis lengths of 2.5 μm and 1 μm respectively. Having in mind that the nano-stripes are arranged in a 1-μm-period array, one can conclude that no more than two adjacent nano-stripes are illuminated by the laser, and therefore, probed in the far-field. However, no Raman signal from the SiGe nano-stripes was observed unless the tip was tunneling right above one of them. With the tip tunneling on the Si substrate, only the peaks at ~300 cm^{-1}, 521 cm^{-1}, and the broad band between 930 and 990 cm^{-1}, still related to bulk silicon (2TO mode) [37] clearly appear. It is worth noting that these spectral features are exactly those present in the spectrum measured with the tip retracted, as it is evident by a comparison with the far-field spectrum presented in the inset of Fig. 13.4.

13.3.2 *SiGe Nano-stripes Raman Mode Assignment*

The upper trace in Fig. 13.4 is the baseline-corrected TERS spectrum obtained with the tip tunneling on a single SiGe nano-stripe. Here new features (designated by arrows) appear, which are totally absent in the far-field spectrum, whereas the most intense bands related to the far-field contribution from the Si bulk substrate remain substantially unmodified. Indeed, the large scattering volume of bulk Si is still probed in the far-field, but the tip-enhanced field also now probes the (much smaller) SiGe volume located just below the tip apex.

In order to perform a correct assignment of the TERS peaks originating from the nano-stripes, the composition of the SiGe nano-stripe must be taken into account since it is one of the factors influencing the Raman peak position. Energy-filtered photo-electron emission microscopy (XPEEM) using soft X-ray excitation was used to measure the Ge concentration of the $Si_{1-x}Ge_x$ alloy filling the nano-stripes providing an average value $x = 0.91 \pm 0.03$ [32]. The intensities and positions of the Ge-Ge and Si-Ge optical phonon modes in Ge-rich $Si_{1-x}Ge_x$ alloys with $1 \geq x \geq 0.5$ are affected as follows: the Ge-Ge peak downshifts from 300 cm^{-1} (bulk Ge) to 285 cm^{-1} ($x \approx 0.5$), while the intensity of the Si-Ge peak increases and its position upshifts from 380 cm^{-1} (for $1 > x \geq 0.9$) to 410 cm^{-1}($x \approx 0.5$) [12, 38, 39]. In the TERS spectrum shown in Fig. 13.4, the contribution of the optical Ge-Ge mode from the nano-stripe and that of the strong Si 2TA mode from the substrate superimpose at ~300 cm^{-1} and cannot be distinguished. Besides, we assign the peak appearing around 380 cm^{-1} to the first-order Si-Ge phonon mode originating from the nano-stripe and being enhanced by the tip. At around 450 cm^{-1}, a very weak feature appears; different authors have observed this band, together with two more subbands (at ~440 cm^{-1} and ~460 cm^{-1}), and have tentatively attributed all three to vibrations of Si impurity-like atoms in Ge-rich specific environments [40–42]. These bands are also sometimes referred to as local Si modes [40, 43]. The doublet structure appearing at 553 cm^{-1} and 575 cm^{-1} is readily attributed to the 2TO(W) and 2TO(L) overtones of the Ge-Ge Raman mode [44–46]. Two more contributions in the TERS spectra remain to be discussed: a feature located at ~814 cm^{-1} can be associated either with the Si overtones 2LO(X) and 2LA(X) (the two being degenerate in frequency [47]) or with a combination of the first-order optical modes of silicon at ~520 cm^{-1} and germanium at ~300 cm^{-1}. The band at ~690 cm^{-1} is also of delicate assignment: according to Uchinokura et al. [47], it can be related to the overtones 2LA(Q) or 2LA(L-K(U)) of either silicon or germanium (or, in our case, even to both of them because of the SiGe alloy). In fact, these overtone states have very close frequencies in both Si and Ge. Considering that the two bands in question (at 814 cm^{-1} and 690 cm^{-1}) appear only when the enhancing tip is located above the embedded nano-stripe, we incline in both cases toward the assignment involving germanium contributions.

13.3.3 *TERS Profile across the SiGe Nano-stripes*

Figures 13.5(c) and 13.5(d) show the integrated TERS intensities of the first-order Si-Ge Raman peak at 380 cm^{-1} and of the Ge-Ge 2TO doublet mode at 553–575 cm^{-1} as a function of the tip position across the nano-stripe during a line scan—the dashed

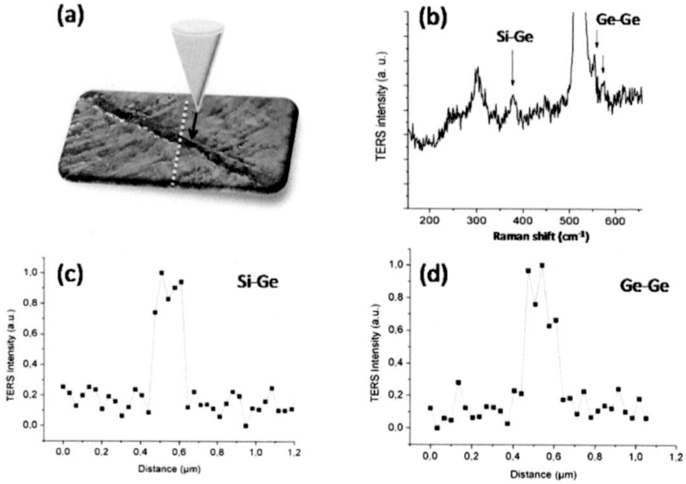

Figure 13.5 (a) Schematic representation of the tip in tunneling position showing the angle between the scan (dotted line) and the nano-stripe direction. (b) TERS spectrum recorded with the tip tunneling on the nano-stripe. Integrated intensities of the first-order Si-Ge (c) and Ge-Ge 2TO (d) peaks as functions of the tip position across the nano-stripe during a line scan with 20-nm step.

line in Fig. 13.5(a)—with a nominal 20-nm step. Figure 13.5(b) shows a typical TERS spectrum obtained on the nano-stripe. The intensity profiles of the Ge-Ge 2TO and the Si-Ge modes exhibit a reproducible experimental behavior and their widths (~200 nm) are fully consistent with the nominal nano-stripe width of 150 nm after taking into account the ~40° angle between the scan and nano-stripe directions (see Fig. 13.5(a)). From these line scans, the experimental spatial resolution can be estimated to be better than 20 nm: the nano-stripe-related TERS peak profiles sharply rise from and fall below the noise level within a single scan step.

For comparison, results obtained on a neighboring nano-stripe are also presented. The TERS spectrum recorded over it is shown in Fig. 13.6(a). Other than the Ge-Ge 2TO and the (up-shifted) Si-Ge peak, the spectrum presents a peak located at 435 cm^{-1}, attributed to a Si-Si local mode in a Ge-rich alloy, like the previously discussed weak feature at ~450 cm^{-1} from the first nano-stripe spectrum in Fig. 13.5(b). It is worth noting that the Si-Ge Raman

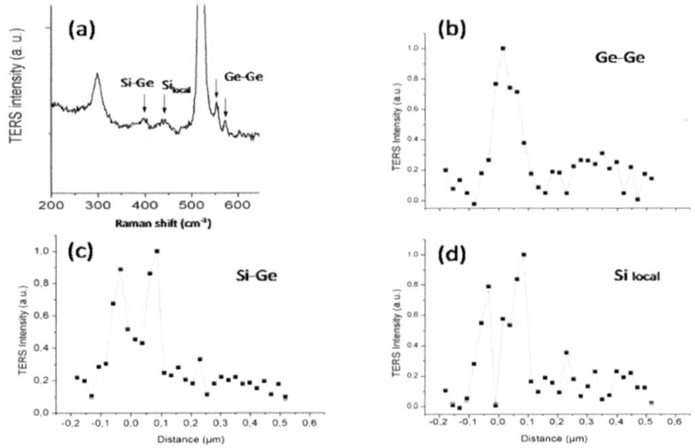

Figure 13.6 (a) TERS spectrum recorded with the tip tunneling on a nano-stripe adjacent to that shown in Fig. 13.5. Integrated TERS intensities of Ge-Ge 2TO (b) the first-order Si-Ge (c) and the local Si (d) modes as functions of the tip position across the nano-stripe during a line scan with a 20-nm step.

mode is now located at higher wavenumbers (~400 cm^{-1} versus ~380 cm^{-1}, compare Figs. 13.6(a) and 13.5(b)). This observation, combined with the clearer spectral signature of the Si local mode at ~435 cm^{-1}, indicates that the Ge concentration of the SiGe alloy in this nano-stripe is locally lower [38, 39, 42]. Figures 13.6(b–d) show the TERS intensity profiles as functions of the tip position for the Si-Ge phonon at ~400 cm^{-1}, the Ge-Ge 2TO doublet and the Si local mode. The intensity profiles of both the local Si and Si-Ge modes correlate with each other, exhibiting a clear dip at the center of the stripe, absent in the profile of the Ge-Ge 2TO mode. The widths of the local Si and SiGe mode intensity profiles are once again ~200 nm, whereas the width of the Ge-Ge mode is only about 150 nm. This possibly indicates a local inhomogeneity in the Si atoms distribution inside the alloy, consistent with a preferential accumulation of Ge atoms at the center rather than near the boundaries. These considerations demonstrate the sensitivity of the TERS technique to the local nano-stripe alloy stoichiometry. Locally measuring composition is of utmost importance since an inhomogeneous Ge distribution in the nano-stripe directly affects the device performance.

13.3.4 *Stress Analysis in Strained SiGe Nano-stripes by TERS*

We now focus on the determination of the local strain present in the nano-stripes. In case of SiGe structures epitaxially grown on Si substrates, the strain results from the lattice mismatch between the SiGe alloy and the bulk Si [15, 48]. For a Ge-rich SiGe-alloy nano-stripe embedded in a Si matrix, it is rather the substantial difference in Young modulus and Poisson ratio values for Ge and Si that are responsible for the induced strain. Irrespectively of its origin, the strain most generally modifies the frequencies and the intensities of the Raman lines.

In cubic crystals such as Si and Ge, the Raman active transverse optical (TO1 and TO2) and longitudinal optical (LO) phonon modes are triply degenerate at $\mathbf{k} = 0$. Lowering of the symmetry by the strain-induced lattice distortion lifts the degeneracy resulting in both mode splitting and modification of the Raman polarizability tensors, $\mathbf{R_i}$, of the three modes [7, 8, 15]. The new phonon frequencies and polarizability tensors are calculated, respectively, from the eigenvalues λ_i and eigenvectors $\mathbf{v}_i$ of the symmetric matrix **K**,

$$\mathrm{K} = \begin{bmatrix} p\varepsilon_{xx} + q(\varepsilon_{yy} + \varepsilon_{zz}) & 2r\varepsilon_{xy} & 2r\varepsilon_{xz} \\ 2r\varepsilon_{xy} & p\varepsilon_{yy} + q(\varepsilon_{xx} + \varepsilon_{zz}) & 2r\varepsilon_{yz} \\ 2r\varepsilon_{xz} & 2r\varepsilon_{yz} & p\varepsilon_{zz} + q(\varepsilon_{yy} + \varepsilon_{xx}) \end{bmatrix} \tag{13.1}$$

in which ε_{ij} are the components of the strain tensor referred to the "crystal" reference frame *xyz* defined by [100], [010], and [001] crystallographic directions and *p*, *q*, and *r* are the so-called phonon deformation potentials describing the elastic properties of the specific phonon mode of the strained crystal. From Ref. [12], one obtains the following values for Si-Ge optical phonon deformation potentials of the $\mathrm{Si_{0.09}Ge_{0.91}}$ alloy: $p = -2.41\ \omega_0^2$ and $q = -2.39\ \omega_0$ where $\omega_0 = 386.9\ \mathrm{cm}^{-1}$ is the triply degenerate "unstrained" frequency of the mode.

The "strained" frequency of the ith optical mode, ω_i, is obtained from the eigenvalues λ_i of the **K** matrix,

$$\omega_i^2 = \omega_0^2 + \lambda_i \tag{13.2}$$

The intensity I_i of each mode is determined from its "strained" polarizability tensor and the Raman selection rule [7] modified to account for the presence of the tip in the TERS experiment [49, 50] (see appendix for the detailed calculation).

In accordance with the nano-stripe morphology and geometry, we consider the following diagonal representation of the stress tensor σ within the nano-stripe, expressed in the "sample" reference frame $x'y'z'$ defined by the [1$\bar{1}$0], [110], and [001] crystallographic directions

$$\boldsymbol{\sigma} = \begin{bmatrix} \sigma_{x'x'} & 0 & 0 \\ 0 & \sigma_{y'y'} & 0 \\ 0 & 0 & \sigma_{z'z'} \end{bmatrix} \tag{13.3}$$

To relate stress to strain, the stress tensor σ must be first rotated by 45° from the sample reference frame $x'y'z'$ to the crystal reference frame xyz and then Hooke's law applied:

$$\begin{bmatrix} \varepsilon_{xx} \\ \varepsilon_{yy} \\ \varepsilon_{zz} \\ 2\varepsilon_{xy} \\ 2\varepsilon_{yz} \\ 2\varepsilon_{zx} \end{bmatrix} = \begin{bmatrix} S_{11} & S_{12} & S_{12} & 0 & 0 & 0 \\ S_{12} & S_{11} & S_{12} & 0 & 0 & 0 \\ S_{12} & S_{12} & S_{11} & 0 & 0 & 0 \\ 0 & 0 & 0 & S_{44} & 0 & 0 \\ 0 & 0 & 0 & 0 & S_{44} & 0 \\ 0 & 0 & 0 & 0 & 0 & S_{44} \end{bmatrix} \begin{bmatrix} \frac{1}{2}\left(\sigma_{x'x'} + \sigma_{y'y'}\right) \\ \frac{1}{2}\left(\sigma_{x'x'} + \sigma_{y'y'}\right) \\ \sigma_{z'z'} \\ \frac{1}{2}\left(\sigma_{x'x'} - \sigma_{y'y'}\right) \\ 0 \\ 0 \end{bmatrix} \tag{13.4}$$

where S_{11}, S_{12}, and S_{44} are the compliance constants of the material. Linearly interpolating the constants of bulk Si and Ge [51] results in $S_{11} = 9.49 \times 10^{-12}$ Pa^{-1} and $S_{12} = -2.60 \times 10^{-12}$ Pa^{-1} for those of the $Si_{0.09}Ge_{0.91}$ alloy.

From Eq. 13.4, it follows that $\varepsilon_{xx} = \varepsilon_{yy}$ and $\varepsilon_{xz} = \varepsilon_{yz} = 0$ and consequently the matrix **K** takes the block-diagonal form:

$$\mathbf{K} = \begin{bmatrix} (p+q)\varepsilon_{xx} + q\varepsilon_{zz} & 2r\varepsilon_{xy} & 0 \\ 2r\varepsilon_{xy} & (p+q)\varepsilon_{xx} + q\varepsilon_{zz} & 0 \\ 0 & 0 & p\varepsilon_{zz} + 2q\varepsilon_{xx} \end{bmatrix} \tag{13.5}$$

K is readily diagonalized providing the following set of eigenvalues:

$$\lambda_{1,2} = (p+q)\varepsilon_{xx} + q\varepsilon_{zz} \pm 2r\varepsilon_{xy} \tag{13.6a}$$

and

$$\lambda_3 = p\varepsilon_{zz} \pm 2q\varepsilon_{xx} \tag{13.6b}$$

Further, one gets from Eq. 13.2, the following expression for the strain-induced frequency shift (or splitting) $\Delta\omega_i = \omega_i - \omega_0$ of the ith mode, in the limit $\lambda_i \ll \omega_0^2$,

$$\Delta\omega_i \approx \frac{\lambda_i}{2\omega_0}. \tag{13.7}$$

In practice, the mode splittings are much smaller than the mode linewidths, so instead of a triplet, one rather observes a single broadened "effective" phonon line approximately shifted at:

$$\Delta\omega_{\text{eff}} \approx \frac{\sum I_i \Delta\omega_i}{\sum I_i} \approx \frac{\sum I_i \lambda_i}{2\omega_0 \sum I_i} \tag{13.8}$$

where I_i is the intensity of the ith mode [15]. Having in mind that, in our TERS experiment the condition $I_1 = I_2 \gg I_3$ holds (see appendix for the detailed derivation), the above expression reduces to

$$\Delta\omega_{\text{eff}} \approx \frac{\lambda_1 + \lambda_2}{4\omega_0} = \frac{(p+q)\varepsilon_{xx} + q\varepsilon_{zz}}{2\omega_0} \tag{13.9}$$

after the substitution of Eq. 13.6a.

Most generally, the stress tensor σ from Eq. (13.3) can be decomposed into hydrostatic, $\boldsymbol{\sigma}_h$, and traceless, $\boldsymbol{\sigma}_t$, parts,

$$\boldsymbol{\sigma} = \boldsymbol{\sigma}_h + \boldsymbol{\sigma}_t = \sigma_h \mathbf{I} + \boldsymbol{\sigma}_t \tag{13.10}$$

where $\sigma_h = \frac{1}{3}(\sigma_{x'x'} + \sigma_{y'y'} + \sigma_{z'z'})$ is the hydrostatic stress component and **I** is the identity matrix [13]. Substituting Eq. 13.4 into Eq. 13.9 while taking into account the stress tensor decomposition 13.10 allows one to express the strain-induced phonon splitting through the stress components,

$$\Delta\omega_{\text{eff}} \approx \frac{(p+2q)(S_{11}+2S_{12})\sigma_h}{2\omega_0} - \frac{(p-q)(S_{11}-S_{22})\sigma_{tz'z'}}{4\omega_0}. \tag{13.11}$$

For the actual values of p, q, S_{11}, and S_{12}, the ratio

$$\frac{(p-q)(S_{11}-S_{12})}{2(p+2q)(S_{11}+2S_{12})} \tag{13.12}$$

is approximately equal to $3.9 \times 10^{-3} \ll 1$ and consequently the second term in Eq. 13.11 may be neglected,

$$\Delta\omega_{\text{eff}} \approx \frac{(p+2q)(S_{11}+2S_{12})\sigma_h}{2\omega_0}, \tag{13.13}$$

which provides a linear relationship between phonon line shift and hydrostatic stress.

To describe the behavior of the hydrostatic stress σ_h across the stripe width, we use the so-called concentrated force or line force model [8, 52]. Within this model, considering the sample reference frame $x'y'z'$ with the origin taken at the middle of a stripe of width w, the stress components along the x' and z' axes in the nano-stripe are given by the following equation:

$$\sigma_{x'x'} = -\frac{2f}{\pi K}\left\{\frac{(x'-\frac{1}{2}w)^3}{\left[(x'-\frac{1}{2}w)^2+z'^2\right]^2} - \frac{(x'+\frac{1}{2}w)^3}{\left[(x'+\frac{1}{2}w)^2+z'^2\right]^2}\right\} \tag{13.14a}$$

$$\sigma_{z'z'} = -\frac{2fz'^2}{\pi}\left\{\frac{(x'-\frac{1}{2}w)}{\left[(x'-\frac{1}{2}w)^2+z'^2\right]^2} - \frac{(x'+\frac{1}{2}w)}{\left[(x'+\frac{1}{2}w)^2+z'^2\right]^2}\right\} \tag{13.14b}$$

where f is the linear density of the force exerted upon the stripe by the substrate parallel to the x' axis and K is a coefficient characteristic of the stripe—substrate interface [52] that depends on the stripe (s) and substrate (S), Young moduli, E_s and E_S, and Poisson ratios, μ_s and μ_S,

$$K = \frac{E_S(1-\mu_s^2)}{E_s(1-\mu_S^2)}. \tag{13.15}$$

Linearly interpolating the values of E and μ for bulk Si and Ge [51] allows one to evaluate those of the $Si_{0.09}Ge_{0.91}$ alloy resulting in $K = 1.25$ by using Eq. 13.15. Finally, the hydrostatic stress is

$$\sigma_h = \frac{1}{3}(\sigma_{x'x'}+\sigma_{z'z'}) + \sigma_{h0} \tag{13.16}$$

where σ_{h_0} is the constant (i.e., coordinate-independent) hydrostatic contribution and the first two terms are given by Eq. 13.14.

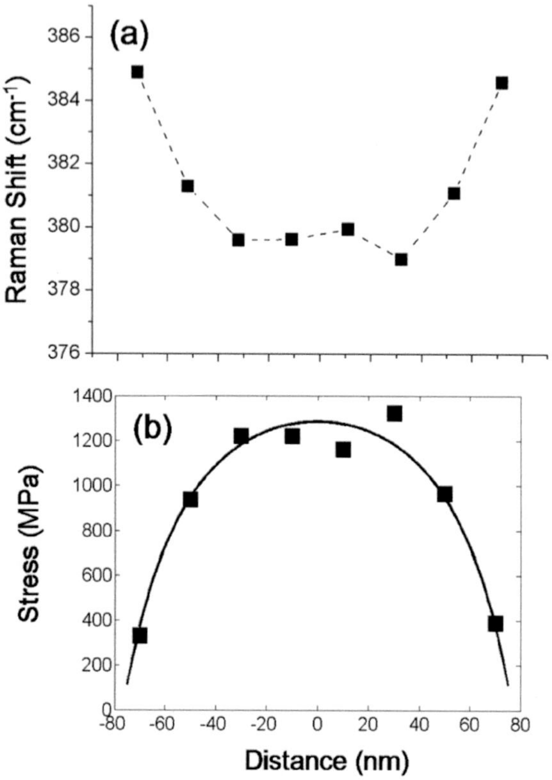

Figure 13.7 (a) Stress-induced Si-Ge phonon shift as a function of the position across the nano-stripe. (b) Experimental profile of the hydrostatic stress component across the nano-stripe (black squares) and the least squares fit (solid line) to the experimental data using a concentrated force model (see text for details).

Figure 13.7(a) shows the stress-induced shift of the Si-Ge optical phonon peak at $\sim$380 cm^{-1} across the SiGe nano-stripe with a 20-nm step. In Fig. 13.7(b), the hydrostatic stress profile across the nano-stripe, deduced from the shift by using Eq. 13.13, is represented. The stress is tensile in nature, attains its maximum value ($\sim$1250 MPa) approximately at the center of the stripe, and decreases symmetrically down to zero at both boundaries. The solid line represents the best fit of the experimentally obtained stress values to Eqs. 13.16 and 13.14 using $\sigma_{h_0} = 2360$ MPa, $f = 320$ N/m,

$w = 200$ nm, and $z' = 10$ nm (the value of w is larger than the nominal width of the nano-stripes because of the "slanted" line scan; see Fig. 13.5(a)). For comparison, stress values of few thousand MPa are typical of intentionally strained epitaxial Si/SiGe [15] and silicon-on-insulator (SOI) structures [48], whereas line force densities of ~260 N/m have been reported in nitride-strained polycrystalline Si pads [8]. The quantity z' is physically related to the near-field penetration depth (~ several tens of nm) in the TERS experiment; varying its value from 5 nm to 20 nm affects the stress-model σ_{h_0} and f values by less than 20 percent. These results clearly demonstrate the ability of the TERS technique to monitor stress in semiconductor structures with nm-scale spatial resolution.

13.4 Conclusion

This chapter presented a spatially-resolved stress analysis with ~20 nm spatial resolution of 150-nm-wide SiGe nano-stripes by means of TERS. The TERS spectra of the nano-stripes revealed the appearance of a number of Ge-Ge, Si-Ge, and Si-Si phonon modes—enhanced owing to the local amplification of the laser radiation by the tip apex—which are absent in the spectra taken from the substrate region in between adjacent nano-stripes. From the frequency profile of the locally enhanced Si-Ge Raman peak monitored across a single nano-stripe during a TERS line scan with a 20-nm step, the profile of the hydrostatic stress component was determined. The tensile stress state monotonously decreases on each side of the nano-stripe center down to a virtually zero value at the boundaries. A concentrated force model from elasticity theory is able to quantitatively reproduce this behavior.

These results unambiguously demonstrate the capability of TERS for the determination of the stress state in SiGe-based semiconductors at the nanometer level. They further contribute to qualify the TERS technique as an advanced characterization tool with nanometer resolution for the stress analysis of modern semiconductor structures whose dimensions are continuously shrinking. This may potentially assist designers at the early stages of device development.

Appendix

The Raman scattering tensors $\mathbf{R}_i$ of the triply degenerate, two transverse optical (TO1 and TO2) and one longitudinal optical (LO) phonon modes of an unstrained cubic crystal are given as follows:

$$\begin{bmatrix} 0 & 0 & 0 \\ 0 & 0 & 1 \\ 0 & 1 & 0 \end{bmatrix}, \quad \begin{bmatrix} 0 & 0 & 1 \\ 0 & 0 & 0 \\ 1 & 0 & 0 \end{bmatrix}, \quad \begin{bmatrix} 0 & 1 & 0 \\ 1 & 0 & 0 \\ 0 & 0 & 0 \end{bmatrix} \tag{13.17}$$

in the {[100], [010], [001]} crystal reference frame. In a strained crystal, the scattering tensors $\mathbf{R}_i^{st}$ generally differ from $\mathbf{R}_i$ and are given by the following relation [8]:

$$\mathbf{R}_i^{st} = v_{i1}\mathbf{R}_1 + v_{i2}\mathbf{R}_2 + v_{i3}\mathbf{R}_3 \tag{13.18}$$

in which v_{ik} is the kth component of the ith normalized eigenvector $\mathbf{v}_i$ of the matrix $\mathbf{K}$ given by Eq. 13.1. For a diagonal stress tensor in the {[1$\bar{1}$0], [110], [001]} sample reference frame, the matrix $\mathbf{K}$ reduces to Eq. 13.5, that is, $\mathbf{K}$ takes the special block-diagonal form, as given below:

$$\mathbf{K} = \begin{bmatrix} A & B & 0 \\ B & A & 0 \\ 0 & 0 & C \end{bmatrix} \tag{13.19}$$

where $A = (p+q)\,\varepsilon_{xx} + q\varepsilon_{zz}$, $B = 2r\varepsilon_{xy}$, and $C = 2q\varepsilon_{xx} + p\varepsilon_{zz}$. The normalized eigenvectors of $\mathbf{K}$ are then readily found to be $\mathbf{v}_1 = {}^1\!/\!\sqrt{2}\begin{bmatrix}1 & 1 & 0\end{bmatrix}^T$, $\mathbf{v}_2 = {}^1\!/\!\sqrt{2}\begin{bmatrix}1 & -1 & 0\end{bmatrix}^T$, and $\mathbf{v}_3 = \begin{bmatrix}0 & 0 & 1\end{bmatrix}^T$ (with associated eigenvalues $A+B$, $A-B$, and C respectively). In accordance with Eq. 13.18, the “strained” scattering tensors, $\mathbf{R}_i^{st}$, become:

$$\frac{1}{\sqrt{2}}\begin{bmatrix} 0 & 0 & 1 \\ 0 & 0 & 1 \\ 1 & 1 & 0 \end{bmatrix}, \quad \frac{1}{\sqrt{2}}\begin{bmatrix} 0 & 0 & -1 \\ 0 & 0 & 1 \\ -1 & 1 & 0 \end{bmatrix}, \quad \begin{bmatrix} 0 & 1 & 0 \\ 1 & 0 & 0 \\ 0 & 0 & 0 \end{bmatrix}. \tag{13.20}$$

Since the TERS experiments in this work were performed in the scattering configuration whereby the stripe direction, laser wavevector, and polarization are coplanar, it is convenient to rotate the “strained” tensors $\mathbf{R}_i^{st}$ by $\pi/4$ about the [001] axis to the {[1$\bar{1}$0], [110], [001]} sample reference frame,

$$\mathbf{R}_i^{st} \leftarrow \mathbf{R}(\pi/4)\,\mathbf{R}_i^{st}\,\mathbf{R}^T(\pi/4) \tag{13.21}$$

in which $\mathbf{R}(\pi/4)$ is the rotation matrix at $\pi/4$ about the [001] axis,

$$\mathbf{R}(\pi/4) = \frac{1}{\sqrt{2}}\begin{bmatrix} 1 & 1 & 0 \\ -1 & 1 & 0 \\ 0 & 0 & \sqrt{2} \end{bmatrix}. \tag{13.22}$$

Substituting Eqs. 13.20 and 13.22 into Eq. 13.21 produces the following "rotated strained" tensors $\mathbf{R}_i^{st}$ for the TO1, TO2, and LO phonon modes respectively,

$$\begin{bmatrix} 0 & 0 & 1 \\ 0 & 0 & 0 \\ 1 & 0 & 0 \end{bmatrix}, \quad \begin{bmatrix} 0 & 0 & 0 \\ 0 & 0 & 1 \\ 0 & 1 & 0 \end{bmatrix}, \quad \begin{bmatrix} 1 & 0 & 0 \\ 0 & -1 & 0 \\ 0 & 0 & 0 \end{bmatrix}. \tag{13.23}$$

Next, the effect of the tip in the TERS experiment can be modeled phenomenologically by introducing the "tip amplification tensor" **A** accounting for the interaction of the electromagnetic field with the tip [49]. For a tip oriented along the z axis, the tensor **A** has the following diagonal representation

$$\mathbf{A} = \begin{bmatrix} b & 0 & 0 \\ 0 & b & 0 \\ 0 & 0 & a \end{bmatrix} \tag{13.24}$$

in which a and b are phenomenological tip-amplification factors ($a>b$) physically related to the longitudinal and transversal polarizability of the tip and depending on its geometry and material dielectric function. The "effective" Raman polarizability tensors $\mathbf{R}_i^{\text{eff}}$ in the presence of the tip are then obtained from the following relation:

$$\mathbf{R}_i^{\text{eff}} = \mathbf{A}^T\,\mathbf{R}_i^{st}\,\mathbf{A} \tag{13.25}$$

representing the action of the tip on both the incident and scattered radiations. They are readily calculated by substituting Eqs. 13.23 and 13.24 into Eq. 13.25,

$$\begin{bmatrix} 0 & 0 & ab \\ 0 & 0 & 0 \\ ab & 0 & 0 \end{bmatrix}, \quad \begin{bmatrix} 0 & 0 & 0 \\ 0 & 0 & ab \\ 0 & ab & 0 \end{bmatrix}, \quad \begin{bmatrix} b^2 & 0 & 0 \\ 0 & -b^2 & 0 \\ 0 & 0 & 0 \end{bmatrix}. \tag{13.26}$$

The intensity of the scattered radiation of the ith mode in the case where no analyzer is present in the output beam (the so-called "depolarized" or, more correctly, "unanalyzed" detection configuration) is given by the classic (far-field) Raman selection rule

$$I_i \propto \left|\mathbf{e}_p^T \mathbf{R}_i^{\text{eff}} \mathbf{e}_i\right|^2 + \left|\mathbf{e}_s^T \mathbf{R}_i^{\text{eff}} \mathbf{e}_i\right|^2 \qquad (13.27)$$

in which

$$\mathbf{e}_p = [0 \cos\theta \sin\theta]^T \text{ and } \mathbf{e}_s = [100]^T \qquad (13.28)$$

are, respectively, the p (parallel to the scattering plane) and s (perpendicular to the scattering plane) polarizations, expressed in the sample reference frame, $\mathbf{e}_i$ is the (incident) laser polarization, and θ is the oblique incidence (off axis) angle (i.e., the angle made by the incident laser and the sample normal). Since in our experiment, $\mathbf{e}_i$ was set to $\mathbf{e}_p$ with the help of the half-wave plate, substitution of Eqs. 13.26 and 13.28 into Eq. 13.27 yields the following expressions:

$$\begin{aligned} I_1 &\propto a^2 b^2 \sin^2\theta \\ I_2 &\propto a^2 b^2 \sin^2 2\theta \\ I_3 &\propto b^4 \cos^4\theta \end{aligned} \qquad (13.29)$$

for the intensities of the three optical modes. In our experiment, $\theta = 60°$ where from the equality $I_1 = I_2$ follows. The ratio I_3/I_1 is as follows:

$$\frac{I_3}{I_1} = \left(\frac{b}{a}\right)^2 \frac{\cos^4\theta}{\sin^2\theta} = \frac{1}{12}\left(\frac{b}{a}\right)^2 \approx 2.8 \times 10^{-3} \qquad (13.30)$$

where a typical STM tip a-to-b ratio value of 5.5 has been used [50]. As a result, within our TERS experiment the transverse modes are excited to equal intensities, whereas the longitudinal mode is practically not excited, that is, the condition $I_1 = I_2 \gg I_3$ holds, as stated in Section 13.3.4.

References

1. Lee M. L., Fitzgerald E. A., Bulsara M. T., Currie M. T., and Lochtefeld A. (2005). Strained Si, SiGe, and Ge channels for high-mobility metal-oxide-semiconductor field-effect transistors, *J. Appl. Phys.* **97**, pp. 011101–011101-27.

2. Bauer M., Weeks D., Zhang Y., and Machkaoutsan V. (2006). Tensile strained selective silicon carbon alloys for recessed source drain areas of devices, *ECS Transactions* **3**, pp. 187–196.
3. Chu M., Sun Y., Aghoram U., and Thomson S. E. (2009). Strain: A solution for higher carrier mobility in nanoscale MOSFETS, *Annu. Rev. Mater. Res.* **39**, pp. 203–229.
4. Sun Y., Thomson S. E., and Nishida T. (2007). Physics of strain effects in semiconductors and metal-oxide-semiconductor field-effect transistors, *J. Appl. Phys.* **101**, pp. 104503–104503-22.
5. Baudot S., Andrieu F., Faynot O., and Eymery J. (2010). Electrical and diffraction characterization of short and narrow MOSFETs on fully depleted strained silicon-on-insulator (SSOI), *Solid State Electron.* **54**, pp. 861–869.
6. Weber W. H. and Merlin R. (2000). *Raman scattering in materials science* (Springer-Verlag, Berlin) and references therein.
7. Anastassakis E. (1997). Selection rules of Raman scattering by optical phonons in strained cubic crystals, *J. Appl. Phys.* **82**, pp. 1582–1591.
8. De Wolf I., Maes H. E., and Jones S. K. (1996). Stress measurements in silicon devices through Raman spectroscopy: Bridging the gap between theory and experiment, *J. Appl. Phys.* **79**, pp. 7148–7156.
9. Tsang J. C., Mooney P. M., Dacol F., and Chu J. O. (1994). Measurements of alloy composition and strain in thin Ge_xSi_{1-x} layers, *J. Appl. Phys.* **75**, pp. 8098–8108.
10. Cazayous M., Groenen J., Demangeot F., Sirvin R., Caumont M., Remmele T., Albrecht M., Christiansen S., Becker M., Strunk H. P., and Wawra H. (2002). Strain and composition in self-assembled SiGe islands by Raman spectroscopy, *J. Appl. Phys.* **91**, pp. 6772–6774.
11. Pezzoli F., Martinelli L., Grilli E., Guzzi M., Sanguinetti S., Bollani M., Chrastina D., Isella G., von Känel H., Wintersberger E., Stangl E. J., and Bauer G. (2005). Raman spectroscopy of $Si_{1-x}Ge_x$ epilayers, *Mat. Sci. Eng. B-Solid* **124**, pp. 127–131.
12. Hossain M. Z. and Johnson H. T. (2010). Effects of composition, strain, and atomic disorder on optical phonon frequencies in $Si_{1-x}Ge_x$, *J. Appl. Phys.* **107**, pp. 073515–073515-6.
13. Bonera E., Fanciulli M., and Batchelder D. N. (2003). Combining high resolution and tensorial analysis in Raman stress measurements of silicon, *J. Appl. Phys.* **94**, pp. 2729–2740.
14. Becker M., Scheel H., Christiansen S., and Strunk H. P. (2007). Grain orientation, texture, and internal stress optically evaluated by micro-Raman spectroscopy, *J. Appl. Phys.* **101**, pp. 063531–063531-10.

15. Ossikovski R., Nguyen Q., Picardi G., and Schreiber J. (2008). Determining the stress tensor in strained semiconductor structures by using polarized micro-Raman spectroscopy in oblique backscattering configuration, *J. Appl. Phys.* **103**, pp. 093525–093525-13.
16. Stadler J., Schmid T., and Zenobi R. (2011). Nanoscale chemical imaging of single-layer grapheme, *ACS Nano* **5**, pp. 8442–8448.
17. Cançado L. G., Hartschuh A., and Novotny L. (2009). Tip-enhanced Raman spectroscopy of carbon nanotubes, J. *Raman Spectrosc.* **40**, pp. 1420–1426.
18. Picardi G., Chaigneau M., Ossikovski R., Licitra C., and Delapierre G. (2009). Tip enhanced Raman spectroscopy on azobenzene thiol self-assembled monolayers on Au(111), *J. Raman Spectrosc.* **40**, pp. 1407–1412.
19. Stadler J., Schmid T., Opilik L., Kuhn P., Dittrich P. S., and Zenobi R. (2011). Tip-enhanced Raman spectroscopic imaging of patterned thiol monolayers, *Beilstein J. Nanotechnol.* **2**, pp. 509–515.
20. Bailo E. and Deckert V. (2008). Tip-enhanced Raman spectroscopy of single RNA strands: Towards a novel direct-sequencing method, *Angew. Chem. Int. Ed.* **47**, pp. 1658–1661.
21. Hayazawa N., Motohashi M., Saito Y., Ishitobi H., Ono A., Ichimura T., Verma P., and Kawata S. (2007). Visualization of localized strain of a crystalline thin layer at the nanoscale by tip-enhanced Raman spectroscopy and microscopy, *J. Raman Spectrosc.* **38**, pp. 684–696.
22. Saito Y., Motohashi M., Hayazawa N., Iyoki M., and Kawata S. (2006). Nanoscale characterization of strained silicon by tip-enhanced Raman spectroscope in reflection mode, *Appl. Phys. Lett.* **88**, pp. 143109–143109-3.
23. Tarun A., Hayazawa N., and Kawata S. (2009). Tip-enhanced Raman spectroscopy for nanoscale strain characterization, *Anal. Bioanal. Chem.* **394**, pp. 1775–1785.
24. Tarun A., Hayazawa N., Motohashi M., and Kawata S. (2008). Highly efficient tip-enhanced Raman spectroscopy and microscopy of strained silicon, *Rev. Sci. Instrum.* **79**, pp. 013706–013706-10.
25. Ramos R. and Gordon M. J. (2012). Reflection-mode, confocal, tip-enhanced Raman spectroscopy system for scanning chemical microscopy of surfaces, *Rev. Sci. Instrum.* **83**, pp. 093706–093706-8.
26. Hanafusa H., Hirose N., Kasamatsu A., Mimura T., Matsui T., Chong H. M. H., Mizuta H., and Suda Y. (2011). Strain distribution analysis of sputter-

formed strained Si by tip-enhanced Raman spectroscopy, *Appl. Phys. Express* **4**, pp. 025701–025701-3.

27. Georgi C., Hecker M., and Zschech E. (2007). Raman intensity enhancement in silicon-on-insulator substrates by laser deflection at atomic force microscopy tips and particles, *Appl. Phys. Lett.* **90**, pp. 171102–171102-3.
28. Merlen A., Valmalette J. C., Gucciardi P. G., Lamy de la Chapelle M., Frigout A., and Ossikovski R. (2009). Depolarization effects in tip-enhanced Raman spectroscopy, *J. Raman Spectrosc.* **40**, pp. 1361–1370.
29. Gucciardi P. G., Lopes M., Détruche R., Julien C., Barchiesi D., and Lamy de la Chapelle M. (2008). Light depolarization induced by metallic tips in apertureless near-field optical microscopy and tip-enhanced Raman spectroscopy, *Nanotechnology* **19**, pp. 215702–215702-7.
30. Ogawa Y., Toizumi T., Minami F., and Baranov A. V. (2011). Nanometer-scale mapping of the strain and Ge content of Ge/Si quantum dots using enhanced Raman scattering by the tip of an atomic force microscope, *Phys. Rev. B* **83**, pp. 081302(R) –081302-4.
31. Hermann P., Hecker M., Chumakov D., Weisheit M., Rinderknecht J., Shelaev A., Dorozhkin P., and Eng L. M. (2011). Imaging and strain analysis of nano-scale SiGe structures by tip-enhanced Raman spectroscopy, *Ultramicroscopy* **111**, pp. 1630–1635.
32. Vanacore G. M., Chaigneau M., Barrett N., Bollani M., Boioli F., Salvalaglio M., Montalenti F., Manini N., Caramella L., Biagioni P., Chrastina D., Isella G., Renault O., Zani M., Sordan R., Onida G., Ossikovski R., Drouin H.-J., and Tagliaferri A. (2013). Hydrostatic strain enhancement in laterally confined SiGe nanostripes, *Phys. Rev. B* **88**, pp. 115309–115309-15.
33. Ren B., Picardi G., and Pettinger B. (2004). Preparation of gold tips suitable for tip-enhanced Raman spectroscopy and light emission by electrochemical etching, *Rev. Sci. Instrum.* **75**, pp. 837–841.
34. Chaigneau M., Picardi G., and Ossikovski R. (2010). Tip enhanced Raman imaging of amorphous carbon contamination on gold surfaces, *Surf. Sci.* **604**, pp. 701–705.
35. Picardi G., Nguyen Q., Schreiber J., and Ossikovski R. (2007). Comparative study of atomic force mode and tunneling mode tip-enhanced Raman spectroscopy, *Eur. Phys. J. Appl. Phys.* **40**, pp. 197–201.
36. Chrastina D., Vanacore G. M., Bollani M., Boye P., Schröder S., Burghammer M., Sordan R., Isella G., Zani M., and Tagliaferri A. (2012). Patterning-induced strain relief in single lithographic SiGe

nanostructures studied by nanobeam x-ray diffraction, *Nanotechnology* **23**, pp. 155702–155702-10.

37. Temple P. A. and Hathaway C. E. (1973). Multiphonon Raman spectrum of silicon, *Phys. Rev. B*7, pp. 3685–3697.
38. Pezzoli F., Grilli E., Guzzi M., Sanguinetti S., Chrastina D., Isella G., von Känel H., Wintersberger E., Stangl J., and Bauer G. (2006). Strain-induced shift of phonon modes in $Si_{1-x}Ge_x$ alloys, *Mater. Sci. Semiconductor Process.* **9**, pp. 541–545.
39. Tsao C.-Y., Liu Z., Hao X., and Green M. A. (2011). In situ growth of Ge-rich poly-SiGe:H thin films on glass by RF magnetron sputtering for photovoltaic applications, *Appl. Surf. Sci.* **257**, pp. 4354–4359.
40. Groenen J., Carles R., Christiansen S., Albrecht M., Dorsch W., Strunk H. P., Wawra H., and Wagner G. (1997). Phonons as probes in self-organized SiGe islands, *Appl. Phys. Lett.* **71**, pp. 3856–3858.
41. Alonso M. I. and Winer K. (1989). Raman spectra of c-$Si_{1-x}Ge_x$ alloys, *Phys. Rev. B* **39**, pp. 10056–10062.
42. Brya W. J. (1973). Raman scattering in Ge-Si alloys, *Solid State Commun.* **12**, pp. 253–257.
43. Qin L., Teo K. L., Shen Z. X., Peng C. S., and Zhou J. M. (2001). Raman scattering of Ge/Si dot superlattices under hydrostatic pressure, *Phys. Rev. B* **64**, pp. 075312–075312-5.
44. Weinstein B. A. and Cardona M. (1973). Second-order Raman spectrum of germanium, *Phys. Rev. B* **7**, pp. 2545–2551.
45. Fuchs H. D., Grein C. H., Thompson C., Cardona M., Hansen W. L., Haller E. E., and Itoh K. (1991). Comparison of the phonon spectra of ^{70}Ge and natural Ge crystals: Effects of isotopic disorder, *Phys. Rev. B* **43**, pp. 4835–4842.
46. Nelin G. and Nilsson G. (1972). Phonon density of states in germanium at 80 K measured by neutron spectrometry, *Phys. Rev. B*5, pp. 3151–3160.
47. Uchinokura K., Sekine T., and Matsuura E. (1974). Critical-point analysis of the two-phonon Raman spectrum of silicon, *J. Phys. Chem. Solids* **35,** pp. 171–180.
48. Hermann P., Hecker M., Renn F., Rölke M., Kolanek K., Rinderknecht J., and Eng L. M. (2011). Effects of patterning induced stress relaxation in strained SOI/SiGe layers and substrate, *J. Appl. Phys.* **109**, pp. 124513–124513-7.
49. Ossikovski R., Nguyen Q., and Picardi G. (2007). Simple model for the polarization effects in tip-enhanced Raman spectroscopy, *Phys. Rev. B* **75**, pp. 045412–045412-9.

50. Picardi G., Nguyen Q., Ossikovski R., and Schreiber J. (2007). Polarization properties of oblique incidence scanning tunneling microscopy-tip-enhanced Raman spectroscopy, *Appl. Spectrosc.* **61**, pp. 1301–1305.
51. Wortman J. J. and Evans R. A. (1965). Young's modulus, Shear modulus, and Poisson's ratio in silicon and germanium, *J. Appl. Phys.* **36**, pp. 153–156.
52. Hu S. M. (1979). Film-edge-induced stress in substrates, *J. Appl. Phys.* **50**, pp. 4661–4666.

Chapter 14

Surface-Enhanced Infrared Spectroscopy

Michele Ortolani[a] and Odeta Limaj[b]

[a]*Sapienza University of Rome, P.le Aldo Moro 2, Rome 00185, Italy*
[b]*École Polytechnique Fédérale de Lausanne, Station 17 Batiment BM, Lausanne 1015, Switzerland*
michele.ortolani@roma1.infn.it

We outline the principles of mid-infrared vibrational spectroscopy of molecular films on a metal surface structured with subwavelength-sized geometric features, both self-assembled and fabricated following electromagnetic design. Molecular chemistry, optical physics, and electromagnetic engineering approaches are presented. We then review molecular film preparation techniques and substrate fabrication technologies specific to surface-enhanced mid-infrared spectroscopy.

14.1 Principles of Infrared Spectroscopy on a Metal Surface

Surface-enhanced infrared (IR) spectroscopy, here defined as vibrational spectroscopy of molecular films in the mid-IR range ($20 < \lambda < 2$ μm) deposited on the surface of a substrate containing subwavelength metal structures, has been the subject of

Handbook of Enhanced Spectroscopy
Edited by Pietro Giuseppe Gucciardi, Marc Lamy de la Chapelle, and Nathalie Lidgi-Guigui

ISBN 978-981-4613-32-3 (Hardcover), 978-981-4613-33-0 (eBook)
www.panstanford.com

discontinuous studies in the last three decades. Different types of substrates have been developed: the initial ones, that we shall call "first-generation" substrates, were based on the self-assembly of metal structures on IR-transparent substrates (e.g., CaF_2 for $\lambda > 12$ μm, ZnSe for $\lambda > 19$ μm, Si for almost any λ). Second-generation substrates rely on lithographic nanopatterning of subwavelength metal structures (metamaterials and antennas). A third generation of substrates is under development, where the metal is substituted by IR-optimized plasmonic materials such as highly doped semiconductors, graphene, or conducting oxides.

In spite of this effort, the observed increase in the molecular sensing capabilities obtained by surface-enhanced IR spectroscopy is in general much lower than that obtained with surface-enhanced Raman spectroscopy (SERS). Consequently, the literature on surface-enhanced IR spectroscopy is less abundant, and both laboratory and commercial applications are still lacking. In this chapter, we review the main results obtained on the different types of substrates. For each reviewed experiment, the different physical, chemical, and electrodynamic mechanisms that can lead to an increase of the IR signal strength are highlighted. The aim of this chapter is to clarify that different phenomena lead to surface-enhanced IR *absorption* (SEIRA) and to surface-enhanced IR *scattering* (SEIRS), although it is not excluded that both effects might cooperate in the same device. In order to decide to what extent each effect is at work in a given experiment, it is required to design and conduct a series of key experiments, where one or the other effect is excluded: such key experiments have seldom been done and/or published. We hope that clarifying the main open points will help motivate future key experiments.

Finally, the results here reviewed have been mainly obtained with Fourier Transform Infrared (FT-IR) spectrometers. However, the recent commercialization of wavelength-tunable quantum cascade lasers (QCL) on one side and mid-IR focal plane arrays for use in mid-IR grating spectrometers on the other side both provide further ingredients for the future development of surface-enhanced mid-IR spectroscopy, with an enormous application potential, if compared to the much more expensive and bulky instrumentation needed for Raman spectroscopy.

14.1.1 *The Chemist's View: Surface-Enhanced Infrared Absorption*

Any molecular bond vibrates around an equilibrium distance at the minimum of a potential well, which can be approximated by a parabola (harmonic oscillator approximation). First quantization of the interatomic potential leads to discrete quantum levels for the vibration of a molecular bond, labeled by the quantum number v. The energy spacing $E_v - E_{v+1} = \hbar\omega_v$ is approximately independent on v in the multi-harmonic oscillator approximation, where the different vibrational resonances ω_ν of a polyatomic molecule are the eigenvalues of a system of forces whose intensity linearly depends on the displacement from the equilibrium position. The value of the electromagnetic frequency ω_ν lies in the mid-IR range for the large majority of existing molecules. IR-active vibrational transitions between states $\langle v \rangle$ and $\langle v+1 \rangle$ can occur, and be experimentally observed, only if the following matrix element is nonvanishing:

$$D_{v,v+1} = \int \psi^*_{v+1} D(R)\, \psi_v \mathrm{d}R \tag{14.1}$$

where $D(R)$ is the matrix element of the electric dipole moment, R is the internuclear distance, and ψs are harmonic-oscillator wavefunctions (Hermite polynomials). Notice that here the rotational degree of freedom is neglected: indeed, in a photon absorption process, momentum conservation must also be fulfilled, therefore for gases vibrational-rotational transitions are required. For soft condensed matter films considered here, the rotational degrees of freedom are blocked and the extra-momentum is absorbed by the aggregate or the substrate. In the small-perturbation approximation, one can then write:

$$D(R) = D(R_{0,v}) + (R - R_{0,v}) \left.\frac{\mathrm{d}D}{\mathrm{d}R}\right|_{R=R_{0,v}} + \ldots. \tag{14.2}$$

where R_0 is the equilibrium value of the internuclear distance and $R_{0,\nu}$ is the same quantity in the excited vibrational state. By substituting Eq. 14.2 into Eq. 14.1, we obtain the IR-vibration selection rule $\Delta\nu = \pm 1$, because two harmonic oscillator wavefunctions with subsequent index display different parity. The following expression for the *first-order line intensity* derived from Fermi's golden rule summarizes this result, as it is different from zero only if $\Delta v = \pm 1$:

$$I(\omega) = \sum_{v} \left.\frac{\mathrm{d}D}{\mathrm{d}R}\right]_{R_{0,v}} \int \psi^*_{v+1}(R - R_0)\psi_v \mathrm{d}R. \quad (14.3)$$

However, the vibrational transition actually occurs because $\mathrm{d}D/\mathrm{d}R \neq 0$, that is, because the dipole moment of the excited state is different from the dipole moment of the ground state. The strength of the absorption line depends therefore on how much the dipole moment changes when the nuclei are in the excited-state position. The absorption line strength is usually considered by IR spectroscopists as an intrinsic parameter of the molecule, which cannot be modified (instead, it is considered as a further fingerprint confirmation beyond the frequency assignment, with IR lines typically classified as "strong," "weak," etc.). We will see that this is not the case in surface-enhanced IR spectroscopy, where the line strength is modified by the proximity to a metal surface. If a molecule is positioned on a metal surface at a distance comparable to that of chemical-bond lengths, it appears from Eq. 14.3 that the intensity of an absorption line can become higher than that of an isolated molecule because of one or both of the following effects induced by the metal surface:

- an **increased average displacement (R–R_0)** of the nuclei when they vibrate. This can happen, for example, because one of the nuclei is blocked by the metal–ion orbital potential, bringing to a static modification of the molecular structure. Notice that this is usually not the case for most molecules, for which binding energies leading to molecule formation are much larger than whatever enhanced electrostatic potential at the metal surface. An exception is given by molecules displaying catalytic reaction rate increase on specific metals. An increased "dynamic" displacement ($R - R_0$) could also arise from an extremely high electric field generated by charge density unbalance at the metal surface oscillating at the radiation frequency (surface plasmon, see below). This high local field can produce an observable effect of the nonlinear susceptibility (i.e., a correction to the linear-oscillator model). It is however usually believed that this nonlinearity correction cannot account for the large signal enhancement reported in some SEIRA experiments.

- an **increased variation of the dipole moment dD/dR** when the nuclei move to the excited-state positions. Indeed, the value of the dipole moment in a given nuclear configuration strongly depends on image-charge effects generated by the proximity to the metal surface (see Fig. 14.1). It is possible that, for symmetry breaking reasons, image-charge effects make the dipole moment in the excited state configuration $D(R)$ to be much larger than $D(R_0)$ leading to a strong absorption line intensity, while for the same molecule in the absence of a metal surface the two values $D(R)$ and $D(R_0)$ are almost identical, and the line is very weak. This effect is expected to be stronger when the dipole direction is perpendicular to the metal surface. Clearly, this assumption has no generality, and it has to be verified by numerical calculations for each specific molecule and vibrational mode. Notice that this effect does not require any static modification of the molecular structure, that is, a static change of the value of $(R - R_0)$, or of D. The molecule is still described by the linear-oscillator model, however, not in vacuum but at the interface between a conducting and a dielectric half-space. This situation adds a high degree of complexity to molecular simulation algorithms if compared to the isolated molecule, making direct experimental proofs a viable faster alternative to understand SEIRA.

The above described effects effectively realize a nonlinear dependence of the single-molecule absorption cross-section on the number of deposited molecules, beyond the Lambert–Beer's law. For very small amount of molecules, such as monolayers, it is then advantageous in terms of signal-to-noise ratio to measure the IR transmittance spectrum of a molecular film on a nanostructured substrate rather than on an unpatterned or a transparent one. Experimental data [1, 2] indicate that the length scale for the onset of nonlinearity is a distance of less than 5 nm from the metal surface, indicating that the direct proximity to the metal surface is crucial in obtaining strong enhancement of the IR signal. This experimental

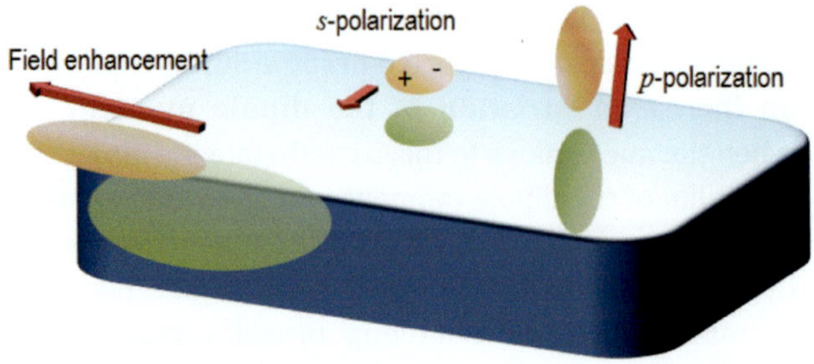

Figure 14.1 Image-charge effects for the vibrational motion of a molecule on a metal surface. The molecular electronic cloud is depicted as a red ellipsoid, while the green ellipsoid is the image-charge region inside the metal structure (blue). The total dipole moment is the sum of that of red and green ellipsoids. Depending on the volume and direction of the image-charge cloud, the variation of the dipole moment with nuclear displacement is suppressed on a flat film with excitation field direction parallel to the surface (s-pol.), it is mildly enhanced for direction normal to the surface (p-pol.), and it can be very strongly enhanced on the edges and/or for subwavelength structures.

fact indicates that there exists a strong interaction between the molecular vibrations and the collective oscillations of conduction electrons at the metal surface at the same frequency, where the radiation field is not screened (actually, it penetrates few tens of nanometers in the mid-IR). These oscillations are called *surface plasmons* and they are responsible for dynamic field enhancement and therefore for the SEIRA effect.

In principle, the SEIRA effect could be generated by the sole presence of a flat, unpatterned metal surface. However, the increase in the first-order line intensity, or the dynamic enhancement of the electric field value, is usually stronger at the edge of a subwavelength metal structure, rather than on a flat metal surface, and therefore SEIRA is identified with IR spectroscopy on nanostructured substrates. The technique called IR reflection-absorption spectroscopy (IRRAS or RAIRS) is performed on molecular monolayers deposited on a flat, thick metal film by measuring the reflection at grazing incidence. Absorption enhancement in the perpendicular polarization is clearly observed (see Fig. 14.1) but, due to the translational symmetry, a maximum enhancement of a factor 2 is expected from the Fresnel relations (recent reviews of this classical technique are available [3]). The absorption signal enhancement higher than 2 sometimes observed in IRRAS is explained in terms of increased optical absorption path length, due to the propagation of a surface plasmon with well-defined wavevector, also called surface plasmon polaritons (SPP). The existence of propagating surface plasmons on flat metal–dielectric interfaces is also responsible for the enhanced surface sensitivity observed in attenuated total reflectance (ATR) prisms when covered by a flat thin metal film, of thickness smaller than the skin depth of the metal (the so-called SPP sensors) [4]. However, it has been thoroughly discussed in the literature [2–5] that SPP sensors are not expected to provide any field confinement in the mid-IR, because the absolute value of the metal dielectric constant is so high, that the field confinement at the surface is of the order of the wavelength, as it happens for standard ATR, and not of the order of the skin depth, as it happens in near-IR surface-enhanced SPP sensors. Finally, we notice that the observation of different absorption enhancement factors when changing the metallic element in SEIRA substrates demonstrates that some kind of "dynamic catalysis" effect exists and modifies the molecular structure [6, 7]. This fact supports the "chemist's view" of SEIRA that the external metal orbitals actively interact with the molecule to produce an effectively higher absorption cross-section.

14.1.2 *The Optical Physicist's View: Surface-Enhanced Infrared Scattering*

The main characteristic of substrates used for surface-enhanced spectroscopy (both Raman and IR) is the presence of metal structures (islands, antennas, holes in a continuous film, corrugations, etc.) of subwavelength size. Subwavelength-sized metal objects scatter the light in a way similar to that of dielectric objects (Rayleigh scattering). However, the scattering cross-section is proportional to the fourth power of the local electric field intensity, therefore it is expected that metallic subwavelength objects scatter light more efficiently than dielectric objects. The Rayleigh scattering cross-section for a nonabsorbing sphere of diameter d and dielectric constant ε is as follows:

$$\sigma_S = \frac{2\pi^5}{3}\frac{d^6}{\lambda^4}\left(\frac{\varepsilon - 1}{\varepsilon + 2}\right) \tag{14.4}$$

which actually diverges for $\varepsilon \rightarrow -2$ if absorption losses can be neglected. This happens in the case of ideal metallic particles close particles close to the plasma frequency. In other words, subwavelength-sized metal objects display an enhanced scattering cross-section in a broad wavelength range defined by both the metal dielectric constant and the object geometry [8], and this effect is at the basis of inelastic surface-enhanced Raman scattering (SERS). The same effect is at work for elastic scattering by subwavelength metal objects with suitable geometry for producing enhanced fields in the mid-IR range, and this phenomenon is called surface-enhanced infrared scattering (SEIRS). We can regard elastic scattering as exchange of momentum between the radiation and the particle, without exchange of energy, as it happens instead in the absorption of a photon from a molecule. Therefore, the two phenomena of SEIRA and SEIRS are physically distinct. The optical physicist point of view of surface-enhanced IR spectroscopy is the presence at the metal surface of a molecule with vibrational frequency $\omega_\nu = \frac{1}{\hbar}(E_{\nu+1} - E_\nu)$ further enhances the scattering cross-section of the subwavelength object at that electromagnetic frequency. Radiation scattering is always a second-order process in perturbation theory, as the photon needs to be first absorbed and then reemitted. Therefore, an increased electric field in the proximity of the metal surface due to plasmonic effects results in an effectively increased scattering cross-section. The enhanced scattering produces higher losses in transmittance (T) experiments, and by a conventional calculation of the absorption $A = \ln(1/T)$, the scattering loss may be visualized as enhanced absorption.

14.1.3 *The Electromagnetic Engineer's View: Dielectric Loading of Infrared Resonators*

The phenomenon of radiation scattering brings us directly to the concept of infrared antennas [9]. In the entire radiofrequency (RF) spectrum from longwave to terahertz, metal antennas are used to convert free-space propagating electromagnetic waves into electrical currents and/or voltages in order to input these into electronic receivers positioned at the antenna feed, and vice versa for transmitters. The antenna works as a resonator that captures (cedes) energy from (to) the free-space, and the device in the feed is called the load of the resonator, as it absorbs (emits) energy from the antenna. Impedance matching between the antenna and the load is necessary to obtain efficient power transfer, and electromagnetic calculations are needed to define the detailed geometrical shape of the antenna, to be then fabricated with deep subwavelength precision. With the advent of nanofabrication techniques, well-designed antennas have been realized in the infrared and visible ranges. They are constituted by metal structures of subwavelength size, precisely defined by lithography on a dielectric substrate. In the electromagnetic engineer's view, the antenna is an infrared resonator, and the molecule represents its load with frequency-dependent impedance given by its vibrational spectrum. Much theoretical work is still needed to bring this analogy to its final consequences for designing and fabricating IR antennas with performances close to ideal for each specific application, as it is currently done for RF antennas.

From the electronic engineer's point of view, the spectroscopic signal enhancement is seen as a modification of the radiation properties of the resonator, due to a specific feature in the impedance of the load. The complex impedance of the load has two main effects on the empty (i.e., vacuum-loaded) resonator spectrum: it red-shifts the resonant frequency and it broadens the linewidth. Therefore, high-quality factor (Q) resonators are preferred for improving the spectroscopic sensing capabilities, as their radiation properties will be more sensitive to slight changes of the load impedance (either changes of the resonant frequency, as in subwavelength hole arrays, or changes of the linewidth, as in dark plasmonic modes, see Section 14.5 for examples). In both cases, the confinement of the electric field close to the metal surface is a key feature, as it results in the spatial overlap of the resonator mode profile with the molecular film. The change in the resonator mode can be described in terms of a complex-valued resonance-frequency variation (if compared to the vacuum-loaded resonator frequency ω_0) due to the deposition of a film of thickness h and complex dielectric tensor $((\epsilon))$ [10], as given below:

$$\frac{\Delta\omega_0}{\omega_0} = -\frac{1}{2}\frac{\int_0^h \mathrm{d}z \int \mathrm{d}x\mathrm{d}y \vec{E}(\mathbf{r}) \cdot [((\epsilon)) - 1] \cdot \vec{E}(\mathbf{r})}{\int \mathrm{d}r E^2(\mathbf{r})}. \quad (14.5)$$

In the case of resonator line broadening, represented by the imaginary part of Eq. 14.5, a much higher shift $\Delta\omega_0$ is expected when ω_0 is close to the frequency ω_v where the molecule has its vibrational transition. Recently, the design and nanofabrication of high-Q mid-IR resonators has led to important advances in substance-specific, label-free optical sensing of molecules by mid-IR radiation (see Sections 14.4 and 14.5).

In summary, we have presented three different views of the class of phenomena usually referred to as surface-enhanced infrared spectroscopy on nanostructured substrates. From the chemist's argument, we have learned that the molecular film under study, or to be detected, should be ultrathin in order to get maximum advantage from the nanostructured substrate, approaching the molecular monolayer limit. From the physicist's view, we have learned that the stronger the enhancement of the radiation electric field in the vicinity of the metal surface, the higher the signal increase. From the electromagnetic engineer's view, we have learned that measuring an increased spectroscopic line intensity is not the only way of exploiting nanostructured substrates: instead, by using a carefully designed substrate with a geometrical resonance frequency that matches the molecular vibration frequency, the complex dielectric constant of the molecule can be quantitatively determined. We shall now describe the experimental techniques used to realize these approaches.

14.2 Preparation of Molecular Films

We have seen that the main characteristic of surface-enhanced IR spectroscopy is the concentration of the electromagnetic field in a region adjacent to the metal surface, with typical dimensions much smaller than one wavelength. This means that the sample should also be located in the close vicinity of the surface. In order to truly exploit the SEIRA/SEIRS effects for spectroscopy of thin molecular layers, it is therefore crucial to use highly controlled sample preparation techniques, with the aim of obtaining a molecular film of controlled thickness, orientation, and packing density. Ideally, the film thickness should approach one molecular monolayer: in these conditions, SEIRA/SEIRS may provide structural information at the single-molecule level. Prior to reviewing in Sections 14.3 and 14.4 the existing types of SEIRA/SEIRS substrates that have been developed so far, here we briefly review a number of well-known techniques for preparing thin molecular films, with particular emphasis on those suitable for depositing molecular monolayers and/or self-assembled monolayers (SAMs). Indeed, the sample preparation technique has considerable impacts on the spectroscopic results, especially when the aim is quantitative comparison of scattering and absorption cross-sections between different types of substrates.

Film cast by drop-drying. A drop of solution containing the molecules as solute is deposited on the substrate, and the film forms as a result of solvent evaporation. Depending on the specific molecule and its interaction with the substrate, films of different thicknesses can be obtained, from a monolayer to several tens of nanometers. Pre-functionalization of the substrate with molecular-specific ligands bound to the metal nanostructure is often used to obtain a monolayer of the desired molecular species, after washing out the unbound species: a typical example of binding agents are thiols on gold [3], often followed by deposition of antibodies or aptamers for biomolecule-specific assays (similarly to what happens in the immobilization step of the ELISA technique [11]). The use of selective binding agents allows for the calibration of the amount of deposited molecules on a metallic surface, therefore enabling quantitative analysis of SEIRA/SEIRS data [4].

Evaporation and sputtering. Synthetic organic molecular films of controlled thickness, including monolayers, can be obtained by thermal evaporation processes or physical sputtering of bulk precursors, such as those routinely employed by the organic electronic industry, directly performed on a SEIRA/SEIRS substrate introduced in the growth chamber. While this type of sample preparation technique is usually not suitable for biomolecules, it is probably the most precise tool available to deposit monolayers of controlled orientation and purity, especially if one resorts to more sophisticated techniques such as molecular-beam epitaxy (MBE) and chemical vapor deposition (CVD). Therefore, this may be the technique of choice for realizing key experiments to validate SEIRA/SEIRS models.

Langmuir films are molecular monolayers of insoluble molecules that form at the liquid–gas interface (typical example: lipids on top of an aqueous solution). The Langmuir balance, or Wilhelmy plate, is a device based on the liquid surface tension, which allows collection of Langmuir–Blodgett films from the so-called Langmuir–Blodgett trough and deposition of these films onto a solid substrate. One or several subsequent monolayers can be deposited onto a solid substrate by vertical dipping of the substrate into the trough, that is, subsequent transitions from the gas phase environment to the liquid phase environment (and/or vice versa). By using a SEIRA/SEIRS substrate as the Wilhelmy plate, it is possible to prepare monolayers suitable for surface-enhanced IR spectroscopy. The possibility to perform SEIRA/SEIRS on Langmuir films strongly depends on the adhesion of the molecules on the nanostructured surface, which may be higher or lower than the adhesion on a flat surface, so that the monolayer nature of the Langmuir film once deposited on a SEIRA/SEIRS substrate may be preserved, or not [2].

Spin-coating is a standard technique to produce thin films of highly controlled thickness basing on the balance between surface adhesion, concentration of the molecule in a volatile solvent, and centrifugal force. However, very few molecules can be spin-coated down to monolayer levels, such as the silk protein [12]. This technique is commonly employed for testing novel SEIRA/SEIRS substrates at a first stage of development, where studies at the molecular monolayer level are not aimed at.

Chemisorption and physisorption of molecules at a liquid–metal or at a gas–metal interface should also be mentioned as a technique for obtaining surface-enhanced spectra of molecules [13]. As well known from studies of catalysis, nanostructuring of metal surfaces has tremendous effects on their adsorption capability. However, as the molecules are trapped inside the crystal lattice of the metal, their absorption spectra may be strongly modified if compared to the free molecule and therefore, strictly speaking, IR absorption measurements on adsorbates cannot be considered a form of molecular spectroscopy, because the target system is physically and chemically different from the free molecule. Moreover, specifications on the exact metal nanostructure geometry are set by the need of maximizing the adsorption rate or concentration, not the spectroscopic cross-section. Nevertheless, this variant of SEIRA is becoming a new analytical tool in the study of catalysis [6, 7].

Deposition of a precipitate from the liquid phase. Although IR spectroscopy cannot be performed through aqueous solutions because of strong mid-IR absorption of liquid water, the ATR technique is commonly employed to obtain indication on the absorption spectra of the solutes (or, more properly, of the precipitate that forms on the prism–solution interface). In this technique, a high-refractive index prism is put in contact with the solution, and an IR beam is guided through the prism so that the incidence angle on the prism–solution interface is beyond the critical angle (total reflection). The evanescent wave that forms on the solution side of the interface probes a region of about one wavelength thickness, where usually only the precipitate is present. The reflected IR beam is only weakly attenuated by water, since it does not propagate inside the solution. If there is no precipitate, the absorption spectrum of the molecules in the solution may be obtained. A well-known plasmonic variant of ATR spectroscopy is the Otto–Kretschmann technique [14], where a metal film of thickness smaller than the skin depth (around 20 nm for most metals in the mid-IR) is grown on the prism surface: the sensitivity to molecules attached to the surface is enhanced as the SPP field extends on the solution side of the interface by few tens of nanometers only. Clearly, the same optical configuration but with ATR prisms covered with nanostructured metal films can be employed to perform SEIRA/SEIRS. The high signal-to-noise ratio and the nonperturbation of the molecule environment typical of ATR techniques, combined with molecular identification capabilities and surface sensitivity, make SEIRA/SEIRS a promising technique to monitor adsorption and reaction dynamics. Biochemistry, electrochemistry, surface photochemistry, and catalysis are all fields where SEIRA/SEIRS could prove useful.

14.3 Bottom-Up Prepared Substrates

In this section, we review some of the results on what we have called "first-generation" substrates for SEIRA, where metallic structures are the result of crystal growth or assembly of aggregates on a given substrate in a controlled physical–chemical environment. As a result, the position, size, shape, and distance relative to each of the nanoparticles can only be defined and controlled "on average", and not by design. These types of substrate preparation processes are usually named "bottom-up" techniques, because here nature assembles the final product from single atoms, ions, or elementary units such as droplets or colloidal particles, and the experimenter can only control the external parameters of the physical process. Bottom-up preparation techniques are opposed to "top-down" fabrication techniques, to be discussed in Section 14.4, where the nanostructure geometry is defined by a priori design (typically, by finite-element electromagnetic simulations) and subsequently realized from bulk raw materials by nanofabrication technologies derived from semiconductor manufacturing processes (lithography, etching, lift-off, ion milling, etc.). The available range of shapes and dimensions of the metal nanoparticles obtained by bottom-up techniques is not arbitrary as it is for top-down techniques. However, in view of possible widespread use of SEIRA/SEIRS as a standard spectroscopic technique, it is also necessary that the substrate preparation process be low-cost and reproducible. Clearly, the future of SEIRA/SEIRS is strongly related to the successful development of substrate preparation techniques with high performances and low cost, therefore it is still unclear whether bottom-up or top-down techniques will prevail. A comprehensive review of the literature on bottom-up substrates has been done by R. F. Aroca and D. J. Ross [15], so here we shall focus on few key experiments only.

Bottom-up prepared substrates can be divided into three main groups: (a) substrates originating from colloidal metal nanoparticles; (b) substrates obtained from evaporation or sputtering of a bulk target; and (c) substrates obtained by electroplating processes. Notice that, for all three groups, the metal nanostructure is obtained by on-purpose modification and control of a standard industrial process, which is otherwise intended for the deposition of smooth metal films. This leads to the clear advantage that the substrate preparation equipment is readily available for future commercialization of the substrates. Instead of proposing an historical review of SEIRA/SEIRS, we mention a number of relevant experiments that were conducted by using each substrate preparation technique. The choice of described experiments is clearly not exhaustive of all designs and ideas proposed in the literature, which actually suffers from the lack of a unitary theoretical framework.

- **Colloids**: metal nanoparticles suspended in colloidal solution are deposited by drop-drying (or more sophisticated drying techniques) on an infrared-transparent substrate. The final result depends on the complex physical interactions between colloidal particles in the high-density regime, which occurs before drying. The interparticle interactions can be tuned in order to obtained a reproducible nanostructure, for example, by using metal–polymer composites (e.g., polymer-coated metal spheres). In the example of Figs. 14.2 and 14.3 [16], gold nanoparticles of 1–2 nm diameter suspended in a solution were employed for seeded-growth

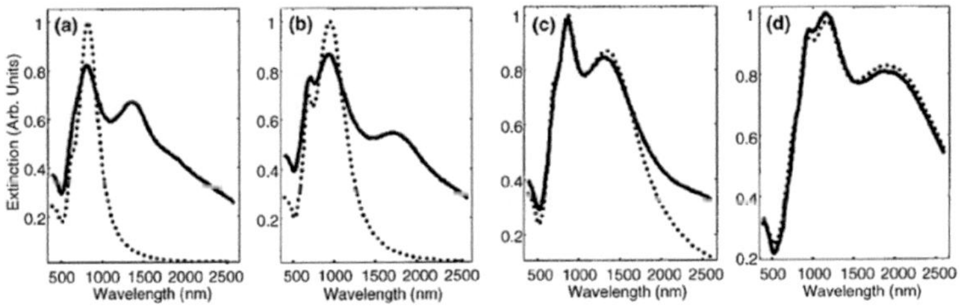

Figure 14.2 Hollow nanospheres (called "nanoshells") make it possible to increase the particle diameter (here 55, 75, 135, and 210 nm from left to right), for shifting the resonance toward the mid-infrared range, while maintaining the metal thickness at 14 nm below the skin-depth limit, for achieving true plasmonic field confinement. Reprinted with permission from Ref. [16]. Copyright 1999, AIP Publishing LLC.

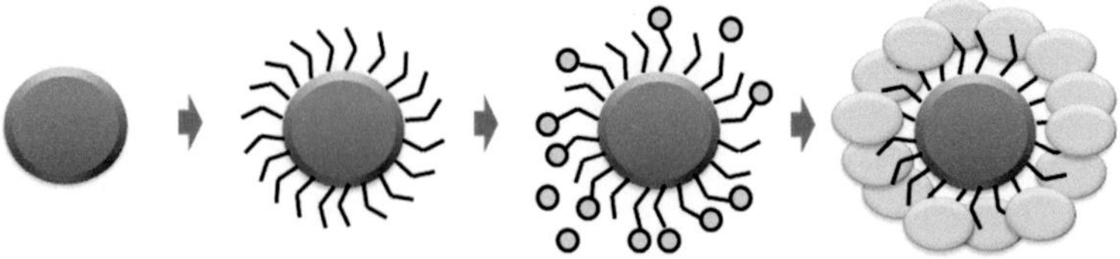

Figure 14.3 Example of bottom-up prepared substrate for infrared sensing. Gold nanoshells are assempled entirely by physico-chemical reactions, the final result being a subwavelength-sized object with resonances in the infrared range. From left to right, a SiO_2 nanosphere is functionalized with amines, then exposed to 2 nm diameter gold nanoparticles. Finally, additional gold is grown to complete the self-assembled gold nanosphere [18].

of complex structures (gold nanoshells grown around a silicon sphere, the surface of the latter terminated with a linker molecule). The gold nanoshell design is crucial for reaching particle sizes of hundreds of nanometers, featuring reasonable scattering cross-sections in the mid-IR, while maintaining the gold film thickness below the skin depth to obtain strong field confinement by resonant interaction of surface plasmons at the two interfaces [17, 18]. In this example of bottom-up technique, the spherical shape of the structure is guaranteed, but the reproducibility of the gold film thickness decreased with decreasing structure size (from 210 nm to 50 nm sphere diameter).

- **Sputtering/evaporation**: a crucible positioned in a high vacuum chamber is loaded with high-purity bulk metal (or alloy), and then heated and/or bombarded by electrons or ions, so to produce liquid metal droplets with sufficient kinetic energy to travel in the free space and reach the target substrate, where they re-assemble to form the final metal structure. Many evaporation and sputtering chambers, routinely employed in the microelectronic industry where they are used to produce smooth metal films, can be used instead to deposit metal nanostructures by controlling the deposition parameters: substrate and crucible temperature, chamber vacuum level, and deposition rates. In one of the first reports of SEIRA [19], the fine tuning of the deposition

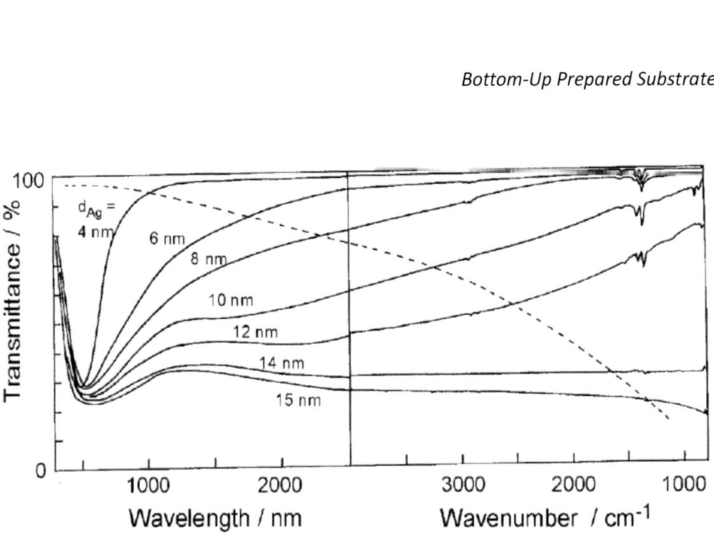

Figure 14.4 Transmission spectra of Ag island films evaporated on CaF2. The effective thickness d_{Ag} indicates the average substrate coverage. The absorption peaks seen in the mid-IR region (right plot) at 1390 cm^{-1} are due to *p*-nitrobenzoate adsorbed on the island surface. The SEIRA peaks disappear when the coverage increases and the islands merge in a continuous film. The dashed line is the calculated transmission of a 10 nm thick continuous Ag film calculated from the Fresnel formulae. Reprinted with permission from Ref. [20], Copyright 1991, American Chemical Society.

rate and time of evaporated silver allowed the authors to produce Ag nano-islands with different degree of substrate coverage, inter-island distance, and connectivity level (see Fig. 14.4). These parameters were only known on average, and they were summarized by an "effective thickness" measured with the standard quartz balance installed inside the evaporation chamber. The increase of substrate coverage by Ag nano-islands increased the inter-island interaction, and resulted in a shift of the surface plasmon resonance from the visible to the mid-IR. Transmission spectra of Ag nano-island films evaporated on CaF_2 with different coverage are shown in Fig. 14.4. The strong band around 500 nm is attributed to the shape resonance of a single Ag nano-island. For increasing substrate coverage, the dipole-coupling among islands increases resulting in the broadening and softening of the 500 nm band. The tail of this band eventually extended into the mid-IR, in agreement with the Bruggeman's theory of the effective medium (see below). Only when the effective thickness surpassed 20 nm, the film became fully connected and opaque in the mid-IR, as expected from a calculation based on the Fresnel relations (dashed line in Fig. 14.4). Note that for $d_{Ag} > 15$ nm, hence for an almost flat non-nanostructured surface, the SEIRA peaks at 1390 cm^{-1} disappeared: therefore, this experiment [19] and its confirmations [20, 21] are considered the first reports of SEIRA.

- **Electroplating and chemical plating**: techniques for metal coating of objects by immersion in ionic solutions containing metal ions can be used to obtain purposely rough coatings, with typical roughness in the range between 100 nanometers and 5 microns, limited by diffusion coefficients in liquid solutions. Therefore, electroplating techniques, even if extremely convenient in terms of costs and reproducibility, are usually considered as not suitable for producing substrates for SERS that require nanostructures of few tens of nanometers. However, substrates made by electroplating have a potential to work well for much longer mid-IR wavelengths. Such electrically connected SEIRA substrates have made possible to perform infrared spectroscopy in situ during an electrochemical reaction. In this way, new intermediate species forming on the electrode surface have been detected and identified. In Fig. 14.5, we report the mid-IR study of the chemisorption of methanol on a platinum electrode in a fuel cell. A peak at 1320 cm^{-1} was observed in the ATR spectrum of the electrode surface, and this peak was absent in the IR spectra of the bulk solution. This fact indicates that the peak is a SEIRA effect and that the new chemical species exist only at the nanostructured Pt surface. As a result, a new intermediate species was identified in the methanol $\rightarrow$ carbon dioxide reaction catalyzed by the Pt surface, and the SEIRA peak intensity was also correlated with the current intensity flowing into the electrochemical cell, clarifying the details of the voltage–current characteristics, which were measured long time before in this type of fuel cells but never fully understood.

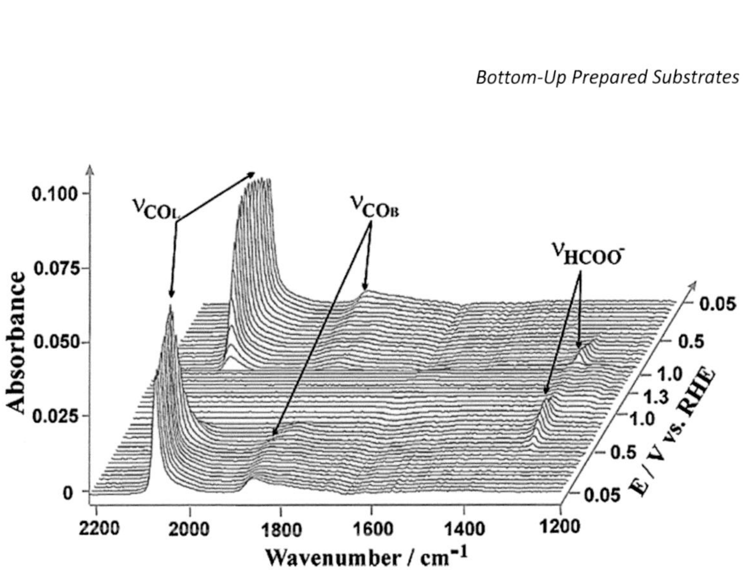

Figure 14.5 Attenuated total reflection spectra of the interface between liquid methanol and a nanostructured platinum electrode. The SEIRA effect allows one to observe the effect of MeOH electro-oxidation on Pt: a new IR peak is observed at around 1320 cm^{-1} when the electrode potential is higher than 0.5 V, the potential at which oxidation of MeOH occurs. This peak has been assigned to the formation of reaction intermediates on the Pt electrode surface. The MeOH electro-oxidation current was observed to be proportional to the IR peak intensity. Reprinted with permission from Ref. [21], Copyright 2003, American Chemical Society.

A very recent biochemical application of SEIRA is the study of solid-supported cell membrane templates [2, 22]. The cell membrane is composed of phospholipid bilayers, and therefore, it is naturally two-molecule thick. Artificial templates can also be grown electrochemically or by Langmuir–Blodgett techniques on metal nanostructures working both as SEIRA substrates and as electrodes. In Fig. 14.6, we report the schematics of a mid-IR spectroscopy setup incorporating both SEIRA and electrochemistry functionalities [22]. SEIRA was exploited to obtain the absorption signal from sparse membrane proteins immersed in the phospholipid matrix. Electric potentials and LED illumination were used to initiate and monitor reactions between the membrane proteins and other functional proteins in solution. It was found that the quantitative interpretation of the experiments critically depends on the orientation of the solid-supported membrane protein with respect to the direction normal to the metal surface: such problems are intrinsic in the concept of

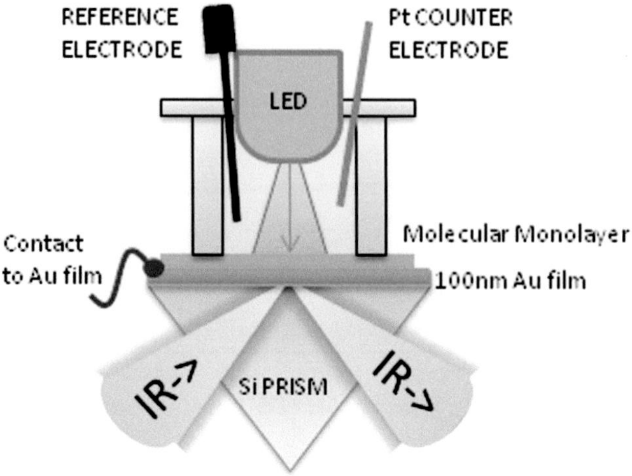

Figure 14.6 Set-up for surface-enhanced spectroscopy in the mid-infrared range with electrochemical and photochemical controls, developed for the study of cell membrane templates with membrane proteins [22].

bottom-up prepared substrates, while in top-down structures it is possible to control surface orientation and field intensity at each position on the substrate.

14.3.1 *Electrodynamic Models*

We conclude our brief discussion of bottom-up prepared substrates by presenting the electrodynamic models that can describe, and eventually predict and reproduce, the transmission and reflection spectra of metal nanoparticle complexes. We shall use the notation ε_h for the dielectric constant of the environment (usually air, $\varepsilon_h = 1$) and $\varepsilon_{\mathrm{M}} = \varepsilon'_{\mathrm{M}} + \iota\varepsilon''_{\mathrm{M}}$ for the complex dielectric function of the metal, usually described by the standard Drude model for the electrodynamic response of metals:

$$\varepsilon_M(\omega) = \varepsilon_\infty - \frac{\omega_p^2}{\omega^2 + i\gamma\omega} \tag{14.6}$$

where ε_∞ is the limit value of the dielectric constant at very high optical frequencies, γ is the free-carrier scattering rate, and Drude spectral weight ω_p^2 is proportional to the free carrier density. The Drude weight is denoted by using the symbol ω_p^2, because ω_p is approximately proportional to the plasma frequency, that is, the frequency where ε'_M changes sign from negative at low frequency to positive at high frequency, and the electrodynamic response of the metal becomes dielectric-like ($\varepsilon'_M > 0$). As we will see below, the exact frequency at which this happens indicates the frequency range of strongest plasmonic behavior of the metal nanoparticles, and it is called the *screened plasma frequency:*

$$\tilde{\omega}_p = \sqrt{\frac{\omega_p^2 - \gamma^2}{\varepsilon_\infty}} \qquad (14.7)$$

The screened plasma frequency of gold is around 2.3 eV [23], and this is the reason why gold nanoparticles work very well for plasmonic applications at long visible wavelengths of 0.5–0.8 microns, such as SERS. Silver has $\tilde{\omega}_p$ ~4 eV and indeed it is often used in experiments involving ultraviolet light. Clearly, highly doped semiconductors [24, 25] or conducting oxides [26] with $\tilde{\omega}_p$ at selectable values below 0.5 eV would be the materials of choice for mid-infrared surface-enhanced spectroscopy, and the development of such materials and substrates (that we have called "third-generation substrates") is now the subject of intense research.

It is worth noticing that in the long-wave infrared range ($\lambda >$ 10 μm), a negative value of $\varepsilon'_{\rm M}$ is reached across narrow frequency ranges in strongly polar crystals with highly symmetric unit cell and in the absence of free charge carriers, such as silicon carbide (SiC). This effect is due to the presence of IR-active transverse-optical phonon modes, followed by longitudinal-optical phonon modes at slightly higher frequency. The equivalent of SPP exists at the interface between such crystals and a conventional dielectric, and they are called surface phonon polaritons. In general, losses are much lower than for metallic nanoparticles, but the field at the surface of nanoparticles made of such crystals is confined only in the narrow, fixed frequency range where $\varepsilon'_{\rm M}$ is negative (few tens of cm^{-1}), and therefore, this effect is not very useful for molecular spectroscopy applications.

The scattering and absorption cross-sections of nanoparticles, and their eventual modifications once they are covered with the molecular film, can be calculated from the first-order electrodynamic polarizability $\alpha_{ij}(\omega)$ defined as the ratio between the i-component of the polarization field generated by the displacement of charges due to the action of the external (radiation) field, and the j-component of the radiation field itself. Here, we restrict to the case of an isolated metal particle of arbitrary shape, cubic crystal symmetry, and subwavelength dimensions. We anticipate that $\alpha(\omega)$ can be calculated only if the particle geometry is precisely known, and often it cannot be written in an explicit form. A special case is the polarizability of a metal sphere in vacuum, which is given by the classical relation $\alpha(\omega) = \frac{\varepsilon_0(\varepsilon_M - 1)}{\varepsilon_M + 2}$. The latter relation provides a clear argument for explaining plasmonic field enhancement: for $\varepsilon_M = -2$ the polarizability diverges. Since $|\varepsilon_M| \gg 1$ for metals, the divergence frequency for spherical nanoparticles is close to the zero-crossing of ε'_M, that is, close to the screened plasma frequency of the metal. In all cases where the particle is not spherically symmetric, it is convenient to model the particles as oblate spheroids with $\eta = a/b$ (ratio of long over short axis diameter length). The polarizability tensor of a spheroid is an implicit relation that can be estimated in several ways [27, 28]. A very approximate explicit relation, that reduces to the classical one for the sphere in vacuum for $\eta \to 1$, $\varepsilon_h \to 1$, is given by the following equation [29]:

$$\alpha(\omega) \approx \varepsilon_0 \frac{\varepsilon_M - \varepsilon_h}{\varepsilon_M + (1+\eta)\varepsilon_h} \tag{14.8}$$

where one sees that a more oblate shape shifts the plasmonic resonance frequency (divergence of $\alpha(\omega)$) more toward the infrared, where ε_M is negative and has a higher absolute value. Indeed, by using strongly elongated nanorods, Neubrech et al. [30] were able to reach the short-wavelength mid-IR (wavelength of 3 μm) with nanoparticles made of gold. In this view, the SEIRA effect is of the same nature of plasmonic field enhancement in the visible, but with very much elongated particles.

Bottom-up prepared SEIRA substrates are made of a system of many subwavelength-sized particles that cannot be considered as isolated objects. If the particles have approximately the same shape, and if their distance is much larger than their size, the mean field Maxwell–Garnett theory of the polarization of inhomogeneous media can be used [31, 32]. The effective dielectric constant of the substrate is given by the following equation:

$$\varepsilon(\omega) \approx \varepsilon_h \frac{3 + 2F\alpha(\omega)}{3 - F\alpha(\omega)} \tag{14.9}$$

where the parameter F is the filling factor of the metal particles randomly distributed inside the hosting matrix of dielectric constant ε_h and $\alpha(\omega)$ can be taken, for example, from the above expression for spheroidal particles. Unfortunately, the mean-field approach can very seldom be used, as the particles are often located within the near-field region of other particles. In this case, more refined approaches based on renormalized polarizability are available [33]. If one now adds the fact that in many bottom-up prepared substrates the shape and the size are not the same for all particles, but are given by a statistical distribution, it turns out that, even if SEIRA effects are clearly observed, it is very difficult to precisely model the substrate optical properties, and to perform an optimization of the geometrical parameters in view of target spectroscopic applications. For this reason, in recent years the research on SEIRA has moved toward top-down fabricated substrates with precisely defined design, which we discuss in the next section.

14.4 Top-Down Fabricated Substrates: Antennas and Metamaterials

In our review of bottom-up first-generation SEIRA/SEIRS substrates, we have seen that they play a crucial role in the application of surface-enhanced IR spectroscopy. Moreover, they feature a relative ease of preparation and low production cost, but they inevitably suffer from severe limitations in terms of control of the direction, strength, and distribution of the enhanced field. Furthermore, in the mid-IR range there is an intrinsic mismatch between the wavelength (several microns) and the characteristic dimensions of bottom-up prepared nanoparticles (few tens of nanometers). Attempts to increase the nanoparticle size to the micron scale, suitable to match mid-IR wavelengths, inevitably result in the particle being much larger than the skin-depth of metals (around 20 to 40 nm for all metals in both the mid-IR and visible range), hence result in the suppression of plasmonic field enhancement.

The top-down approach instead, when realized with modern high-resolution lithographic technologies capable of nanometric precision, offers the possibility of engineering the electromagnetic response of the nanostructured substrate in a controlled and versatile way. The general idea is to fabricate nanostructured substrates by repetition of a single element, after having calculated its local electromagnetic response on a subwavelength scale by means of finite- difference time-domain (FDTD) and/or finite-element simulation methods. Nowadays, this task can be performed (with some care and theoretical know-how) with a standard personal computer equipped with commercial software packages initially developed for high-frequency integrated circuit simulation. In the last decade, a myriad of different shapes of the single element have been explored by joint simulation and fabrication efforts, with the aim of generating the highest possible field enhancement at the so-called *hot-spots*, that is, the high field locations where one plans to mechanically or chemically immobilize the molecules. Two physical mechanisms originate the hot-spots: (i) the lightning effect (or compression of electric field) when two metallic structures are placed at subwavelength distance, and (ii) the excitation of surface plasmons, which also works for a single metal–dielectric interface. The first mechanism is based on the concept of perfect electric conductor, which works well, for example, for gold in the mid-IR, and therefore it is purely geometrical and almost wavelength independent. The second mechanism indeed depends on the value of the dielectric constants of the conductor and of the dielectric to obtain a sharp frequency resonance. In second-generation metal-made substrates described in this section, the fine tuning of the resonances is still obtained by means of geometrical shaping of the nanostructures. Forthcoming third-generation substrates made of nonmetallic materials will instead allow for active tuning of the resonances through external control of the material properties.

What is actually measured in second-generation substrates with sharp frequency resonances is not directly the (enhanced) absorption spectrum of the molecule, but rather a perturbation of radiation scattering by the subwavelength element resulting from the plasmon–molecule coupling (see Section 14.5 for details). For this reason, the concept of SEIRS has been introduced [34]. Indeed, the spectral profile obtained by SEIRS is somehow different from the absorption spectrum, due to the near-field interactions involved in the scattering process [35]. The observed SEIRS signal strongly depends on the relative value of the real part of the dielectric constant of the molecule and that of the environment (host and conductor). As a result, the plasmon resonance lineshape, upon interaction with the target molecule, experiences both a redshift (that is the key mechanism in refractive-index surface plasmon resonance sensors) and a line broadening, which can be used to evaluate the true absorption enhancement [1]. In the literature, different definition of "SEIRA enhancement factors" are used, sometimes multiplying the absorption line strength by the ratio between the unit cell area and the "hot-spot area", a fact that leads to extremely high numbers for the SEIRA enhancement factor. We prefer not to report any numerical value of "SEIRA enhancement factors" taken from the literature, instead we report maps of the field enhancement value relative to the modulus of the incident radiation field, which facilitate the comparison among different nanoparticle designs.

14.4.1 *The Role of the Nanoparticle Shape*

We now consider the role of the geometrical shape of the single nanoparticle, assuming that its details were optimized by electromagnetic design. If we consider a nanoparticle of elongated shape, the length L along the electric field polarization direction and the average refractive index of the surrounding environment n_{eff} are the main parameters defining the resonance frequency of its first-order plasmonic excitation. The simplest infrared antenna geometry one can think about is the optical translation of the half-wave printed dipole antenna as described by radio-frequency engineers:

Ref.	Structure	Molecule	Exp. results
Neubrech *et al.*, 2008		Octadecanthiol (ODT) $[CH_3(CH_2)_{16}CH_2SH]$	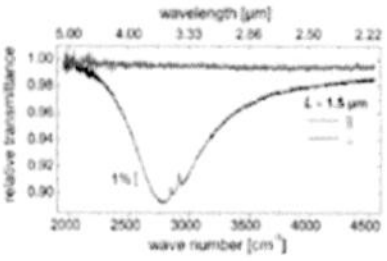
Adato *et al.*, 2009	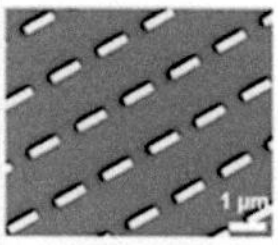	Silk fibroin	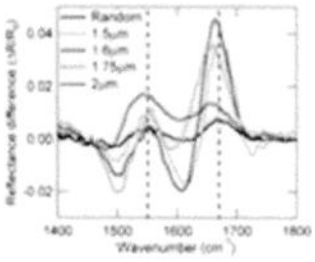
Aouani *et al.*, 2013		SAM of carboxyl-terminated alkanethiols [AT-EG6-COOH] and EDC-NHS	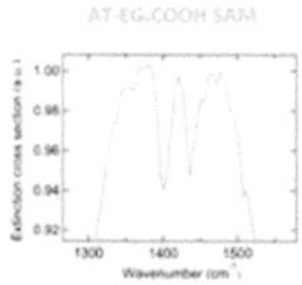
Aksu *et al.*, 2013	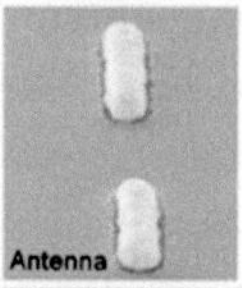	protein A/G monolayer and antibody (IgG)	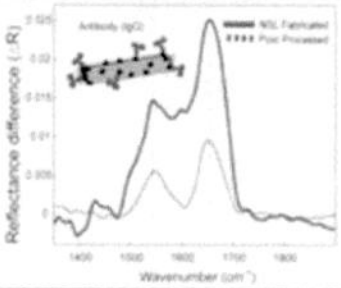
Brown *et al.*, 2013		Octadecanthiol (ODT) Hemoglobin Silica	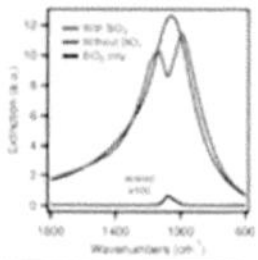
Pryce *et al.*, 2011		p-mercaptoaniline (pMA)	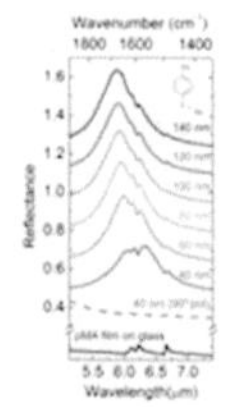

Figure 14.7 Here summarized some works recently published in the literature. The unit cell structure of the nanostructured arrays proposed for SEIRA measurements and the target molecule detected are reported. In the right column, the observed signal from reflection or transmission measurements is reported. Reference to these works can be found throughout the text in Section 14.4.

a rectangle defined by lithography on an infrared-transparent substrate. When conventional lithographic technologies are used, the metal film thickness (hundreds of nanometers) will inevitably compare with the rectangle width, and therefore they are referred to as *nanorods* [30, 36]. For electric field polarized along the long side, the nanorod structure has a plasmonic resonance at $\lambda_m = (2/m)n_{\text{eff}} L$, where m is an odd integer indicating the mode number (even integers would correspond to symmetric field profiles with no net dipole to couple with electromagnetic radiation). For electric field polarized along the short side of the rectangle, a different resonance exists at much shorter wavelengths. This short-side resonance is weaker, because there is no polarization enhancement effect due to the elongated shape. Indeed, the intrinsic double-energy scale present in planar rectangular dipole antennas has been exploited to obtain both mid-IR and Raman signal enhancement for the same molecules on the same substrate [37]. The next step is to place two nanorods adjacent to one another to form an antenna with a nanometric gap. At odds with the prediction of RF antenna theory that, in the case of impedance-matched load, the resonance wavelength should double, two adjacent nanorods display only a slight redshift of the resonance with respect to a single nanorod, due to strong impedance mismatch between metal and vacuum [38]. However, two adjacent nanorods feature a strong enhancement of the electric field in the gap at the single-rod resonance frequency, which can be exploited for SEIRA. Moreover, as the high field is confined in the gap, the rest of the nanorod can be shaped to obtain different functions, as for example Raman signal enhancement on the same substrate [39]. Infrared nanorods are relatively simple to fabricate, and they were also obtained through nanostencil lithography, suitable for low-cost mass production of substrates. The latter technique even provided further electric field enhancement at the nanorod surface, due to molecular aggregates forming in its proximity [40].

In order to relax the linear polarization requirement of nanorod antennas, the next step is the cross-shape nanostructure formed by two orthogonal nanorod pairs. If the two arms constituting the cross nanoparticle have the same length, the resulting response will be polarization independent [41]. If, instead, one of the two arms has a different length, the cross becomes a dual-band resonator for multispectral analysis. Additional solutions in this direction have been made possible by advancements in the precision of nanofabrication techniques that allowed the realization of mid-IR log-periodic antennas featuring an operating wavelength window of approximately 3 μm in the mid-IR [42].

A different approach beyond scaling rectangular dipoles is to exploit ultrahigh-resolution lithography to create sharp edges and tips to obtain stronger charge accumulation and hence stronger fields in a highly confined volume. This approach works well in three dimensions with metalized scanning-probe tips, but this topic is outside the subject of the present chapter (see, e.g., Ref. [35] and references therein). However, a planar triangular shape, such as a planar bow-tie antenna, will also confine the field to some extent. In this case, the unit cell is made of two triangles with a nanometric gap in between their opposed tips. Because of the lack of a true resonance frequency in the bow-tie design, the highest field confinement is reached at the frequency matching the plasma frequency of the metal, and therefore bow-tie metallic antennas do not display strong resonances in the mid-IR range. However, by fabricating a sharp tip on the front surface of cross dipole antennas, very strong field enhancement at specific resonant mid-IR wavelengths (the C–H stretch at 3.4 μm and the amide-I band of hemoglobin at 6.05 μ) was obtained [43].

As stated above, the main disadvantage of geometrical resonances is that they cannot be tuned once the substrate has been fabricated. A proposed approach toward tunability is offered by integrating the nanostructures with a compliant substrate and use mechanical deformation of the substrate to tune the resonance frequency. This was realized for mid-IR split-ring resonators [44] whose resonance frequency is given by the classical LC resonance formula $\omega_0 \sim (LC)^{-1/2}$, where L is the inductance and C is the capacitance across the split in the resonator. By stretching the polymeric substrate, the horizontal arm can be moved away from the split-ring resonator changing the interparticle interaction, hence their resonant frequency. However, frequency tunability was obtained at the expenses of the field enhancement at the gap, which decays as the interparticle distance is increased.

14.4.2 *The Role of the Array Periodicity*

In order to produce suitable substrates for far-field FT-IR spectroscopy, the subwavelength units need to be repeated, typically on a two-dimensional array. Periodic arrays of antennas that extend over millimeter-sized areas are typically required. Antenna design based on electromagnetic simulations may also exploit the additional degree of freedom coming from the array periodicity, which tunes the nearest-neighbor interaction. When the periodicity becomes dominant over the single-object geometry in defining the resonance frequencies, a wavevector-space description based on Bloch's theorem (similar to that developed for photonic crystals) is more suitable than the single-antenna analysis. In the case of connected metallic structures (i.e., nanohole arrays), SPP propagate at the interface between metal and dielectric medium and their resonance frequency is set mainly from the array periodicity. A notable effect is the extraordinary transmission through subwavelength hole arrays [45]. Subwavelength hole arrays feature grating resonances with extremely high-quality factor, which can be exploited for mid-IR mass-sensing applications [2, 5]. However, no significant mid-IR vibrational absorption enhancement is expected in such grating-mode substrates because of the lack of field confinement, as indeed observed in Ref. [2] (Fig. 14.8). Therefore, we will not discuss subwavelength hole arrays, but we shall rather focus our attention on nanostructured substrates made of disconnected nanoantennas arranged in a periodic array.

The array periodicity plays an important role, since it sets the distance between neighboring nanoantennas that tunes the mutual interaction between charge oscillations sustained by two different structures. The near-field interaction between plasmonic modes on adjacent nanostructures is described by using the quantum mechanics concept of wavefunction hybridization [46, 47]. Although the concept of physically disconnected metal nanoparticles exhibiting a collective response if placed at near-field distances may seem not intuitive, indeed periodicity defines the collective response of the array and the transition between radiative and diffraction regimes. The interparticle interaction can be treated using coupled dipole methods, taking into account that the total electric field at a

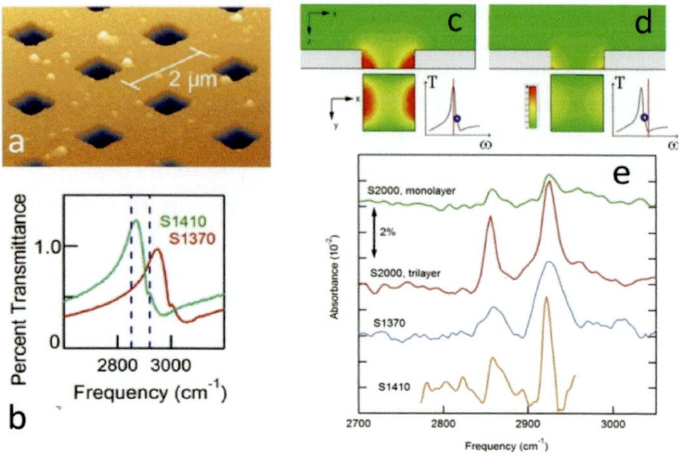

Figure 14.8 Surface plasmon polariton sensor for mid-IR detection of lipid bilayer mimicking cell membrane. (a) Three-dimensional rendering based on AFM data of square hole array with period 2.00 μm (S2000) (metal surface roughness below 10 nm). (b) Experimental transmittance spectrum of square hole array with period 1.37 μm (S1370, red line) and S2000 (green line). The frequency position of the main phospholipid vibrational lines (CH_2 stretching doublet) is marked by blue dashed lines. Numerical calculation of field profiles at the maximum (c) and minimum (d) far-field transmittance frequency, as indicated by the red marking line in the insets; blue circle indicates SPP resonance frequency. (e) Absorbance spectra calculated from experimental data, for monolayer/trilayer lipid formation of different arrays. Reprinted with permission from Ref. [2], Copyright 2013, American Chemical Society.

certain position results from the sum of all retarded electric fields of other dipoles. The Rayleigh wavevectors $k_m = 2\pi m/d \sin\theta$ (with m integer and θ incidence angle) set the grating resonances: at this wavelength, all electric fields sum up in phase, while for all shorter wavelengths the single-particle geometry dominates the response.

14.4.3 *The Role of the Substrate Polarizability*

When dealing with metal nanoparticles, one generally assumes a homogeneous host medium with given refractive index n_h. However, in an ordinary SEIRA/SEIRS sensing chip, the metallic nanostructures lie on a rigid, IR-transparent dielectric substrate with a different refractive index n_s (e.g., silicon, $n_s = 3.42$, or CaF_2, $n_s = 1.40$ for mid-IR frequencies) and are exposed to air ($n_h = 1.00$) or water ($n_h = 1.4$–1.7 depending on wavelength). The refractive index mismatch results in different modal indexes at the two interfaces, with consequent splitting of the normal modes of oscillation. Sometimes, when designing the resonance of a SEIRA substrate, an approximate effective refractive index is considered:

$$n_{\text{eff}} = \sqrt{\frac{n_h^2 + n_s^2}{2}} \tag{14.10}$$

However, the fine tuning of the resonance frequencies to match molecular vibrational lines, which is required in the top-down approach, makes this approximation too rough. One trick when dealing with aqueous environments is to use CaF_2, which has almost the same value of the refractive index as water. More generally, *under-cut* and *buried* nanoantennas have been proposed with the purpose of reducing refractive index mismatch (see Fig. 14.9).

Under-cut geometry: a wet-etching step is used to remove part of the substrate around and under the edges of the nanoantenna, in order to maximize its exposure to target molecule in air/water. As a result n_{eff} will shift toward the limit value n_h, and the resonance frequency will shift to higher frequencies, unless the nanoparticles size is made larger to preserve molecular fingerprints matching. The possibility to work with larger nanostructures offers the advantage of improved nanofabrication output, sharper tips and edges, and higher area exposed to surface chemistry processes for molecule immobilization.

Buried geometry: the opposite approach consists of burying nanoparticles inside the substrate. A dry-etching step is used in the fabrication procedure before the metal deposition, so as to

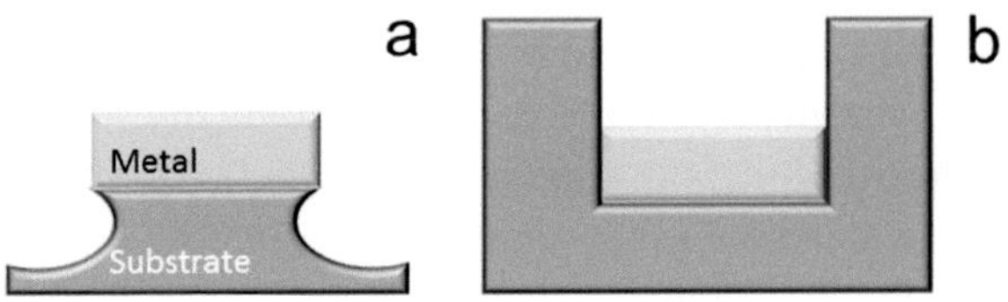

Figure 14.9 Sketch of the under-cut (a) and buried (b) geometry for metallic nanostructures onto a dielectric substrate.

obtain etch pits of specific geometry where the metal nanoparticles are then evaporated in the subsequent process step. In this case, n_{eff} will be closer to n_s, therefore the plasmonic resonance will move to lower frequency, but one can take advantage of the higher polarization field in the dielectric substrate. The high-field volume is distributed both in the dielectric and in the air gap on the top of the nanostructure, but the air gap can be made such a way that it does not significantly affect the near-field distribution according to FDTD simulations. Despite the fact that the accessible area for sensing applications is only the one in correspondence of the air gap, the buried geometry features a field enhancement more than two times higher than a corresponding nanoparticle–on-substrate [48].

14.4.4 *Real-Time and In situ*

We conclude the review of top-down approaches by describing a recent demonstration of real-time in situ monitoring and identification of various chemically distinct substances in aqueous environment by SEIRA/SEIRS [36]. In this work, the dynamic capabilities of SPR refractive index sensors (typically designed to work in the near-IR) are added to the chemical recognition capability of mid-IR spectroscopy. Here, the sensing mechanism is based on surface-enhanced absorption signal due to the spectral overlap of the plasmon resonance and the molecule absorption. In addition, the integrated spectral weight of the absorbance peaks corresponding to specific chemical fingerprints of the target molecule provides chemical identification, at odds with SPR instruments where only a substance-independent variation of refractive index is detected. The aim of the experiment was to monitor the secondary structure of proteins by detecting changes in the amide-I and amide-II bands around 1530 and 1660 cm^{-1}. An ATR configuration, as schematized in Fig. 14.10, was employed in order to overcome the strong water absorption due to OH-bending line at 1650 cm^{-1}. Gold nano-antenna arrays were realized on a CaF_2 chip inserted in a fluidic cell with capability of loading different analytes through inlet and outlet ports. The gold nanoantennas were functionalized with a SAM of a biotin-labeled alkanethiol (BAT) and spotted with IgG antibody.

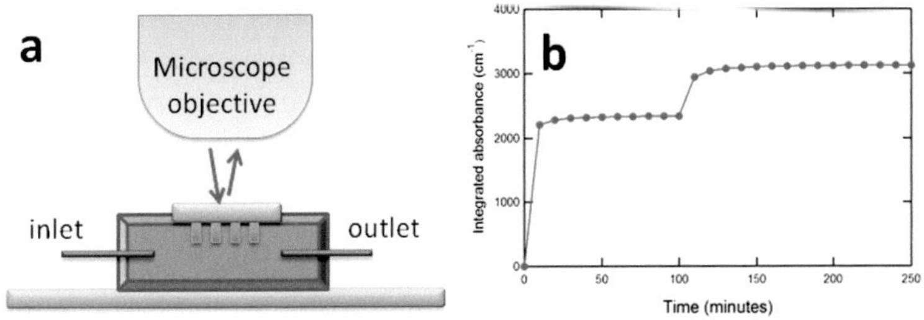

Figure 14.10 Plasmonic-induced absorption (PIR) platform for the quantitative detection of proteins bound to a self-assembled monolayer covering gold nanoantennas resonating at the amide I infrared vibration band (1630–1670 cm^{-1}). Protein samples with high-, low-, and zero-binding efficiency to SAM are introduced. (a) Experimental configuration for the PIR measurements. (b) Sketch of the integrated amide I absorbance signal used to monitor protein binding onto the nanostructures similarly to surface plasmon resonance sensors in the near-infrared or visible range, with the advantage of chemical specificity [36].

SEIRA signal was monitored in time upon introduction in the flow cell of three different protein samples with high-, low-, and zero-binding efficiency to the IgG antibody. The frequency-integrated absorbance under the amide-I and amide-II bands depended on the number of amide bonds in the given protein, multiplied by the protein coverage. Monitoring the integrated spectral weight over time then provided the dynamics of protein accumulation on the nanoantennas. In addition to near-IR SPR setups, which detect the target molecules but cannot identify them, in this case the integrated absorbance signal monitors the protein-specific chemical fingerprint.

14.5 Plasmon–Molecule Interaction

In this last section, we provide the elements of a phenomenological electrodynamic model of the plasmon–molecule interaction, which lies behind all phenomena described here. The absorption lines associated with molecular bond excitation in general exhibit the symmetric Lorentzian lineshape characteristic of the response of forced damped harmonic oscillators. However, the absorption lines observed in resonant SEIRA/SEIRS experiments (i.e., when the geometrical resonance frequency matches the molecular vibration frequency) exhibit a remarkable asymmetry (see spectra in Fig. 14.6). The asymmetric lineshape can be described as a Fano interference profile, named after the physicist Ugo Fano. This interference model describes the (quantum or classical) overlap between two possible excitation pathways in a system displaying a discrete transition and a continuum of excited states encompassing the energy of the excited discrete state. The continuum interferes in a constructive or destructive way with the discrete excitation depending on whether the energy of the final state is higher or lower than that of the discrete state, hence producing an asymmetric lineshape. In our case, the (broad bandwidth) plasmonic excitation constitutes the quasi-continuum background and the (narrow bandwidth) molecular vibration is the discrete transition. In this scenario, the extinction coefficient, that sums absorption and scattering losses, can be reproduced by an expression given by the product of two analytical terms, whose numerical parameters can be optimized by electromagnetic design [49–51]. The first term is the plasmon lineshape represented by a standard Lorentzian:

$$L(E) = \frac{1}{1 + \left(\frac{E - E_P}{\Gamma_P/2}\right)^2} \tag{14.11}$$

where E_{P} is the resonance energy of the plasmonic mode and Γ_{P} its width, usually only a factor of 5 smaller than E_{P} due to radiative losses. The second is the interference term given by the following equation:

$$\sigma(\varepsilon) = \frac{(\varepsilon + q)^2}{\varepsilon^2 + 1} \tag{14.12}$$

where q is the Fano parameter determining the asymmetry of the lineshape and ε is the reduced energy, which depends on the energy of the incident radiation E, the energy of the plasmonic (E_P) and molecular (E_d) modes and their linewidths Γ_P and Γ_d, respectively:

$$\varepsilon = \frac{E - E_d}{\Gamma_d(E)/2} - \frac{E - E_P}{\Gamma_P(E)/2}. \tag{14.13}$$

The product of the Lorentzian lineshape and the interference term reproduces the experimentally measured extinction coefficient of many SEIRA/SEIRS nanostructured substrates (see Fig. 14.8), and it has been used as a powerful tool for extracting the parameters describing the interaction between the plasmonic mode and the molecular absorption. One may think that an increased coupling between the plasmonic and the molecular mode necessarily leads to a stronger signal in the correspondent SEIRA/SEIRS spectrum, but this is not always the case. Instead, a careful design of the nanostructure has to be performed in order to optimize the molecular detection signal.

An intuitive approach to the plasmon–molecule system is offered by coupled classical harmonic oscillators [47]. The plasmonic resonance, which has a high cross-section for the interaction with the electromagnetic field, is represented as a mass-and-spring system driven by an external stimulus. The molecular absorption, which instead has a very small cross-section, is represented by a mass not coupled to the external radiation, but weakly interacting with the "plasmonic mass". In this model, the molecular vibration is not directly coupled to electromagnetic radiation (the so-called *dark mode*). In other words, in the absence of coupling to the plasmonic mode, the molecular vibration would not be observable since without surface enhancement it would be too weak to be detected. By definition, the damping rate γ_{P} of the narrow-bandwidth molecular vibration is much smaller than that of the broad-bandwidth plasmonic mode γ_{A}. The latter term is given by the sum of external and intrinsic damping rates as $\gamma_{\mathrm{A}} = \gamma_{\mathrm{Ae}} + \gamma_{\mathrm{A0}}$ where γ_{Ae} are radiative losses (external) and γ_{A0} is the energy ceded to the walls, or the other mass and springs. It can be demonstrated, and hopefully also visualized in the spring-and-mass model, that the maximum movement of the "molecular mass" is obtained through a complex impedance-matching problem, which results in equal distribution of intrinsic and external damping of the plasmonic mode (see Fig. 14.11). Moving below or above the critical impedance-matching point, the system is driven into one of the two following regimes: electromagnetic-induced transparency (EIT, i.e., the molecular absorption superimposes the plasmonic absorption resulting in a dip, a transparency window) and electromagnetically induced absorption (EIA, i.e., the molecular vibration appears as an absorption peak in the plasmonic band).

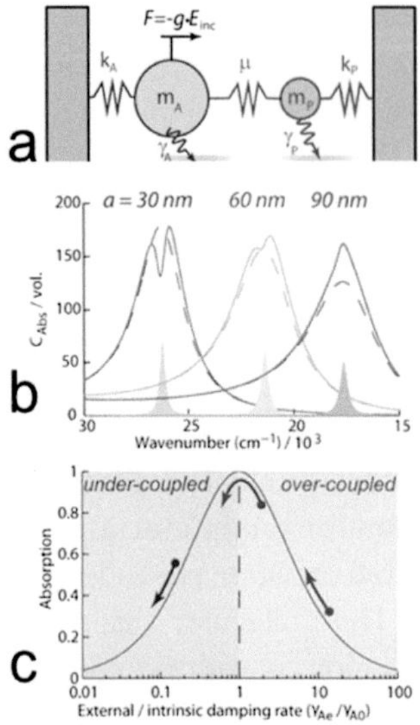

Figure 14.11 (a) Schematics of the mass-and-spring model. A classical harmonic oscillator with mass m_A, force constant k_A, and damping γ_A is driven by the external radiation field E_{inc} with coupling constant g. A second harmonic oscillator (corresponding parameters m_P, k_P, and γ_P) representing the molecular vibration with damping does not interact directly with the external radiation but it is coupled to the first oscillator by a third spring with force constant μ. The third spring represents the plasmon–molecule interaction. (b) Absorption cross-section renormalized to particle volume for a spheroidal nanoparticle before (dashed line) and after (continuous line) coating with a 5 nm-thick molecular layer displaying a vibrational mode in the same frequency region (filled peaks). The three curves represent the under-coupled (UC, violet), critically-coupled (CC, green), and over-coupled (OC, orange) regime according to (c). FDTD simulations (d) and experimental data (e) of nanorod array before (blue) and after (red) spin-coating an 8-nm-thick PMMA layer on top. Reprinted with permission from Ref. [52], Copyright 2013, American Chemical Society.

References

1. Wu, C., Khanikaev, A. B., Adato, R., Arju, N., Yanik, A. A., Altug, H., and Shvets, G. (2012). Fano-resonant asymmetric metamaterials for ultrasensitive spectroscopy and identification of molecular monolayers, *Nat. Mater.*, **11**, pp. 69–75.

2. Limaj, O., D'Apuzzo, F., Di Gaspare, A., Giliberti, V., Domenici, F., Sennato, S., Bordi, F., Lupi, S., and Ortolani, M. (2013). Mid-infrared surface plasmon polariton sensors resonant with the vibrational modes of phospholipid layers, *J. Phys. Chem. C*, **117**, pp. 19119–19126.
3. Kudelski, A. (2005). Characterization of thiolate-based mono- and bilayers by vibrational spectroscopy: A review, *Vib. Spectrosc.,* **39**, pp. 200–213.
4. Liedberg, B., Nylander, C., and Lunström, I. (1983). Surface plasmon resonance for gas detection and biosensing, *Sens. Actuators*, **4**, pp. 299–304.
5. Coe, J. V., Rodriguez, K. R., Teeters-Kennedy, S., Cilwa, K., Heer, J., Tian, H., and Williams, S. M. (2007). Metal films with arrays of tiny holes: Spectroscopy with infrared plasmonic scaffolding, *J. Phys. Chem. C*, **111**, pp. 17459–17472.
6. Olave, C. R., Carrasco, E. A. F., Campos-Vallette, M., Saavedra, M. S., Diaz, G. F., Clavijo, R. E., Figueroa, W., Garcia-Ramos, J. V., Sanchez-Cortes, S., Domingo, C., Costamagna, J., and Rios, A. (2002). Vibrational study of the interaction of dinaphthalenic Ni(II) and Cu(II) azamacrocycle complexes methyl and phenyl substituted with different metal surfaces, *Vib. Spectrosc.,* **28**, pp. 287–297.
7. Nakamura, R., Ueda, K., and Sato, S. (2001). In situ observation of the photoenhanced adsorption of water on TiO_2 films by surface-enhanced IR absorption spectroscopy, *Langmuir,* **17**, pp. 2298–2300.
8. P. Alonso-González, P. Albella, M. Schnell, J. Chen, F. Huth, A. García-Etxarri, F. Casanova, F. Golmar, L. Arzubiaga, L. E. Hueso, J. Aizpurua, and R. Hillenbrand. (2012). Resolving the electromagnetic mechanism of surface-enhanced light scattering at single hot spots, *Nat. Comm.,* **3**, p. 684.
9. Novotny, L. and van Hulst, N. (2011). Antennas for light, *Nat. Photon.*, **5**, pp. 83–90.
10. Joannopoulos, J. D., Johnson, S. G., Winn, J. N., and Meade, R. D. (2008). *Photonic Crystals: Molding the Flow of Light,* 2nd ed. (Princeton University Press, USA).
11. De la Rica, R. and Stevens, M. M. (2012). Plasmonic ELISA for the ultrasensitive detection of disease biomarkers with the naked eye, *Nat. Nanotech.,* **7**, pp. 821–824.
12. Adato, R., Yanik, A. A., Amsden, J. J., Kaplan, D. L., Omenetto, F. G., Hong, M. K., Erramilli, S., and Altug, H. (2009). Ultra-sensitive vibrational spectroscopy of protein monolayers with plasmonic nanoantenna arrays, *Proc. Nat. Acad. Soc.*, **106,** pp. 19227–19232.

13. Herminjard, S., Sirigu, L., Herzig, H. P., Studemann, E., Crottini, A., Pellaux, J.-P., Gresch, T., Fischer, M., and Faist, J. (2009). Surface Plasmon Resonance sensor showing enhanced sensitivity for CO2 detection in the mid-infrared range, *Optics Express,* **17**, pp. 293–303.

14. Kretschmann, E. (1971). Die bestimmung optischer Konstanten von Metallen durch Anregung von Oberflachenplasmaschwingungen, *Z. Phys.*, **241**, pp. 313–324.

15. Aroca, R. F. and Ross, D. J. (2004). Surface-Enhanced Infrared Spectroscopy, *Appl. Spectrosc.*, **58**, pp. 324A–338A.

16. Oldenburg, S. J., Jackson, J. B., Westcott, S. L., and Halas, N. J. (1999). Infrared extinction properties of gold nanoshells, *Appl. Phys. Lett.*, **75**, pp. 2897–2899.

17. Oldenburg, S. J., Averitt, R. D., Westcott, S. L., and Halas, N. J. (1998). Nanoengineering of optical resonances, *Chemical Physics Letters*, **288**, pp. 243–247.

18. Halas, N. J. (2002). The optical properties of nanoshells, *Optics Photonics News*, **13**, pp. 26–30.

19. Osawa, M. and Ikeda, M. (1991). Surface-enhanced infrared absorption of p-nitrobenzoic acid deposited on silver island films: contributions of electromagnetic and chemical mechanisms, *J. Phys. Chem.*, **95**, pp. 9914–9919.

20. Nishikawa, Y., Fujiwara, K., Ataka, K., and Osawa, M. (1993). Surface-enhanced infrared external reflection spectroscopy at low reflective surfaces and its application to surface analysis of semiconductors, glasses, and polymers, *Anal. Chem.*, **65**, pp. 556–562.

21. Chen, Y. X., Miki, A., Ye, S., Sakai, H., and Osawa, M. (2003). Formate, an active intermediate for direct oxidation of methanol on Pt electrode, *J. Am. Chem. Soc.*, **125**, pp. 3680–3681.

22. Ataka, K., Stripp, S. T., and Heberle, J. (2013). Surface-enhanced infrared absorption spectroscopy (SEIRAS) to probe monolayers of membrane proteins, *Biochim. Biophys. Acta*, **1828**, pp. 2283–2293.

23. Johnson, P. B. and Christy, R. W. (1972). Optical constants of the noble metals, *Phys. Rev. B*, **6**, pp. 4370–4379.

24. Ortolani, M., Baldassarre, L., Nucara, A., Finazzi, M., Frigerio, J., Isella, G., Samarelli, A., Paul, D. J., and Biagioni, P. (2013). Mid-infrared plasmonic antennas made of electron-doped epitaxial germanium-on-silicon, *IEEE Proceedings of Int. Conf. on IR, millimeter and THz waves (IRMMW-THz 2013)*, Mainz, Germany.

25. Law, S., Yu, L., Rosenberg, A., and Wasserman, D. (2013). All-semiconductor plasmonic nanoantennas for infrared sensing, *Nano Lett.*, **13**, pp. 4569–4574.

26. Naik, G. V., Kim, J., and Boltasseva, A. (2011). Oxides and nitrides as alternative plasmonic materials in the optical range, *Opt. Mater. Express*, **1**, pp. 1090–1099.

27. Roseler, A. and Korte, E.-H. (1998). The optical constant of metallic island films as used for surface enhanced infrared absorption, *Thin Solid Films,* **313–314**, pp. 732–736.

28. Eagen, C. F. (1981). Nature of the enhanced optical absorption of dye-coated Ag island films, *Appl. Opt.*, **20**, pp. 3035–3042.

29. Kang, H. and Kim, K. (2007). Anisotropic polarization tensors for ellipses and ellipsoids, *J. Comput. Math.*, **25**, pp. 157–168.

30. Neubrech, F., Pucci, A., Cornelius, T. W., Karim, S., García-Etxarri, A., and Aizpurua, J. (2008). Resonant plasmonic and vibrational coupling in a tailored nanoantenna for infrared detection, *Phys. Rev. Lett.*, **101**, pp. 157403–157407.

31. Maxwell–Garnett, B. A. (1904). Colours in metal glasses and metallic films, *Philos. Trans. Royal. Soc. A*, **203**, pp. 385–420.

32. Niklasson, G. A. and Granqvist, C. G. (1984). Optical properties and solar selectivity of coevaporated Co–Al2O3 composite films, *J. Appl. Phys.*, **55**, pp. 3382–3410.

33. Bruggeman, D. A. G. (1935). Berechnung verschiedener physikalischer Konstanten von heterogenen Substanzen, *Ann. Phys. (Leipzig)*, **24**, pp. 636–664.

34. Neubrech, F. and Pucci, A. (2012). Surface-enhanced infrared spectroscopy, *Spectrosc. Europe,* **24**, p. 6.

35. Alonso-González, P., Albella, P., Neubrech, F., Huck, C., Chen, J., Golmar, F., Casanova, F., Hueso, L. E., Pucci, A., Aizpurua, J., and Hillenbrand, R. (2013). Experimental verification of the spectral shift between near-and far-field peak intensities of plasmonic infrared nanoantennas, *Phys. Rev. Lett.*, **110**, pp. 203902–203906.

36. Adato, R. and Altug, H. (2013). In-situ ultra-sensitive infrared absorption spectroscopy of biomolecule interactions in real time with plasmonic nanoantennas, *Nat. Comm.*, **4**, p. 2154.

37. Le, F., Brandl, D. W., Urzhumov, Y. A., Wang, H., Kundu, J., Halas, N. J., Aizpurua, J., and Nordlander, P. (2008). Metallic nanoparticle arrays: a common substrate for both surface-enhanced Raman scattering and surface-enhanced infrared absorption, *ACS Nano,* **2**, pp. 707–718.

38. Jer-Shing Huang, Thorsten Feichtner, Paolo Biagioni, and Bert Hecht (2009). Impedance matching and emission properties of nanoantennas in an optical nanocircuit, *Nano Lett.*, **9**, pp. 1897–1902.
39. D'Andrea, C., Bochterle, J., Toma, A., Huck, C., Neubrech, F., Messina, E., Fazio, B., Marago, O. M., Di Fabrizio, E., de La Chapelle, M. L., Gucciardi, P. G., and Pucci, A. (2013). Optical nanoantennas for multiband surface-enhanced infrared and Raman spectroscopy, *ACS Nano,* **7**, pp. 3522–3531.
40. Aksu, S., Cetin, A. E., Adato, R., and Altug, H. (2013). Plasmonically enhanced vibrational biospectroscopy using low-cost infrared antenna arrays by nanostencil lithography, *Adv. Opt. Mater.*, DOI: 10.1002/adom.201300133.
41. Businaro, L., Limaj, O., Giliberti, V., Ortolani, M., Di Gaspare, A., Grenci, G., Ciasca, G., Gerardino, A., de Ninno, A., and Lupi, S. (2012). Mid-infrared nanoantenna arrays on silicon and CaF2 substrates for sensing applications, *Microelect. Engineer.*, **97**, pp. 197–200.
42. Aouani, H., Sipova, H., Rahmani, M., Navarro-Cia, M., Hegnerova, K., Homola, J., Hong, M., and Maier, S. A. (2013). Ultrasensitive broadband probing of molecular vibrational modes with multifrequency optical antennas, *ACS Nano,* **7**, pp. 669–675.
43. Brown, L. V., Zhao, K., King, N., Sobhani, H., Nordlander, P. and Halas, N. J. (2013). Surface-enhanced infrared absorption using individual cross antennas tailored to chemical moieties, *J. Am. Chem. Soc.*, **135**, pp. 3688–3695.
44. Pryce, I. M., Kelaita, Y. A., Aydin, K., and Atwater, H. A. (2011). Compliant metamaterials for resonantly enhanced infrared absorption spectroscopy and refractive index sensing, *ACS Nano,* **5**, pp. 8167–8174.
45. Ebbesen, T. W., Lezec, H. J., Ghaemi, H. F., Thio, T., and Wolff, P. A. (1998). Extraordinary optical transmission through sub-wavelength hole arrays, *Nature*, **391**, pp. 667–669.
46. Prodan, E., Radloff, C., Halas, N. J., and Nordlander, P. (2003). A hybridization model for the plasmon response of complex nanostructures, *Science*, **302**, pp. 419–422.
47. Biagioni, P., Huang, J. S., and Hecht, B. (2012). Nanoantennas for visible and infrared radiation, *Rep. Prog. Phys.*, **75**, pp. 024402.
48. Adato, R., Yanik, A. A., Wu, C., Shvets, G., and Altug, H. (2010). Radiative engineering of plasmon lifetimes in embedded nanoantenna arrays, *Opt. Express*, **18**, pp. 4526– 4537.

49. Giannini, V., Francescato, Y., Amrania, H., Phillips, C. C., and Maier, S. A. (2011). Fano resonances in nanoscale plasmonic systems: A parameter-free modeling approach, *Nano Lett.*, **11**, pp. 2835–2840.

50. Gallinet, B. and Martin, O. J. F. (2011). Influence of electromagnetic interactions on the line shape of plasmonic Fano resonances, *ACS Nano*, **5**, p. 8999.

51. Gallinet, B. and Martin, O. J. F. (2011). Ab initio theory of Fano resonances in plasmonic nanostructures and metamaterials, *Phys. Rev. B*, **83**, p. 235427.

52. Adato, R., Artar, A., Erramilli, S., and Altug, H. (2013). Engineered absorption enhancement and induced transparency in coupled molecular and plasmonic resonator systems, *Nano Lett.*, **13**, pp. 2584–2591.

Chapter 15

Enhanced Spectroscopies Based on Nonmetallic Nanoparticles

Shinji Hayashi[a,b]

[a] *Optics and Photonics Center, Moroccan Foundation for Advanced Science, Innovation and Research (MAScIR), Madinat Al Irfane, 10100 Rabat, Morocco*
[b] *Graduate School of Engineering, Kobe University, Rokko, Nada, Kobe 657-8501, Japan*
s.hayashi@dragon.kobe-u.ac.jp

Although metallic nanostructures have widely been used to enhance spectroscopic signals of molecules placed very close to the nanostructures, relatively large losses in metals impose some limit in their applications, in particular, in the fluorescence spectroscopy. In recent years, nonmetallic nanostrucures made of low-loss materials have attracted much interest to further extend enhanced spectroscopies. This chapter is aimed to give an overview of the near-field enhancement brought by nonmetalic nanostructures. On the basis of Mie theory, the near-field enhancement brought by a spherical nonmetallic nanoparticle is first discussed. Experimental results demonstrating the Raman and fluorescence enhancement brought by GaP nanoparticles are also presented.

Handbook of Enhanced Spectroscopy
Edited by Pietro Giuseppe Gucciardi, Marc Lamy de la Chapelle, and Nathalie Lidgi-Guigui

ISBN 978-981-4613-32-3 (Hardcover), 978-981-4613-33-0 (eBook)
www.panstanford.com

15.1 Introduction

It is well known that metallic nanoparticles, such as Ag and Au nanoparticles, support localized surface plasmons (LSPs), which are collective oscillations of free electrons localized in the nanoparticles [6, 13, 18, 23, 37]. When LSPs are excited by light, electric fields strongly enhanced relative to those of incident light are generated in the vicinity of the surfaces of nanoparticles. Until now, a tremendous amount of work has been done on the enhancement of Raman [22, 33] and fluorescence [14] signals of molecules placed very close to metallic nanoparticles and the enhancement has been attributed to enhanced electric fields associated with the excitation of LSPs. The enhancement has also been extended to other linear and nonlinear spectroscopies, such as surface-enhanced infrared absorption spectroscopy (SEIRA), coherent anti-Stokes Raman scattering (CARS), second harmonic generation (SHG), *etc.*, as described in other chapters of this book.

Although enhanced spectroscopies based on metallic nanoparticles find a variety of potential applications, especially in biosensing and bioimaging, a good choice of metal and high-precision controls of size, shape, and arrangement of nanoparticles are required to achieve high enhancements of spectroscopic signals in a desired wavelength region. In the metal-enhanced fluorescence spectroscopy, it is well known that when the molecules are placed very close to metal surfaces (at a distance smaller than ~10 nm), the fluorescence signal is quenched due to the energy transfer from the excited molecules to the metals. The quenching can be avoided by inserting a spacer layer between molecules and metal surfaces, but one loses the enhancement factor (EF), since the field enhancement is maximal at the surfaces.

One of the approaches to further extend the enhanced spectroscopies is to use nonmetallic nanoparticles, instead of metallic nanoparticles. A great advantage of using nonmetallic nanoparticles is the absence of, or very small, quenching in fluorescence signals. In 1976, two years after the first observation of surface-enhanced Raman scattering (SERS) by Fleischman et al. [10], Chew et al. [7] have developed an electromagnetic (EM) theory of Raman scattering and fluorescence of molecules embedded in dielectric particles.

It was in 1982 that Inoue and Ohtaka theoretically predicted the enhancement of Raman signals from molecules adsorbed on a dielectric sphere [32] and an array of dielectric spheres [21]. Clear experimental evidence for enhanced Raman scattering on nonmetalic nanoparticles has been reported by Hayashi et al. [16] for copper phthalocyanine molecules adsorbed on GaP nanoparticles. Very recently, quenching-free enhancement of fluorescence from Rhodamine B molecules caused by GaP nanoparticles has been reported by Hayashi et al. [19]. In spite of these theoretical and experimental studies, enhanced spectroscopies based on nonmetallic nanoparticles received little attention, being hidden by the extensive development of enhanced spectroscopies based on metallic nanoparticles and the rapid growth of *plasmonics*. However, enhanced spectroscopies based on nonmetalllic nanostructures have gained renewed interest in recent years and much effort has been devoted to build up counterparts of metallic nanostructures.

The purpose of this chapter is to give an overview of the EM enhancement and enhanced spectroscopies based on nonmetalic particles. This chapter is organized as follows. After a brief summary of the EM mechanism of enhanced spectroscopies given in the next section, in Section 15.3 the near-field efficiency Q_{NF} for a spherical particle is introduced and numerical results of Q_{NF} suggesting large near-field enhancements caused by nonmetallic spheres are presented. In Section 15.4, experimental results of Raman and fluorescence enhancement caused by GaP particles are discussed. Finally, concluding remarks are given in Section 15.5.

15.2 Electromagnetic Mechanism of Enhanced Spectroscopies

Before entering into the details of enhanced spectroscopies based on nonmetallic particles, the EM mechanism of enhancement in Raman scattering [12, 33, 42] and fluorescence spectroscopy [3, 14, 24, 26] is summarized here. For simplicity let us consider a single molecule placed very close to a single nanoparticle schematically shown in Fig. 15.1. The nanoparticle can be either metallic or nonmetallic (dielectric). We assume that the model system is excited by incident

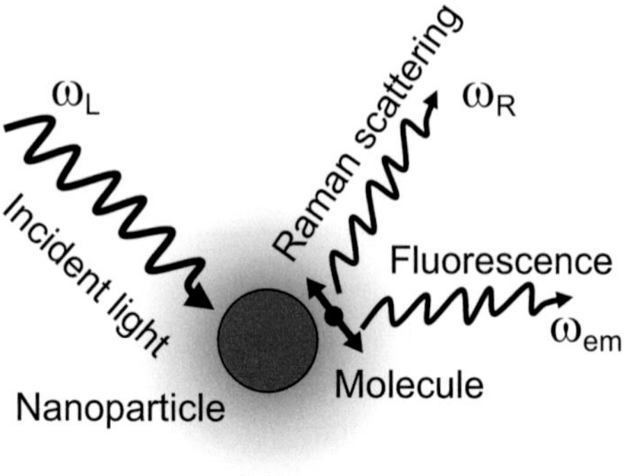

Figure 15.1 Simple model for discussing EM mechanisms of enhanced spectroscopies.

light with a frequency ω_L. In the presence of the nanoparticle, spectroscopic signals from the molecule can be enhanced relative to those from the molecule placed in a free space. In general, the enhancement of the spectroscopic signals takes place in two processes, that is, the excitation and light emission processes.

In a classical picture, Raman scattering process can be described by a dipole oscillating at Raman frequency ω_R, which is shifted from ω_L by a Raman shift. Raman scattered light is radiated by the Raman dipole. For the model system shown in Fig. 15.1, the Raman dipole of the molecule can be written as $\mathbf{P}_R(\omega_R) = \alpha_R \mathbf{E}(\mathbf{r}_0, \omega_L)$, where α_R is a Raman tensor, which we consider isotropic (scalar) for simplicity, and $\mathbf{E}(\mathbf{r}_0, \omega_L)$ is the local electric field induced at the position of the molecule $\mathbf{r}_0$. In the presence of the nanoparticle, the magnitude of $\mathbf{E}(\mathbf{r}_0, \omega_L)$ can be larger than that of the incident electric field $\mathbf{E}_i(\omega_L)$. The EF in the excitation process $g(\omega_L)$ is thus given by,

$$g(\omega_L) = \frac{|\mathbf{E}(\mathbf{r}_0, \omega_L)|^2}{|\mathbf{E}_i(\omega_L)|^2}. \tag{15.1}$$

$g(\omega_L)$ represents the local-field or near-field EF at ω_L.

In the presence of the nanoparticle, the intensity of Raman scattered light radiated by the Raman dipole can be enhanced due to the antenna effect of the nanoparticle. Although several different expressions have been derived theoretically [12, 33, 42], it is very common to use the following approximate expression for the EF in the Raman radiation process,

$$g(\omega_R) = \frac{|\,\mathbf{E}(\mathbf{r}_0, \omega_R)\,|^2}{|\,\mathbf{E}_i(\omega_L)\,|^2}, \tag{15.2}$$

where $\mathbf{E}(\mathbf{r}_0, \omega_R)$ is the local electric field at the Raman frequency ω_R.

By combining Eqs. 15.1 and 15.2, the total EF in the enhanced Raman scattering is given by,

$$F_R = g(\omega_L)g(\omega_R). \tag{15.3}$$

In general, the Raman shift $(\omega_L - \omega_R)$ is small and we can assume $\omega_L \approx \omega_R$ and $g(\omega_L) \approx g(\omega_R)$ to get,

$$F_R \approx g^2(\omega_L) = \frac{|\,\mathbf{E}(\mathbf{r}_0, \omega_L)\,|^4}{|\,\mathbf{E}_i(\omega_L)\,|^4}. \tag{15.4}$$

This is a well-known fourth-power law in SERS.

We now turn to the enhanced fluorescence spectroscopy. As described in the literatures [3, 14, 24, 26], the fluorescence emission rate γ_{em} can be expressed as,

$$\gamma_{em} = \gamma_{exc} q_e, \tag{15.5}$$

where γ_{exc} is the excitation rate, and q_e is the quantum efficiency given by $q_e = \gamma_r/\gamma$ with γ_r and γ being the radiative decay rate and the total decay rate, respectively. When the molecule has an absorption dipole $\mathbf{p}_a$, then $\gamma_{exc} \propto |\,\mathbf{p}_a \cdot \mathbf{E}(\mathbf{r}_0, \omega_L)\,|^2$, where $\mathbf{E}(\mathbf{r}_0, \omega_L)$ is the local electric field at the position of the molecule. Therefore, the enhancement in the local electric field leads directly to the enhancement in γ_{exc}.

The fluorescence emission process can be described by an emission dipole $\mathbf{p}_e$ oscillating at the emission frequency ω_{em}. Based on Fermi's golden rule, the spontaneous decay rate of the dipole can be expressed as (in SI units),

$$\gamma = \frac{2\omega_{em}}{3\hbar\varepsilon_0} |\,\mathbf{p}_e\,|^2 \rho(\mathbf{r}_0, \omega_{em}), \tag{15.6}$$

where $\rho(\mathbf{r}_0, \omega_{em})$ denotes the photonic mode density, which depends on the environment where the dipole is placed. The above equation indicates that the spontaneous decay rate is not an intrinsic property of the emitter, but is modified by the environment, as first pointed out by Purcell [36]. The enhancement in the radiative decay rate mediated by the enhancement in $\rho(\mathbf{r}_0, \omega_{em})$ is often called the Purcell effect.

In general, the total decay rate is given by the sum of the radiative decay rate γ_r and the nonradiative decay rate γ_{nr} ($\gamma = \gamma_r + \gamma_{nr}$). The presence of the nanoparticle in the vicinity of the molecule can lead to changes in both γ_r and γ_{nr}. Therefore, q_e, in the presence of the nanoparticle, is determined by the competition between γ_r and γ_{nr} and can be either larger or smaller than q_{e0} of a free molecule.

As mentioned above, the enhancement of the local electric fields or near fields at the position of the molecule plays a crucial role in the enhanced Raman and fluorescence spectroscopy. Although a single molecule placed near a nanoparticle is considered here, the discussion can be extended to many molecules placed close to more complicated nanostructures. In recent years, much attention has been paid on hot spots of enhancement, which are the points of extremely large local electric fields, in a variety of nanostructures such as dimers, arrays of nanoparticles, complicated assemblies of nonspherical particles, etc. [22]. Since the Raman scattering and fluorescence are incoherent processes, the intensities of emitted EM waves from molecules, and not their amplitudes, are additive. Therefore, for an ensemble of molecules, the enhanced intensity can be discussed by taking an appropriate summation or average of individual intensities over the molecules.

15.3 Near-Field Enhancement Predicted by Mie Theory

In this section, we discuss the possibility of achieving the near-field enhancement by nonmetallic particles. The discussion is based on the Mie theory, which is valid for a sphere of arbitrary size. It was in 1908 that Gustav Mie developed a general EM theory of scattering and absorption of a plane wave (light) incident on a spherical particle of an arbitrary size [30]. The Mie theory allows us to calculate the extinction efficiency Q_{ext}, which is a sum of the scattering efficiency Q_{sca} and the absorption efficiency Q_{abs}. In 1981, modifying the expression of Q_{sca}, Messinger et al. [29] introduced a near-field efficiency Q_{NF} and examined the near-field enhancement brought by various metallic particles. Since Q_{NF} can be applied to any material, we apply here to nonmetallic particles. In the following, we first compare the behaviors of Q_{ext}, Q_{sca}, and Q_{abs} for metallic and nonmetallic particles. As typical examples of metallic and nonmetallic particles, we consider Ag and GaP spherical particles. After presenting calculated results of Q_{ext}, Q_{sca}, and Q_{abs}, we present calculated results of Q_{NF} that demonstrate the near-field enhancement by nonmetallic particles.

Figures 15.2 (a), (b), and (c) show spectra of Q_{ext} calculated for an Ag sphere surrounded by air for different radii of $R = 100$, 60, and 20 nm, respectively. Figures 15.2 (d), (e), and (f) show the spectra obtained for a GaP sphere of $R = 100$, 60, and 40 nm surrounded by air, respectively. In these figures, Q_{sca} and Q_{abs} are also shown to clarify their relative contributions to Q_{ext}. To obtain the spectra, literature values of the dielectric functions for Ag and GaP [34] were used. It should be noted that the real part of the dielectric function of Ag takes negative values for wavelengths longer than 324 nm, while that of GaP takes positive values in the whole wavelength range presented in Fig. 15.2. Figure 15.2 demonstrates that the Ag sphere exhibits relatively broad extinction peaks, while the GaP sphere exhibits sharp and broad extinction peaks.

In general, the appearance of the peaks in Q_{ext} is physically understood as the consequence of the resonances of incident light to EM normal modes of a sphere, often referred to as Mie resonances. Here, EM normal modes of a sphere imply EM oscillations specific to the sphere, which satisfy Maxwell's equations and EM boundary conditions at the sphere surface [38, 39]. The EM normal modes can be classified into transverse electric (TE) and transverse magnetic (TM) modes [6, 37–39], and specified by an order number l. Radial components of electric fields inside and outside the sphere are zero for the TE_l modes, while those of magnetic fields are zero for the TM_l modes.

The pronounced extinction peaks (Mie resonances) of the GaP sphere with $R = 100$ nm appearing in Fig. 15.2 (d) can be attributed to both the TE_l and TM_l modes with $l = 1$ and 2 as indicated in the figure. For a sphere composed of material with positive real part of the dielectric function, which is the case for GaP, the resonances are also called morphology-dependent resonances or whispering-

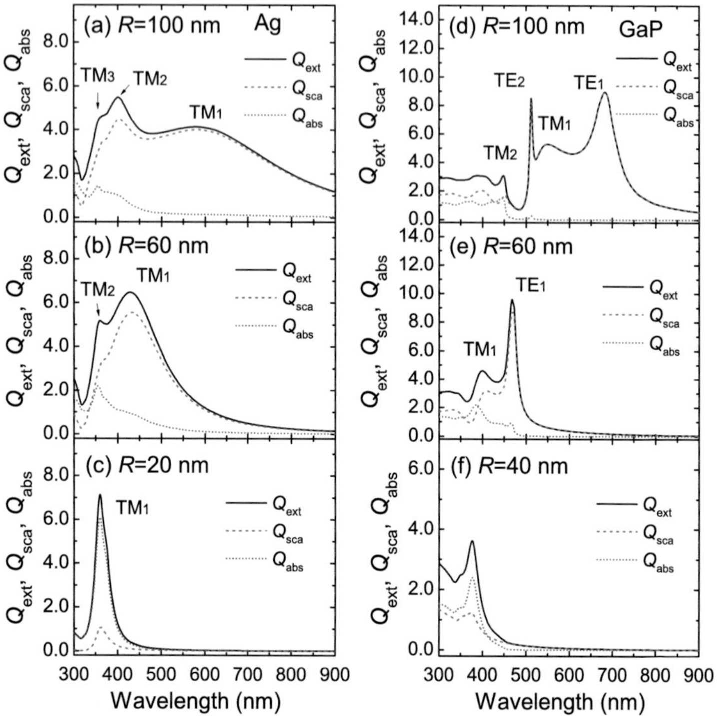

Figure 15.2 Q_{ext}, Q_{sca} and Q_{abs} calculated for an Ag sphere with $R = 100$ nm (a), 60 nm (b), and 20 nm (c), and for a GaP sphere with $R = 100$ nm (d), 60 nm (e), and 40 nm (f).

gallery mode resonances [37]. The latter term was first used by Lord Rayleigh for sound waves guided along a cathedral ceiling [28].

Figures 15.2 (d), (e), and (f) demonstrate that as R decreases, the extinction peaks of the GaP particle shift to shorter wavelengths and only the peaks of TE_1 and TM_1 modes remain in the spectra, when the particle is sufficiently small. The peaks of the higher-order modes are merged to form a continuum in the wavelength region of $\lambda < \sim 400$ nm. Figures 15.2 (d), (e), and (f) also show that the relative contributions of Q_{sca} and Q_{abs} change depending on R. For $R = 100$ nm, TE_1, TM_1, and TE_2 extinction peaks are dominated by Q_{sca}. However, as R decreases, the contribution of Q_{sca} decreases and that of Q_{abs} increases. For the extinction peak appearing around 380 nm for $R = 40$ nm (Fig. 15.2(f)), the contribution of Q_{abs} to Q_{ext} is about twice larger than that of Q_{sca}.

For a sphere composed of material with negative real part of the dielectric function, which is the case for Ag, only the TM_l modes exist and they lead to the extinction peaks. Indeed, the Ag sphere with $R = 100$ nm (Fig. 15.2(a)) shows extinction peaks attributable only to the TM_l modes with $l = 1$, 2, and 3. In this case, the extinction peaks are attributed to the LSP resonances. The TM_1, TM_2, and TM_3 modes correspond to a dipolar, quadrapolar and higher-order multipolar LSP modes. We see that as R decreases, the peaks in Q_{ext} shift to shorter wavelengths and the higher-order mode peaks tend to disappear, finally leaving a single peak of the TM_1 mode located at $\lambda = 360$ nm. Figures 15.2(a), (b), and (c) also demonstrate that the relative contributions of Q_{sca} and Q_{abs} greatly change depending on the radius. Namely, as R decreases, the relative contribution of Q_{abs} increases and finally becomes dominant for a small enough sphere. For isolated Ag nanoparticles such as those in Ag colloidal solutions, an LSP resonance peak very similar to that in Fig. 15.2(c) is commonly observed by absorption measurements [13, 23, 37, 48].

In the Mie theory, the expression of Q_{sca} is derived from electric and magnetic fields of scattered wave at the positions very far from the sphere (far fields). In the derivation, the EM energy (Poynting vector) flowing out of a hypothetical sphere with a radius much larger than the particle radius is calculated. Therefore, Q_{sca} gives the efficiency of converting the incident wave into the scattered wave. In the framework of the Mie theory, it is also possible to calculate fields in the vicinity of the particle surface (near fields). Messinger et al. [29] introduced a near-field efficiency Q_{NF} defined as,

$$Q_{NF} = \frac{r^2}{\pi R^2} \int_0^{2\pi} \int_0^{\pi} \mathbf{E}_s \cdot \mathbf{E}_s^* \sin\theta d\theta d\phi \Big|_{r=R} \tag{15.7}$$

where $\mathbf{E}_s$ is the electric field of the scattered wave. They also introduced Q_r, which takes into account only the radial component of $\mathbf{E}_s$. From the above expression, it is clear that Q_{NF} is the near-field intensity (square of the absolute value of the electric field) averaged over the sphere surface ($r = R$). Therefore, Q_{NF} and Q_r offer very convenient measures of near-field enhancement when we discuss the enhancement of optical signals caused by the Mie resonances.

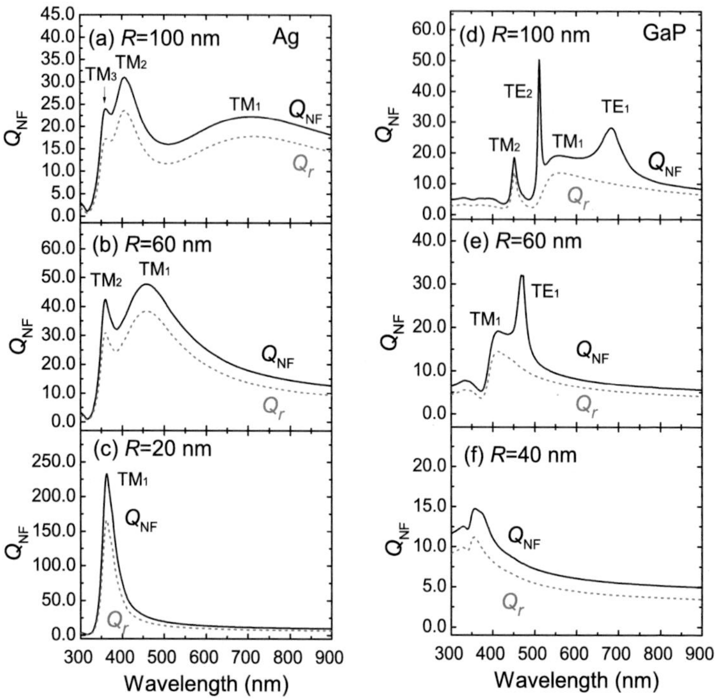

Figure 15.3 Q_{NF} and Q_r calculated for an Ag sphere with $R = 100$ nm (a), 60 nm (b), and 20 nm (c), and for a GaP sphere with $R = 100$ nm (d), 60 nm (e), and 40 nm (f).

Spectra of Q_{NF} and Q_r calculated for Ag and GaP spheres are presented in Fig. 15.3. For the TM_l mode resonances, we see that a large part of the near-field enhancement is supported by Q_r, that is, the radial component of the electric field. For the TE_l mode resonances, there is no contribution of Q_r, since the radial component of electric field is zero for the TE_l modes. A comparison between Figs. 15.2 and 15.3 reveals that the resonance peaks seen in Q_{NF} spectra correlate fairly well with those in Q_{ext} spectra, although positions, shapes, and relative intensities of the resonance peaks in Q_{NF} do not coincide exactly with those in Q_{ext}. This implies that the resonances detected by the far fields correlate well with those of the near fields.

An important fact seen for Q_{NF} spectra of the Ag particle (Figs. 15.3(a), (b), and (c)) is a dramatic increase in the maximum value with decreasing R. Although the EF depends on various factors in actual experiments of enhanced spectroscopies, the results presented here suggest that larger EFs can be achieved by smaller Ag nanoparticles. In contrast to the case of the Ag particle, the maximum value in the Q_{NF} spectra of the GaP particle (Fig. 15.3(d), (e), and (f)) decreases as R decreases. However, it should be noted that the GaP sphere exhibits considerably large values of Q_{NF}. In fact, the GaP particle with $R = 100$ nm can generate a near-field enhancement as large as or even larger than that generated by an Ag sphere of the same radius. Although the EM normal modes supported by the GaP sphere are not the LSP modes, relatively large particles can leads to the large near-field enhancement. The Q_{NF} spectra of the GaP particle presented here strongly suggests the possibility of realizing enhanced spectroscopies based on nonmetallic particles.

From Q_{NF} spectra calculated for various nonmetallic particles, it turns out that large EFs can be achieved when the real part of dielectric function of the particle is large and its imaginary part is small. In terms of the complex refractive index $(n + ik)$, this criteria corresponds to a large refractive index n (preferably larger than $\sim$2.5) and a small extinction coefficient k. In Fig. 15.4, n and k of crystalline GaP [34] are plotted as a function of the wavelength. This figure clearly demonstrates that the crystalline GaP satisfies the above criteria for $\lambda > \sim 350$ nm. In Fig. 15.4, n and k of crystalline Si [34], which is another candidate of near-field enhancer, are also plotted. The wavelength dependence of n and k is very similar to that of the crystalline GaP and satisfies the above criteria for $\lambda > \sim 400$ nm. Although not shown here, spectra of Q_{NF} calculated for Si spheres are very similar to those for GaP spheres presented in Figs. 15.3(d), (e), and (f). The near-field enhancement caused by Si particles has been pointed out also by Tanaka and Obara [44]. They performed numerical simulations based on the Mie theory and a three-dimensional finite-difference time-domain (3D FDTD) method to demonstrate the applicability of Si particles to laser-induced nanopatterning mediated by enhanced near fields. In qualitative agreement with our results presented in Fig. 15.3, they found that

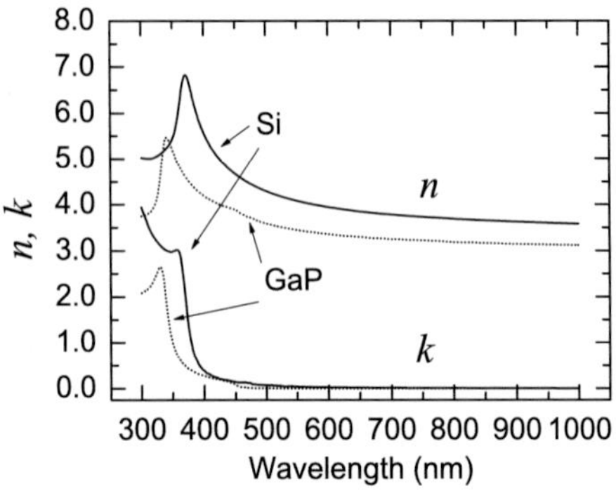

Figure 15.4 n and k for crystalline GaP and Si.

under illumination of light with $\lambda = 800$ nm, a Si particle 100 nm in radius can lead to near-field EFs considerably larger than those brought by a gold particle.

15.4 Raman and Fluorescence Enhancement of Molecules on GaP Particles

The results of theoretical calculations presented in the preceding section predict that GaP particles are good candidates of enhancers for enhanced spectroscopies, even though they are not metallic and do not support LSPs. The enhancement in Raman and fluorescence from dye layers deposited on GaP particles has indeed been observed by the author's group [16, 19]. The experimental results are briefly summarized in this section.

15.4.1 *Sample Preparation*

The structure of samples used for Raman and fluorescence measurements is schematically shown in Fig. 15.5. To prepare the samples, GaP particles were first deposited onto substrates by a gas-evaporation technique described in our previous papers [15, 16, 19]. A vacuum chamber was first evacuated to $\sim 1.0 \times 10^{-3}$ Pa and then filled with Ar gas. Small pieces of GaP wafers were charged in a tungsten basket fixed in the chamber and heated by

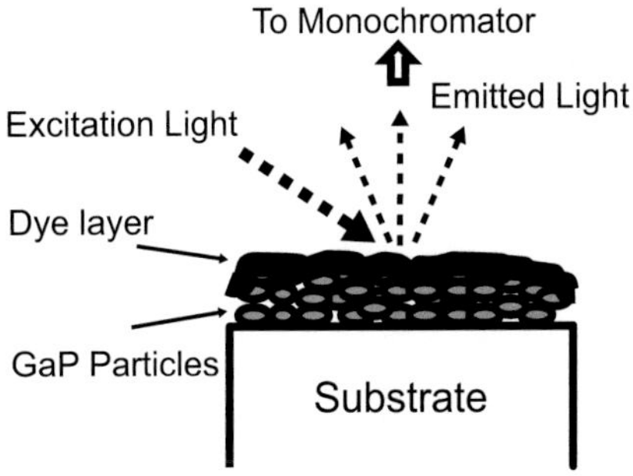

Figure 15.5 Structure of samples used for Raman and fluorescence measurements.

injecting ac current. Particles grown along the convection of Ar gas were collected by substrates placed above the heater. In the gas-evaporation technique, the average size of the particles can be controlled by the gas pressure, distance between the heater and the substrate, and heater temperature. The pressure of Ar gas was set at 0.2 to 2×10^4 Pa. The substrate was placed 4 to 10 cm above the heater. The heater temperature was set at a temperature slightly higher than the melting point of GaP ($\sim$1800K).

For transmission electron microscopic (TEM) observation, a part of the particle layers was scrapped off and dispersed in ethanol. A drop of ethanol was put on TEM grids. After the evaporation of ethanol, the grids were examined by an electron microscope. Figure 15.6(a) shows a typical TEM image of GaP particles prepared. The electron diffraction pattern given in the inset exhibits sharp Debye-Scheller rings, demonstrating that the GaP particles have good crystallinity. The particles seen in the image are ellipsoidal rather than spherical and randomly oriented. Figure 15.6(b) is a distribution of the linear dimension of particles measured along one direction of the image. The average size of the sample determined from the histogram is $d_{\text{ave}} = 126$ nm. The average determined in this manner implies the rotational average as well as the average over the size distribution.

To perform Raman and fluorescence measurements, thin layers of dye molecules were deposited by a vacuum deposition technique on top of the GaP particle layers. Reference samples were also

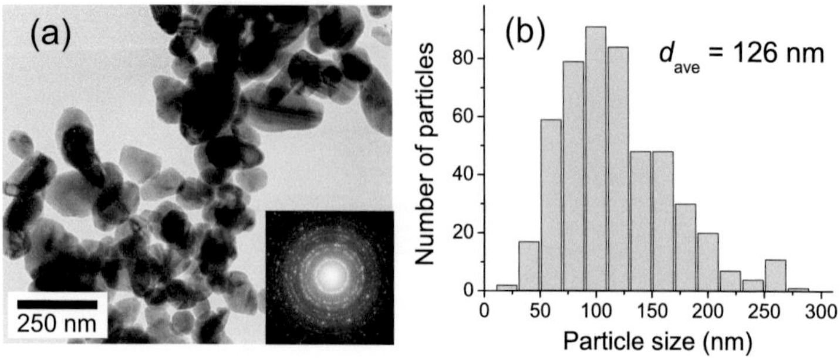

Figure 15.6 (a) Typical TEM image of GaP particles prepared by gas-evaporation technique and (b) size distribution.

prepared by simultaneously depositing the dye layers onto glass substrates. The thickness of the dye layers was monitored by a quartz microbalance. Pieces of Si wafers were used as substrates for Raman samples, while glass slides were used for fluorescence samples.

15.4.2 *Raman Enhancement*

Evidence for Raman enhancement caused by GaP nanoparticles has been reported in 1988 by the author's group [16]. In the experiments, copper phthalocyanine (CuPc) molecules were used as test molecules. As schematically shown in Fig. 15.5, CuPc dye layers 1 nm in thickness were deposited onto layers of gas-evaporated GaP particles. Raman spectra of the samples corresponding to $d_{\text{ave}} =$ 106, 60, and 40 nm are shown in Fig. 15.7. The spectra show many Raman lines corresponding to the molecular vibrations in CuPc as well as the lines of transverse optical (TO) and longitudinal optical (LO) phonons in GaP particles. It should be noted that the intensities of the lines increase as the particle size increases. The Raman EF, determined by taking the ratio of the 1530 cm^{-1} line of the sample to that of the reference, is 700, 300, and 200 for the spectra (a), (b), and (c), respectively.

According to the EM mechanism of SERS summarized in Section 15.2 and the definition of Q_{NF} given by Eq. 15.7, the intensity of SERS is considered to be proportional to Q_{NF}^2. The dependence of

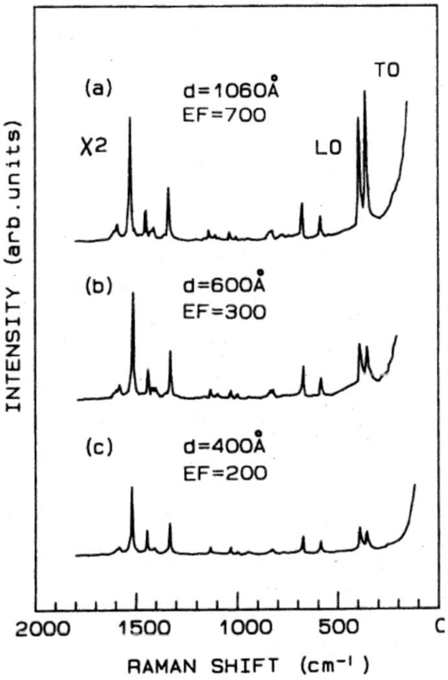

Figure 15.7 Raman spectra of 1 nm-thick CuPc layers deposited on GaP particle layers with various average sizes. Figure reprinted with permission from Ref. [16], Copyright 1998 by the American Physical Society.

Q_{NF} on the particle diameter ($2R$) calculated at the wavelength of the incident light ($\lambda_i = 514.5$ nm) is shown in Fig. 15.8(a). In Fig. 15.8(b), the experimental dependence of the Raman EF on the average particle size is compared with a calculated dependence of Q_{NF}^2 on $2R$. Many sharp peaks of Q_{NF} appearing in Fig. 15.8(a) are attributed to the Mie resonances of a GaP sphere; the assignment of some strong peaks to the TE_l and TM_l modes is given in the figure. The average size of GaP particles used in the experiment was smaller than 120 nm. This size range corresponds to the tail of the TE_1 resonance in Fig. 15.8(a), where Q_{NF} increases monotonously as the particle diameter increases. As seen in Fig. 15.8(b), Q_{NF}^2 also increases resulting in a good qualitative agreement with the experimentally observed increase in the Raman EF. The experimental results presented above clearly demonstrate

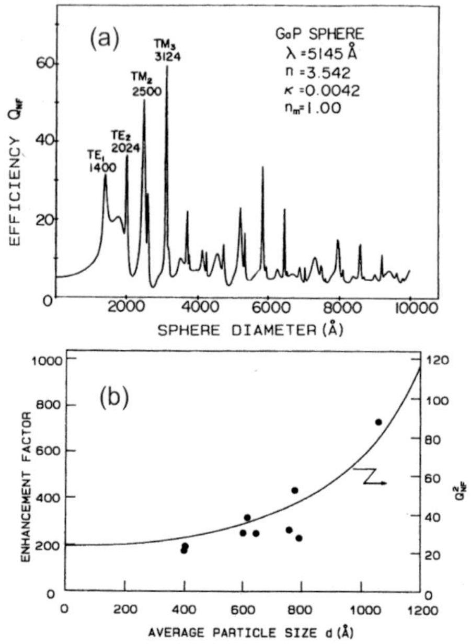

Figure 15.8 (a) Calculated size dependence of Q_{NF} at $\lambda_i = 514.5$ nm and (b) comparison between experimental Raman EFs and Q_{NF}^2. Figure reprinted with permission from Ref. [16], Copyright 1998 by the American Physical Society.

the SERS activity of GaP particle layers. Furthermore, a good qualitative agreement with calculated results of Q_{NF} suggests that the enhancement is of EM origin.

15.4.3 *Fluorescence Enhancement*

For fluorescence measurements, we used layers of well known dye molecules, Rhodamine B[a] and DCM[b]. Results obtained with Rhodamine B is found in our previous paper [19]. We describe here results obtained with DCM. Figure 15.9 compares a fluorescence spectrum of a 30 nm thick DCM layer deposited on a GaP particle layer (sample) with that deposited on a glass substrate (reference).

[a] [9-(2-carboxyphenyl)-6-diethylamino-3-xanthenylidene]-diethylammonium chloride

[b] 4-(dicyanomethylene) -2-methyl-6-(p-dimethylaminostyryl)-4H-pyran

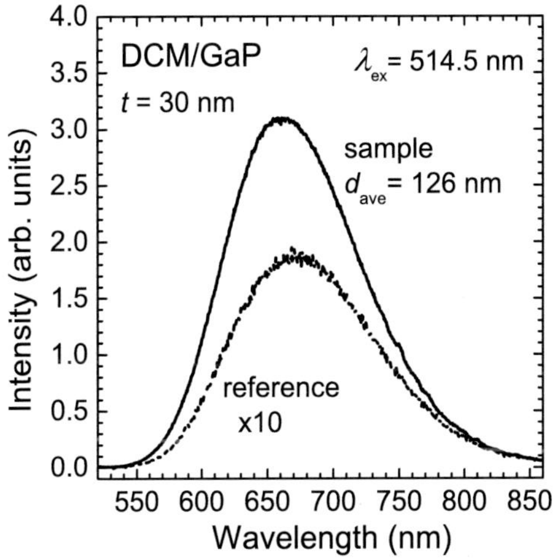

Figure 15.9 Comparison of fluorescence spectrum of a 30-nm thick DCM layer on a GaP particle layer (sample) with that on a glass substrate (reference).

The TEM image of the GaP particles corresponding to this sample was previously shown in Fig. 15.6. To excite the spectra, the 514.5 nm line of an Ar^+ laser was used. A slightly focused laser beam was incident on the sample surface at an angle of 45°. Emitted light was collected by a lens from the normal to the sample surface and focused on a slit of 25 cm monochromator equipped with a Si charge-coupled device. The fluorescence measurements were performed under the identical condition for the sample and reference. In Fig. 15.9, we see that the fluorescence intensity of the sample is greatly enhanced relative to that of the reference. The EF is around 30.

We have examined the dependence of the fluorescence EF on the dye layer thickness. The result obtained is plotted in Fig. 15.10. In the figure, the filled squares represent the average values of EF determined from the data of at least five identical samples and the error bars represent the ranges of fluctuations. We see that the EF increases rapidly as the thickness decreases in the thickness region of $t < \sim 10$ nm. It should be noted that the result presented in Fig. 15.10 is very similar to the previous result obtained with Rhodamine B layers (Fig. 5 in [19]).

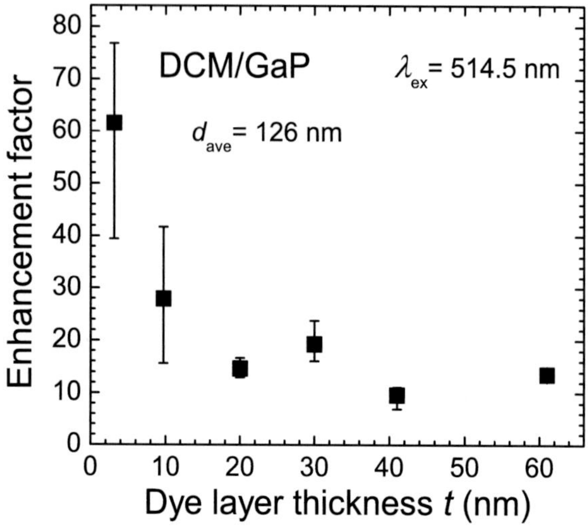

Figure 15.10 Dependence of fluorescence EF on the thickness of DCM layer.

Anger et al. [3] explained theoretically and demonstrated experimentally that when a single molecule is placed close to a metallic nanoparticle and the distance between the particle and the molecule decreases, the fluorescence intensity first increases, takes a maximum, and finally decreases rather rapidly (quenching). The initial increase is due to the increase in the excitation rate γ_{exc} given in Eq. 15.5, reflecting the increase in the local electric fields. The rapid decrease in the fluorescence intensity at very short distances is due to a rapid increase in the nonradiative decay rate γ_{nr} caused by the energy transfer from the molecule to the metallic particle. If the quenching process is absent, the fluorescence intensity is expected to increase monotonously as the distance decreases. In our experiments, we have a lot of molecules in the dye layers. Therefore, EFs plotted in Fig. 15.10 represent those averaged over the layers and the decrease in the layer thickness can be reinterpreted as the decrease in the average distance between molecules and GaP nanoparticles. The rapid increase in the fluorescence EF for very thin layers thus implies that the nonradiative decay due to the energy transfer is absent or negligibly small. To the best of author's knowledge, the fluorescence data described above are the first demonstration of quenching-free fluorescence enhancement based on nonmetallic nanoparticles.

As pointed out in the paper of Anger et al. [3], the nonradiative decay rate of the molecule near the metal surface is determined by Ohmic losses in the metal, which is determined by the imaginary part of dielectric function of the metal. The larger the imaginary part, the larger the nonradiative decay rate. Note that the imaginary parts for Ag and Au are relatively large in the wavelength regions of LSPs and lead to large nanoradiative decay rates. On the other hand, the imaginary part of dielectric function of the GaP crystal is practically zero (loss-less) in the visible spectral region,[a] since the GaP crystal is a semiconductor with an indirect band gap. This is the reason why the fluorescence quenching is absent on GaP particle layers.

To gain more insight into the enhancement caused by the GaP nanoparticles, we have measured the fluorescence excitation spectra. For the measurements, light from a Xe lamp was monochromatized by a monochromator and focused on the sample. The intensity of the emitted light at the wavelength of 665 nm (around the peak of the fluorescence spectra) was monitored by varying the wavelength of incident light. Figure 15.11(a) compares a raw excitation spectrum of a sample with that of a reference. The DCM thickness was less than 1 nm and the average size of the GaP nanoparticles was 143 nm. The observed excitation spectrum of the reference is found to be very similar to its absorption spectrum, as it should be. However, the excitation spectrum of the sample is shifted to longer wavelengths and broadened compared to that of the reference. Furthermore, the monitored intensity is very much enhanced. The great changes in the excitation spectrum is of course due to the presence of the GaP nanoparticle layer. The dependence of the EF on the excitation wavelength, that is, excitation spectrum of EF, is shown in Fig. 15.11(b). We see that the EF increases monotonously as the excitation wavelength increases and reaches a value as large as 160 at around 580 nm.

For the same sample, we also measured the spectrum of light scattering using the same setup as the fluorescence excitation

[a]The imaginary part of complex refractive index k is also practically zero as seen in Fig. 15.4.

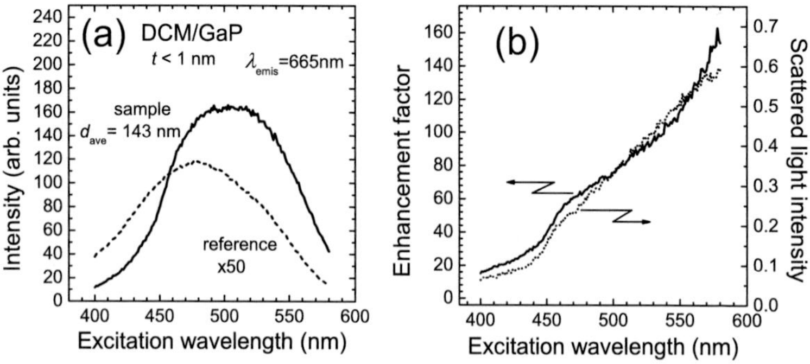

Figure 15.11 (a) Fluorescence excitation spectra monitored at the emission wavelength of 665 nm for a sample with $d_{\text{ave}} = 143$ nm and a reference. (b) Excitation spectrum of the EF.

measurements. The intensity of the light scattered at the same wavelength as the incident light was monitored and normalized to the scattering intensity of a standard sample, that is, a $BaSO_4$ particle layer. The scattering spectrum obtained is compared with the excitation spectrum of EF in Fig. 15.11(b). We see that the excitation spectrum of EF and the scattering spectrum show almost the same dependence on the excitation wavelength. This good correlation between the two spectra can be understood on the basis of the EM mechanism as explained below.

The excitation spectrum of fluorescence EF should reflect the dependence of the excitation rate γ_{exc} in Eq. 15.5 on the excitation wavelength and thus that of local electric fields. In Section 15.3, we pointed out that for GaP particles, the spectra of Q_{NF} correlate rather well with those of Q_{ext}. Furthermore, Q_{ext} for relatively large particles is dominated by Q_{sca}. Since the magnitude of Q_{NF} increases with increasing the particle size, larger particles make larger contributions to the fluorescence EF. Therefore, for an ensemble of particles with a relatively large average particle size, we can expect a good correlation between averaged Q_{NF} and averaged Q_{sca}. If we admit the same argument for an ensemble of nonspherical particles, the observed good correlation between the excitation spectrum of EF and the scattering spectrum can be well explained.

15.5 Concluding Remarks

In this chapter, after a brief summary of the EM mechanism of Raman and fluorescence enhancement, numerical results of Q_{NF} for Ag and GaP nanoparticles are presented. The results indicate that relatively large GaP nanoparticles (R ~100 nm) can induce the near-field enhancement comparable to or even larger than that induced by Ag nanoparticles. Experimental results demonstrating the Raman and fluorescence enhancement of dye layers deposited onto gas-evaporated GaP particles layers are also presented. In particular, the fluorescence enhancement has been shown to be free from quenching for the first time.

It should be stressed here that the near-field enhancement described in this chapter is brought by relatively large nanoparticles of high n and low k materials, such as GaP ($n = 3.59$ at $\lambda = 500$ nm) and Si ($n = 4.30$ at $\lambda = 500$ nm). In this regime of n and k ($n >$ ~2.5 is preferable), the lowest-order TE_l and TM_l modes of a sphere with $l = 1$ (the lowest-order Mie resonances) are located in the visible spectral region and the corresponding resonance bands are relatively broad. Furthermore, the near-field enhancement achieved at the resonances is considerably high. These behaviors of the resonances make it possible to use the nonmetallic nanoparticles as near-field enhancers. In recent years, much attention has been paid on these aspects brought by various nonmetallic nanostructures [43, 45]. In particular, Si nanostructures have attracted great interest [1, 8, 9, 41, 44]. Unlike the case of metals, the low-loss nature of Si, its abundancy in nature, and its lower conductivity are encouraging features to develop not only novel enhanced spectroscopies but also highly efficient photonic devices such as light-emitting and photovoltaic devices.

It is well known that the near-field enhancement can be obtained using particles with low n and low k materials, such as SiO_2 ($n = 1.46$ at $\lambda = 500$ nm) and polystyrene ($n = 1.60$ at $\lambda = 500$ nm). To get a large enough enhancement with these materials, particles as large as a few μm or more have commonly been used. Such particles are known to support higher-order TE_l and TM_l resonances with narrow bandwidths in the visible spectral region. A large enough sphere can be regarded as a high-Q microcavity, and the enhanced fluorescence of dye molecules embedded in [5, 11] or adsorbed on [31] the spherical microcavities has been studied extensively. Laser oscillation in such systems has also been studied intensively [25, 46]. In recent years, the enhancement in Raman signals caused by large dielectric particles has been the subject of theoretical [4] and experimental [2] studies.

As described in a recent review paper of Wang et al. ([47] and references therein), there exist a lot of reports on SERS using nonmetallic nanostructures, including those of metal oxides (NiO, ZnO, etc.), semiconductors (ZnS, CdS, CdTe, Si, Ge, etc.), and silver halides (AgCl, AgBr, etc.). The SERS activity of graphene has also been demonstrated [27]. Although the details of enhancement mechanism are not clear at present, the observed SERS phenomena have commonly been attributed to the charge-transfer mechanism, which is outside the scope of this chapter.

It should be noted finally that purely dielectric multilayer systems can be used to generate the enhanced local electric fields. Enhanced spectroscopies based on various structures, such as waveguide structures [20], photonic double barrier structures [17, 40], and multilayer structures supporting Bloch surface waves [35], are currently under investigation.

Acknowledgments

This work was supported by JSPS KAKENHI Grant Number 23310064. The author is grateful to C. Yoshikawa for assistance in fluorescence experiments with DCM dye layers.

References

1. Albella, P., Poyli, M. A., Schmidt, M. K., Maier, S. A., Moreno, F., Sáenz, J. J. and Aizpurua, J. (2010). *Appl. Phys. Lett.* **97**, p. 131116.
2. Anderson, M. S. (2010). *Appl. Phys. Lett.* **97**, p. 131116.
3. Anger, P., Bharadwaj, P. and Novotny, L. (2006). *Phys. Rev. Lett.* **96**, p. 113002.

4. Ausman, L. K. and Schatz, G. C. (2008). *J. Chem. Phys.* **129**, p. 054704.
5. Benner, R. E., Barber, P. W., Owen, J. F. and Chang, R. K. (1980). *Phys. Rev. Lett.* **44**, p. 475.
6. Bohren, C. F. and Huffman, D. R. (1983). *Absorption and Scattering of Light by Small Prticles* (John Wiley & Sons, New York).
7. Chew, H., McNulty, P. J. and Kerker, M. (1976). *Phys. Rev. A* **13**, p. 396.
8. Evlyukhin, A. B., Novikov, S. M., Zywietz, U., Eriksen, R. L., Reinhardt, C., Bozhevolnyi, S. I. and N., C. B. (2012). *Nano Lett.* **12**, p. 3749.
9. Evlyukhin, A. B., Reinhardt, C., Seidel, A., Luk'yanchuk, B. S. and Chichkov, B. N. (2010). *Phys. Rev. B* **82**, p. 045404.
10. Fleischmann, M., Hendra, J. P. and McQuillan, A. J. (1974). *Chem. Phys. Lett.* **26**, p. 163.
11. Fujiwara, H., Sasaki, K. and Masuhara, H. (1999). *J. Appl. Phys.* **85**, p. 2052.
12. Gaponenko, S. V. (2010). *Introduction to Nanophotonics* (Cambridge University Press, Cambridge).
13. Garcia, M. A. (2011). *J. Phys. D: Appl. Phys.* **44**, p. 283001.
14. Geddes, C. D. (ed.) (2010). *Metal-Enhanced Fluorescence* (John Wiley & Sons, New Jersey).
15. Hayashi, S. and Kanomori, H. (1982). *Phys. Rev. B* **60**, p. 7079.
16. Hayashi, S., Koh, R., Ichiyama, Y. and Yamamoto, K. (1988). *Phys. Rev. Lett.* **60**, p. 1085.
17. Hayashi, S., Kurokawa, H. and Oga, H. (1999). *Opt. Rev.* **6**, p. 204.
18. Hayashi, S. and Okamoto, T. (2012). *J. Phys. D: Appl. Phys.* **45**, p. 433001.
19. Hayashi, S., Takeuchi, Y., Hayashi, S. and Fujii, M. (2009). *Chem. Phys. Lett.* **480**, p. 100.
20. Hu, D. B. and Qi, Z. M. (2013). *J. Phys. Chem. C* **117**, p. 16175.
21. Inoue, M. and Ohtaka, K. (1982). *Phys. Rev. B* **26**, p. 3487.
22. Kneipp, K., Moskovits, M. and Kneipp, H. (eds.) (2006). *Surface-Enhanced Raman Scattering, Topics in Applied Physics*, Vol. 103 (Springer, Berlin).
23. Kreibig, U. and Vollmer, M. (1995). *Optical Properties of Metal Clusters, Springer Series in Materials Science*, Vol. 25 (Springer, Berlin).
24. Kümmerlen, J., Leitner, A., Brunner, H., Aussenegg, F. R. and Wokaun, A. (1993). *Mol. Phys.* **80**, p. 1031.
25. Kuwata-Gonokami, M., Takeda, K., Yasuda, H. and Ema, K. (1992). *Jpn. J. Appl, Phys.* **31**, p. L99.

26. Lakowicz, J. R. (2001). *Anal. Biochem.* **298**, p. 1.
27. Ling, X., Xie, L., Fang, Y., Xu, H., Zhang, H., Kong, J., Dresselhaus, M. S., Zhang, J. and Liu, Z. (2010). *Nano Lett.* **10**, p. 553.
28. Lord Rayleigh (1910). *Philos. Mag.* **20**, p. 1001.
29. Messinger, B. J., von Raben, K. U., Chang, R. K. and Barber, P. W. (1981). *Phys. Rev. B* **24**, p. 649.
30. Mie, G. (1908). *Ann. Phys. (Leipzig)* **25**, p. 377.
31. Norris, D. J., Kuwata-Gonokami, M. and Moemer, W. E. (1997). *Appl. Phys. Lett.* **71**, p. 297.
32. Ohtaka, K. and Inoue, M. (1982). *J. Phys. C: Solid State Phys.* **15**, p. 6463.
33. Otto, A. (1984). *Surface-Enhanced Raman Scattering: "Classical" and "Chemical" Origins, Topics in Applied Physics*, Vol. 54, chap. 6 (Springer-Verlag, Berlin), pp. 289–418.
34. Palik, E. D. (ed.) (1985). *Handbook of Optical Constants of Solids* (Academic Press, Orlando).
35. Pirotta, S., Xu, X. G., Delfan, A., Mysore, S., Maiti, S., Dacarro, G., Patrini, M., Galli, M., Guizetti, G., Bajoni, D., Sipe, J. E., Walker, G. C. and Liscidini, M. (2013). *J. Phys. Chem. C* **117**, p. 6821.
36. Purcell, E. M. (1946). *Phys. Rev.* **69**, p. 681.
37. Quinten, M. (2011). *Optical Properties of Nanoparticle Systems* (WILEY-VCH, Weinheim).
38. Ruppin, R. (1982). Chapter 9: *Spherical and cylindrical polaritons in solids*, In *Electromagnetic Surface Modes*, ed. Boardman, A. D. (John Wiley, Chichester), p. 345.
39. Ruppin, R. and Englman, R. (1970). *Rep. Prog. Phys.* **33**, p. 149.
40. Sainidou, R., Renger, J., Teperik, T. V., González, M. U., Quidant, R. and Javier Garcia de Abajo, F. (2010). *Nano Lett.* **10**, p. 4450.
41. Sigalas, M. M., Fattal, D. A., Williams, R. S., Wang, S. Y. and Beausoleil, R. G. (2007). *Opt. Express* **15**, p. 14711.
42. Stockman, M. I. (2006). *Electromagnetic Theory of SERS, Topics in Applied Physics*, Vol. 103 (Springer), pp. 47–65.
43. Tanaka, Y., Obara, G., Zenidaka, A., Terakawa, M. and Obara, M. (2010). *Appl. Phys. Lett.* **96**, p. 261103.
44. Tanaka, Y. and Obara, M. (2009). *Jpn. J. Appl. Phys.* **48**, p. 122002.
45. Terakawa, M., Tanaka, Y., Obara, G., Sakano, T. and Obara, M. (2011). *Appl. Phys. A* **102**, p. 661.

46. Tzeng, H. M., Wall, K. F., Long, M. B. and Chang, R. K. (1984). *Opt. Lett.* **9**, p. 499.
47. Wang, X., Shi, W., She, G. and Mu, L. (2012). *Phys. Chem. Chem. Phys.* **14**, p. 5891.
48. Xia, Y. and Halas, N. J. (2005). *MRS Bull.* **30**, p. 338.

Index

For Product Safety Concerns and Information please contact our EU representative GPSR@taylorandfrancis.com Taylor & Francis Verlag GmbH, Kaufingerstraße 24, 80331 München, Germany

Batch number: 10793311

Printed by Printforce, the Netherlands